Relativity Made Relatively Easy, Volume 2

Relativity Made Relatively Easy

Volume 2

General Relativity and Cosmology

A course for physics undergraduates

Andrew M. Steane

OXFORD

UNIVERSITY PRESS

OXFORD
UNIVERSITY PRESS

Great Clarendon Street, Oxford, OX2 6DP,
United Kingdom

Oxford University Press is a department of the University of Oxford.
It furthers the University's objective of excellence in research, scholarship,
and education by publishing worldwide. Oxford is a registered trade mark of
Oxford University Press in the UK and in certain other countries

First Edition published in 2021

Impression: 1

Published in the United States of America by Oxford University Press
198 Madison Avenue, New York, NY 10016, United States of America

British Library Cataloguing in Publication Data

Data available

Cataloging-in-Publication data is on file at Library of Congress

ISBN 978–0–19–289564–6 (hbk.)
ISBN 978–0–19–289354–3 (pbk.)

DOI: 10.1093/oso/9780192895646.001.0001

Printed and bound by
CPI Group (UK) Ltd, Croydon, CR0 4YY

Preface

This book is motivated by a desire to make General Relativity and cosmology accessible to physicists, either as part of an undergraduate degree, or as a starting point for a graduate. The treatment of cosmology is sufficiently thorough to present an introductory course in its own right on that subject. The final part is a short introduction to classical field theory, which could serve as a preparation for a course on particle physics or quantum field theory.

The book follows a previous volume on Special Relativity (Steane, 2012). The present volume assumes knowledge of Special Relativity and a modest familiarity with index notation and the stress-energy tensor, and that is all. For a reader with such a foundation this book stands alone and does not require the first volume. Having said that, some introductory material on the meaning of Gaussian curvature, geodesics and parallel transport which was presented in the first volume is not repeated here; a simple presentation of those ideas would smooth the way for readers of the present book but is not required.

The subtitle announces that the book aims to be 'relatively easy'. This is not to say that the subject is straightforward—it is not—but the reader will be eased into it, and in writing I have been alert to needless confusions that can be avoided simply by choosing notation carefully and ordering the ideas in a helpful way. The level is intended to be accessible to a student studying for a first degree in physics, who would like to answer such questions as: How are gravitational waves generated and detected? What do we mean by the horizon of a black hole? What exactly is a wormhole and can they exist? What can and cannot physics tell us about the origins of the cosmos? What is dark energy? Can galaxies travel faster than light? The book aims to bring as much clarity and physical insight as possible to these and other questions, and to support self-study. There are 248 exercises.

General Relativity can be loosely divided into four areas:

1. Tensor analysis and differential geometry (manifolds, metric, covariant derivative).
2. Finding particular solutions (e.g. weak field, Schwarzschild, Kerr, FLRW) and their physical implications.
3. Properties of the field equation (stability, existence of solutions, singularities).
4. Causality and global properties (topological methods, global spacetime diagrams).

This book is mostly focussed on the first two areas in this list, but some basic points from the other two are also introduced. If a phenomenon is included, then it is calculated in sufficient detail to allow precise comparison with experimental data, as far as possible without use of the phrase 'it can be shown that'. For example, we display and interpret gravitational wave data as precisely as possible without recourse to the advanced numerical methods that are required for strong-field dynamical simulations. Birkhoff's theorem is derived but Buchdahl's theorem in quoted. We derive the metric of the charged non-rotating black hole (Reissner–Nordström) but quote the metric of the rotating neutral black hole (Kerr). The Friedmann equations of cosmology are derived. The meaning of cosmological expansion and the Big Bang is carefully explained. There is an emphasis on physical understanding throughout.

Should students wish to avail themselves of computer algebra packages, they are welcome to do so, but none are needed for the problems treated here. I think it is good to solve the basic problems by hand, just as it is good to learn how to do arithmetic before picking up a calculator. (This does not preclude using a computer to check your working.)

The ordering of material is itself part of the decision process when teaching and learning. The text starts with applications of General Relativity that can be understood without the full mathematical apparatus, and then develops the mathematics of manifolds and differential geometry. We thus arrive at the complete theory half way through the book. The second half of the book then treats exact applications. This ordering helps to build up confidence and familiarity, and to keep the subject in contact with physical observations.

General Relativity lives in close connection with the mathematics of manifolds and differential geometry, which is a large subject in its own right. Every writer in this area has to form a judgement about what parts of the mathematical background to include, and at what stage to include them. I have deliberately avoided tools such as the wedge product which are not needed in the first instance, if one's goal is to make basic calculations in astrophysics and cosmology. The mathematician looking to learn differential geometry more generally will need further ideas, but I hope the book will be useful even for such readers, in providing insight and physical examples. For the physicist the book is more than sufficient for a first course, giving a base from which advanced texts become accessible.

Acknowledgements

I learned General Relativity by reading books, papers and lecture notes and doing large numbers of calculations. In preparing this text I consulted existing textbooks and many sets of lecture notes generously made available on the internet. The following books were especially helpful: M. P. Hobson, G. Efstathiou and A. N. Lasenby, *General Relativity, An Introduction for Physicists*, Cambridge University Press; W. Rindler, *Relativity: Special, General and Cosmological*, Oxford University Press; B. Schutz, *A First Course in General Relativity*, Cambridge University Press; S. Carroll, *Spacetime and Geometry*, Pearson; Lightman, Press, Price and Teukolsky, *Problem Book in Relativity and Gravitation*, Princeton University Press.(Hobson, Efstathiou and Lasenby, 2006; Rindler, 2006; Schutz, 2009; Carroll, 2014a; Lightman, Press, Price and Teukolsky, 1975) I also learned specific points or useful approaches from: B. Crowell, *General Relativity*, www.lightandmatter.com; P. A. M. Dirac, *General Theory of Relativity*, Princeton University Press; R. D'Inverno, *Introducing Einstein's Relativity*, Clarendon Press; C. W. Misner, K. S. Thorne, and J. A. Wheeler, *Gravitation*, W. H. Freeman and Co.; E. Poisson, *A Relativist's Toolkit*, Cambridge University Press; R. Wald, *General Relativity*, University of Chicago Press, (Crowell, 2018; Dirac, 1996; d'Inverno, 1992; Misne, Thorne and Wheeler, 1975; Poisson, 2004; Wald, 1984; Peebles, 1993).

Some of the calculations and physical pictures in this book are original, but the central ideas and the mathematical methods are not. Writing involved first disentangling a number of concepts that range over mathematics and physics, and then bringing together ideas and approaches. My role has been first to glean, or to sieve, like a prospector looking for gold, and then to purify and shape. The gold was whichever approach seemed to me to be clearest. It may be possible to tell, in parts of the book, where I am following one predecessor, and where another. I can only acknowledge my indebtedness.

I thank all those lecturers who have provided lecture notes, and writers of review articles. I would particularly like to acknowledge comprehensive lecture notes by Steven Balbus, University of Oxford; Matthias Blau, University of Bern, Sean Carroll, California Institute of Technology and David Tong, University of Cambridge (Carroll, 1997; Balbus, 2019; Blau, 2018; Tong, 2019).

I thank Steven Balbus and Henry Davies for reading parts of the text and giving feedback, and Marlan Scully for indicating and kindly giving me a copy of the text by Lightman *et al.*. I thank Sonke Adlung and all others at Oxford University Press who worked on the book.

Many of the diagrams were prepared using GNU Octave and TikZ (Eaton, Bateman, Hauberg and Wehbring, 2018; Tantau, 2013). For the text I used LaTeX via MikTeX. Much highly skilled, creative and generous labour goes into the building and sharing of such software tools.

I thank the leadership of Oxford University for its commitment to academic values and its willingness to take a long-term view of the development of scholarly work. I thank David Lucas, Tom Harty and Chris Ballance for their friendly supportive attitude. Finally, my special thanks go to Emma, Joseph, Wilfred and Elsie for together creating a home full of wide interests and generous instincts, in which books can be written.

Contents

1

Terminology and notation

This chapter should be skimmed at first reading; it is here as a reference to which the reader may refer when needed. The main thing to say is that some familiarity with index notation is assumed and the Einstein summation convention is adopted throughout.

$\det(m)$	determinant of matrix m
$\lvert x \rvert$	absolute value of x
$[m_{ab}]$	the matrix whose components are m_{ab}
g_{ab}	(set of covariant components of) metric tensor
g	$\equiv \det([g_{ab}])$
η_{ab}	Minkowski metric
$\mathbf{v}, \mathbf{t}$	bold font is adopted for 4-vectors and also for tensors of higher rank
$\mathbf{v}, \mathbf{E}$	in some sections the bold font is used for 3-vectors
$\vec{p}, \vec{g}$	3-vectors
u^a	(set of contravariant components of) 4-velocity
u	a parameter
$\mathrm{d}s$; $\mathrm{d}\tau$	invariant interval; proper time
$\mathbb{F}$	the electromagnetic field tensor
T_{ab}	the (covariant components of the) energy-momentum tensor
ρc^2, p	scalar invariant energy density and pressure
c; G	speed of light; gravitational constant
Γ^a_{bc}	Christoffel symbol; also called connection coefficient; affine connection
$\Box^2$	$= \eta^{\mu\nu}\partial_\mu\partial_\nu$ d'Alembertian
$\nabla^\mu\nabla_\mu$	$= g^{\mu\nu}\nabla_\mu\nabla_\nu$ covariant d'Alembertian
Φ	gravitational potential, (dimensions $[L^2/T^2]$)
Λ	cosmological constant, (dimensions $[1/L^2]$)
$\mathrm{d}\Omega^2$	$= \mathrm{d}\theta^2 + \sin^2\!\theta\,\mathrm{d}\phi^2$ solid angle line element

4-vectors and their components. In discussions of Special Relativity, including in the companion Volume 1 of the present book, it is common to write

$$\mathsf{P'} = \Lambda\mathsf{P} \tag{1.1}$$

Relativity Made Relatively Easy: General Relativity and Cosmology. Volume 2. Andrew M. Steane,
Oxford University Press. © Andrew M. Steane 2021. DOI: 10.1093/oso/9780192895646.003.0001

Table 1.1 Some physical constants. $M_{\odot}$ is the solar mass. The value of cG can serve as a mnemonic for the value of G (since everyone knows the value of c).

G	6.67428×10^{-11} J m/kg^2
c	$299\,792\,458$ m/s
cG	0.02 m^{4}s^{-3}kg^{-1}
$M_{\odot}$	1.98847×10^{30} kg
$GM_{\odot}$	1.3271244×10^{20} m^{3}s^{-2}
$R_{\odot}$	6.957×10^8 metres
AU	$149\,597\,870\,700$ metres
year	$31\,557\,600$ seconds
AU $/c$	499.005 seconds
Mpc	$3.08567758149137 \times 10^{22}$ metres
Mpc	$6.48 \times 10^{11}/\pi$ AU
$8\pi G/c^2$	111.491 mm per Earth mass
$8\pi G/c^4$	2.07664×10^{-43} metres per joule
$5c^5/G$	1.8141×10^{53} watts

to show the Lorentz transformation of a 4-vector such as the energy-momentum 4-vector, where Λ is the transformation matrix. One may also write

$$\mathsf{P}' \cdot \mathsf{U}' = \mathsf{P} \cdot \mathsf{U} \tag{1.2}$$

to show the invariance of the scalar product. However, many a modern-day general relativist will react to that notation with a sort of horror or indignation, as if something morally wrong has been done. The reason for this reaction is that the 4-vector *itself* does not change at all when one switches attention from one reference frame to another. Rather, its *components relative to one frame differ from its components relative to the other frame.* To express this one may write

$$p'^a = \Lambda^a{}_\mu p^\mu. \tag{1.3}$$

This shows what happens to the components. If one wishes to express what has happened to the 4-vector itself during a passive transformation, one refuses to admit into the notation any change. The 4-vector itself has not changed.

In the present book we will adopt the general relativist's correct insistence that 4-vectors do not change when one considers a passive transformation of the coordinate system, leading to a change of basis vectors. We will adopt the bold font, as in **p**, to write 4-vectors, and when we wish to refer to 3-vectors we will put an arrow, such as $\vec{p}$ (with exceptions which will be clear). However, we will not adopt a state of moral indignation towards notation such as (1.1), (1.2); all we need to do is be clear that in those expressions P is not 'the 4-vector'; rather it is 'the list of contravariant components of the 4-vector relative to the unprimed reference frame'. Similarly, P' is not 'the 4-vector'; rather it is 'the list of contravariant components of the 4-vector relative to the primed reference frame'.

That is all that needs to be said on the matter, and that is both the beginning and the end of the use of the notation P in the present book. That notation was an index-free method to write about lists of contravariant components. In the present book we will always use indices to refer to such lists, because this is the clearer way to approach the more general subject. We will still be able to make much use of index-free notation, because we will let bold symbols like **p** and **u** refer to 4-vectors, and more generally we adopt the bold font for tensors (to be defined later) of any rank above zero. To report a result such as (1.2) we shall write some such assertion as

$$\text{`}\mathbf{p}\cdot\mathbf{u}\text{ is a scalar invariant quantity.'} \tag{1.4}$$

Those awkward signs. Sign conventions have been notoriously awkward to settle in General Relativity, and in differential geometry in general. The issue can be summarized as follows. Let $S1$, $S2$, $S3$ refer to three values which may be $+1$ or -1. Then we have

$$\eta_{ab} = [S1]\,\text{diag}(-1,\,+1,\,+1,\,+1)$$
$$R^a_{bcd} = [S2]\,\left(\partial_c\Gamma^a_{db} - \partial_d\Gamma^a_{cb} + \Gamma^a_{c\lambda}\Gamma^\lambda_{db} - \Gamma^a_{d\lambda}\Gamma^\lambda_{cb}\right)$$
$$G_{ab} = R_{ab} - \frac{1}{2}g_{ab}R \;=\; [S3]\,\frac{8\pi G}{c^4}T_{ab}$$

and to be consistent one must then have (assuming $T_{00} > 0$)

$$R_{ab} = [S2][S3]\,R^\lambda_{a\lambda b}, \tag{1.5}$$

and, in the absence of torsion,

$$R^a_{\lambda cd}v^\lambda = [S2]\,(\nabla_c\nabla_d - \nabla_d\nabla_c)v^a. \tag{1.6}$$

The energy tensor for an ideal fluid is then

$$T_{ab} = (\rho + p/c^2)u_a u_b + [S1]\,pg_{ab} \tag{1.7}$$

which takes the form $\text{diag}(\rho c^2,\,p,\,p,\,p)$ in the instantaneous rest frame of a fluid element. Finally, in a local inertial frame and Cartesian coordinates the action of the d'Alembertian operator on a scalar field is

$$\partial^\lambda\partial_\lambda\phi \overset{\text{LIF}}{=} [S1](-c^2\partial_t^2 + \partial_x^2 + \partial_y^2 + \partial_z^2)\phi. \tag{1.8}$$

In order to commit to memory the sign in the equation for R^a_{bcd}, notice the placement of the differential operators ∂_c and ∂_d.

The choices made in this book compare as follows with some other books:

	Present text	MTW	HEL	d'Inverno	Rindler	Schutz
$[S1]$	+	+	−	−	−	+
$[S2]$	+	+	+	+	+	+
$[S3]$	+	+	−	−	−	+

The above choices were not made merely out of a desire to copy esteemed predecessors. The positive sign in front of T_{ab} on the right-hand side of our Einstein field equation is a reminder of the positive sign in the Poisson equation for Newtonian gravity, $\nabla^2 \Phi = 4\pi G \rho$. Also, I like the fact that these choices make spatial matters come out largely positive for the 2-sphere. That is to say, if one considers the 2-sphere (the surface of a sphere in three-dimensional Euclidean space), then it is universally agreed that the Gaussian curvature K is positive, and with the above choices, the eigenvalues of R_{ab} are also positive for the 2-sphere, and so is $R \equiv R^\sigma_\sigma$.

On the 2-sphere it is obvious that geodesics converge, so since we have $K > 0$ for this case, the sign in the geodesic deviation equation must be

$$\ddot{\eta} = -K\eta \tag{1.9}$$

and more generally, for an isotropic space of positive curvature, geodesic flow results in reducing volumes. This can help in checking the sign when writing the more general geodesic deviation equation. When the choices are made as indicated above one gets

$$\frac{\mathrm{D}^2 s^a}{\mathrm{d}\tau^2} = -R^a_{\lambda\mu\nu} \dot{x}^\lambda s^\mu \dot{x}^\nu. \tag{1.10}$$

When comparing work from different authors, it is helpful to note that the connection coefficients Γ^a_{bc} are quadratic in the metric, and therefore do not change sign when the metric does.

Alphabets and indices.   When writing equations involving indices, the question arises whether the relation is correct in any basis, or in only one or some. One way to keep this distinction clear is to adopt the *abstract index notation*, in which an index does not refer to a component, but merely signals that the quantity is a vector (or tensor, as the case may be). Then any expression involving components can be readily converted to abstract indices as long as the expression holds in all bases. However, for the purposes of the present text, abstract index notation would merely add a further layer of notation which will not offer any significant help. Therefore in the present text we will **not** adopt abstract index notation, but we **will** be concerned almost entirely with expressions that are valid in any basis.

Some authors choose to use Greek letters for components in one particular basis, and Roman letters otherwise. The reverse practice is also widespread. In the present text, when a particular basis is being assumed, this will be indicated by an annotated equals sign, as in, for example,

$$g_{ab} \overset{\mathrm{LIF}}{=} \eta_{ab}. \tag{1.11}$$

With this method we are free to use as indices the letters of any alphabet, and we mostly prefer the Roman alphabet for an index which is not being summed over, and the Greek alphabet (especially λ, μ, ν) for dummy indices (i.e. those which are being summed over). But we do not always adopt that practice. We adopt whatever practice seemed to the author to be clearest for any given equation or sequence of derivations.

The indices i, j, k are special: they range over the values $1, 2, 3$ (not 0), and in almost all circumstances where i, j, k are used, these values correspond to spatial coordinates.

Ordering of indices on a second- or higher-rank tensor does matter. It often happens that the first index is up and the others down. We adopt the following useful shorthand:

$$\text{for any } \mathbf{Q}, \quad Q^a_{bcd} \equiv Q^a{}_{bcd} \tag{1.12}$$

That is, when the notation leaves it ambiguous as to which of two indices is first, then it is understood that the upper index comes first.

Naming the flat space. A local inertial frame or **LIF** is both a physical and a mathematical idea. Physically, it is a small set of standard rods and clocks, near to some event P in spacetime, each in free fall and all having (momentarily at least) the same velocity.[1] One can determine the absence of relative motion through radar signalling. Mathematically, a LIF is a coordinate chart and associated basis in the vicinity of P such that the metric tensor has no first-order dependence on any of the coordinates at P: $(\partial_c g_{ab})|_P = 0$. Typically one also chooses the coordinates such that they are orthogonal at P, and scales them such that $g_{ab} = \eta_{ab}$ at P. Then, near P we have

$$g_{ab}(x) \stackrel{\text{LIF}}{=} \eta_{ab} + \mathcal{O}(x - x_P)^2 \tag{1.13}$$

The following are synonyms:

> local inertial frame, local Lorentz frame, local Minkowski frame, freely falling frame

and they are all closely related to the synonyms:

> Riemann normal coordinates, geodesic coordinates, local Cartesian coordinates

These terms are occasionally employed in non-synonymous ways in other books.

One interested purely in geometry might prefer to say 'local Lorentz frame'[2] rather than 'local inertial frame'; one interested in physics may prefer to say 'local inertial frame'. In this book we will mostly say *local inertial frame* or LIF.

If P is an event, then the phrase 'in a LIF at P' can be translated to mean any or all of the following:

1. 'if we consider what is observed relative to a small light frame at P, all the parts of which are in free fall.'
2. 'if we adopt coordinates in which $\partial_c g_{ab} = 0$ at P'
3. 'if we adopt coordinates in which $\Gamma^a_{bc}(P) = 0$' (this amounts to the same condition as (2))

---

[1] Which is to say, more formally, they are on neighbouring parallel geodesics.

[2] Since Lorentz already has the transformation named after him, and Minkowski explicitly emphasized the geometric point of view, I think it would be appropriate to use the terminology 'local Minkowski frame' rather than 'local Lorentz frame', but the latter appears to have won. I shall stick to LIF.

4. 'if we adopt coordinates in which $\Gamma^a_{bc}(P) = 0$ AND $g_{ab}(P) = \eta_{ab}$'
5. 'using Riemann normal coordinates with origin at P'.

Note that these statements are not all exactly the same, because on the one hand one can find a metric with zero gradient which is not the Minkowski metric, so (2) does not necessarily imply (4), and on the other hand, one does not have to use Cartesian coordinates to map a flat space, so (1) does not necessarily imply any of the others. However, when we wish to consider LIFs we almost always wish to adopt the Minkowski metric, and this will be assumed unless it is explicitly stated otherwise. This implies that in a LIF, the coordinates are Cartesian, not polar, unless it is explicitly stated otherwise.

Riemann normal coordinates is a system of coordinates in which the equations for geodesics passing through the origin have the same form as the equations of straight lines in Cartesian coordinates in Euclidean space. In the context of GR, this captures the same notion as local inertial frame. A general way to construct these coordinates near any point is given in Section 8.3. Using this construction one finds that, in the case of four dimensions and Minkowski signature, (1.13) holds.

Parentheses and integrals.   Standard mathematical notation is sometimes ambiguous. For example $g_{ab}(x - x_0)$ might mean either a quantity g_{ab} multiplied by $(x - x_0)$, or it might signify that g_{ab} is a function of one parameter, and this parameter has the value $(x - x_0)$. This ambiguity is avoided in most cases by the context, but occasionally it is not, and in that case we shall employ the notation $g_{ab}(x - x_0)$ for the second meaning, that is, a functional dependence.

Indefinite and definite integrals are written

$$\int f(x)\, \mathrm{d}x, \qquad \int_a^b f(x)\, \mathrm{d}x. \qquad (1.14)$$

It often happens that one wishes to find a definite integral with upper limit equal to x. This could for example be written

$$\int_a^x f(x')\, \mathrm{d}x' \qquad (1.15)$$

where the prime is introduced in order to distinguish the 'dummy' integration variable from the quantity x that gives the upper end of the range of integration. However, in the context of GR this notation is distracting because it suggests we have introduced a new coordinate called x'. To avoid giving that false impression, I will in this book employ the notation

$$\int_a^x f(x)\, \mathrm{d}x. \qquad (1.16)$$

If one were performing multiple nested integrals then such a notation would be liable to break down. For an integral over a single variable, however, it is clear and unambiguous and we will adopt it without further comment.

Exercises

The reader will require sufficient familiarity with index notation to be able to prove the following concerning correctly constructed tensorial expressions.

1.1 If one term in an equation has all indices down (or no indices at all), then up-indices elsewhere in the equation must be dummies.

1.2 If S_{ab} is symmetric and A_{ab} is antisymmetric then the double-contraction $S_{\mu\nu}A^{\mu\nu}$

must give zero, and this generalizes to index pairs at any rank.

1.3 If S_{ab} is symmetric and T_{ab} is any tensor then $S_{\mu\nu}T^{\mu\nu}$ only depends on the symmetric part of **T**.

1.4 $(T_{abc} + T_{cab} + T_{bca})X^a X^b X^c = 3T_{abc}X^a X^b X^c$. [Hint: index relabelling]

Part I

2

The elements of General Relativity

We now begin our study. In this chapter some central ideas are introduced, and the main equations of General Relativity are quoted with a view to getting a sense of where we are heading. The sequence of ideas in the book is such that we view these equations now somewhat as mountains on the skyline; we discern their overall shape but we will not arrive at fully understanding and thus surmounting them till half way through the book. On the journey we will learn what gravitation is and how spacetime behaves by studying physical effects and by developing mathematical tools.

General Relativity (hereafter GR) deals with a subtle interplay between 'stuff' and 'arena', between 'a thing' and 'that which is not quite a thing but is not nothing'. The 'stuff' is *matter and radiation* (excluding gravitational radiation). The 'arena' is spacetime. In tackling GR we are trying to study physics while walking on the water—we have to accept a certain level of abstraction right from the start. In General Relativity the subject tells us, 'ok; here is your map (chart; frame; system of coordinates), but the map is not the territory! Those straight lines on the map are not straight lines on the ground!' When we reply 'ok, so show me the ground itself' then the answer comes back, 'well there are tensor fields, and here they take this value, and there they take that value', and so we enquire 'yes, but where is 'here' and 'there'?', and the subject replies 'just exactly where I showed you on your map ...'.

Thus spacetime is elusive. Nonetheless, I would like to encourage the reader to adopt, from the start, the notion that spacetime is 'a something'. Spacetime is a 'thing', a manifold, something that can warp and vibrate. This perspective has been resisted, for good reasons, by some very able physicists, but I think it is safe, and helpful, to adopt it, as long as one retains a sense that spacetime is a very special kind of thing because you cannot tell whether you are moving relative to it. You cannot even tell if you are accelerating relative to it (but you can tell if you are moving non-inertially; more on this later). You cannot dip your finger in spacetime in order to watch it go by. Indeed, you cannot measure or detect the least thing about spacetime at any single point, not even if there is a huge gravitational wave passing by. This is an informal statement of the strong equivalence principle. But when we compare information gathered from different points, spacetime makes its presence felt.

This notion that spacetime is a 'thing', if a rather subtle thing, is born out by quantum field theory, where we can if we like regard gravity as a field, and the 'thing' I have been referring to

Relativity Made Relatively Easy: General Relativity and Cosmology. Volume 2. Andrew M. Steane,
Oxford University Press. © Andrew M. Steane 2021. DOI: 10.1093/oso/9780192895646.003.0002

is the gravitational field. However, there remain deep and unresolved mathematical difficulties when this combination is pursued in depth. Both quantum field theory and General Relativity, as we know them today, are inadequate to capture fully the nature of the physical world at the level of fundamental physics. Rather, they provide a set of tools, and a starting point, from which to gain both a rich amount of good insight, and encouragement to find out more.

2.1   The gravitational field equations: first view

One way to start to study GR is to quote a set of differential equations which describe spacetime. There is more than one way to present the sequence of ideas. We shall begin by introducing a second-rank tensor **g** called the *metric*. The metric is most commonly treated by dealing with its components, written g_{ab}, where the indices a and b are in the range 0 to 3. The fundamental equation of GR is the Einstein field equation, which is a differential equation (or, if you prefer, a set of differential equations) relating the metric to the stress-energy tensor T_{ab}:

Einstein field equation, first version

$$R_{ab} - \tfrac{1}{2} R\, g_{ab} = \frac{8\pi G}{c^4} T_{ab}. \tag{2.1}$$

Here G and c are constants, the stress-energy tensor T_{ab} describes the energy, pressure and stress of whatever matter and electromagnetic fields may be present, $R = R^\mu_\mu$ and R_{ab}, called the **Ricci tensor**, is related to g_{ab} by a complicated sequence of differential operations:

$$R_{bd} \equiv R^\mu_{b\mu d} = \partial_\mu \Gamma^\mu_{db} - \partial_d \Gamma^\mu_{\mu b} + \Gamma^\mu_{\mu\lambda}\Gamma^\lambda_{db} - \Gamma^\mu_{d\lambda}\Gamma^\lambda_{\mu b} \tag{2.2}$$

where

$$\Gamma^a_{bc} = \frac{1}{2} g^{a\lambda}(\partial_b g_{c\lambda} + \partial_c g_{b\lambda} - \partial_\lambda g_{bc}). \tag{2.3}$$

Einstein's field equation reduces to Poisson's equation (the equation for Newtonian gravity) in the limit of weak fields and ordinary forms of matter. It remains to say how g_{ab} is physically manifested, so that we can use the theory to make predictions about observable phenomena. The **geodesic equation** is one way to do this. It is the equation of motion of a test particle subject to only gravitational influences:

Geodesic equation

$$\frac{du^a}{d\tau} + \Gamma^a_{\mu\nu} u^\mu u^\nu = 0 \qquad \text{where} \qquad u^a = \frac{dx^a}{d\tau}. \tag{2.4}$$

Here $x^a(\tau)$ is the set of numbers giving the coordinate location of the particle, as a function of the elapsed proper time τ along its worldline. By solving the field equation for g_{ab} and then the geodesic equation for given initial conditions, one can, for example, find the orbits of planets about a star of any size. The geodesic equation reduces to Newton's second law of motion in the weak field limit.

So far we have presented the field equation and the geodesic equation merely in order to have a logically complete set of ideas, and in order to show that GR is difficult! The field equation is non-linear: note the product of Γ terms in the equation for the Ricci tensor. This means that we cannot employ the method of superposition in order to make new solutions by combining those already found. Each problem has to be started afresh. There is also the issue of how to specify T_{ab} for some given configuration of matter. This is another difficult task, in general, because the arrangement of matter, and therefore its properties such as pressure and flow, will itself depend on the gravitational environment, so one has to seek a self-consistent solution in which g_{ab} and T_{ab} each depend on the other. No wonder, then, that most of this book is concerned with solutions in empty space where $T_{ab} = 0$!

Through the use of the terms 'proper time' and 'worldline' we have assumed that the reader is familiar with Special Relativity, and we have brought in some very reasonable general assumptions about the nature of spacetime—we assume that a worldline is a well-defined idea, for example, which amounts to assuming that spacetime is a kind of continuous 'space' known to mathematicians as a *differentiable manifold*. The word 'manifold' here is a reference to a kind of fabric, a generalization to many dimensions (in GR, four dimensions) of a smooth sheet or cloth, but one which may warp and bend in interesting ways. This leads us to the very important *geometric interpretation of gravity* which is described further in the next section.

By manipulating the field equation a little, we can bring it to a form which is often more convenient in practice. First contract both sides to obtain

$$R_\lambda^\lambda - \tfrac{1}{2}R\delta_\lambda^\lambda = -R = \frac{8\pi G}{c^4}T_\lambda^\lambda \tag{2.5}$$

(where we used that $g^{a\mu}g_{\mu b} = \delta_b^a$ and $\delta_\lambda^\lambda = 4$ in four dimensions). Hence

$$R = -\frac{8\pi G}{c^4}T_\lambda^\lambda. \tag{2.6}$$

Substituting this result for R into (2.1) we have

Einstein field equation, second version (equivalent to first)

$$R_{ab} = \frac{8\pi G}{c^4}\left(T_{ab} - \tfrac{1}{2}g_{ab}T_\lambda^\lambda\right) \tag{2.7}$$

2.1.1 Field and geometric interpretations of gravity

The equations of GR can be interpreted by us in more than one way, as suits our own intuition. Whichever way helps to solve, or get correct insight into, any given physical scenario is valuable. In this section we will examine two ways: GR as a *field theory* and GR as a *geometric theory*.

We will set out the field theory approach first in order to examine it briefly, and then move on. The geometric approach will be adopted in the rest of the book. The field theory approach can be useful as a kind of sanity check, whenever the geometric ideas become confusing. In the field theory approach one simply asserts that the quantity Γ^a_{bc} which appears in (2.3) and (2.4) is itself 'the gravitational field'. The three indices indicate that this field has, at any given location, sixty-four components (not all independent), and its role in the geodesic equation shows that it can be interpreted as a force per unit momentum per unit velocity. The set of ideas can be presented as follows:

Gravitational field theory
Field (N.B. not a tensor field)
$$\Gamma^a_{\mu\nu}$$

Relationship of field to potential

$$\Gamma^a_{bc} = \frac{1}{2}g^{a\lambda}(\partial_b g_{c\lambda} + \partial_c g_{b\lambda} - \partial_\lambda g_{bc}) \tag{2.8}$$

Field equation

$$\partial_\mu \Gamma^\mu_{ab} - \partial_b \Gamma^\mu_{\mu a} + \Gamma^\mu_{\lambda\mu}\Gamma^\lambda_{ab} - \Gamma^\mu_{b\lambda}\Gamma^\lambda_{a\mu} = \frac{8\pi G}{c^4}\left(T_{ab} - \tfrac{1}{2}g_{ab}T^\lambda_\lambda\right) \tag{2.9}$$

Equation of motion of test particle

$$\frac{\mathrm{d}u^a}{\mathrm{d}\tau} = -\Gamma^a_{\mu\nu}u^\mu u^\nu + \frac{q}{m}\mathbb{F}^a_\mu u^\mu \tag{2.10}$$

where

$$u^a = \frac{\mathrm{d}x^a}{\mathrm{d}\tau}. \tag{2.11}$$

Useful facts:
$$g_{ab} = g_{ba}, \qquad \Gamma^a_{bc} = \Gamma^a_{cb}, \qquad \Gamma^\lambda_{\lambda c} = \partial_c \ln\sqrt{|g|} \tag{2.12}$$

When $\Gamma^a_{bc} = \Gamma^a_{cb}$ we say the connection has no *torsion*; this is assumed in General Relativity.

The cosmological constant term in the field equation has been absorbed into T_{ab}. Eq (2.10) can be derived from the field equations; see Section 16.4.

In this approach we regard the metric tensor as a kind of 'potential'; the field is related to the gradients of this potential in various directions. The field equation (2.9) is precisely the same as (2.7) (which you should check). We included in the equation of motion of test particles a term giving the influence of electromagnetic forces on a particle of charge q and mass m. The electromagnetic force is obtained from the electromagnetic field tensor $\mathbb{F}$ which we shall describe more fully in a moment.

Notice that in the field theory approach it has not been necessary to mention spacetime as such; we simply take it for granted that the coordinates x^a are coordinates of events in a four-dimensional region, and we do not need to know much about that region. It simply supports the gravitational field $\mathbf{\Gamma}$ and the electromagnetic field $\mathbb{F}$ and the stress-energy tensor $\mathbf{T}$. Our task

---

Symmetries of the curvature tensor

$$\nabla_e R_{ab\,cd} + \nabla_c R_{ab\,de} + \nabla_d R_{ab\,ec} = 0 \qquad \text{`Bianchi identity'} \qquad (2.14)$$

$$R_{ab\,cd} = R_{cd\,ab}, \quad R_{ab\,cd} = -R_{ab\,dc} = -R_{ba\,cd}, \qquad (2.15)$$

$$R_{a\,bcd} + R_{a\,cdb} + R_{a\,dbc} = 0, \qquad (2.16)$$

$$R_{ab} = R_{ba}, \qquad G_{ab} = G_{ba}. \qquad (2.17)$$

Small gaps in the list of indices have been used to draw attention to the pattern in some results, and have no other significance. For the action of the covariant derivative operator ∇ on a fourth-rank tensor see Section 12.4.

---

is to solve all these differential equations, and eventually we will be able to make predictions such as whether or not particles on two trajectories will hit one another, and how much proper time will have elapsed for each of them when they do. Notice, however, that by interpreting τ as proper time we are making a physical statement which is far from obvious in the field theory approach, but which becomes self-evident in the geometric approach. So one may say that the field theory approach looks to the geometric approach for a convenient justification of some of the basic concepts. But the field theory approach has the useful feature that one never needs to ask whether spacetime is curved, or what that could mean.

Now let us turn to the **geometric interpretation**. The equations are all the same, but now we get a lot more insight into what is going on.

In the geometric interpretation the quantities Γ^a_{bc} are called *Christoffel symbols* or *connection coefficients*. The name 'gravitational field' is not needed, but if it is adopted then it can be used for the **Riemann curvature tensor**, which is related to the Christoffel symbols by

$$R^a_{bcd} \equiv \partial_c \Gamma^a_{db} - \partial_d \Gamma^a_{cb} + \Gamma^a_{c\lambda}\Gamma^\lambda_{db} - \Gamma^a_{d\lambda}\Gamma^\lambda_{cb}. \qquad (2.13)$$

This fourth-rank tensor is something of a 'monster', but it has many symmetries which are described in the box. The symmetry properties are not postulates but mathematical properties which follow from the definition (you will be invited to obtain them in Chapter 15 using eqns (15.4), (15.5)). In a two-dimensional spacetime, R^a_{bcd} is still of fourth rank, but the indices only range over two dimensions, say $0, 1$. In this case the symmetries of the tensor are such that it is completely specified by a single component such as R^0_{101}.

The Riemann curvature tensor expresses the degree of **Gaussian curvature** in each direction at each point in any given manifold. Exactly how or why it does that is a long story which will be told in Chapters 8–15. Gaussian curvature is a measure of the amount by which geometry on a surface departs from Euclidean geometry; see box for some further information. This was discussed in Volume 1 and we will revisit it in Chapter 15.

The Ricci tensor is obtained directly from the Riemann curvature tensor:

$$R_{bd} \equiv R^\mu_{b\mu d} \qquad (2.19)$$

Gaussian curvature

The Gaussian curvature K of a two-dimensional manifold (i.e. a surface) quantifies the degree to which geometry on the surface departs from Euclidean geometry. Define a circle on the surface as a locus of points at the same distance r from some given point, as measured along the surface. Let C be the circumference of such a circle, again as measured along the surface. Then at any point on the surface the Gaussian curvature is defined to be

$$K \equiv \lim_{r \to 0} 3\frac{(2\pi r - C)}{\pi r^3}. \tag{2.18}$$

Let a triangle be defined as three points connected by lines of minimal length on the surface. It is shown in Volume 1 that the internal angles of such a triangle sum to $\pi + K\sigma$ radians in the limit as $\sigma \to 0$, where σ is the area of the triangle.

If a three-dimensional sphere in three-dimensional Euclidean space has radius d, then one finds that the Gaussian curvature of its surface is equal to $K = 1/d^2$.

(this was previewed without comment in eqn (2.2)). You should check that (2.2) has been obtained correctly from (2.13).

Sometimes a further tensor called the Einstein tensor is introduced:

$$G_{ab} \equiv R_{ab} - \tfrac{1}{2}R\,g_{ab} \tag{2.20}$$

and then one can write the field equation as

$$G_{ab} = \frac{8\pi G}{c^4}T_{ab}. \tag{2.21}$$

In practice one deals mostly with the Ricci tensor.

The Ricci tensor, and therefore also the Einstein tensor, may be interpreted as an average of the curvature in different directions. The physical interpretation of the Einstein field equation is that this average curvature is zero in empty space, and in the presence of matter and other (non-gravitational) fields it is proportional to the stress-energy tensor. The proportionality constant is contained in the field equation, and also in (2.6).

The heart of the geometric interpretation is to note that spacetime is a differentiable manifold and g_{ab} is the metric tensor of this manifold. What this means is that by combining g_{ab} with infinitesimal coordinate changes dx^a one can find out the spacetime distance, called *interval*, between any two neighbouring events. The formula for this interval ds is

$$ds^2 = g_{\mu\nu}dx^\mu dx^\nu \tag{2.22}$$

There is a huge amount of physics hidden in this simple equation. For when we combine this with the assertion that the whole influence of gravity can be expressed as an influence on g_{ab}, then we obtain the strong equivalence principle—the principle which says that ordinary (special relativistic) physics, without any mention of gravity, holds in any sufficiently small region of spacetime mapped in appropriate coordinates. This can be deduced because the mathematics of manifolds will enable us to show that in the vicinity of any given event it is always possible to find coordinates in which g_{ab} takes a simple form, namely $g_{ab} = \eta_{ab}$ where η_{ab} is the Minkowski metric. This tells us a huge amount of information about gravitational phenomena! It means that ordinary materials, and ordinary processes such as clocks, chemical reactions and so on, all behave, on small scales, just as they would in the absence of gravity. It would take a large amount of calculation to deduce this directly from the field equations and geodesic equation without the benefit of geometric insight!

Another reason behind the powerful role of (2.22) is that when we write the equation we make two further assertions: we assert that g_{ab} is the set of covariant components of a tensor, and $\mathrm{d}x^a$ is the set of contravariant components of a vector (that is, a 4-vector). It follows, by the rules of tensor analysis to be described in subsequent chapters, that $\mathrm{d}s$ is a *scalar invariant* quantity, and this is a remarkable statement. It means that, for a given pair of neighbouring events P and Q, the value of $\mathrm{d}s$ will not depend on the particular way the coordinates x^a are assigned. But coordinate systems are almost arbitrary! We can imagine laying out coordinates across any given region of space and time almost however we like (the *almost* here is to do with smoothness, more on this later). What happens is that if one changes the coordinate assignments from one system, say x^a, to another system, say x'^a, then the values of the components g_{ab} will also change, according to a regular rule, namely

$$g'_{ab} = \frac{\partial x^\mu}{\partial x'^a}\frac{\partial x^\nu}{\partial x'^b}g_{\mu\nu} \tag{2.23}$$

and one will find that

$$g'_{\mu\nu}\mathrm{d}x'^\mu\mathrm{d}x'^\nu = g_{\mu\nu}\mathrm{d}x^\mu\mathrm{d}x^\nu. \tag{2.24}$$

In other words the interval is invariant, as claimed.

It is an important property of (2.22) that the right-hand side is quadratic in the coordinate displacements. It differs from other possibilities such as $g_{\mu\nu}|\mathrm{d}x^\mu||\mathrm{d}x^\nu|$ or $(-\mathrm{d}t^4 + \mathrm{d}x^4 + \mathrm{d}y^4 + \mathrm{d}z^4)^{1/2}$; this is a further way in which the equation has mathematical and physical content. The type of space in which distance measurements can be expressed in the form (2.22) (a sum of quadratic dependences on coordinate separations) is called a *Riemannian* or a *pseudo-Riemannian* manifold. If intervals are always positive we have the Riemannian case. GR asserts that spacetime is a pseudo-Riemannian manifold.

2.1.2   The indirect nature of gravity

We will next take a look at the way gravitational physics bears on other areas of physics. We are particularly interested in the subtle way in which gravity often works 'behind the scenes', as we will explain.

Let us first consider how Newtonian physics proceeds. We typically take an interest in how blobs of matter called 'bodies', or small blobs called 'particles', move about and impinge on one another. It is very convenient, in order to keep track of these bodies or particles, to introduce the notion of space and time. We set up a system of coordinates in space, and a way to keep track of time, and then we can define things like displacement, velocity, acceleration, and eventually force and mass. When Einstein invites us to go back and think about what we did, we come to realize that our system of coordinates was really a shorthand for the layout of a set of rigid objects such as a scaffolding of iron bars, and our notion of time was a reference to physical phenomena involving regularly repeating oscillation. The 'background' to our study of physical stuff thus turns out to be just more physical stuff!

After noticing this, we might ask ourselves, is there any such thing as space after all, then? Is there any such thing as time? Is not the whole story of the world just things hitting other things, following patterns set out in various mathematical relationships, and we can forget about spacetime?

The answer provided by General Relativity is 'no: spacetime is real enough, but its reality is of a subtle kind. Here is my proof that spacetime is real: it does not have just any old metric. It only ever has, at each location, a metric which is locally Minkowskian. Also, its curvature is tensorial.[1] Furthermore, it can transmit energy and momentum via gravitational waves.' Now that we have settled that we have a spacetime to think about, which can serve as the arena in which blobs of matter (and other things such as electromagnetic fields) can exist, we can get on with the job of finding out how those blobs and fields affect one another. For example, electromagnetic phenomena are captured by the equations called Maxwell's equations. We know how those equations are expressed in the absence of gravity. It turns out that in the presence of gravity we can express them as follows:

$$\nabla_\lambda \mathbb{F}^{\lambda b} = -\mu_0 j^{\cdot b} \tag{2.25}$$

$$\partial_c \mathbb{F}^{ab} + \partial_a \mathbb{F}^{bc} + \partial_b \mathbb{F}^{ca} = 0. \tag{2.26}$$

where $\mathbb{F}$ is a second-rank tensor field, **j** is the electric 4-current density and the operator ∇ denotes the **covariant derivative**, to be developed in Chapter 10 (it will feature prominently in the rest of the book). These equations look a lot like Maxwell's equations! In fact they are Maxwell's equations, but written in a more general way. The field $\mathbb{F}$ is called the electromagnetic field, and we can aim to find the motion of particles subject to its force, and the evolution of the field itself, and thus predict electromagnetic phenomena. Since there is no direct mention of gravity in these equations, it might look as though electromagnetic phenomena are not influenced by gravity. Does light not bend on its journey around the Sun then?

In fact it does. Hiding in that innocent-looking ∇_λ there is a bunch of Christoffel symbols Γ:

$$\nabla_c \mathbb{F}^{ab} \equiv \partial_c \mathbb{F}^{ab} + \Gamma^a_{c\mu} \mathbb{F}^{\mu b} + \Gamma^b_{c\mu} \mathbb{F}^{a\mu}. \tag{2.27}$$

Gravity influences electromagnetic phenomena by affecting the Christoffel symbols. Thus we find that the electromagnetic field equation (2.25) turns out to involve gravity after all.

---

[1]That is to say, the curvature is what it is, independent of what system of coordinates may have been adopted.

A **test particle**, in the context of eqn (2.10), is a particle of small enough mass and electric charge that it does not itself significantly disturb the gravitational and electromagnetic environment, and having sufficiently small intrinsic angular momentum (spin) that the gravitational coupling to the latter can be neglected. We will often be interested in cases where the electromagnetic field tensor $\mathbb{F}$ is zero, and then we will be studying motion under gravity alone. In that case equation (2.10) reduces to the geodesic equation which we repeat here:

Geodesic equation again

$$\frac{\mathrm{d}u^a}{\mathrm{d}\tau} + \Gamma^a_{\mu\nu} u^\mu u^\nu = 0 \qquad \Leftrightarrow \qquad u^\lambda \nabla_\lambda u^a = 0 \qquad (2.28)$$

The second version shows how the equation is written using the covariant derivative operator; this will be presented more fully in Chapter 10.

The name 'geodesic equation' indicates that this equation has a geometric interpretation: it is a differential equation describing those curves in spacetime which are geodesics of spacetime. The notion of a geodesic was introduced in Volume 1 and will be expounded more fully in Chapter 13. It is essentially a path which is as straight as it can be in the given spacetime. The equation asserts that test particles in free fall have geodesic worldlines.

The geodesic equation can also be written

$$\frac{\mathrm{d}\mathbf{u}}{\mathrm{d}\tau} = 0 \qquad (2.29)$$

where the bold font indicates that we are giving a statement about the 4-velocity $\mathbf{u}$ as a tensorial mathematical quantity (here, a first-rank tensor) as opposed to a list of components, and we are invoking a type of differentiation which has to be defined carefully when applied to such an object (Chapter 10). Equations (2.28) and (2.29) demonstrate neatly the two ways of looking at gravity. In eqn (2.28) the role of $\Gamma^a_{\mu\nu}$ is a force per unit velocity per unit momentum, but in eqn (2.29) there is no 'force of gravity' at all! How can these equations possibly be saying the same thing? The second version of (2.28) gives a hint at how the mathematics works. We can regard gravity as affecting the very meaning of change and the lack of it as one moves from one event in spacetime to another. This is the geometric interpretation; it is all about how one defines the notion of a straight line in a curved space. One of the aims of the book is to make that concept crystal clear, so that the student can move between (2.28) and (2.29) at will.

2.2 The strong equivalence principle

A naive take on GR would suggest that all one can do with it is calculate worldlines. But of course a physicist would never be satisfied with that. We want to get physical understanding, and we want to calculate more than just isolated particles. We want to know what happens to things such as solid bodies with 10^{24} particles in them, and processes such as chemical

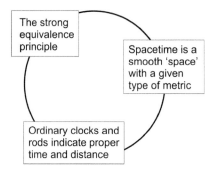

Fig. 2.1 Three mutually interacting ideas that underpin General Relativity.

reactions and radioactive decay. It turns out that much of the behaviour of ordinary physical objects can be deduced by noting that *spacetime is locally flat*. This statement is a geometric way of stating the *strong equivalence principle* which was introduced in Volume 1 and already mentioned above. One of the implications is:

> Any physical law which can be expressed as a tensor equation in Special Relativity has exactly the same form in a local inertial frame in a curved spacetime.

This is a useful statement, but it should be read carefully. It does *not* assert that Special Relativity itself suffices to find physical laws; it asserts merely that if there is a physical law that can be expressed without making direct appeal to concepts not applicable to Special Relativity, then we can expect such a law to apply locally in each region of spacetime, with the mathematics of tensors taking care of the details. We will develop tensor methods in Chapter 12; here we shall illustrate the results for some simple phenomena.

Fig. 2.1 presents three related ideas. The statement that spacetime is a smooth 'space' asserts that we can accurately describe or represent spacetime as the type of mathematical entity called *differentiable manifold*, mentioned already. This matches our everyday intuition. The statement about a 'metric' was discussed in connection with eqn (2.22). The other two boxes in Fig. 2.1 contain assertions that could (just possibly) be regarded as following from the Einstein field equation and other equations of fundamental physics, but it makes equally good sense to regard them as axioms by which such equations can be deduced or constrained. Both are statements about physics. If we have found or been given the metric applicable to some region of spacetime, then mathematics alone will allow us to determine, from the metric, mathematical distances and times in spacetime. The statement about clocks and rods is then a further statement, concerning physical behaviour. It asserts that a small (somewhat idealized, but largely ordinary) clock will 'match', by its internal evolution, what the metric says about time: it will evolve in step with the accumulated proper time along its own worldline. Similarly, the internal forces which result in the equilibrium length of a small rod (again somewhat idealized) will result in a fixed proper length of the rod, as indicated by the metric. This statement about rods and clocks could also be regarded as following from the strong equivalence principle and the statement about spacetime, but it is useful to single it out for special attention nonetheless, as we have done in Fig. 2.1.

The **local inertial frame** (LIF) is an important idea in GR. It can be thought of as an idealized set of rods and clocks all freely falling alongside each other in some region of spacetime, without disturbing that region. A LIF can be used to construct a coordinate labelling in which, locally, the metric tensor takes a very simple form, as we will prove in Chapter 8.

General Relativity predicts (or, if you prefer, assumes) the strong equivalence principle. This principle asserts that local physics always goes the same in local freely falling frames, and therefore if our measuring rods and clocks were in free fall, then for short enough rods and short enough elapsed time, gravity will have no effect, and everything will transpire just as in Special Relativity. In particular, the distance measurements will be consistent with Euclidean geometry, and released objects will move in straight lines relative to any LIF, in the absence of ambient electromagnetic fields etc. Again, this is a substantial prediction. It includes the prediction that in these circumstances the circumference of small circles, as measured by these rods, will be equal to 2π times their radius, and triangles will have internal angles summing to $180°$, and all other features of Euclidean geometry will hold, and it also implies the symmetries which result in conservation laws for energy and momentum in small enough regions of spacetime. Finally, we may also deduce that the **causal structure** of spacetime is delineated by **light cones**, just as in Special Relativity. For at each event on a worldline, we can always consider a LIF at that event, and we must find that in the LIF the worldlines of massive entities are timelike, and those of massless entities are null. The new feature introduced by GR is that now the light cones (each defined as the surface formed by all null geodesics passing through a given event) can weave through spacetime in non-trivial ways, because spacetime itself is warped.

2.2.1 All frames are equal, but some are more equal than others

The title of this section is an oblique reference to Orwell's *Animal Farm*. In the present context the phrase 'all frames are equal' indicates that we can, if we wish, track position, time and motion relative to a reference body that is in any state of motion (even a living jellyfish could be used, as Einstein liked to point out), and GR provides the mathematical tools to do this. Nevertheless, some reference bodies are more useful than others in practice, which is indicated by the second half of the title. The slightly paradoxical or enigmatic tone is serving a further purpose. It is to signal that GR has an ambiguous relationship with the notion that acceleration and rotation cannot be determined absolutely. The relationship can be succinctly stated as follows:

> One *can* determine, without looking out of the window, whether or not one is in a local inertial frame, but one *cannot* determine, without looking out of the window, the amount of acceleration (including rotation) of this frame relative to the average matter distribution of the universe.

The strong equivalence principle implies both parts of this statement. The first half of the statement considers a physically constructed chamber around oneself, relative to which one may make position and timing measurements. In order to determine whether or not the chamber is a local inertial frame (LIF), just note whether objects released or thrown move in straight lines relative to the chamber and each other (and be sure to try more than one speed and at least three directions of travel). Or, for a more sophisticated measurement, observe the chamber using radar echos and a single clock.

What the above statement means is that if one is floating free, like an astronaut in a space suit, then one *can* determine whether or not there are inertial forces, such as centrifugal forces, acting on ones hands or feet, *without* the need to look around at the distant stars, but one cannot assume that a frame with no such forces will be one that shows no rotation relative to the distant stars. The latter is the more surprising fact, and it is so because a large enough nearby mass that is itself rotating can so disturb spacetime that the LIFs near to it are in a state of rotation relative to LIFs far away. This is called *frame dragging*.

There is an important distinction here between the concept 'inertial' and the concept 'accelerating'. The concept 'inertial' refers to motion in which no force other than gravity is acting; the concept 'acceleration' refers to rate of change of velocity. The 3-velocity is always relative—we can never define a 3-velocity except in terms of relative motion. The concept of inertial motion, on the other hand, is absolute—one can say absolutely whether or not some given entity is moving inertially. It is quite common, in the context of GR, to define a 4-acceleration 4-vector such that its value is zero for inertial motion. Such a quantity is, obviously, zero for an object moving inertially because that is how it has been defined. All its components are zero, and therefore they remain zero after a change of coordinate system or reference frame. This is unlike 4-velocity, which always has at least one non-zero component (the temporal one). So we can define a notion of absolute acceleration. However, when discussing some specific physical situation, such as motion of planets or rockets near a star, it is often useful to track their motion relative to a coordinate frame which is itself fixed relative to the star. The word 'acceleration' may then refer to rate of change of 3-velocity relative to the coordinate frame, and this usage more closely matches the use of the word 'acceleration' in everyday speech. We may say, for example, that a falling apple accelerates towards the ground. But we can also assert that the 4-acceleration of the apple is zero and the 4-acceleration of the ground is not zero—the whole surface of the Earth is being accelerated outwards (relative to any LIF) by electromagnetic forces in the rocks of the Earth. It is not a matter of right or wrong usage; it is a matter of being clear about what terminology has been adopted.

In order to connect GR to our everyday experience of gravity, it is helpful to define a quantity called 'the acceleration due to gravity'. This can be defined as the relative acceleration between two bodies owing to their mutual gravitation. For example, planet Earth moves in freefall on its journey around the Sun, and an object released above its surface also moves in freefall, with a motion that orbits the Sun along with Earth, and also accelerates towards Earth. The two trajectories (two geodesics) approach one another more and more rapidly—a fact whose geometric name is *geodesic deviation*. The confusing fact is that both these objects have zero 4-acceleration, and yet there is this non-zero 3-acceleration between them. To avoid the confusion, learn to regard 'acceleration due to gravity' not as an acceleration of one entity, but as a relative motion of two entities (in our example, the centre of planet Earth and the falling rock).

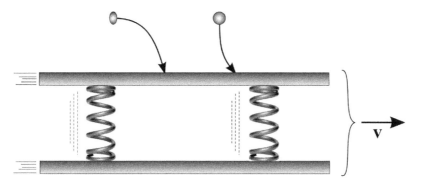

Fig. 2.2 An argument to show that gravity relates in part to flow of matter. The figure shows a pair of massive rods which attract each other by gravitation, and which are held at some equilibrium separation by a pair of springs. In the rest frame S of rods and springs, the springs exert a force f. Special Relativity tells us that the force exerted by the springs is smaller in the frame S' depicted in the diagram, in which the springs are moving to the right. It follows that the gravitational attraction of the rods must be reduced also, even though their energy density is larger in S'. We deduce, therefore, that *the motion of these rods must reduce their mutual gravitation*. The figure also shows two small balls falling towards the rods. The ball on the left was released from rest in S. Let a be its initial acceleration relative to S. Special Relativity tells us that the vertical component of its acceleration is smaller than a in S'; this is consistent with the reduced gravitation already noted. The ball on the right is released from rest in S'. It accelerates towards the rods, and there is also a slight gravitational acceleration in the direction of motion of the rods. We will show in Chapter 5 that this is consistent with the other properties. The above reasoning can be made precise in the weak field limit, and gives a good qualitative guide more generally.

One may also regard 'acceleration due to gravity' as a reference to the upwards acceleration, relative to a LIF, which an object has to be given by non-gravitational forces in order to prevent it from moving relative to some given reference body (such as planet Earth). What is 'due to gravity' here is not the non-gravitational forces themselves, but the fact that they have to act in order to prevent the supported body from moving towards the centre of the Earth. In this scenario the supported object has a non-zero 4-acceleration. In a suitable coordinate system the spatial part of this 4-acceleration is equal to the acceleration due to gravity, and the *weight* of an object is the size of the force that is required to keep it at a fixed coordinate position. This concept is only useful when there is a natural choice of such coordinates, which may not happen when spacetime is itself dynamic.

The announcement 'some frames are more equal than others' draws attention to the fact that local inertial frames play a sufficiently useful role in GR that they deserve recognition as having special status. Einstein was determined to pursue a theory in which the mathematical form of the equations of physics would be the same in *all* frames, including ones where all the parts wobble and accelerate in general ways, and he succeeded. Nevertheless, in order to *understand* what the theory is telling us, the LIFs are, again and again, the most helpful systems of reference.

2.3   The source of gravity

The mathematical form of the field equation is one in which the stress-energy tensor is the *source* of the gravitational field. In electromagnetism we had a source in the form of a 4-vector; here we see the source of the gravitational field is a second-rank tensor. We should have expected to see the T^{00} component (energy density)—we know that mass causes gravitational attraction—and equally we know that this quantity alone is not a Lorentz scalar so cannot be the whole story. We might have held out the (forlorn) hope that some 4-vector related to the Poynting vector could act as source; we have to reconcile ourselves to the fact that the whole stress-energy appears. It follows that not only does mass gravitate, but so does energy flux and momentum density. Just as a current of electrical charge will generate a magnetic field, similarly, a gravity field may be generated in part by a current of matter—see Fig. 2.2. Two gases having the same internal energy but different internal pressure will generate different gravitational fields.

To construct a stress-energy tensor, let it be *defined* through its energy and momentum components in some convenient LIF:

$$T^{ab} \overset{\mathrm{LIF}}{=} \begin{pmatrix} \rho c^2 & c\mathbf{g} \\ c\mathbf{g} & \sigma^{ij} \end{pmatrix} \tag{2.30}$$

where ρc^2 is energy density, $\mathbf{g}$ is momentum density and σ^{ij} is stress. Then we assert that T^{ab} is a tensor, which tells us how to express it in all coordinate systems (Chapter 12). Note that this method works owing to the strong equivalence principle. We can use ordinary quantum mechanics and materials science to figure out T^{ab} at any one event in a LIF. There is a difficulty, however. When the gravitational effects are extreme, the metric may vary so rapidly as a function of distance and time that the LIF approximation (which amounts to taking a first-order approximation for the dependence of g_{ab} on position and time) is not sufficient to allow ordinary materials science to be used. In this case all one can do is resort to a new form of materials science in curved spacetime. However, it often happens that the details are not needed in order to perform accurate calculations. For example, in the case of spherical symmetry we can assert that Schwarzschild and Droste's solution (Chapter 17) gives the metric outside the gravitating body, and all we need to do is measure one parameter (the mass of the central body such as a star) in order to find out everything we need in order to find the particle orbits etc. It will not be necessary to calculate the mass from a model of the stellar interior; one simply infers it by measuring any one orbit or by observing the ticking of clocks at two different distances from the star (exercise 2.6).

A further issue arises from a contribution called by either of two names: the *cosmological constant*, or *dark energy*. These are two names for the same mathematical contribution to the field equation.[2] Let us introduce a term Λg_{ab} into the field equation, where Λ is a constant, independent of time and space. If we place this term on the right-hand side, then it can be absorbed into the stress-energy by writing

$$T_{ab} = T_{ab}^{(\text{everything else})} - \frac{c^4}{8\pi G}\Lambda g_{ab} \tag{2.31}$$

---

[2]Strictly speaking, the term *dark energy* is slightly more general. A cosmological constant is one form that dark energy may take; this will be discussed in Chapter 22.

and this is what we have done in writing eqn (2.9). The reason to consider such a term is two-fold. We will show later that $\nabla_c g_{ab} = 0$, and this means that the equation with the new term still makes sense as a model of gravity; the second reason is that cosmological measurements suggest $\Lambda \neq 0$. The observations suggest $\Lambda \sim \mathcal{O}(10^{-52})\,\mathrm{m}^{-2}$, which implies that $c^2\Lambda/(8\pi G)$ has a value equal to the mass density corresponding to a few hydrogen atoms per cubic metre. Therefore this term is negligible compared to ordinary matter, when ordinary matter is present, and even in the vacuum outside ordinary matter its influence on spacetime will remain negligible up to some considerable distance from a gravitating body. However, the Λ contribution appears *throughout* space, including in the vast voids between galaxies, and therefore on the global cosmological scale it can be the dominant contribution. We will return to this in Chapters 22–25 and mostly ignore it until then.

Students who have heard of the idea that gravitation has something to do with curved spacetime might guess that spacetime is curved only at points where matter is present. This is wrong. The Einstein equation (2.1) is like Poisson's equation $\nabla^2\phi = \rho$: it sets a condition on the second derivatives of the quantity of interest (g_{ab}). In the presence of matter, the relevant derivatives sum to a non-zero value. Elsewhere they sum to a zero value, but this does not require that they are all zero individually. In short, G^{ab} is not the gravitational field, nor is it the curvature (it is a measure of the orientation-averaged curvature). In free space outside a finite gravitating body $G^{ab} = 0$, but the curvature tensor R^a_{bcd} is not zero. Rather, it falls gracefully to zero as one moves away from the body. The summary is:

- If $R^a_{bcd} \neq 0$ then spacetime is curved (and consequently there are tidal effects owing to gravity), and there may or may not be matter present.
- If $R_{ab} \neq 0$ then there is matter present, and spacetime is curved.
- If $R_{ab} = 0$ then there is no matter present, and spacetime may or may not be curved.
- If $R^a_{bcd} = 0$ then spacetime is not curved.

In this list of statements we have used the word 'matter' as a catch-all for 'gravitating stuff', which includes electromagnetic fields and all other non-gravitational fields, as well as ordinary matter, and the dark energy or cosmological constant.

We will be much concerned with the metric tensor defined implicitly by eqn (2.22). A given metric may be written down by displaying, as a 4×4 matrix, all of its components, as they appear in some given system of coordinates. Here is an example (the Schwarzschild–Droste metric, expressed in a spherical system of coordinates):

$$[g_{ab}] = \begin{pmatrix} -(1 - 2GM/c^2r)c^2 & 0 & 0 & 0 \\ 0 & (1 - 2GM/c^2r)^{-1} & 0 & 0 \\ 0 & 0 & r^2 & 0 \\ 0 & 0 & 0 & r^2\sin^2\theta \end{pmatrix}. \tag{2.32}$$

The same information can be furnished by writing down the line element equation (2.22):

$$\mathrm{d}s^2 = -\left(1 - \frac{2GM}{c^2r}\right)c^2\mathrm{d}t^2 + \frac{1}{1 - 2GM/c^2r}\mathrm{d}r^2 + r^2\mathrm{d}\theta^2 + r^2\sin^2(\theta)\mathrm{d}\phi^2 \tag{2.33}$$

For this reason, the equation for the line element is often introduced with the phrase 'here is the metric'. It is understood that the reader will be able to extract the components of the metric by noting the factors appearing in the various terms in the line element equation.

Exercises

2.1 Prove from equations (2.8)–(2.10) that if $\mathbb{F} = 0$ and g_{ab} is independent of position and time, then the 4-velocity of a test particle $(\mathrm{d}x^a/\mathrm{d}\tau)$ will be constant.

2.2 We will gain some physical intuition about Γ^i_{00} by relating it to the acceleration due to gravity in a simple case. Consider a stationary spacetime treated by adopting coordinates x^a. Let $\vec{x}(\tau)$ be the set of three functions describing the trajectory of a particle in free fall as a function of elapsed proper time along the worldline. Consider a particle released from rest at some event A. Define the 3-acceleration due to gravity as the value of $\mathrm{d}^2 l/\mathrm{d}\tau^2$ when the particle is momentarily at rest relative to the coordinates, where $\mathrm{d}l = (g_{ij}\mathrm{d}x^i\mathrm{d}x^j)^{1/2}$ is the proper distance between adjacent points on the trajectory (here i and j range over spatial coordinates 1 to 3). For simplicity suppose that only the first component x^1 has a non-zero second-derivative at A, so that we can take $\mathrm{d}l = \sqrt{g_{11}}\mathrm{d}x^1$ in the vicinity of A. (i) Let $u^a = \mathrm{d}x^a/\mathrm{d}\tau$ be the particle's 4-velocity. The norm of u^a is given by $g_{\mu\nu}u^\mu u^\nu$ and it will be shown later that this must be equal to $-c^2$. At A we have that u^a has zero spatial part. Show that at A we must have

$$u^a = (c/\sqrt{-g_{00}}, \, 0, 0, 0). \qquad (2.34)$$

(ii) Use the geodesic equation to relate Γ^a_{00} to $\mathrm{d}^2 l/\mathrm{d}\tau^2$. [*Ans.* $\Gamma^1_{00} = (g_{00}/\sqrt{g_{11}})c^{-2}\,\mathrm{d}^2 l/\mathrm{d}\tau^2$] (c.f. (11.51))

2.3 Let $\vec{g}(r)$ be the acceleration due to gravity at distance r from the centre of the Earth.

This is the relative 3-acceleration of Earth and nearby objects in free fall. Show that if we adopt ordinary coordinates then the geodesic equation implies that the non-zero Christoffel symbols must be of order $|\vec{g}|/c^2$ and hence the non-zero components of the curvature tensor must be of order $|\vec{g}|/rc^2$ (detailed calculation is not required).

2.4 For a two-dimensional spacetime, let the indices on coordinates be $0, 1$. Write down all sixteen possibilities for a string of four such indices. If $k = R_{0101}$ find the values of all the other components of R_{abcd} in terms of k using the symmetry properties (2.15). Confirm that (2.16) holds automatically for this case. [*Ans.* $R_{1010} = k$, $R_{0110} = R_{1001} = -k$, others zero]

2.5 Repeat the previous exercise, now for three dimensions, but do not write out all the eighty-one combinations; focus attention on the thirty-six combinations for which the first two indices differ and the last two indices differ. Hence show that there are six independent elements.

2.6 Find the mass of the Sun using the periods of orbits treated by Newtonian physics, where you are given that two circular orbits with periods close to 88 days differ in period by 197 seconds when their radii differ by 1000 km. You may find it useful first to obtain the formula

$$GM = \frac{27\pi^2}{2}T\left(\frac{\mathrm{d}T}{\mathrm{d}r}\right)^{-3}$$

where T is the period.

3

An introductory example: the uniform static field

In this short chapter we show how the equations of the previous chapter are used to extract information about curvature and the stress-energy tensor for a simple but non-trivial form of metric. We then apply the geodesic equation to find the motion of test particles in a region of spacetime described by the given metric.

We will explore a metric having the following form:

$$ds^2 = -\alpha^2(x)dt^2 + dx^2 + dy^2 + dz^2 \tag{3.1}$$

where $\alpha(x)$ is a function of x and each of the coordinates t, x, y, z has the range $-\infty$ to ∞. We shall take particular interest in the two cases

$$\alpha(x) = ax \qquad\qquad\qquad \text{Rindler metric} \tag{3.2}$$
$$\alpha(x) = ce^{kx} \qquad\qquad\qquad \text{uniform static field} \tag{3.3}$$

where a and k are constants. The gravitational field in both of these cases may be said to be spatially uniform in some respects and not in others, as we shall see. Nonetheless we refer to the second case as 'uniform' because if one makes a shift in the spatial coordinates and multiplies the time coordinate by an appropriate factor then the metric is unchanged. To be precise, if one introduces new coordinates T, X defined by

$$X = x - x_0, \qquad T = e^{kx_0}t \tag{3.4}$$

where x_0 is a constant then one has

$$ds^2 = -c^2 e^{2kX} dT^2 + dX^2 + dy^2 + dz^2 \tag{3.5}$$

which is the same metric as before, apart from a coordinate relabelling. Consequently all the local gravitational effects are independent of position in the uniform static field, which is not the case for the Rindler metric.

Faced with a line element of the form (3.1), our first task is to try to interpret the coordinates a little. You should suspect from the names that t has something to do with time, and x, y, z form

Relativity Made Relatively Easy: General Relativity and Cosmology. Volume 2. Andrew M. Steane,
Oxford University Press. © Andrew M. Steane 2021. DOI: 10.1093/oso/9780192895646.003.0003

perhaps a rectangular or Cartesian system of coordinates. To confirm these guesses, note that when $dx = dy = dz = 0$ then $ds^2 < 0$ which shows that the t direction is timelike. Similarly, the other directions are spacelike. You can also note that at any given t the region of spacetime mapped out by the other coordinates has a Euclidean metric. This confirms, for example, that we will never encounter a place where many different values of a coordinate all refer to the same event (contrast this with the situation in some polar coordinate systems, where at $\theta = 0$ or $\theta = \pi$ all terms in the line element involving $d\phi$ vanish, which implies that different values of ϕ refer to the same event, c.f. eqn (2.33)).

We have now said enough to settle in our minds that it makes sense to regard x, y, z as mapping space in Cartesian coordinates, with t indicating a measure of time (but it is not proper time). Note that, more generally, we are not guaranteed always to get such a neat separation of spatial and temporal directions via the coordinate system.

From (3.1) we deduce that the metric and its inverse are

$$g_{ab} = \begin{pmatrix} -\alpha^2 & & & \\ & 1 & & \\ & & 1 & \\ & & & 1 \end{pmatrix}, \qquad g^{ab} = \begin{pmatrix} -\frac{1}{\alpha^2} & & & \\ & 1 & & \\ & & 1 & \\ & & & 1 \end{pmatrix}. \tag{3.6}$$

The only non-zero derivative of g_{ab} is then

$$\partial_1 g_{00} = -2\alpha \frac{d\alpha}{dx}.$$

This makes the equation (2.8) for the connection symbols manageable (later we will show how to obtain them an easier way). One finds that the only non-zero connection symbols are

$$\Gamma^0_{01} = \Gamma^0_{10} = \frac{1}{\alpha}\frac{d\alpha}{dx}, \qquad \Gamma^1_{00} = \alpha\frac{d\alpha}{dx}. \tag{3.7}$$

The only non-zero derivatives of these are

$$\partial_1 \Gamma^0_{01} = \frac{1}{\alpha}\frac{d^2\alpha}{dx^2} - \frac{1}{\alpha^2}\left(\frac{d\alpha}{dx}\right)^2, \qquad \partial_1 \Gamma^0_{10} = \partial_1 \Gamma^0_{01},$$

$$\partial_1 \Gamma^1_{00} = \alpha\frac{d^2\alpha}{dx^2} + \left(\frac{d\alpha}{dx}\right)^2.$$

We can now find the Ricci tensor. A moment's investigation will teach you that this is essentially a two-dimensional problem because all elements of R_{abcd} with any index not equal to 0 or 1 vanish. We therefore only have to work out the elements $00, 01, 11$ of R_{ab} (since $R_{10} = R_{01}$ and other elements are zero). Using (2.2) we find (exercise 3.2)

$$R_{bd} = \begin{pmatrix} \alpha\alpha'' & & & \\ & -\alpha''/\alpha & & \\ & & 0 & \\ & & & 0 \end{pmatrix} \tag{3.8}$$

where $\alpha'' \equiv \mathrm{d}^2\alpha/\mathrm{d}x^2$. Pre-multiplying this by g^{ab} in order to raise the first index, we find $R^a_{\ d} = \mathrm{diag}(-\alpha''/\alpha, -\alpha''/\alpha, 0, 0)$. Taking the trace gives

$$R = -\frac{2}{\alpha}\alpha''.$$

Therefore

$$G_{ab} = \alpha'' \begin{pmatrix} \alpha & & & \\ & -1/\alpha & & \\ & & 0 & \\ & & & 0 \end{pmatrix} + \frac{\alpha''}{\alpha} \begin{pmatrix} -\alpha^2 & & & \\ & 1 & & \\ & & 1 & \\ & & & 1 \end{pmatrix} = \begin{cases} 0 & \text{Rindler metric} \\ \mathrm{diag}(0, 0, k^2, k^2) & \text{uniform field} \end{cases}$$

If we now substitute this into the Einstein field equation, we discover that the stress-energy tensor of the matter producing such a field must be zero for the Rindler case (i.e. a vacuum— no matter at all), and must have pressure (in the y and z directions) but no energy-density in the uniform case. No such state of matter exists, therefore the second gravitational field under discussion is just a theoretical device useful for teaching purposes, not a field which could be physically realized.

The motion of test particles in this hypothetical field can be obtained from the geodesic equation (2.28), or from an Euler–Lagrange method to find a curve of maximum proper time for a given metric. In practice the second method is algebraically the more simple. We will present the details in Chapter 13. At this stage we will merely present the answer, as an illustration of the fact that the metric leads to the worldlines of test particles in free fall. For light rays one finds

$$t = t_0 \pm (1/ck)e^{-kx} \tag{3.9}$$

and for massive particles one finds

$$\frac{\mathrm{d}^2t}{\mathrm{d}\tau^2} = -2k\frac{\mathrm{d}t}{\mathrm{d}\tau}\frac{\mathrm{d}x}{\mathrm{d}\tau}, \tag{3.10}$$

$$\frac{\mathrm{d}^2x}{\mathrm{d}\tau^2} = -kc^2 e^{2kx}\left(\frac{\mathrm{d}t}{\mathrm{d}\tau}\right)^2. \tag{3.11}$$

where τ is the proper time along the worldline. The solution of these is discussed in the exercises. Fig. 3.1 illustrates the result by showing example freefall wordlines of photons and massive particles in the xt coordinate plane. Notice that the photons propagating upwards in the field accelerate as they go, as measured by coordinate speed. This does not change the fact that the rate at which they traverse a given ruler distance (as determined by the local metric),[1] per unit local proper time (again, indicated by the metric), is everywhere fixed at c (exercise 3.3).

The example massive particle worldlines in Fig. 3.1 indicate that, when measured by coordinate distance and time, the acceleration of a particle released from rest at a given height x increases

---

[1] *ruler distance* refers to distance on a spatial hypersurface; this is a more complex idea than proper time because it requires one to specify a hypersurface, not just a worldline. The word 'ruler' serves as a reminder of this difference, but the term 'proper distance' is also widely used.

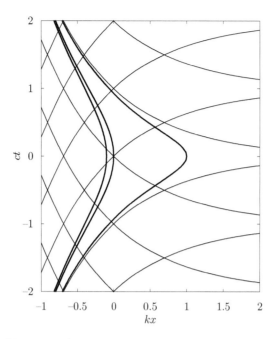

Fig. 3.1 Null geodesics (thin curves) and example free fall trajectories for massive particles (thick curves) in the gravitational field described by (3.1), (3.3) (see (3.9)–(3.11)).

with x. However, we may define 'the acceleration due to gravity' more usefully as the value of $d^2l/d\tau^2$ for a particle released from rest, where l is ruler distance (i.e. spatial separation as indicated by the metric) and τ is proper time (not coordinate time). This is the relative 3-acceleration, as observed in a LIF, between two entities: a particle in free fall and a particle whose position in the x^a coordinates is maintained constant by the forces acting on it. The spatial part of the metric under discussion is Euclidean, so $dl = dx$ here, and for adjacent events on the worldline of a particle momentarily at rest relative to the spatial coordinates, the metric (3.1) gives $dt/d\tau = \exp(-kx)$ (using $dx = dy = dz = 0$). Substituting this into eqn (3.11) we find the acceleration due to gravity is $d^2x/d\tau^2 = -kc^2$. This is independent of both position and time. This is an example of the sense in which the metric (3.1) with $\alpha = ce^{kx}$ represents a uniform gravitational field.

Sometimes the phrase 'uniform gravitational field' is used to refer to the Rindler metric (3.2), on the grounds that then there are no tidal effects, that is, no squeezing or stretching of an object in free fall. This is in contrast to (3.3) which gives a non-zero R^a_{bcd} and consequently tidal effects. But does not this mean the latter case is non-uniform after all? This is a matter of terminology. We already exhibited in (3.5) the sense in which there is spatial uniformity in the case $\alpha(x) = ce^{kx}$. It follows that for this case the amount of squeezing or stretching owing to the tidal effect of gravity is independent of position. Hence a given static string extended between given masses will be stretched by the same amount, in terms of proper length, no matter where it happens to be.

Exercises

3.1 Find the Christoffel symbols for the Rindler metric, and show that in this case $R_{ab} = 0$. What does this tell us about the physical situation? [*Ans.* $\Gamma^0_{01} = \Gamma^0_{10} = 1/x$, $\Gamma^1_{00} = a^2 x$; it is a vacuum solution]

3.2 Show that in the case where the only non-zero Christoffel symbols are Γ^0_{01}, Γ^0_{10}, Γ^1_{00} and they only depend on the coordinate x^1, then (2.2) gives

$$R_{bd} = \partial_1 \Gamma^1_{db} - \partial_d \Gamma^0_{0b} + \Gamma^0_{01} \Gamma^1_{db}$$
$$- (\Gamma^0_{d1}\Gamma^1_{0b} + \Gamma^1_{d0}\Gamma^0_{1b} + \Gamma^0_{d0}\Gamma^0_{0b}). \quad (3.12)$$

Hence obtain (3.8).

3.3 Consider neighbouring events A, B on the worldline described by (3.9). Let the co-ordinate separation between A and B be $(\delta t,\ \delta x)$. How much time would a clock in free fall, momentarily at rest relative to the spatial coordinates, register between A and B? What distance would a rigid measuring rod in free fall, momentarily at rest relative to the spatial coordinates, indicate between A and B? Hence prove that the speed of particles having the worldline (3.9), as indicated by these instruments, is equal to $\pm c$ for all events on the worldline. [*Ans.* $\exp(kx)\delta t,\ \delta x,\ \delta x/(\exp(kx)\delta t) = \pm c$]

3.4 Solve equations (3.10), (3.11), as follows. First eliminate τ, e.g. by considering $(\mathrm{d}/\mathrm{d}\tau)(\dot{t}/\dot{x})$ where the dot signifies differentiation with respect to τ. Hence obtain

$$\frac{\mathrm{d}^2 x}{\mathrm{d}t^2} = 2k \left(\frac{\mathrm{d}x}{\mathrm{d}t}\right)^2 - kc^2 e^{2kx}. \quad (3.13)$$

To solve this, introduce the change of variable $u = \exp(-2kx)$. Hence obtain

$$x = -(2k)^{-1}\ln(A + Bt + k^2 c^2 t^2) \quad (3.14)$$

where A and B are constants of integration.

3.5 Use your analysis of this example to illustrate the answer to exercise 2.2 of the previous chapter.

3.6 Use eqn (2.13) to obtain R^0_{101} and R^1_{001} for a metric of the form (3.1). Confirm that $R_{1001} = -R_{0101}$ for this case. [*Ans.* $-\alpha''/\alpha$, $-\alpha\alpha''$] *We will show in Chapter 15 how to relate this quantity to the geodesic deviation, that is, the way in which neighbouring geodesics approach or swerve away from one another. In the present example the negative value implies that a string attached to two objects, released into free fall from rest one above the other, will be stretched.*

3.7 **The Rindler fish.** Consider the force required at some location B in order to support in static conditions an object located at some other location A by dangling it on a string whose mass is negligible. It was shown in Chapter 8 of Volume 1 that in order to avoid unphysical scenarios such as an infinite energy source, the magnitude of this force must be given by

$$f_B = m_B |\vec{g}_A| e^{(\Phi_A - \Phi_B)/c^2} \quad (3.15)$$

where f_B is the force applied in a LIF at B, m_B is the inertial mass which the dangled object would have if it were located at B, $\vec{g}_A$ is the acceleration due to gravity at A, and Φ is the gravitational potential, which for the metric (3.1) is $\Phi(x) = c^2 \ln(\alpha/c)$. Equation (3.15) will also be derived (by you) at the end of Chapter 14. Assuming it for now, show that for the Rindler metric, f_B is independent of A, whereas for the uniform field f_B depends on A. More colloquially, then: a fish supported at x in the Rindler metric finds that its weight depends on x, but the fisher finds that the force on his fishing rod does not change as the fish is slowly raised.

4

Life in a rotating world

The aim of this chapter is to convey some of the flavour of GR and to develop physical intuition, by showing how phenomena involving space and time can be looked at in more than one way. We will begin to learn about gravitational effects associated with rotation, by considering in the first instance a rotating reference frame in flat spacetime. The phrase 'flat spacetime' implies that there exists a set of coordinates in which the solution of the Einstein field equation takes a form such that g_{ab} is everywhere equal to the Minkowski metric. In other words we are treating a situation that can be treated with perfect precision using Special Relativity. By doing so, we will nevertheless pick up some ideas and themes that also apply more generally, as we will show in the later parts of the chapter. The treatment will be exact where possible, and approximate where appropriate. Most of the mathematical treatment will concern 3-vectors not 4-vectors; we will adopt the bold font for 3-vectors in this chapter.

Consider, then, a race of people who live on the inside surface of a cylinder of radius 6371 km which rotates once every 1.407 hours (Fig.4.1).Their experience would be, to first approximation,

Fig. 4.1 The cylinder world. [Image Steve Bowers; https://www.orionsarm.com/eg-article/460db7f55a8d3]

Relativity Made Relatively Easy: General Relativity and Cosmology. Volume 2. Andrew M. Steane, Oxford University Press. © Andrew M. Steane 2021. DOI: 10.1093/oso/9780192895646.003.0004

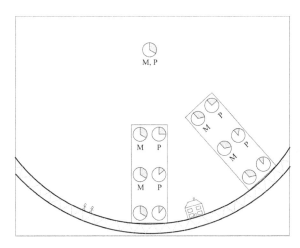

Fig. 4.2 A part of the cylinder world, showing master clocks (M) and proper clocks (P) on a pair of skyscrapers. The proper clocks run slow compared to the master clocks.

much like that of ourselves on Earth. They experience a 'gravitational' acceleration of $9.8\,\mathrm{ms}^{-2}$ towards their ground (i.e. the inside surface of the cylinder). They can build houses and skyscrapers, they can play frisbee and tennis. A plum bob suspended from the top of a skyscraper will hang straight down to the ground. They are well aware that their world has a cylindrical global shape. We onlookers, observing the situation from the perspective of an inertial frame of reference, may wish to say that there is no gravitational acceleration here, but the material of the cylinder has a centripetal acceleration towards the cylinder's axis. However, we should also acknowledge that the cylinder inhabitants' experiences are not illusory nor are their claims wrong. It is simply another way of describing relative acceleration.

As they develop more accurate time and distance measurements, the inhabitants come to realize that their standard clocks and rulers have to be interpreted with care. They install at each point two clocks: a standard clock and a 'master' clock.[1] By radar signalling up and down a skyscraper, they decide it makes sense to calibrate their master clocks such that all master clocks on a given skyscraper (i.e. all those on a given line out from the rotation axis) tick at the same rate—see Fig. 4.2. It follows that all master clocks throughout their world tick at the same rate. They then find that proper clocks on the ground run slow compared to master clocks; proper clocks at the top of a skyscraper run somewhat less slow. They call this 'gravitational time dilation'. We (onlookers inhabiting an inertial frame) see the same discrepancy in their two types of clock and attribute it to motional time dilation (the master clocks are keeping step with a clock at the axis, i.e. a non-moving clock from our point of view, the proper clocks are moving and consequently slowed).

To calculate this effect we can use Special Relativity. In the inertial frame S at rest relative to the cylinder's axis of rotation, the speed of a point at radius r is $v = r\omega$ where ω is the angular velocity of rotation of the cylinder. The time dilation factor is therefore

---

[1]The inhabitants of the rocket described in Volume 1 similarly investigated timing by using a pair of clocks at each location.

$$\gamma = \frac{1}{\sqrt{1 - v^2/c^2}} = \frac{1}{\sqrt{1 - r^2\omega^2/c^2}}.$$

Writing e^{Φ/c^2} for this 'gravitational time dilation' factor we find

$$\Phi = \frac{c^2}{2} \log(1 - r^2\omega^2/c^2) \tag{4.1}$$

where the radial distance r is agreed between the frames because transverse distances are not contracted.

A particle moving around a circle at radius r has acceleration in S given exactly by $a = r\omega^2$. Jumping now into the particle's instantaneous (inertial) rest frame we find this is a transverse acceleration so the proper acceleration is $a_0 = \gamma^2 a = \gamma^2 r\omega^2$ (Volume 1 eqn (2.61)). This is the 'acceleration due to gravity' in the cylinder world, which we shall call α:

$$\alpha = \gamma^2 r\omega^2 = \frac{r\omega^2}{1 - r^2\omega^2/c^2} \tag{4.2}$$

Hence we have the relation

$$\boldsymbol{\alpha} = -\boldsymbol{\nabla}\Phi. \tag{4.3}$$

This is an exact relation between gravitational acceleration and gravitational time dilation in the case under consideration.

Another feature of the cylinder world is that the times for light to complete a given round trip in the two possible directions are not always equal. For example, a light pulse sent all the way around the circumference (using a set of mirrors to reflect it) takes a shorter time to return to the source if it travels in one sense than if it travels in the other. From the inertial frame point of view, this is no surprise: the light propagating one way has to travel first a distance $2\pi r$ and then a further $r\omega t$, because the cylinder has rotated on, making a total distance $ct = r(2\pi + \omega t)$. Solving for t gives $t = 2\pi r/(c - r\omega)$. The light travelling the other way has a distance reduced by $r\omega t$, so the conclusion in the inertial frame is

$$t_\pm = \frac{2\pi r}{c(1 \mp r\omega/c)}.$$

The proper clocks in the cylinder have time dilation relative to this, so register travels times smaller by a factor of γ:

$$\tau_\pm = \frac{t_\pm}{\gamma} = \frac{2\pi r}{c} \sqrt{\frac{1 + v_\pm/c}{1 - v_\pm/c}}$$

where $v_\pm = \pm r\omega$. This is exact. To first approximation, the difference in travel time is (in either frame)

$$\tau_+ - \tau_- \simeq 4\pi r^2\omega^2/c^2.$$

This is called the **Sagnac effect**. More generally, for any trajectory, one finds a time difference proportional to the area enclosed by the trajectory.

The inhabitants of the cylinder world also notice that rulers laid on the ground behave strangely (exercise 4.1). They eventually attribute this to a direction-dependent 'gravitational space contraction' (we call it Lorentz contraction along the direction of motion). However, long before they notice either this or time dilation, they have observed another crucial feature of their world: dropped objects do not fall straight down! This is the Coriolis effect. Consider for example a Galileo dropping balls from a 50-metre high tower. The tower is strictly vertical: a plum bob hangs straight down it. However, the balls hit the ground 20 cm from the bottom of the plum bob. From an inertial point of view this is easy to explain: the tower top is moving at $v_1 = (r - h)\omega = 7902.918$ m/s, the tower bottom is moving at $v_2 = r\omega = 7902.980$ m/s; the balls take 3.2 s to hit the ground, so they fall $3.2(v_2 - v_1) = 0.20$ m behind the tower.

The classical (Newtonian) expression for the Coriolis acceleration is

$$a_C = 2\mathbf{u} \times \boldsymbol{\omega}. \tag{4.4}$$

Ignoring relativistic corrections for a moment, we find that the cylinder inhabitants must write the complete expression

$$\mathbf{f} = m\boldsymbol{\alpha} + 2m\mathbf{u} \times \boldsymbol{\omega} \tag{4.5}$$

for the 'gravitational' force on an object of mass m and velocity $\mathbf{u}$. Here m and $\mathbf{u}$ are properties of the body experiencing the force, but $\boldsymbol{\alpha}$ and $\boldsymbol{\omega}$ are not: they must describe the local gravity. An analogy with the Lorentz force of electromagnetism is beginning to emerge. It looks as though gravity includes a field $\boldsymbol{\alpha}$ that acts on all objects whatever their state of motion, and also a field

$$\mathbf{B_g} = 2\boldsymbol{\omega} \tag{4.6}$$

that only acts on moving objects. We shall call the latter a 'gravimagnetic field'. The gravimagnetic field in the cylinder world is uniform and parallel to the axis of the cylinder.

Here are a couple of easy gravimagnetic effects. A ball thrown horizontally from a skyscraper at speed $r\omega \simeq 7.9$ km/s in a direction directly opposite to the rotation of the cylinder will be found to stay at fixed height, orbiting the world. We onlookers are not surprised: we say the ball is not moving and the cylinder rotates under it. The cylinder inhabitants say the gravimagnetic field is producing a force in the opposite direction to $m\boldsymbol{\alpha}$ and exceeding it by a factor 2, thus providing enough centripetal force to keep the ball in orbit, tracing what is to them a circular path around their world. A child standing on the ground and watching the ball fly past overhead is not in any doubt about this gravimagnetic force, because after all it is preventing the ball from falling.

Next, gyroscopes experience a torque tending to align their axis of rotation with the gravimagnetic field direction. They are acting like a 'magnetic compass', except that unlike the magnetic compasses customarily used on Earth they possess (and cannot help but possess) angular momentum as well as gravimagnetic dipole moment. Therefore, when they experience a torque, they precess.

A gyroscope with its axis initially 'vertical' (i.e. along a radius of the cylinder) and mounted in a gimbal, so that it is free to rotate about any axis, is observed by the cylinder inhabitants to

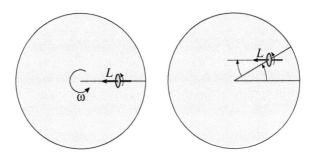

Fig. 4.3 A view down the cylinder, as observed in the inertial frame, in the non-relativistic limit $v \ll c$. The cylinder rotates anticlockwise at rate ω. A gyroscope mounted in the cylinder has constant $\mathbf{L}$ relative to an inertial reference frame. Therefore $\mathbf{L}$ precesses in the clockwise sense relative to a line fixed in the cylinder frame, at rate $-\boldsymbol{\omega}$ for $v \ll c$. The relativistic treatment gives $-\gamma^2\boldsymbol{\omega}$ in the disc frame and $-\gamma\boldsymbol{\omega}$ when the timing is relative to a LIF.

precess, completing approximately 2π radians once every 1.4 hours—see Fig. 4.3 (we shall make a more precise statement in a moment). We say the gyroscope axis maintains a constant direction (apart from a small Thomas precession); the cylinder inhabitants say the gravimagnetic field acts on it, as follows. Eqn (4.5) implies that the gravitational force on a rotating ring of matter is exactly analogous to the Lorentz force on a ring of electric current. In a uniform field a current ring experiences a torque $\boldsymbol{\mu} \times \mathbf{B}$ where $\boldsymbol{\mu}$ (the dipole moment) is proportional to the angular momentum $\mathbf{L}$ of the ring. In the electromagnetic case the relation is[2] $\boldsymbol{\mu} = q\mathbf{L}/2m$. For the inertial force under consideration here the 'charge' is the inertial mass so we have $\boldsymbol{\mu} = \mathbf{L}/2$. Hence the torque is

$$\boldsymbol{\mu} \times \mathbf{B}_{\text{g}} = \frac{\mathbf{L}}{2} \times (2\boldsymbol{\omega}) = \mathbf{L} \times \boldsymbol{\omega}. \tag{4.7}$$

Therefore the precession rate is $-\omega$. This is just enough to complete a revolution as the cylinder does, but in the opposite sense. In terms of the gravimagnetic field, the result is

$$\boldsymbol{\Omega}_{\text{precess}} = -\frac{1}{2}\mathbf{B}_{\text{g}}. \tag{4.8}$$

Now let us make these statements more precise by bringing in Special Relativity. In order to do this it is helpful to consider a rigidly rotating flat disc (such as a disc closing either end of the rotating cylindrical world). We already saw that the first term in (4.5) has to be multiplied by a factor γ^2 compared to the classical result; now we claim that the whole of the right-hand side receives this same factor, with $\mathbf{u}$ referring to velocity relative to the rotating disc. Thus the total acceleration, relative to the disc, of an inertially moving object, measured using standard rods and clocks on the disc, is

---

Free-fall acceleration, in standard (not coordinate) measures

$$\mathbf{a} = \gamma^2 \left(r\omega^2 \hat{\mathbf{r}} + 2\mathbf{u} \wedge \boldsymbol{\omega} \right). \tag{4.9}$$

---

[2]If a particle of mass m, charge q moves at speed v around a circle of radius r then $\mu = \pi r^2 qv/(2\pi r) = qvr/2$ and $L = mrv$.

This can be proved by invoking the full apparatus of coordinate transformations developed later in the book. The calculation is not simple because the *spatial* geometry of the disc's rest frame is not Euclidean. For example, owing to contraction of the rulers, the ratio of circumference to diameter of a circle is not always 2π. This does not mean spacetime is curved—it is still Minkowskian—but from a spacetime point of view we have (deliberately) adopted a non-trivial way of specifying which three-dimensional hypersurface within spacetime we choose to call 'space' at any given time. We offer two example cases to motivate the result. First consider an object at rest in the inertial frame. Its velocity relative to the disc is clearly $u = r\omega$ in the tangential direction, and its proper acceleration is zero. Its acceleration relative to the disc is sufficient to allow it to complete circular orbits. Because this relative acceleration is in the radial direction, the argument we gave before for $\mathbf{a}$ applies again (radial rods are not contracted; proper clocks give two factors of γ compared with the classical result), hence the result is $\mathbf{a} = -\gamma^2 r\omega^2 \hat{\mathbf{r}}$, which is consistent with (4.9). Next consider an object moving with $\mathbf{u}$ in the radial direction. The Newtonian formula (4.4) applies in the lab frame, as long as it is interpreted as a statement about the rate of change of *closing velocity* of particles on the disc with the object in question (because then it follows from Euclidean geometry). Passing now to an instantaneous rest frame on the disc, we have two effects from the Lorentz transformation of acceleration and velocity: $\mathbf{a} \to \gamma^3 \mathbf{a}$ and $\mathbf{u} \to \gamma \mathbf{u}$ (since $\mathbf{a}$ is along the boost and $\mathbf{u}$ is transverse to it). The result agrees with (4.9) again.

Now let us interpret the second term in (4.9) as a gravimagnetic field. We find

$$\mathbf{B}_{\mathrm{g}} = 2\gamma^2 \boldsymbol{\omega}. \tag{4.10}$$

Thus all gravimagnetic effects are somewhat larger than we said before. In particular, the factor of γ^2 means the gyroscope we considered above must precess *faster* than we thought, by a factor γ^2. Transforming back to the lab (inertial) frame, whose clocks are running faster than local proper clocks on the disc, we find the precession rate is $\gamma\omega$ (instead of the Newtonian result ω). After subtracting the rate of rotation of a line painted on the disc, i.e. ω, we have that a gyroscope supported torque-free on the disc and carried around by the latter will precess in the inertial frame at a rate $(\gamma - 1)\omega$. This is the Thomas precession! (c.f. eqn (6.50) of Volume 1).

4.1 The canonical form of the stationary metric

The metric of flat spacetime (Minkowski spacetime) expressed in cylindrical coordinates $(r, z, \bar{\phi})$ is

$$\mathrm{d}s^2 = -c^2 \mathrm{d}t^2 + \mathrm{d}r^2 + \mathrm{d}z^2 + r^2 \mathrm{d}\bar{\phi}^2.$$

In order to describe the rotating disc in flat spacetime, we introduce the coordinate

$$\phi \equiv \bar{\phi} - \omega t \qquad \Rightarrow \quad \mathrm{d}\bar{\phi} = \mathrm{d}\phi + \omega \mathrm{d}t. \tag{4.11}$$

This ϕ is the clearly the angle from some radial line painted on the disc, at the time indicated by the inertial clocks. We define coordinate time (the 'master clock' time for our cylinder world) to agree with t, so no further changes are needed. The metric in these coordinates is

$$ds^2 = -c^2dt^2 + dr^2 + dz^2 + r^2(d\phi + \omega dt)^2 \tag{4.12}$$

$$= -\left(1 - \frac{r^2\omega^2}{c^2}\right)\left(c\,dt - \frac{r^2\omega/c}{1 - r^2\omega^2/c^2}d\phi\right)^2 + dr^2 + dz^2 + \frac{r^2}{1 - r^2\omega^2/c^2}d\phi^2, \tag{4.13}$$

where the second version follows after a little algebra. Exercise 15.8 of Chapter 15 describes a way to extract a purely spatial metric in order to study spatial relativistic effects on a rigidly rotating disc. The 3-surface so defined has a corkscrew-like shape in spacetime and consequently is not flat.

In order to interpret what a given metric is telling us, it can be useful to introduce the concept of a **lattice**. By a lattice we mean a physical entity, which one might imagine to be built out of a collection of steel bars bolted together, constructed in such a way that each point on the lattice is at a fixed coordinate location. Then motion of other things can be interpreted as motion relative to such a lattice. The metric (4.13) is time-independent. This implies that lattice points at fixed (r, z, ϕ) will have constant separations, a fact that could be checked by radar signalling or by other methods. But in the terminology of GR we do not say the lattice is static, because, owing to the time-space cross terms in the metric, it will be found that for some closed circuits (say, around the edges of a polygon fixed in the lattice), the time for a light signal to complete the circuit depends on the direction of travel—the Sagnac effect. This reveals the non-static nature of the lattice. Lattices with these properties are said to be *stationary* but not static.

In the stationary but non-static metric (4.13), the time dilation factor can be still be 'read off' from the g_{tt} term. By using $g_{00} = -c^2 \exp(2\Phi/c^2)$ (eqn (5.46)) we obtain (4.1). More generally this is true for any metric that can be expressed in the 'canonical' form

$$ds^2 = -e^{2\Phi/c^2}\left(cdt - W_i dx^i\right)^2 + k_{ij}dx^i dx^j \tag{4.14}$$

where i, j run over spatial indices, so $W_i dx^i = \mathbf{W} \cdot d\mathbf{x}$. You can easily show that any time-independent metric can be brought exactly to this form by 'completing the square' to bring the cross-terms into the bracket with dt. It is shown in exercise 11.2 of Chapter 11 that in this case the 4-acceleration of a lattice point (i.e. a point at fixed spatial coordinates) is given by

$$a^a = g^{a\mu}\partial_\mu \Phi = k^{aj}\partial_j \Phi \tag{4.15}$$

and therefore the magnitude of the acceleration due to gravity (that is, the relative acceleration between the lattice and a LIF) is

$$|\vec{a}| = \left(k^{ij}(\partial_i \Phi)(\partial_j \Phi)\right)^{1/2}. \tag{4.16}$$

4.2   The lessons of the rotating cylinder

The effects that arise in the treatment of the rotating cylinder do not have to be called gravitational—they can be called an effect of acceleration—but in view of the close relationship between gravity and inertial forces, we expect similar effects to be possible in a reference frame fixed in a gravitational field. We will show in subsequent chapters that the notion of a magnetism-like contribution to gravity is a useful way to understand various gravitational effects, and allows them to be calculated to high accuracy in the weak field limit, and exactly in some cases. The statements made in connection with (4.14) are also exact.

There are two main gravitational effects related to precession of gyroscopes. The precession of a gyroscope in free fall, such as one in orbit around a planet or star, is called *geodetic precession*, or *de Sitter* precession. An extra precession owing to the gravitation of a nearby rotating body is called *Lense-Thirring* precession. Owing to the fact that one can always adopt a frame moving with a body in free fall, the two effects are not altogether different, but in simple cases one can maintain the distinction by asserting that the de Sitter effect is due simply to the presence of a central mass, whereas the Lense-Thirring effect is due to the rotation of the central mass. For motion around a rotating body, the two effects must be added.

Geodetic precession is so named because a gyroscope in free fall follows, very nearly, a spacetime geodesic.[3] Its interpretation need not invoke the magnetic analogy, but a calculation of it may. An exact calculation for circular orbits in Schwarzschild spacetime is presented in Section 17.2.4. The second case (Lense-Thirring) is treated in Chapter 6, where we shall make use of eqn (4.8).

Exercises

4.1 A disc of proper radius $R = 1000\,$km rotates rigidly at 29.22 revolutions per second in Minkowski spacetime. The disc is measured by means of a large collection of steel rulers, each of proper length 1 metre, which can be glued in place anywhere on the disc without changing their proper length. For the purpose of this question, let a line of minimal length on the disc be defined as one joining two points on the disc using the smallest number of such rulers laid end-to-end.

(i) Prove that radial lines on the disc are lines of minimal length (a brief argument in words is sufficient)

(ii) Two radial lines are painted on the disc, such that at radius $R/4$ they are separated by the length of one steel ruler. How many steel rulers can be laid end-to-end between these lines at radius $R/2$ and at radius R? [*Ans.* 2.08, 5]

(iii) Qualitatively describe a line of minimal length between non-neighbouring points at the same radius.

(iv) A triangle is formed from three lines of minimal length. Is the sum of its internal angles equal to, greater than, or less than 180°? Exercise 15.8 of Chapter 15 gives further information.

4.2 An inhabitant of the cylinder world is first standing on the ground and then jumps upwards (that is, in the direction of the force they experience on their feet). While off the ground they are in zero gravity are they not? So why do they fall back down? Explain.

4.3 Write down a completely general time-independent metric by writing the line element in terms of unspecified functions, and show that it can be brought to the canonical form (4.14) by a change of coordinates.

---

[3]A spinning object in freefall in a non-uniform field does not follow quite the same path as a non-spinning; therefore its worldline is very slightly non-geodesic, but this correction can typically be neglected when calculating the precession called geodetic precession.

5

Linearized General Relativity

Many interesting problems in gravitational physics arise in the weak field limit, in which effects such as weak gravitational waves can occur which are not treatable by Newtonian gravity but which can be treated with high accuracy by a linear approximation to GR. This *linearized* theory is a well-constructed, accurate and very useful theory, and mathematically it is very much simpler than the full theory. It can treat many gravitational effects with very high precision, including lensing, low-amplitude gravitational waves, and many gyroscopic precession and timing measurements. It does not treat the precession of the perihelion of Mercury, since this is a cumulative effect built up over very long times, but with minor additions it can treat that too.

Let η_{ab} be the Minkowski metric,

$$\eta_{ab} \equiv \begin{pmatrix} -1 & & & \\ & 1 & & \\ & & 1 & \\ & & & 1 \end{pmatrix}.$$

The basic idea is to introduce a quantity h_{ab} such that the metric in curved spacetime can be written

$$g_{ab} = \eta_{ab} + h_{ab}. \tag{5.1}$$

This can be done exactly, of course, but we shall be interested in cases where h_{ab} is small, with components of magnitude $\ll 1$, and we shall ignore terms of order h_{ab}^2 until the final section of this chapter, where second-order terms are judiciously included in order to discuss the energy and momentum associated with the gravitational field.

The main results of the chapter are summarized in Table 5.1, where the gravitational results to be described are presented alongside similar equations from electromagnetism.

We are introducing linearized General Relativity in this chapter before developing all the mathematical ideas we shall need in the rest of the book. This is because it is possible to do this and it gives a sense of grounding before we continue to more subtle material. However, we will require a few basic ideas concerning tensors and index notation which we state here and develop more fully in Chapters 8–12.

Relativity Made Relatively Easy: General Relativity and Cosmology. Volume 2. Andrew M. Steane, Oxford University Press. © Andrew M. Steane 2021. DOI: 10.1093/oso/9780192895646.003.0005

Table 5.1 The main results to be derived in the chapter.

Linearized GR: a complete theory of weak-field gravity

	electromagnetism	linearized GR
source	j^a	T^{ab}
conservation law	$\partial_\lambda j^\lambda = 0$	$\partial_\lambda T^{\lambda b} = 0$
potential	A^a	h_{ab}
		and $\bar{h}_{ab} \equiv h_{ab} - \frac{1}{2}\eta_{ab}h$
gauge transformation	$A_a \to A_a + \partial_a\chi$	$h_{ab} \to h_{ab} + \partial_a\chi_b + \partial_b\chi_a$
preferred gauge	$\partial_\lambda A^\lambda = 0$	$\partial^\lambda \bar{h}_{\lambda b} = 0$
Field equation	$\Box^2 A_a = -(1/c^2\epsilon_0)j_a$	$\Box^2\bar{h}_{ab} = -(16\pi G/c^4)(T_{ab} + t_{ab}),$ (5.2)

where in order to find $\bar{h}_{ab}$, t_{ab} is set to zero in the first instance.

Equation of motion of test particles

$$\frac{\mathrm{d}u_a}{\mathrm{d}\tau} = \frac{q}{m}\left(\partial_a A_\lambda - \partial_\lambda A_a\right)u^\lambda \qquad \frac{\mathrm{d}u_a}{\mathrm{d}\tau} = -\frac{1}{2}(\partial_\mu h_{a\nu} + \partial_\nu h_{a\mu} - \partial_a h_{\mu\nu})u^\mu u^\nu \quad (5.3)$$

$$\text{where} \quad h_{ab} = \bar{h}_{ab} - \tfrac{1}{2}\eta_{ab}\bar{h}^\lambda_\lambda$$

The above assume the linear (weak field) approximation but have no other restriction. The following make also the approximation of slow stationary sources with negligible stress.

$$\phi = \int \frac{\rho\,\mathrm{d}V}{4\pi\epsilon_0 r} \qquad\qquad \Phi = -\int \frac{G\rho\,\mathrm{d}V}{r} \qquad (5.4)$$

$$\mathbf{A} = \int \frac{\mathbf{j}\,\mathrm{d}V}{4\pi\epsilon_0 c^2 r} \qquad\qquad \mathbf{W} = -4\int \frac{G\rho\mathbf{u}\,\mathrm{d}V}{c^2 r} \qquad (5.5)$$

$$\mathbf{E} = -\boldsymbol{\nabla}\phi \qquad\qquad \mathbf{E}_\mathrm{g} = -\boldsymbol{\nabla}\Phi \qquad (5.6)$$

$$\mathbf{B} = \boldsymbol{\nabla}\times\mathbf{A} \qquad\qquad \mathbf{B}_\mathrm{g} = \boldsymbol{\nabla}\times\mathbf{W} \qquad (5.7)$$

$$\mathbf{a} = \frac{q}{m}(\mathbf{E} + \mathbf{v}\times\mathbf{B}) \qquad\qquad \mathbf{a} = \mathbf{E}_\mathrm{g} + \mathbf{v}\times\mathbf{B}_\mathrm{g} \qquad (5.8)$$

In General Relativity we can develop scalar invariants, 4-vectors and tensors of all ranks, just as in Special Relativity, except that now one must be especially careful not to assume that tensors at different locations in spacetime can be added to one another. In fact, forming such sums is already dubious in Special Relativity, so this is not a complete change. We shall have contravariant, covariant and mixed tensor components, and we will use superscript and subscript indices accordingly. The metric tensor g_{ab} can be used to lower indices, and its inverse g^{ab} can be used to raise indices. g^{ab} is **defined** as that set of quantities which gives

$$g^{a\lambda}g_{\lambda b} \equiv \delta^a_b. \qquad (5.9)$$

> **A few notes on tensors**. In GR, the *trace* of a second-rank tensor S^{ab} is not the sum of the diagonal elements, but rather S^{μ}_{μ} (which is equal to $g_{\mu\nu}S^{\mu\nu}$). Tensors can be symmetric, in which case $S^{ab} = S^{ba}$ and $S_{ab} = S_{ba}$ and $S^{a}_{\ b} = S_{b}^{\ a}$ but one should resist the notion of swapping up and down indices. Rather, use g_{ab} to lower and g^{ab} to raise. The Kronecker delta δ^{a}_{b} is a legitimate tensor and is written this way because this indicates correctly how it behaves under coordinate transformations.

In other words, when expressed as matrices, g_{ab} and g^{ab} are inverses of one another. This can be used to show that the 'see-saw' rule applies; that is, $a_{\mu}b^{\mu} = a^{\mu}b_{\mu}$ and similarly at higher rank (since higher-rank tensors behave like outer products of first-rank tensors). See box for some further basic points.

When changing coordinate system, the tensors themselves, as geometric objects, do not change, but their set of components with respect to the coordinate system do change. We shall indicate this by attaching a prime, as for example in

$$x'^{a} = f^{a}(x^{0}, x^{1}, x^{2}, x^{3}) \tag{5.10}$$

where the functions f^{a} express some coordinate transformation. In Volume 1 we attached such a prime to the index, rather than the underlying symbol (or kernel). In the present volume we attach a prime, as shown, to the symbol itself. This is because no ambiguity arises and the choice made in the present volume is consistent with our notation which was briefly introduced in Chapter 1 and will be set out more fully in Chapter 9. The reader should bear with us and understand that these choices are not being made arbitrarily.

Having defined h_{ab} using (5.1), it is easy to show (exercise 5.3) that

$$g^{ab} = \eta^{ab} - h^{ab} + O(h^2) \tag{5.11}$$

where

$$h^{ab} \equiv \eta^{a\mu}\eta^{\nu b}h_{\mu\nu}. \tag{5.12}$$

Note that in this equation η^{ab} is playing the role of 'raiser of indices'. This is *not* the standard situation in General Relativity, where g^{ab} has that role. However, one can see that in the linearized theory η^{ab} can stand in for g^{ab} when working with quantities of first order in h. This suggests a perspective in which we regard η_{ab} as the metric, in which case we are dealing with a flat spacetime and we can invoke all the ideas of Special Relativity. We shall pursue this idea further once we have developed it more fully.

5.1 Global Lorentz transformations

Lorentz transformations are that set of transformations which preserve the Minkowski metric. A transformation Λ is a Lorentz transformation if and only if (in matrix notation), $\Lambda^{T}\eta\Lambda = \eta$ (Volume 1 eqn (2.48)). In index notation this is written

$$\Lambda^{\mu}{}_{a}\Lambda^{\nu}{}_{b}\,\eta_{\mu\nu} = \eta_{ab}. \tag{5.13}$$

Let us suppose we have two coordinate systems x^a and x'^a, related by a Lorentz transformation, and the coefficients Λ^a_b give the Lorentz transformation from unprimed to primed symbols, as in

$$x'^a = \Lambda^a{}_{\lambda}x^{\lambda}. \tag{5.14}$$

We shall then have

$$x^a = K^a{}_{\mu}x'^{\mu} \tag{5.15}$$

where K^a_b is the inverse transformation:

$$K^a{}_{\mu}\Lambda^{\mu}{}_{b} = \delta^a_b. \tag{5.16}$$

We shall now explain why this K^a_b is also written $\Lambda_b{}^a$ (note the placement of the indices). This is not a major issue; it is simply a clarification of notation which might otherwise be confusing.

In Special Relativity, the quantity $\Lambda_b{}^a$ fits the rules of index placement as long as it is defined as

$$\Lambda_b{}^a \equiv \eta_{b\mu}\eta^{a\nu}\Lambda^{\mu}{}_{\nu}. \tag{5.17}$$

One finds that this $\Lambda_b{}^a$ is the inverse transformation to $\Lambda^c{}_d$, since

$$\Lambda_{\mu}{}^{a}\Lambda^{\mu}{}_{b} = \eta_{\mu\nu}\eta^{a\lambda}\Lambda^{\nu}{}_{\lambda}\Lambda^{\mu}{}_{b} = \left(\Lambda^{\mu}{}_{b}\Lambda^{\nu}{}_{\lambda}\eta_{\mu\nu}\right)\eta^{a\lambda} = \eta_{b\lambda}\eta^{a\lambda} = \delta^a_b, \tag{5.18}$$

where we used (5.13). By comparing this with (5.16) we deduce

$$\Lambda_b{}^a = K^a{}_b. \tag{5.19}$$

In matrix terminology one would say that the matrix whose elements are $\Lambda_b{}^a$ is the inverse transpose of the matrix whose elements are $\Lambda^a{}_b$.

It turns out that $\Lambda_b{}^a$ plays two roles. First, it is the inverse Lorentz transformation, as we just showed. Secondly, it can be used to transform covariant quantities, since, starting from (5.14), we have

$$\eta_{ab}x'^a = \eta_{ab}\Lambda^a{}_{\lambda}\eta^{\lambda\mu}x_{\mu}$$
$$\implies \quad x'_b = \Lambda_b{}^{\mu}x_{\mu} \tag{5.20}$$

where we have used η_{ab} and η^{ab} to lower and raise indices exactly as one would in Special Relativity. This dual role is the one described in Volume 1 Section 12.2.2, where we noted that, in matrix notation, covariant quantities transform as $(\Lambda^{-1})^T$. The transpose operation is taken care of in index notation through the fact that in (5.20) the sum is over the second index, whereas in (5.18) the sum is over the first index. Equations (5.14) and (5.20) are examples of more general coordinate transformation equations which we will present in Chapter 8.

In GR we distinguish between global Lorentz transformations and local Lorentz transformations. A local Lorentz transformation is understood to refer to a change of coordinates which is only applied (and in some cases only makes sense) in the vicinity of some particular event. It could be used to switch between LIFs both local to some given event, for example. A global Lorentz transformation is understood to refer to a change in coordinates throughout spacetime. Such a global transformation will not always result in a valid (i.e. smooth) coordinate assignment everywhere. Therefore one should not assume that global Lorentz transformations can always be adopted. However, in the weak field limit there is no problem.

Applying a global Lorentz transformation to the metric tensor expressed as in (5.1), we find

$$g'_{ab} = \Lambda_a{}^\mu \Lambda_b{}^\nu \left(\eta_{\mu\nu} + h_{\mu\nu} \right) = \eta_{ab} + \Lambda_a{}^\mu \Lambda_b{}^\nu h_{\mu\nu} \tag{5.21}$$

This is an interesting result, because it shows that the metric thus obtained also has the generic form of (5.1), i.e. $g'_{ab} = \eta_{ab} + h'_{ab}$, but now with the gravitational term given by

$$h'_{ab} = \Lambda_a{}^\mu \Lambda_b{}^\nu h_{\mu\nu}. \tag{5.22}$$

This is the very same expression that would be used to transform a second-rank tensor in Special Relativity. Thus we have a further suggestion of how to interpret GR in the weak field limit. Instead of saying that we have a slightly curved spacetime, we can, if we like, consider that the gravitational effects are given by a symmetric second-rank tensor field h_{ab} in flat spacetime. Thus we can regard the linearized theory from either of two points of view. In the first point of view, the spacetime is slightly curved and h_{ab} describes its departure from flatness. In the second point of view, spacetime is *exactly flat*, i.e. with metric η_{ab}, and h_{ab} is a tensor field living in this flat spacetime. We will make use of both points of view, as it suits us.

In the second perspective h_{ab} plays the role for gravitational phenomena analogous to the role played by the vector potential A_a for electromagnetic phenomena, and we can proceed to calculate the results just as we would in Special Relativity. This is a powerful observation; it enables the development of ideas in the present chapter to proceed smoothly and efficiently.

5.2 Coordinate transformations and gauge transformations

We need one more idea before we can derive the results listed in Table 5.1. Recall first of all the concept of *gauge transformation* which was introduced in connection with electromagnetism in Volume 1. In that context, we have a field tensor $\mathbb{F}$ which is related to the 4-vector potential through a derivative:

$$\mathbb{F}^{ab} = \partial^a A^b - \partial^b A^a \tag{5.23}$$

The physical effects, such as forces on particles, are related directly to $\mathbb{F}$ not the potential, and, owing to the derivatives in (5.23), more than one potential can give rise to the same $\mathbb{F}$ and hence the same physical effects. This situation is called *gauge freedom*. It is an interesting and important idea. If, in the context of Special Relativity (i.e. flat spacetime) we replace A^a by

$$\tilde{A}^a \equiv A^a + \partial^a \chi \tag{5.24}$$

for any scalar invariant function $\chi(t, x, y, z)$, then we shall find

$$\partial^a \tilde{A}^b - \partial^b \tilde{A}^a = \partial^a A^b - \partial^b A^a + \partial^a \partial^b \chi - \partial^b \partial^a \chi = \partial^a A^b - \partial^b A^a. \tag{5.25}$$

Therefore, in eqn (5.23) we can use either A or $\tilde{A}$ and get the same $\mathbb{F}$.

We now propose that a coordinate transformation is itself a form of gauge transformation. For, when we change our system of coordinates, nothing has really changed. That is to say, if we have a set of vectors, for example, representing physical properties such as force and momentum, and then we rotate the coordinate system, then the forces and momenta are unchanged, and so are all those physical predictions which do not refer directly or indirectly to the system of coordinates. For example, the prediction 'the momentum of this electron is equal to difference between the momenta of the incoming and outgoing photons' remains true, and so does the prediction 'the pointer on the voltmeter will be located next to the painted number "2"'. The prediction 'the electron will move along the first axis of my coordinate system' will not necessarily remain true. *As long as we adopt the perspective that the system of coordinates is not itself part of the physical apparatus to be discussed*, then we can say that a change of coordinates has no physical effect, and is a form of gauge transformation.[1]

It requires some careful thought to deduce what we are and are not claiming by such a statement. After all, effects such as Doppler shift are real enough, and those are often calculated by using the Lorentz transformation. The idea is that the Doppler effect is really a statement about two sets of physical interactions: one set at the emitter and one set at the receiver. We do not need to bind our coordinates to those emitting and receiving physical entities, but sometimes it is convenient to do so.

Again, by insisting that 'nothing has changed' when we change coordinates, we are adopting a perspective that takes a little getting used to. Since switching between inertial frames in Special Relativity is itself no more and no less than a coordinate change, it might appear as if we are suggesting that there is no such thing as time dilation, Lorentz contraction, the Doppler effect, the headlight effect, Thomas precession and so on. Not so. We are merely saying that what goes on in spacetime is what it is, irrespective of how we slice up spacetime by choosing one frame or another. In GR it often happens that the coordinates themselves do not have a straightforward relationship with physical devices such as particle detectors. In such a case we are thrown back on the metric tensor and the line element equation (2.22), which gives a strictly invariant (coordinate independent) result.

We will use the concept of gauge invariance in the following in order to simplify the Einstein field equation in the weak field limit.

5.3 The linearized field equations

Since η_{ab} is constant, the Christoffel symbols (eqn (2.8)) are

---

[1]The laws of physics as we know them (the Standard Model plus GR) have gauge transformations that divide into two separate, independent classes: transformations of the fields and transformations of the coordinate system. Supersymmetry, which is widely expected to be a feature of a quantum theory of gravity, uses a gauge transformation that combines coordinate and field transformations together.

$$\Gamma^a_{bc} = \frac{1}{2} g^{a\lambda} (\partial_b h_{c\lambda} + \partial_c h_{\lambda b} - \partial_\lambda h_{bc}) \tag{5.26}$$

$$\simeq \frac{1}{2} \eta^{a\lambda} (\partial_b h_{\lambda c} + \partial_c h_{\lambda b} - \partial_\lambda h_{bc}) \tag{5.27}$$

where the first version is exact and the second is correct to first order.

Using (2.2) we find the linearized Ricci tensor is (exercise 5.4)

$$R_{ab} = \partial_\mu \Gamma^\mu_{ab} - \partial_b \Gamma^\mu_{\mu a} + O(h^2)$$

$$= \frac{1}{2} \left(\partial_a \partial_\lambda h^\lambda{}_b + \partial_b \partial_\lambda h^\lambda{}_a - \partial_a \partial_b h^\lambda_\lambda - \Box^2 h_{ab} \right) \tag{5.28}$$

where $\Box^2 \equiv \partial^\lambda \partial_\lambda$ is the d'Alembertian, and we now adopt the practice of using the equality sign for results that are correct up to first order in h_{ab}. By substituting this into the Einstein field equation we obtain the linearized field equation. In vacuum, the field equation is $R_{ab} = 0$.

For the sake of completeness, let us also note that to first order in h_{ab},

$$R^a_{bcd} = \tfrac{1}{2} (\partial_b \partial_c h^a_d - \partial_b \partial_d h^a_c + \partial_d \partial^a h_{bc} - \partial_c \partial^a h_{db}) \tag{5.29}$$

(the reader may wish to verify this as an exercise after completing this chapter).

The linearized Ricci tensor (5.28) has a $\Box^2 h_{ab}$ term and various other terms. Sensing the hint of a wave equation, we ask ourselves, just as in electromagnetic theory (Volume 1 Section 7.4), whether we can make the other terms go away. To this end, it turns out to be helpful to introduce another way of looking at h_{ab}, namely the 'trace reversed' form

$$\bar{h}_{ab} \equiv h_{ab} - \tfrac{1}{2} \eta_{ab} h \tag{5.30}$$

where $h \equiv h^\lambda_\lambda$. The tensor is 'trace reversed' since $\bar{h}^\lambda_\lambda = -h^\lambda_\lambda$ (the Einstein tensor is similarly related to the Ricci tensor, eqn (2.20)). Then we have

$$h_{ab} = \bar{h}_{ab} - \tfrac{1}{2} \eta_{ab} \bar{h}^\lambda_\lambda \tag{5.31}$$

and substituting this into (5.28) gives

$$R_{ab} = \frac{1}{2} \left(\partial_a \partial_\lambda \bar{h}^\lambda_b + \partial_b \partial_\lambda \bar{h}^\lambda_a - \Box^2 h_{ab} \right). \tag{5.32}$$

Now recall that h_{ab} plays the role of 'tensor potential', which is related through a derivative to effects such as tidal effects which cannot be transformed away by adopting a LIF. This suggests that the concept of gauge transformation might be applicable—that is, a transformation which changes the 'potential' h_{ab} without changing the 'field' R_{ab}. If we can thus achieve that $\partial_\lambda \bar{h}^\lambda_a = 0$ then we shall get rid of both unwanted terms and gain the wave equation we would like to have.

With the above in mind, consider what happens when we replace h_{ab} by

$$\tilde{h}_{ab} = h_{ab} + \epsilon (\partial_a \xi_b + \partial_b \xi_a), \tag{5.33}$$

where $\xi^a(x)$ is some set of arbitrary functions of the coordinates,[2] and ϵ is small. The change can be shown to follow from a slight adjustment of the coordinates (see exercise 5.6), but all we need for the moment is to make the change and see what happens. This is reminiscent of $A_a \rightarrow A_a + \partial_a \chi$ in electromagnetism.

The components of the new Ricci tensor $\tilde{R}_{ab}$ are obtained by replacing h_{ab} with $\tilde{h}_{ab}$ on the right-hand side of (5.28). Therefore we have

$$
\begin{aligned}
\tilde{R}_{ab} &= \tfrac{1}{2} \left(\partial_a \partial_\lambda \tilde{h}^\lambda{}_b + \partial_b \partial_\lambda \tilde{h}^\lambda{}_a - \partial_a \partial_b \tilde{h}^\lambda_\lambda - \Box^2 \tilde{h}_{ab} \right) \\
&= R_{ab} + \frac{\epsilon}{2} \left[\partial_a \partial_\lambda (\partial^\lambda \xi_b + \partial_b \xi^\lambda) + \partial_b \partial_\lambda (\partial^\lambda \xi_a + \partial_a \xi^\lambda) - \partial_a \partial_b (2\partial_\lambda \xi^\lambda) - \partial^\lambda \partial_\lambda (\partial_a \xi_b + \partial_b \xi_a) \right] \\
&= R_{ab}.
\end{aligned}
\tag{5.34}
$$

All the effects of the change in h_{ab} cancel out: the components $\tilde{R}_{ab}$ are the same as R_{ab}. Note, this will not happen in the exact theory; it is owing to the linear approximation which has been involved when we wrote the Christoffel symbols in terms of h_{ab}. This is why we require ϵ in (5.33) to be small. It is of the same order of smallness as h_{ab}.

The result $\tilde{R}_{ab} = R_{ab}$ is reminiscent of the effect of a gauge change: namely, no change at all in the quantity of interest. It follows that the Ricci scalar R will not change either, and then the Einstein field equation tells us that neither will T_{ab}. In short, we can change the metric tensor as in (5.33) and get no change at all in the linearized theory in any part of the field equation when it is written in terms of the Ricci tensor and the stress-energy tensor. For this reason the metric change (5.33) is called a *gauge transformation*.

The move from h_{ab} to $\tilde{h}_{ab}$ is possible because in any given spacetime there is more than one way to find coordinates in which g_{ab} takes the form (5.1). We can adopt one coordinate choice, or a slightly different one, and accordingly the difference between the exact metric and the Minkowski metric will take one form or another. It is sometimes asserted that this gauge change is not a coordinate change, because one does not adopt new tensors R_{ab}, T_{ab} but just continues with the old ones. However, this point of view misses the significant fact that the components of the metric tensor have changed, and this is a significant change. It is not true that the metric reveals its influence only through R_{ab}, because we can and will sometimes use the metric tensor directly, in order to calculate the invariant interval between given events, and in order to write the geodesic equation. The summary is that *under the change of gauge (5.33), the functional forms of all scalars, vectors and tensors are unaltered,* with the exception of *the metric tensor.* See exercise 5.6 for further information.

We will now adopt the standard terminology, and refer to (5.33) as a gauge transformation. We note that tensors other than h_{ab} do not change, so we will not attach a tilde to them.

The $\bar{h}_{ab}$ tensor is changed to

$$
\tilde{\bar{h}}_{ab} = \bar{h}_{ab} + \epsilon \left(\partial_a \xi_b + \partial_b \xi_a - \eta_{ab} \partial_\lambda \xi^\lambda \right).
\tag{5.35}
$$

---

[2] The notation $\xi^a(x)$ is a shorthand for $\xi^a(x^0, x^1, x^2, x^3)$, i.e. each ξ^a is a function of the four coordinates.

Recall now that we would like the divergence of this tensor[3] to be zero, if possible. From (5.35) we have

$$\partial^\mu \bar{\bar{h}}_{\mu b} = \partial^\mu \bar{h}_{\mu b} + \epsilon \left(\partial^\mu \partial_\mu \xi_b + \partial^\mu \partial_b \xi_\mu - \partial^\mu \eta_{\mu b} \partial_\lambda \xi^\lambda \right) = \partial^\mu \bar{h}_{\mu b} + \epsilon \partial^\mu \partial_\mu \xi_b \tag{5.36}$$

It follows that, in order for this to be zero, it is sufficient that

$$\epsilon \Box^2 \xi_b = -\partial^\mu \bar{h}_{\mu b} \tag{5.37}$$

(recall that $\Box^2 \equiv \partial^\mu \partial_\mu$). This is a wave equation with source term, which always has a solution for any source, so it will always be possible to make a gauge choice satisfying (5.37). After making such a choice, we have

$$\partial^\mu \bar{\bar{h}}_{\mu b} = 0. \tag{5.38}$$

In other words, we can always find a gauge for which there holds (now dropping the tilde) the

---

Lorenz gauge condition in weak field

$$\partial^\lambda \bar{h}_{\lambda b} = 0 \qquad \text{which may also be written } \partial^\lambda h_{\lambda b} - \frac{1}{2} \partial_b h = 0 \tag{5.39}$$

This is like the Lorenz gauge condition of electromagnetism, hence the name. It is also called the harmonic gauge (motivated by eqn (5.37)) and the de Donder gauge.

With this gauge choice we have achieved the condition we wanted in order to simplify eqn (5.32), so now we have

$$R_{ab} = -\tfrac{1}{2} \Box^2 h_{ab} \tag{5.40}$$

$$R = -\tfrac{1}{2} \Box^2 h \tag{5.41}$$

$$G_{ab} = -\frac{1}{2} \left(\Box^2 h_{ab} - \tfrac{1}{2} \eta_{ab} \Box^2 h \right) = -\tfrac{1}{2} \Box^2 \bar{h}_{ab} \tag{5.42}$$

and the Einstein field equation (2.1) is

Linearized field equation in Lorenz gauge

$$\Box^2 \bar{h}_{ab} = \frac{-16\pi G}{c^4} T_{ab}. \tag{5.43}$$

This is a remarkable and very useful result, because it is a wave equation. With our knowledge of electromagnetism to guide us, we can immediately see in our mind's eye the possibility of inverse-square laws, waves propagating at the speed of light and other familiar ideas. Not only is (5.43) a wave equation, but also it is one that is separated into ten independent equations for the ten independent elements of the symmetric $\bar{h}_{ab}$. The solution of each wave equation in

---

[3] h_{ab} is only a tensor in the flat spacetime sense, adopted in the linearized theory, not a tensor in the full theory; for this reason it is often called a pseudo-tensor. We shall not adopt that terminology.

the set, for any given T_{ab}, proceeds exactly as in Volume 1 Section 8.2. The method (Green's method) can also be found in texts on mathematics or electromagnetism. The solution is:

Solution of field equations in Lorenz gauge

$$\bar{h}_{ab}(t, \mathbf{x}) = \frac{4G}{c^4} \int \frac{[T_{ab}]}{r_{\text{sf}}} \mathrm{d}^3 \mathbf{x}_s \tag{5.44}$$

---

where $(t, \mathbf{x})$ is the field event, r_{sf} is the length of the null vector from each source event to the field event, and $[T_{ab}] = T_{ab}(t_s, \mathbf{x}_s) = T_{ab}(t - r_{\text{sf}}/c, \mathbf{x}_s)$ is the energy tensor at the retarded source event (hence the integral is over the past light cone of the field event). In particular,

$$\bar{h}_{00} = \frac{4G}{c^2} \int \frac{[\rho]}{r_{\text{sf}}} \mathrm{d}^3 \mathbf{x}_s \tag{5.45}$$

where ρ is the mass density in a LIF. This reminds us of the Newtonian solution, and indeed it reduces to that solution in the limit of static conditions and low velocities. The present result is much more general, however. It can be used to treat the trajectories of photons, for example, and it is not restricted to static conditions.

Among the mathematical tools available to perform the integral on the right-hand side of (5.44) is the multipole expansion; this is presented in Section 7.3.

We now have the tools to calculate many gravitational problems; it is a good moment to look back at Table 5.1 where the results are brought together. The field equation in the table includes a term t_{ab} which will be discussed at the end of the chapter; it is a second-order term that is not needed for the calculation of geodesics. The table also includes the equation of motion for test particles (geodesic equation) which follows immediately from substituting (5.27) (the Christoffel symbols) into (2.28). This completes the theory. The further results in the table show how the comparison with electromagnetism can be drawn out more fully in the case of slow stationary sources; this is presented in Chapter 6.

5.4   Newtonian limit

In Volume 1 we showed how Newtonian gravitation can be obtained purely from gravitational time dilation, for slowly moving particles. Writing

$$g_{00} = -e^{2\Phi/c^2} \simeq -(1 + 2\Phi/c^2) \tag{5.46}$$

we found that in the limit $\Phi \ll c^2$, $v^2 \ll c^2$ the principle of most proper time allows us to interpret the gravitational time dilation function Φ as the Newtonian gravitational potential function. We shall now show that the geodesic equation leads to the same conclusion (as it must, since this is essentially the same calculation). The geodesic equation (2.28) gives

$$\frac{\mathrm{d}u^a}{\mathrm{d}\tau} = -\Gamma^a_{\mu\nu} u^\mu u^\nu \simeq -c^2 \Gamma^a_{00} = -\frac{c^2}{2} \left(2\partial_0 h^a_0 - \partial^a h_{00} \right) \simeq \frac{c^2}{2} \partial^a h_{00} \tag{5.47}$$

where in the first approximation we used that $u^a = (\gamma c, \gamma \vec{u})$ so the zeroth term dominates when $u \ll c$ (giving also $\gamma \simeq 1$) and in the last we assumed the time-dependence of h_{00} is small compared to c times its gradient (using that dx^0 is $c dt$ for a Minkowskian metric). Replacing now $d\tau$ by dt, which is valid when $u \ll c$, we have that the spatial part of this equation reads

$$\frac{d\vec{u}}{dt} = \tfrac{1}{2} c^2 \, \mathrm{grad}\,(h_{00}) \,,$$

where we have adopted the arrow notation for 3-vectors, to avoid confusion with results to be obtained in later chapters. Now, (5.46) asserts that $h_{00} = -2\Phi/c^2$ so we have

$$\frac{d\vec{u}}{dt} = -\,\mathrm{grad}(\Phi) \tag{5.48}$$

which confirms that Φ acts as the Newtonian potential. Note that Φ is related directly to the metric perturbation h_{00} without reference to the source.

By substituting $h_{00} = -2\Phi/c^2$ into (5.47) we also find that, in the case of a static weak field,

$$\Gamma^a_{00} = \frac{1}{c^2} \partial^a \Phi. \tag{5.49}$$

which illustrates the sense in which Γ^a_{bc} is 'the gravitational field' (and recall exercise 2.2 of Chapter 2).

Now we shall extract the Newtonian field equation—i.e. the formula for how Φ behaves in free space (Laplace's equation), and how it is related to its source (Poisson's equation).

Picking out the 00 term from the linearized Ricci tensor (5.28), we find[4]

$$R_{00} \simeq \frac{1}{2} \nabla^2 h_{00}$$

since the other terms involve derivatives with respect to time so are negligible in the stationary limit. Now we apply the Einstein field equation (2.7), setting on the right-hand side the energy tensor for a perfect fluid[5] (1.7), which gives

$$R_{00} = -\frac{4\pi G}{c^2}(\rho + 3p/c^2).$$

Therefore

$$\nabla^2 h_{00} = -\frac{8\pi G}{c^2}(\rho + 3p/c^2).$$

---

[4] Here $\nabla^2 = \mathrm{div}\,\mathrm{grad}$ is the Laplacian in three dimensions.

[5] The quantities ρ and p here are scalar invariants; ρc^2 and p are, respectively, the energy density and the pressure in the instantaneous rest frame of the fluid element.

and after using (5.46) one obtains:

Gravity in weak field limit

$$\nabla^2 \Phi = 4\pi G(\rho + 3p/c^2). \qquad (5.50)$$

This is Poisson's equation for the gravitational potential. It is in agreement with Newton's law of gravitation since $p \ll \rho c^2$ for ordinary matter, but displays also the effect of pressure which is a prediction of General Relativity. A source at high pressure gravitates more strongly than another of lower pressure but the same proper mass. Pressure enters alongside energy density also in Euler's equation for fluid flow, noted already in Volume 1 Section 16.3. The combination $(\rho + 3p/c^2)$ is sometimes called the *active gravitational mass* (per unit volume). However, it does not follow that the total active gravitational mass of a star, as observed from far away, is given merely by the volume integral of this quantity, because one also has to take into account the influence of the gravitational potential well; see Section 18.1 for more information.

5.5   Field energy and the gravity of gravity

So far in this chapter we have found a good way to treat the first approximation to the metric perturbation h_{ab}, and this enabled us to find Christoffel symbols, also to linear approximation, which can then be used in the geodesic equation to find the motion of test particles. However, owing to the higher-order terms which have been so far neglected, there is an inconsistency which we now need to address. A central feature of the linearized theory, which is both a great bonus and, as we shall now see, a weakness, is that it is linear!

The linearity of the linearized field equation (5.43) means that it satisfies the superposition principle: we can superpose existing solutions to find new solutions. This is very useful; it enabled us to write down solutions such as (5.44) which take advantage of this feature through Green's method. However, the non-linear terms in Einstein's full field equation contain some of the important physics. For example, if two balls stick together by gravitational attraction, then their energy content is smaller than when they are far apart, because we would need to provide energy in order to separate them. Equally, when bringing them together, we could extract energy by lowering one towards the other on the end of a rope, and using the rope to turn an electrical generator or whatever. This implies that the joint field when two balls are close to one another must be less than the sum of their separate fields (this is explored more fully in exercise 5.8). No allowance is made for this on the right-hand side of the exact Einstein field equation. Therefore it is the *non-linearity* of the left-hand side which takes care of this issue. It is through this non-linearity that the equation captures or expresses the fact that 'gravity gravitates' and the field of two balls is not twice that of one ball. This effect corresponds loosely to the notion of binding energy, but the attempt to ascribe an energy directly to the gravitational field is not straightforward. For example, the field Γ can be transformed away at a moment's notice by going to a LIF. In view of this, the energy associated with gravity must be something to do with curvature, not merely a non-zero Christoffel symbol.

It will be admitted by the reader that there is some important physics in the non-linear terms which have been neglected in the linearized theory, but does this amount to an inconsistency? It does, for the reason we now expound.

The problem has to do with energy and momentum conservation. In Special Relativity, the conservation of energy and momentum is expressed by $\partial_\lambda T^{\lambda b} = 0$. If an electromagnetic field has an energy tensor $T^{ab}_{(\text{em})}$ with $\partial_\lambda T^{\lambda b}_{(\text{em})} \neq 0$, for example, then we deduce that energy and momentum is moving between the field and the charged matter which it is pushing. After taking this into account we get a total energy tensor T^{ab} of matter and e-m. field together which satisfies $\partial_\lambda T^{\lambda b} = 0$. However, if we take the divergence of the linearized field equation (5.43) then, using (5.39), we find the left-hand side vanishes, and we are forced to conclude that

$$\partial_\lambda T^{\lambda b} = 0.$$

'Is that not that what we just claimed,' you ask? Well yes it is, but this is telling us that the energy and momentum of the matter-and-radiation in the universe is being conserved (in the linearized theory), so what is the gravitational field doing? Where are its effects? It appears that it is having no influence on the energy and momentum of the stuff it is squeezing and stretching. The equation $\partial_\lambda T^{\lambda b} = 0$ is an equation of motion for matter which makes no mention of gravity, so it asserts that matter will not accelerate towards nor be squeezed by massive gravitating bodies: it will just ignore them! This is inconsistent. We avoided this incorrect prediction up till now in the linearized theory by adopting the geodesic equation for the motion of matter, but now we need a more general approach that will allow for the possibility of energy and momentum movements carried by the gravitational field itself. Or if you prefer, carried by spacetime itself.

In the full theory one finds $\partial_\lambda T^{\lambda b} \neq 0$, but the departure from zero only appears at second order in h_{ab} so we have ignored it. We would like to fix this problem, while retaining the benefits of linearization as much as we can. This will enable us to calculate the energy radiated in weak gravitational waves, and things like that. We proceed through a series expansion, as follows.

Write

$$G_{ab} = G^{(1)}_{ab} + G^{(2)}_{ab} + \dots$$

where G_{ab} is the exact Einstein tensor, $G^{(1)}_{ab} = -\Box^2 \bar{h}_{ab}/2$ (eqn (5.42)) and $G^{(2)}_{ab}$ is the second-order term. Then the Einstein field equation is (to second order)

$$G^{(1)}_{ab} + G^{(2)}_{ab} = \frac{8\pi G}{c^4} T_{ab}$$

$$\implies \qquad G^{(1)}_{ab} = \frac{8\pi G}{c^4} T_{ab} - G^{(2)}_{ab} = \frac{8\pi G}{c^4} \left(T_{ab} + t_{ab} \right) \tag{5.51}$$

where

$$t_{ab} \equiv -\frac{c^4}{8\pi G} G^{(2)}_{ab}. \tag{5.52}$$

Equation (5.51) is our new linearized Einstein field equation. In Lorenz gauge it can be written

$$\Box^2 \bar{h}_{ab} = -\frac{16\pi G}{c^4} \left(T_{ab} + t_{ab} \right) \tag{5.53}$$

where it is understood that t_{ab} may be set to zero in the first instance in order to find $\bar{h}_{ab}$ for given T_{ab}, and then t_{ab} is calculated from (5.52). We can find $G_{ab}^{(2)}$ by substituting the first-order solution $g_{ab} = \eta_{ab} + h_{ab}$ into the exact equation for G_{ab} and extracting the $O(h^2)$ term. This yields the full $G_{ab}^{(2)}$ with nothing omitted; see appendix E.

The above is mathematically correct, but it does not quite succeed in yielding a self-consistent theory. It can be shown, after considerable algebra, that t_{ab} as given by (5.52) does not behave in a fully tensorial manner. It transforms correctly under global Lorentz transformations, but it does not under general infinitesimal coordinate transformations (5.61) and hence it is not guaranteed to be invariant under the gauge transformation involved in choosing a gauge such as the Lorenz gauge. This causes its status and interpretation to become ambiguous. The problem is related to the fact that at any single point the gravitational field can be transformed away by adopting a LIF, which suggests that one cannot associate an energy with the gravitational field at a point. This is consistent with the general idea that gravitational effects are to do with curvature. The problem can be circumvented by replacing (5.52) by

$$t_{ab} \equiv -\frac{c^4}{8\pi G} \left\langle G_{ab}^{(2)} \right\rangle \tag{5.54}$$

where the angle brackets denote an average over a region of spacetime. The width L of the region should be larger than the length-scale of local ripples in spacetime, such as the wavelength of gravitational waves, but small compared to any radius of curvature of the background on which the ripples are imposed. For waves of wavelength λ on a background with maximum Gaussian curvature K, this averaging is always possible if the background is static, and more generally it is possible if $\lambda \ll |K|^{-1/2}$. A suitable length-scale on which to take the average is then $L \sim |\lambda^2/K|^{1/4}$. It can be shown that one thus arrives at a gauge-invariant result.

It is shown in appendix E that the result is, up to a fractional error of order $\lambda|K|^{1/2}$,

$$t_{ab} = \frac{c^4}{32\pi G} \left\langle A_{ab} + B_{ab} + C_{ab} + D_{ab} \right\rangle \tag{5.55}$$

where

$$A_{ab} = (\partial_a \bar{h}_{\mu\nu})(\partial_b \bar{h}^{\mu\nu}) \tag{5.56}$$

$$B_{ab} = -(\partial_\nu \bar{h}^{\mu\nu})(\partial_a \bar{h}_{b\mu} + \partial_b \bar{h}_{a\mu}) \tag{5.57}$$

$$C_{ab} = -\frac{1}{2}(\partial_a \bar{h})(\partial_b \bar{h}) \tag{5.58}$$

$$D_{ab} = -\frac{8\pi G}{c^4} \left(2\bar{h}_{a\mu} T_b^{\mu} + 2\bar{h}_{b\mu} T_a^{\mu} + \eta_{ab} h^{\mu\nu} T_{\mu\nu} \right). \tag{5.59}$$

This lengthy expression simplifies considerably in many situations of practical interest. In Lorenz gauge, $B_{ab} = 0$. Also $C_{ab} = 0$ when $\bar{h} = 0$, and $D_{ab} = 0$ in vacuum. Hence A_{ab} is the most important term.

We now have a definition that is gauge invariant, and the tensor t_{ab} succeeds in capturing that part of the non-linearity of the original (exact) equation that is needed in order to allow for the gravity of gravity—the fact that gravitational fields themselves act as sources of gravity.

The new field equation (5.53) has a natural physical interpretation. The quantity t_{ab} is playing the role of an energy tensor. It can be interpreted as describing the energy and stress of the gravitational field, because if we regard the equation as one describing the gravitational field as a tensor field in exactly flat (Minkowskian) spacetime, then energy-momentum conservation is handled just as in Special Relativity, by using partial derivatives not covariant derivatives.

It is valuable to reflect on the physical justification for this procedure of moving $G_{ab}^{(2)}$ to the right-hand side of the equation and giving it a new name. The procedure is justified by thinking about the very definition of energy and momentum. What is energy? It is a concept we invent, or a property we draw attention to, because of its *usefulness*, which stems chiefly from the fact that it is conserved. How do we define field energy? One way is to examine *what happens when the field pushes on particles*—whatever energy goes to the particles we say must have come from the field. That is the strategy we adopt in electromagnetism to relate the energy of the electromagnetic field to the energy of the charged matter it interacts with. Equation (5.53) adopts the same strategy. With (5.39), it says the 4-divergence of T_{ab} and t_{ab} are equal and opposite, the former being associated purely with the sources, the latter purely with the field. This is all we can or need ask of a field energy tensor; it follows that t_{ab} is the energy tensor of the gravitational field in the linearized theory.

We shall apply (5.55) to the study of gravitational radiation in Chapter 7, where an alternative derivation of the important A_{ab} term is also provided. Field energy in the full theory is addressed in Chapter 16.

Exercises

5.1 *We shall obtain the gravitational interaction between two parallel beams of light by using the equations gathered in Table 5.1.*
   (i) Adopting rectangular coordinates (t, x, y, z), and using your knowledge of electromagnetism, satisfy yourself that the energy tensor for an electromagnetic plane wave propagating in the x direction is

$$T^{ab} = \begin{pmatrix} 1 & 1 & 0 & 0 \\ 1 & 1 & 0 & 0 \\ 0 & 0 & 0 & 0 \\ 0 & 0 & 0 & 0 \end{pmatrix} \epsilon_0 E^2 \cos^2(kx - \omega t).$$

We shall consider a pencil beam of light. It is a cylinder centred on the x axis with sufficient radius that diffraction is negligible, and composed of a few frequencies close together with no fixed phase relationship, so that the $\cos^2$ function averages, leaving a smooth time-independent energy tensor in-

side the cylinder. The energy tensor outside the cylinder is zero.
   (ii) Setting $t_{ab} = 0$, show that the field equation (5.2) gives $\vec{\nabla} \cdot (\vec{\nabla} \bar{h}_{ab}) = -(16\pi G/c^4) T_{ab}$.
   (iii) Using Gauss' theorem and the cylindrical symmetry, or otherwise, show that $\vec{\nabla} \bar{h}_{ab}$ (for each a, b) is in the radial direction, and is given by

$$\vec{\nabla} \bar{h}_{ab} = -\frac{8G}{rc^4} T_{ab} \, \hat{r} \qquad (5.60)$$

where $r = (y^2 + z^2)^{1/2}$ and $\hat{r}$ is a unit vector in the outwards radial direction.
   (iv) Consider another beam of light, initially parallel to the first, located in the xy plane at some distance y. To treat a null geodesic one should take as the parameter τ in the geodesic equation not proper time

(which is zero), but some other parameter which increments uniformly along the worldline. We can always normalize this 4-velocity in such a way that the first component is initially 1, and then the whole 4-velocity for the beam in question is, initially, $u^a = (1, 1, 0, 0)$. Show from (5.3) that the rate of change of this 4-velocity with respect to the parameter τ is given by

$$\frac{\mathrm{d}u_a}{\mathrm{d}\tau} = \frac{1}{2}\partial_a \left(h_{00} + h_{01} + h_{10} + h_{11} \right).$$

Hence show that these two beams of light do not attract one another. Notice the interesting implication: a propagating beam of light does not self-focus by its own gravitation.

(v) Show that counter-propagating beams attract one another with four times the acceleration that one might have expected from a Newtonian argument. (Tolman, Ehrenfest and Podolsky, 1931; Scully, 1979; Rätzel, Wilkens and Menzel, 2016)

5.2 Prove that a spacetime which looks nearly flat to one observer still looks nearly flat to any other observer in uniform motion relative to the first (where by 'looks nearly flat' we mean $|g_{ab} - \eta_{ab}| \ll 1$). [Hint: this is simple]. Note the implication: you cannot 'boost up' the curvature merely by running fast. Is this obvious? What bearing does the Ricci scalar have on such questions?

5.3 Write $g^{ab} = \alpha\eta^{ab} + \beta h^{ab} + O(h^2)$ where h^{ab} is defined in (5.12) and α and β are constants to be discovered. Hence obtain (5.11) by using the definition (5.9).

5.4 For Γ^a_{bc} given by (5.27) show that

$$\partial_\mu \Gamma^\mu_{bc} = \tfrac{1}{2}\eta^{\mu\lambda}(\partial_\mu \partial_b h_{\lambda c} + \partial_\mu \partial_c h_{\lambda b} - \partial_\mu \partial_\lambda h_{bc})$$
$$\partial_c \Gamma^\mu_{b\mu} = \tfrac{1}{2}\eta^{\mu\lambda}(\partial_c \partial_b h_{\lambda\mu} + \partial_c \partial_\mu h_{\lambda b} - \partial_c \partial_\lambda h_{b\mu})$$

Hence obtain (5.28) (you may find it useful to obtain R_{bc} and then change the names of the indices at the end).

5.5 x^a and x'^a are two sets of coordinates for the same region of spacetime. g_{ab} is the metric tensor in terms of unprimed coordinates; g'_{ab} is the metric tensor in terms of primed coordinates. Using

$$\mathrm{d}s^2 = g_{ab}\mathrm{d}x^a\mathrm{d}x^b = g'_{ab}\mathrm{d}x'^a\mathrm{d}x'^b$$

and $(\partial x^a/\partial x^b) = \delta^a_b$ (which you should confirm if you do not consider it to be obvious; c.f. eqn (8.10)) deduce that

$$g'_{cd} = g_{ab}\frac{\partial x^a}{\partial x'^c}\frac{\partial x^b}{\partial x'^d} .$$

5.6 Consider the coordinate transformation given by

$$x'^a = x^a - \epsilon\xi^a(x) \qquad (5.61)$$

where $\xi^a(x)$ is some set of functions, and ϵ is of the same order of smallness as h_{ab}. Using the result of the previous exercise, show that the metric tensor becomes

$$g'_{cd} = g_{ab}(\delta^a_c + \epsilon\partial_c\xi^a)(\delta^b_d + \epsilon\partial_d\xi^b). \quad (5.62)$$

Hence, by writing $g_{ab} = \eta_{ab} + h_{ab}$ and $g'_{ab} = \eta_{ab} + h'_{ab}$, obtain

$$h'_{ab} = h_{ab} + \epsilon(\partial_a\xi_b + \partial_b\xi_a). \qquad (5.63)$$

Note that it follows that the gauge change (5.33) can be regarded as the result of a small adjustment in the coordinates.

5.7 Obtain an expression for the covariant components of the linearized curvature tensor R_{abcd}, and show that it is invariant under an infinitesimal coordinate transformation.

5.8 Explore the gravitational impact of gravitational binding energy, as follows. Suppose there is a spherical planet of mass M_1 and radius R_1, surrounded by a shell of mass M_2 and radius $R_2 \gg R_1$. Lower the shell onto the planet using light ropes, thus accumulating some energy E at the top of the ropes, where it is stored at R_2 (retaining spherically symmetry). Argue (carefully!) that this procedure cannot change the field outside R_2 because the whole system is isolated. Then argue that this implies the gravitation of the combined planet and shell must have become smaller than it was, at any location $r > R_2$, by an amount equal to the gravitation associated with E.

5.9 (Not for the faint-hearted.) Obtain eqn (E.23) for $G^{(2)}_{ab}$ and show that this quantity is not invariant under a gauge transformation of the form (5.33).

6

Slow stationary sources

In GR the situation is called *stationary* if the metric does not depend on time in the coordinates which have been adopted.

In this chapter we consider the linearized theory applied to the case of sources in stationary motion (for example, a rotating ball or ring), where the *source* moves at speeds small compared to the speed of light. The speed of test particles or light waves moving near the source will be unrestricted. We further specialize to the case where the stress T^{ij} in the source is negligible compared with the energy density ρc^2; this is a very good approximation for ordinary matter such as ordinary stars and planets (but not neutron stars). Under these assumptions the energy tensor takes the form

$$T^{ab} = \begin{pmatrix} \rho c^2 & \rho \mathbf{v} c \\ \rho \mathbf{v} c & \mathbf{0} \end{pmatrix} \quad \Leftrightarrow \quad T_{ab} = \begin{pmatrix} \rho c^2 & -\rho \mathbf{v} c \\ -\rho \mathbf{v} c & \mathbf{0} \end{pmatrix} \tag{6.1}$$

where $\mathbf{v}$ is the 3-velocity of the various parts of the source, and we are employing the linearized theory where the Minkowski metric suffices to relate T_{ab} to T^{ab}. Both ρ and $\mathbf{v}$ are functions of position, but we are treating the case where they are not time-dependent.[1]

6.1 Gravitational 'Maxwell's equations'

The metric can be written conveniently by introducing scalar and vector potentials Φ, $\mathbf{W}$ defined by

$$\Phi \equiv -\frac{c^2}{4}\bar{h}_{00} = -G \int \frac{[\rho]}{r_{\mathrm{sf}}}\mathrm{d}V \tag{6.2}$$

$$\mathbf{W} \equiv c\bar{h}_{0i} = \frac{-4G}{c^2} \int \frac{[\rho\mathbf{v}]}{r_{\mathrm{sf}}}\mathrm{d}V. \tag{6.3}$$

where the relation of these integrals to h_{ab} follows from (5.44). We include factors of c in these definitions so that Φ matches the Newtonian potential (it has physical dimensions of squared

---

[1]Throughout this chapter we use the bold font for 3-vectors such as velocity, acceleration, vector potential and gravi-magnetic field.

Relativity Made Relatively Easy: General Relativity and Cosmology. Volume 2. Andrew M. Steane, Oxford University Press. © Andrew M. Steane 2021. DOI: 10.1093/oso/9780192895646.003.0006

velocity) and so that $\mathbf{W}$ (dimensions of velocity) makes the analogy with electromagnetism closer (c.f. eqn (6.14)).

For $T_{ij} = 0$ we have $\bar{h}_{ij} = 0$ so $\bar{h}_{00} = -4\Phi/c^2$ is the only diagonal element and therefore $\bar{h}^{\lambda}_{\lambda} = -\bar{h}_{00} = 4\Phi/c^2$. Using (5.31) we then find $h_{00} = -2\Phi/c^2$. Also,

$$h_{11} = h_{22} = h_{33} = -\bar{h}^{\lambda}_{\lambda}/2 = h_{00} \quad \text{and} \quad h_{0i} = \bar{h}_{0i} = W_i/c. \tag{6.4}$$

Thus

$$[h_{ab}] = \begin{pmatrix} -2\Phi/c^2 & \mathbf{W}/c \\ \mathbf{W}/c & -2\Phi\delta_{ij}/c^2 \end{pmatrix} \tag{6.5}$$

and therefore the metric (in a rectangular system of coordinates) is

$$ds^2 = -\left(1 + \frac{2\Phi}{c^2}\right)c^2 dt^2 + 2W_i dx^i dt + \left(1 - \frac{2\Phi}{c^2}\right)\left[(dx^1)^2 + (dx^2)^2 + (dx^3)^2\right]. \tag{6.6}$$

This is the linearized version, so we are restricted to the weak field limit, but we have significant information, namely expressions for the potentials $\Phi, \mathbf{W}$ in terms of the source properties, and, in terms of them, the complete form of the metric. In the static case, where $W_i = 0$, we find that the spatial metric takes the form of the factor $(1 - 2\Phi/c^2)$ multiplying a Euclidean metric. This makes it easy to state the result in terms of other coordinates such as polar coordinates.

The metric (6.6) produces the linear approximation to the Schwarzschild and the Kerr metrics, and can handle more general cases such as a rotating oblate spheroid—for example, a star. We will use it in Chapter 19 to confirm the connection between parameters of the Kerr metric and physical properties (angular momentum and mass) of the source and surrounding spacetime. It can be used to calculate planetary orbits, and also photon trajectories (gravitational lensing) and the Shapiro effect (see Volume 1) to high accuracy in all but extreme cases such as neutron stars and black holes. It can be used to obtain de Sitter and Lense-Thirring precession, described in the next section.

The relationships (6.2) and (6.3) can equally well be expressed

$$\nabla^2 \Phi = 4\pi G\rho, \qquad \nabla^2 \mathbf{W} = \frac{16\pi G}{c^2}\mathbf{j} \tag{6.7}$$

and this suggests an analogy with electromagnetism, as we have already hinted. We make the analogy more complete by introducing fields $\mathbf{E}_g$ and $\mathbf{B}_g$ **defined** by

$$\mathbf{E}_g \equiv -\boldsymbol{\nabla}\Phi, \qquad \mathbf{B}_g \equiv \boldsymbol{\nabla}\times\mathbf{W} \tag{6.8}$$

(but do not forget, we can only handle situations without time-varying potentials and fields; see appendix D for the more general case). We thus obtain:

Gravitoelectromagnetic field equations

$$\begin{aligned} \boldsymbol{\nabla}\cdot\mathbf{E}_g &= -4\pi G\rho, & \boldsymbol{\nabla}\cdot\mathbf{B}_g &= 0 \\ \boldsymbol{\nabla}\times\mathbf{E}_g &= 0, & \boldsymbol{\nabla}\times\mathbf{B}_g &= -16\pi G\mathbf{j}/c^2 \end{aligned} \tag{6.9}$$

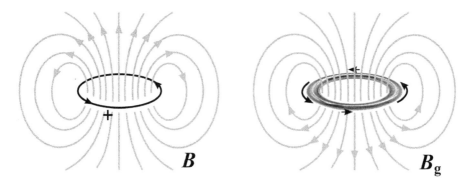

Fig. 6.1 The left diagram shows the *electromagnetic* **B** field owing to a current loop. The right-hand diagram shows the *gravimagnetic* field $\mathbf{B}_{\mathrm{g}}$ owing to a rotating massive ring. The two sets of field lines agree except that they are in opposite directions.

Note the signs in front of the integrals for Φ and $\mathbf{W}$, which correspond to the fact that like 'charges' (i.e. masses) *attract* in GR, and note the factor of 4 in the expression for $\mathbf{W}$. This factor can also be set up another way, for example by using a factor 2 in front of the integral and putting the other factor 2 in the relation between $\mathbf{B}_{\mathrm{g}}$ and $\mathbf{W}$. The main point is that we can now use familiar results from electromagnetism to do calculations in GR, as long as we respect the limitations of the linear and stationary approximation. It is because of such respect that one should treat with caution the above equations for grad, div, curl of $\mathbf{E}_{\mathrm{g}}$ and $\mathbf{B}_{\mathrm{g}}$—there is a danger of losing track of what has been assumed. Nevertheless, familiarity with such things as the magnetic field due to a current-carrying wire or a rotating ball of charge serves us well here, and we can immediately get some good insight into the gravitational effects of the corresponding movements of mass. Fig. 6.1 shows the gravimagnetic field around a uniform rotating ring, and compares it with the electromagnetic case.

In the case of a rigidly rotating axisymmetric mass distribution, one can show (exercise 6.2) that the lowest order contribution to $\mathbf{W}$ is

$$\mathbf{W} = -\frac{4G}{c^2}\frac{J\sin\theta}{2r^2}\hat{\phi} = -\frac{2G}{c^2}\frac{\mathbf{J}\wedge\mathbf{r}}{r^3} \tag{6.10}$$

where $\mathbf{J} = \int \mathbf{r} \wedge \mathbf{p}\,\mathrm{d}V$ is the angular momentum of the source. This has the same functional form as the expression for the vector potential of a magnetic dipole $\mathbf{m}$ in electromagnetism $(\mathbf{A} = (\mu_0/4\pi)\mathbf{m}\wedge\mathbf{r}/r^3)$, therefore we can deduce that (6.10) gives rise to a dipole field. We can substitute this into (6.6) in order to find the metric. The term involving $\mathbf{W}$ is

$$2(W_x\mathrm{d}x + W_y\mathrm{d}y + 0)\mathrm{d}t = 2Wr\sin\theta\,\mathrm{d}\phi\mathrm{d}t \tag{6.11}$$

where the second version adopts spherical polar coordinates. We then find the complete expression (6.6) is

$$\mathrm{d}s^2 = -\left[1 + \frac{2\Phi}{c^2}\right]c^2\mathrm{d}t^2 - \frac{4GJ\sin^2\theta}{c^2 r}\mathrm{d}t\mathrm{d}\phi + \left[1 - \frac{2\Phi}{c^2}\right]\left(\mathrm{d}r^2 + r^2\mathrm{d}\theta^2 + r^2\sin^2\theta\,\mathrm{d}\phi^2\right). \tag{6.12}$$

The physical significance of $\mathbf{E}_g$ and $\mathbf{B}_g$ remains unclear until we have developed a formula for the motion of test particles in the assumed conditions. We employ the equation of motion given in Table 5.1 (it is the geodesic equation for the linearized theory). We are interested in the 3-acceleration $\mathbf{a}$, which in the low-velocity limit is given by the spatial components of $du^a/d\tau$. Using (6.5), dropping ∂_0 terms (stationary source), keeping only up to linear terms in v, and paying careful attention to signs, one finds (exercise 6.1)

$$u^\mu u^\nu \partial_\mu h^i_\nu \simeq \mathbf{u} \cdot \boldsymbol{\nabla} W^i,$$
$$u^\mu u^\nu \partial^i h_{\mu\nu} \simeq -2\partial_i \Phi + 2\partial_i (\mathbf{u} \cdot \mathbf{W}). \tag{6.13}$$

Hence, after noting that the x-component of $\mathbf{u} \times (\boldsymbol{\nabla} \times \mathbf{W})$ is $\partial_x (\mathbf{u} \cdot \mathbf{W}) - (\mathbf{u} \cdot \boldsymbol{\nabla}) W_x$, we obtain

$$\mathbf{a} = -\boldsymbol{\nabla}\Phi + \mathbf{u} \times \operatorname{curl} \mathbf{W} = \mathbf{E}_g + \mathbf{u} \times \mathbf{B}_g, \tag{6.14}$$

which you can, if you like, write

$$\mathbf{f} = m(\mathbf{E}_g + \mathbf{u} \times \mathbf{B}_g) \tag{6.15}$$

where $\mathbf{f}$ is force and m is inertial mass. This confirms that it is completely legitimate to regard $\mathbf{E}_g$ as the 'gravitational field' (in the Newtonian sense) and $\mathbf{B}_g$ as the 'gravimagnetic' field, which combines with the Newtonian result to give the total field of force per unit mass. Note also that the Newtonian limit for low-velocity low-pressure sources also immediately follows (and c.f. Section 5.4 for more general sources).

You should now be able to understand the example of the two rods and springs illustrated in Fig. 2.2. Each rod produces a radial field $\mathbf{E}_g$ and also a gravimagnetic field $\mathbf{B}_g$ which circulates around it in the left-handed sense. The latter field, when it interacts with the other rod, *reduces* the net attraction between the rods. By contrast, if the rods were moving in opposite directions then the gravimagnetic effect would increase the attraction (the opposite of what happens in the electromagnetic case). The gravimagnetic force on the falling balls also explains the reduced acceleration of the first ball, and the component of acceleration along the rods for the second ball.

One can similarly deduce that particles falling in the gravitational field of a rotating ball such as the Earth will not fall straight down (i.e. in a purely radial direction), owing to the gravimagnetic term which pushes them off to one side, and furthermore we can calculate this to high precision. Further measurable effects are described in the next section.

The generalization of the subject of this section to the case of sources which are neither slow nor stationary is described in appendix D.

6.2 Lense-Thirring precession

Lense-Thirring precession was briefly mentioned at the end of Chapter 4. We are now ready to calculate this effect in the weak field limit.

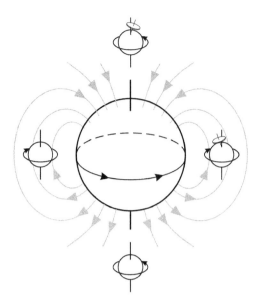

Fig. 6.2 Effect of a large rotating sphere on spacetime around it. The lines show the gravimagnetic field, the smaller spheres show the sense of rotation of LIFs and thus the sense of precession of gyroscopes near the large body. They rotate in the same sense that they would if the collection of spheres were immersed in a very slightly viscous liquid.

The Lense-Thirring precession is most naturally interpreted in terms of gravimagnetism, but it is useful also to understand it as an example of what should be expected in a rotating frame. Near a rotating body such as Earth, spacetime is warped in such a way that each of the LIFs is slowly rotating relative to the distant stars. Therefore a frame fixed in space relative to the distant stars is, in an absolute sense, a rotating frame, because it is rotating (very slowly) relative to any local inertial frame. Consequently one should expect precession of a gyroscope whose centre of mass is fixed in this frame, as illustrated in Fig. 4.3. The Lense-Thirring precession is analogous to the *whole* of the precession observed in a rotating frame, not just the residual Thomas precession, but it is nevertheless a GR effect because if any LIF is in a state of rotation relative to the distant stars then this must be owing to the warping of spacetime.

We shall describe some exact results in the presence of a rotating black hole in Chapter 19; here we content ourselves with the weak field case. To calculate the Lense-Thirring effect in the weak field limit, much the easiest and most intuitive way is to appeal to gravimagnetism. First one needs to know the gravimagnetic field due to the gravitating system in question. For the case of an isotropic spinning ball the gravimagnetic field outside it is a dipole field (exercise 6.6), given by

$$\mathbf{B}_{\mathrm{g}} = \frac{-2GJ}{c^2 r^3}\left(2\cos\theta\,\hat{\mathbf{r}} + \sin\theta\,\hat{\boldsymbol{\theta}}\right) = \frac{-4G}{c^2 r^3}\left(\frac{3(\boldsymbol{\mu}\cdot\mathbf{r})\mathbf{r}}{r^2} - \boldsymbol{\mu}\right) \tag{6.16}$$

where $\boldsymbol{\mu} = \mathbf{J}/2$ is the dipole moment and $\mathbf{J}$ is the angular momentum of the ball. Using (4.8) we find

$$\Omega_{\text{LT}} = -\frac{1}{2}\mathbf{B}_{\text{g}} = \frac{G}{c^2 r^3}\left(\frac{3(\mathbf{J}\cdot\mathbf{r})\mathbf{r}}{r^2} - \mathbf{J}\right). \tag{6.17}$$

This is the formula for the Lense-Thirring precession rate in the weak field limit. For points along the axis of rotation, $\mathbf{r}$ is aligned with $\pm\mathbf{J}$ so the bracket evaluates to $2\mathbf{J}$. At the equator it evaluates to $-\mathbf{J}$. Thus a gyroscope at either pole precesses in the same sense as the Earth's rotation, one at the equator in the opposite sense (like a cog loosely engaged by one fixed to the Earth) and at half the rate; see Fig. 6.2.

The field lines of the Earth's gravimagnetic field emerge from the south pole and follow the familiar dipole pattern around and up, to finish near the north pole. Modeling the Earth as a uniform ball, we find that a gyroscope hovering near the equator would precess by 0.46 milliarc-sec per day (0.168 arcsec per year). A gyroscope in near-Earth orbit would precess by the sum of this and the de Sitter precession, the latter with a sign depending on the sense of the orbit. A gyroscope in free fall along the axis of rotation (i.e. in the radial direction at the poles) would precess by the same amount as one fixed near the poles, as calculated above, which you can confirm using Lorentz transformations.

For the canonical stationary metric (4.14) the exact result can be written as follows. The proper precession rate (i.e. the rate according to local standard clocks) of a gyroscope fixed in the lattice is

$$|\mathbf{\Omega}_{\text{LT}}| = \frac{1}{2\sqrt{2}}e^{\Phi/c^2}\left[k^{ik}k^{jl}(\partial_j W_i - \partial_i W_j)(\partial_k W_l - \partial_l W_k)\right]^{1/2}. \tag{6.18}$$

This is essentially a covariant expression for '$\mathbf{\Omega}_{\text{LT}} = \text{curl}\,\mathbf{W}/2$'. The most common case is where $\mathbf{W}$ has only one component, W_3, which depends on only one coordinate, x^1. For this case you can check that the only terms appearing in the sum are 1313 and 3131, and the expression simplifies to

$$|\mathbf{\Omega}_{\text{LT}}| = \frac{1}{2}e^{\Phi/c^2}\sqrt{k^{11}k^{33}}\,\partial_1 W_3. \tag{6.19}$$

6.2.1 Frame dragging?

The term 'frame dragging' or 'space dragging' ('Mitführung') was coined by early workers such as Einstein and Thirring in order to give an insight into the spacetime physics underlying effects such as Lense-Thirring precession. The idea is that a rotating body drags the space around it, like a kind of whirlpool. This idea has some appeal and gives a rough suggestion of what is going on inside the ergosphere of the Kerr metric (Section 19.2). However, the concept of frame dragging is rather vague in practice, and hard to use for quantitative calculations. (Rindler, 1997) By contrast, gravimagnetism is both clear and quantitatively precise.

For example, we saw above that a gyroscope near the Earth will Lense-Thirring precess in a different sense at the equator than it would at the poles. The notion of frame dragging does not predict this clearly; rather it gives a convenient way to think about the result once it has been calculated by other means.

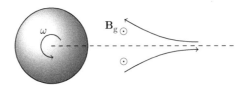

Fig. 6.3 Example infalling and outgoing trajectory (arrows) for a particle moving near a rotating body, in the equatorial plane. The gravimagnetic field is out of the page as shown.

Inside a rotating spherical cavity the ordinary gravitational field is zero and the gravimagnetic field is uniform (the corresponding electromagnetic case is a standard problem in undergraduate physics). This means that in this situation the gravimagnetic field can be transformed away by adopting a rotating reference frame. One then has, to first approximation, a Minkowski metric for a lattice which is rotating. However, the metric is not exactly Minkowski because the centrifugal force remains (it appears at second order in ω in the metric, see (4.13), (4.9)). Note also that the rotation of the lattice is very slow; exercise 6.5. This is NOT Mach's principle.

Finally, let us consider a particle falling towards the equator of a rotating body. Using the gravimagnetic calculation, we find that its velocity is swept to the side a little, in the direction which agrees with the sense of rotation of the body—see Fig. 6.3. This is consistent with what the 'frame dragging' idea would suggest. But now consider a particle moving outwards. It experiences the same field $\mathbf{B_g}$ and therefore the force $\mathbf{v} \times \mathbf{B_g}$ is now in the opposite direction, *against* the sense of rotation of the massive body. Therefore it is *not* true to say that timelike geodesics are all swept around a rotating body in the same sense, as a naive interpretation of frame dragging might suggest. This does not rule out that one can construct interpretations of 'frame dragging' such that correct answers are obtained. However, in order to do this one requires first to know the right answer by some other method.

The phrase 'frame dragging' has persisted in physics terminology because the physical effects discussed above are real enough. However, the phrase does not always succeed in giving a correct intuition about either the size or the sign of the various effects, and therefore it should be used with caution.

Exercises

6.1 Let $u^a = (\gamma c, \gamma \vec{u})$ be the 4-velocity of a particle moving in a weak gravitational field. Show that for a static metric and $u \ll c$,

$$u^\mu u^\nu \partial_\mu h_\nu^i \simeq u^j u^0 \partial_j h_0^i$$
$$u^\mu u^\nu \partial^i h_{\mu\nu} \simeq 2u^0 u^\nu \partial^i h_{0\nu}.$$

Hence obtain (6.13).

6.2 (i) Show from (6.3) that at positions $r \gg a$, the leading order contribution to the gravimagnetic potential produced by a ring of mass M, radius a, rotating at angular velocity ω is $\mathbf{W} = -(2GM\omega a^2/c^2 r^2)\sin\theta\,\hat{\phi}$. (ii) Derive (6.10).

6.3 Replace the electric current loop on the left of Fig. 6.1 by a *negatively* charged ring, rotating in the same sense as the massive ring on the right. Comment on the directions of both the 'gravi-electric' and 'gravimagnetic' parts of the acceleration due to gravity compared to the electromagnetic case, making reference also to the equations in Table 5.1.

6.4 A sphere of mass M travels down the z axis at speed v, and a particle travels in the opposite direction along the line $y = 0$, $x = r$ so as to pass the sphere. At the moment when the two are abreast, show that the ratio between the ordinary and the gravimagnetic contributions to the acceleration of the particle is $4v^2/c^2$. (For the Hulse–Taylor binary star system described in Section 7.5, this effect is of order 10^{-5} and is detectable.)

6.5 Using your knowledge of the magnetic field inside a solenoid of infinite length, or otherwise, find the gravimagnetic field inside a long cylindrical shell of mass μ per unit length, rotating at angular velocity ω. Deduce that, to first order approximation, spacetime inside the cylinder is Minkowskian and local inertial frames are rotating at angular velocity $4G\mu\boldsymbol{\omega}/c^2$ relative to infinity. [*Ans.* $\mathbf{B_g} = -8G\mu\boldsymbol{\omega}/c^2$]

6.6 Show that the gravimagnetic field outside an isotropic spinning ball is as given by (6.16) (c.f. exercise 6.2 of Chapter 5).

6.7 Relabel ϕ as $\bar{\phi}$ in (6.12), and then introduce the coordinate $\phi = \bar{\phi} - \omega t$ as in (4.13). Hence show that in the new coordinates, the metric outside a rotating axisymmetric mass distribution takes a form such that $g_{t\phi} = 0$ when $\omega = 2GJ/c^2r^3$. What does this tell us about Lense–Thirring precession? [Hint: note that since ω is here a function of r, there is no simple sense in which we have adopted a rotating frame, but regions near to the rotation axis do bear that interpretation—why?]

6.8 Using the same change of coordinates as in the previous exercise, show that the resulting metric has, after setting $c = 1$,

$$g_{00} = -(1 + 2\bar{\Phi}) - \frac{4GJ\sin^2\theta}{r}\omega + (1 - 2\bar{\Phi})\omega^2 r^2 \sin^2\theta \quad (6.20)$$

where $\bar{\Phi}$ is the potential before the coordinate change. Show that, for points in the equatorial plane ($\theta = \pi/2$), this corresponds to the canonical form (4.14) with

$$\Phi = \bar{\Phi} + \frac{2GJ\omega}{r} - \tfrac{1}{2}(1 - 2\bar{\Phi})\omega^2 r^2 + O(\bar{\Phi}^2/c^4).$$

Now if the acceleration due to gravity given by (4.15) is equal to zero, then we must have that ω is the angular velocity for a circular orbit, since a particle in free fall then suffers no acceleration relative to the lattice. Use this to show that the angular velocity for such an orbit is

$$\omega = -\frac{GJ}{c^2r^3} \pm \sqrt{\frac{GM}{r^3}} + O(J^2)$$

where we reinstated c and used $\bar{\Phi} = -GM/r$. Confirm that this can also be written

$$\omega^2 = \frac{GM}{r^3} \mp \frac{2GJ}{c^2r^4}\sqrt{\frac{GM}{r}} + O(J^2). \quad (6.21)$$

(This method to find ω via a change of coordinates can also be used without approximation, in appropriate circumstances, in order to find a geodesic without solving any differential equations.)

6.9 Obtain (6.21) via gravimagnetism. Give an intuitive explanation of why it is that the retrograde orbit has the shorter period. Find the difference in periods between circular orbits in opposite senses above Earth's equator at altitude 2000 km.

6.10 Will a clock fixed at Earth's equator run slow compared to one fixed at the north pole? In order to answer this question, take into account the fact that the Earth is not spherical but slightly ellipsoidal owing to its rotation. More precisely, there is a surface called the *geoid* which is the surface of constant potential Φ, where Φ includes the effects of gravitation and global rotation. In the absence of other external forces the surface of a body of water not moving relative to the ground will follow the geoid. Now the question becomes, will clocks fixed at different places on the geoid run at different rates? [Hint: adopt a rotating frame and do not introduce any approximation.]

6.11 If you stand still for a year on Earth's surface, by how much more will your head age than your feet? Is the answer different at the poles compared to the equator?

6.12 The Lense-Thirring precession rate for an object near Earth's north pole is 4×10^{-14} rad/s. Therefore an observer located there who took pains to construct an accurate LIF must find that his LIF rotates relative to the distant stars at this rate. For example, the LIF would complete a revolution once every 5 million years, which is equally to say that the distant stars would orbit such a LIF at this rate. The galaxy Centaurus A is at a distance of 12 million lightyears. Calculate the speed at which Centaurus A is moving relative to a LIF at Earth's north pole, if one extends the LIF all the way to Centaurus A. What is the meaning of this calculation? You may postpone a complete answer until you have finished the book!

7

Gravitational waves

The linearized field equation (5.43) obviously has plane wave solutions. One can find plane wave solutions of the exact field equation too, but it is easier to start with the weak field limit, which is sufficient to describe many interesting phenomena. In this chapter we consider such gravitational waves. We discuss production and detection, amplitude and energy. Some of the derivations are somewhat lengthy but are presented in full; the reader wishing to avoid these algebraic details can skim through them and receive the main results on trust.

7.1 Identifying and simplifying the plane wave solutions

In vacuum we have $\Box^2 \bar{h}_{ab} = 0$ and also $\Box^2 h = 0$ so

$$\Box^2 h_{ab} = 0. \tag{7.1}$$

Consider plane wave solutions of the form

$$h_{ab} = \epsilon_{ab} e^{i k_\mu x^\mu} \tag{7.2}$$

where ϵ_{ab} (symmetric) is a constant polarization tensor and k^a is the 4-wave-vector. Here ϵ_{ab} may be complex, and it is understood that the physical solution is obtained by taking the real part of h_{ab}. Upon substituting this into (7.1) we find that it is a solution and the phase velocity is c. In order to allow us to use (7.1) the solution must also satisfy the Lorenz gauge condition (5.39). Anticipating that $h = 0$ (the traceless gauge to be described shortly) this can be written $\partial^\lambda h_{\lambda b} = 0$ and therefore

$$k^\lambda \epsilon_{\lambda b} = 0 \tag{7.3}$$

that is, each row or column of ϵ_{ab} is orthogonal to k^a: the wave is 'transverse' in the space-time sense when $h = 0$. For example, if the wave is travelling in the z direction, then $[k^a] = (k, 0, 0, k)$ and then we have that $\epsilon^{a3} = \epsilon^{a0}$, i.e. for any given row the z-component matches the t-component. After using that ϵ^{ab} is symmetric, this reduces the number of independent components of ϵ^{ab} from ten to six.

We can exploit some further gauge freedom within the Lorenz gauge. Further changes of h_{ab} using a gauge function satisfying

Relativity Made Relatively Easy: General Relativity and Cosmology. Volume 2. Andrew M. Steane,
Oxford University Press. © Andrew M. Steane 2021. DOI: 10.1093/oso/9780192895646.003.0007

$$\Box^2 \xi_a = 0 \qquad (7.4)$$

will remain in Lorenz gauge, since they will not change (5.37). A useful choice is to adopt a coordinate change that itself oscillates in step with the wave. That is, we propose

$$\xi^a = \rho^a e^{ik_\mu x^\mu} \qquad (7.5)$$

and we adjust the constant parameters ρ^a such that the polarization tensor simplifies. It is shown in exercise 7.1 how to do this so as to arrive at a polarization tensor that is both traceless and purely spatial. That is,

$$\epsilon^\lambda_\lambda = 0 \qquad \text{and} \qquad \epsilon_{0a} = \epsilon_{a0} = 0. \qquad (7.6)$$

The Lorenz gauge condition now shows that the wave must be transverse in the spatial sense, i.e. having only components orthogonal to the direction of travel of the wave. Consider for example such a wave travelling in the z direction. Then the metric perturbation h_{ab} must have the general form

$$h_{ab}(z,t) = \begin{pmatrix} 0 & 0 & 0 & 0 \\ 0 & h_+ & h_\times & 0 \\ 0 & h_\times & -h_+ & 0 \\ 0 & 0 & 0 & 0 \end{pmatrix} e^{i(kz-\omega t)} \qquad (7.7)$$

since this is the only way to construct a traceless symmetric purely spatial matrix which is transverse to the z direction. The variables h_+ and $h_\times$ are constant 'wave amplitudes'. General phases of oscillation can be treated by allowing h_+ and $h_\times$ to be complex numbers. There are thus two independent polarization states which can be taken to be

$$\epsilon^{(+)}_{ab} = \begin{pmatrix} 0 & 0 & 0 & 0 \\ 0 & 1 & 0 & 0 \\ 0 & 0 & -1 & 0 \\ 0 & 0 & 0 & 0 \end{pmatrix}, \qquad \epsilon^{(\times)}_{ab} = \begin{pmatrix} 0 & 0 & 0 & 0 \\ 0 & 0 & 1 & 0 \\ 0 & 1 & 0 & 0 \\ 0 & 0 & 0 & 0 \end{pmatrix}. \qquad (7.8)$$

The gauge in which the plane wave has this form is called the **transverse traceless** gauge, or TT gauge.

The movement from an arbitrary form of h_{ab} to one respecting both the Lorenz gauge condition and our chosen further conditions has used up all the gauge freedom. The starting point was a symmetric matrix having ten independent components. Since we imposed first four constraints and then another four, there are now two parameters remaining, in terms of which h_{ab} is completely specified within the TT gauge.

7.1.1 General method to adopt transverse traceless gauge

A useful feature of the TT gauge is that there is a method, starting from some plane wave solution h_{ab} not in that gauge, to quickly obtain the form which that particular h_{ab} takes in TT gauge, without repeating the sequence of steps outlined above and in exercise 7.1. All one needs to do is project h_{ab} onto the right direction in spacetime and then adjust the trace. Let **n** be a spacelike unit vector; one with no temporal component at all (in the coordinates we are adopting) and whose spatial part is along the direction of propagation of the waves:

$$n_0 = 0, \qquad n_j = ck_j/\omega. \tag{7.9}$$

Now introduce a tensor P_{ab} whose effect is to project onto the spatial surface orthogonal to **n**:

$$P_{ab} = \eta_{ab} - n_a n_b. \tag{7.10}$$

In order to understand the sense in which this is a projection tensor, note its effect on vectors. For any **v**, one finds $n_\mu P^\mu_\nu v^\nu = 0$ and $P^a_\mu P^\mu_\nu v^\nu = P^a_\nu v^\nu$, in other words the action of **P** on a vector gives a result orthogonal to **n** and leaves unchanged items which are already orthogonal to **n**. We can now find the transverse part of h_{ab} as simply the projection $P^\mu_a P^\nu_b h_{\mu\nu}$ and the transverse traceless part is obtained by subtracting off the trace:

$$h^{\mathrm{TT}}_{ab} = P^\mu_a P^\nu_b h_{\mu\nu} - \tfrac{1}{2} P_{ab} P^{\mu\nu} h_{\mu\nu}. \tag{7.11}$$

In order to prove that this procedure does indeed give the correct h^{TT}_{ab} corresponding to h_{ab}, and not some multiple of it for example, one mainly needs to note that the h_{12} component survives intact in the case of propagation along the z direction.

Applied to the case of propagation along a coordinate axis, the projection amounts to a very simple method: to convert to TT gauge, all we need to do is set to zero the components of h_{ab} that are not transverse (e.g. t and z components for propagation along z) and then subtract from each of the diagonal elements half their sum.

7.2 The physical impact of a gravitational wave

Consider a free particle before its encounter with a gravitational wave, at rest with 4-velocity $u^a = (c, 0, 0, 0)$. The motion of the particle in the presence of the wave is determined by the geodesic equation

$$\frac{\mathrm{d}u^a}{\mathrm{d}\tau} + \Gamma^a_{\mu\nu} u^\mu u^\nu = 0.$$

The situation initially is that only u^0 is nonzero, so the initial acceleration is

$$\left.\frac{\mathrm{d}u^a}{\mathrm{d}\tau}\right|_{\tau=0} = -c^2 \Gamma^a_{00} = 0 \tag{7.12}$$

where the four Christoffel symbols vanish because in TT gauge $\epsilon_{\mu 0} = \epsilon_{0\nu} = 0$. Since the initial acceleration is zero, the particle remains at rest, and the same conclusion must then hold at all subsequent times. Thus, in TT gauge, a free particle initially at rest *stays motionless with respect to the coordinate system* as a gravitational wave passes over it.

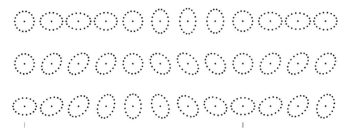

Fig. 7.1 The effect of a gravitational wave on a ring of free-floating particles as time goes on from left to right. The three lines of images show the case of $+$, $\times$ and circularly polarized waves, respectively. In the latter case the particles do not travel around the ring, but each one performs an epicycle. Also, the ring is circular before the wave arrives, and is then squeezed into an elliptical shape along a direction which depends on the initial phase of the wave. The scale of the diagram is such that proper distances are constant on the page.

Now consider two points on the x axis of the coordinate system, separated in the x direction by a coordinate distance d, in the presence of a 'plus-polarized' ($\epsilon_{ab}^{(+)}$) wave travelling in the z direction. The ruler distance[1] between the points is

$$s = \sqrt{g_{11}}\,d \simeq \left(1 + \tfrac{1}{2}h_{11}\right) d = \left(1 + \tfrac{1}{2}h_+ \cos(kz - \omega t)\right) d. \tag{7.13}$$

This clearly fluctuates with time. Similarly, the ruler distance between two points on the y axis separated by coordinate distance d varies as

$$s = \left(1 - \tfrac{1}{2}h_+ \cos(kz - \omega t)\right) d. \tag{7.14}$$

Thus when the ruler distance on the x axis is elongated, that on the y axis is compressed. Fig. 7.1 shows this ruler-distance effect on a ring of free-floating particles.

We can now interpret the above results more physically. We saw that freely moving test particles do not change their *coordinate* locations as the wave passes, but this means they *do* change their *proper separations* (ruler distances from one another). Suppose a ring of free dust particles is floating above a flat metal plate. The particles in the plate also try to change their proper separations as the wave passes—this is what the geodesic equation is telling them to do—but they are not in free fall because they are subject to electromagnetic and other forces within the material of the plate (c.f. eqn (2.10)), and compared to those forces the gravitational tidal force is miniscule. The plate therefore maintains its shape and proper distances. Such a metal plate acts, to very good approximation, like an ideal measuring rod or ruler. If we now adopt the reference frame natural to a laboratory measurement—which is a reference frame in which the *metal plate* does not get squeezed or stretched—then we must conclude that the floating dust particles move, relative to such a laboratory, exactly as shown in Fig. 7.1. They track the changing ruler separations of coordinate points, whereas a scale of fixed ruler distances could be literally marked on the metal plate by a series of scored lines, and would stay fixed in the

---

[1]Ruler distance, also called proper distance, is the integral of spacetime interval along some appropriately chosen spacelike line.

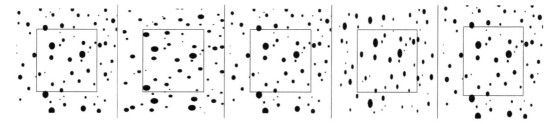

Fig. 7.2 The effect of a + polarized gravitational wave. The situation at five instants of time is shown. The black square is a rigid steel frame. The dots represent blobs of matter floating in space (in free fall), made of material with a very small or vanishing coefficient of elasticity (Young's modulus). The images illustrate the 'breathing' motion caused by the wave throughout any given plane perpendicular to the direction of propagation. This serves to show that there is no centre to the plane wavefronts, and nothing special about the point at the centre of one of the ellipses shown in Fig. 7.1.

laboratory. If there was friction between the dust and the plate then the gravitational wave would drag the dust across the plate and generate (a tiny amount of) heat.

Note that the effect of the wave is to move the dust particles by an amount independent of their mass and in proportion to their separation. A cloud of particles would undergo a sort of 'breathing' motion (with no unique centre—exercise 7.3), as shown in Fig. 7.2. To detect this motion one would do well to use not a tiny dust particle but a heavy object, in order to give the gravitational forces a chance to be as large as possible compared to other forces. A gravitational wave with a long enough wavelength will shift the whole Earth and Moon just as readily as it would shift the smallest mote of dust!

7.2.1 Detection

With the above in mind, one deduces that a good way to detect gravitational waves is to suspend a pair of large, massive objects (called test masses) from threads attached to two ends of a long rigid bar, and seek to detect motion of the objects relative to one another, in the coordinate frame in which the bar itself does not move. From (7.13) the fractional change in separation of the test masses as a wave passes by is given by

$$\frac{\delta s}{s} = \frac{h_+}{2} \tag{7.15}$$

(and more generally see exercise 7.5). The largest gravitational waves we can expect to observe on Earth come from astrophysical sources such as binary stars, and are expected to produce values of h_+ of order 10^{-21}, therefore this is the amount of strain (i.e. $\delta s/s$) one is trying to detect in a gravitational wave detector. Even for s of order kilometres, the change in s is smaller than the size of an atomic nucleus, so this is an extremely demanding experimental challenge.

In the *laser interferometric* method, the test masses are large heavy mirrors suspended from highly sophisticated vibration isolation platforms in high vacuum, and illuminated by high-power laser beams in a Michelson interferometer arrangement (Fig. 7.3). There are four test

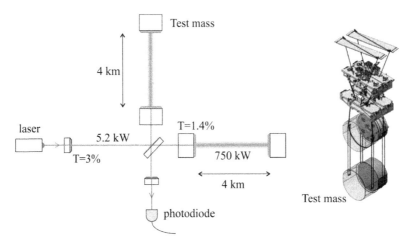

Fig. 7.3 A laser interferometric gravitational wave detector. The design is essentially a large Michelson interferometer with resonant optical cavities in each arm. All the optics are vibration-isolated; the test masses (heavy mirrors) are suspended from multi-stage vibrational isolation platforms. The parameters are shown for the LIGO detector, USA. [Image courtesy Caltech/MIT/LIGO Laboratory]

masses. The most significant additional element compared to the traditional interferometer design is the pair of 1.4% transmission mirrors in Fig. 7.3. These act to form a pair of optical cavities with the fully reflecting mirrors, in which the light bounces too and fro. This builds up the sensitivity to displacements much as if the interferometer arms were many times longer.

In the future it may be possible to detect gravitational waves by two further methods. One is the timing of pulses emitted by a collection of millisecond pulsars distributed across the galaxy. Such measurements are sensitive to gravity waves of period of order years. In the more distant future interferometers based on matter waves, as opposed to light waves, may be able to access frequencies of order kHz with sufficient sensitivity.

A simple linear detector. Equations (7.13) and (7.14) are already perfectly adequate as a basis for analysing a laser interferometric gravitational wave detector, if the restoring and damping forces offered by the mirror supports are negligible. In this section we will develop the ideas a little further, in order to obtain a result that will be useful later when considering the energy in gravitational waves.

We consider the effect of a gravitational wave on a pair of masses m supported above a rigid metal plate, see Fig. 7.4. We suppose the masses are free to move in the x-direction, and are located initially at coordinates x_A^{TT}, x_B^{TT} in a coordinate system where the wave is in TT gauge and is +-polarized as it travels in the z direction. As we discussed after (7.12), the values of x_A^{TT} and x_B^{TT} do not change as the wave passes by, but the proper distance between x_A^{TT} and x_B^{TT} does, and therefore the location of one or both oscillates relative to the metal plate.

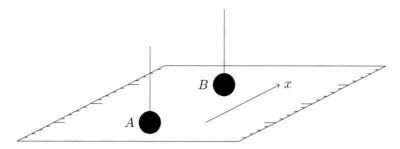

Fig. 7.4 A pair of masses supported above a steel plate.

Let x be a coordinate which increases in proportion to proper distance along the metal plate, and assume the gravitational wave has a negligible effect on the plate. If we place the origin $x = 0$ at a location half way between A and B before the wave arrives, then the change in proper distance between the two masses implies that their locations expressed in the x coordinate must be

$$x_A = -s/2, \qquad x_B = s/2 \qquad (7.16)$$

where s is given by (7.13) and we can take $z = 0$. Therefore

$$\begin{aligned} \frac{\mathrm{d}^2 x_A}{\mathrm{d}t^2} &= -\frac{\ddot{s}}{2} = \frac{1}{4}\omega^2 d\, h_+ \cos\omega t, \\ \frac{\mathrm{d}^2 x_B}{\mathrm{d}t^2} &= \frac{\ddot{s}}{2} = -\frac{1}{4}\omega^2 d\, h_+ \cos\omega t. \end{aligned} \qquad (7.17)$$

The velocities of the masses are of order $\omega d\, h_+$. If this is small compared to c then we can now use Newtonian mechanics to interpret what this motion is telling us about the forces on the masses. From Newton's second law we determine that the tidal force from the gravitational wave is such as to give forces $m\ddot{x}_A$, $m\ddot{x}_B$ respectively on the two masses.

With this result in hand, we can analyse the case shown in Fig. 7.5. We now treat the case where each mass is also subject to a restoring force and a damping force owing to the pendulum or spring which is holding it in place. The equations of motion (Newton's second law) for the two masses are

$$\begin{aligned} m\ddot{x}_A &= -kx_A - \nu\dot{x}_A + \tfrac{1}{4}m\omega^2 d\, h_+ \cos\omega t, \\ m\ddot{x}_B &= -kx_B - \nu\dot{x}_B - \tfrac{1}{4}m\omega^2 d\, h_+ \cos\omega t, \end{aligned} \qquad (7.18)$$

where k and ν are spring and damping constants. All terms linear in h_+ have been accounted for correctly. Subtracting the first equation from the second gives

$$m\ddot{s} = -ks - \nu\dot{s} - \tfrac{1}{2}m\omega^2 d\, h_+ \cos\omega t \qquad (7.19)$$

and therefore

$$\ddot{s} + 2\gamma\dot{s} + \omega_0^2 s = -\tfrac{1}{2}\omega^2 d\, h_+ \cos\omega t \qquad (7.20)$$

where $\omega_0^2 = k/m$ and $\gamma = \nu/2m$. This is the equation for driven damped harmonic motion. The solution (after a transient has died away) is

$$s = R\cos(\omega t + \phi) \qquad (7.21)$$

Fig. 7.5 A pair of masses joined by a lightly damped spring.

where

$$R = -\tfrac{1}{2}\omega^2 d\, h_+ / ((\omega_0^2 - \omega^2)^2 + 4\gamma^2\omega^2)^{1/2} \tag{7.22}$$

$$\tan\phi = -2\gamma\omega/(\omega_0^2 - \omega^2). \tag{7.23}$$

7.2.2   An interplanetary warp drive?

We are used to the idea that faster-than-light travel and communication is not possible. This is inescapable in GR because the metric is always locally Minkowskian and worldlines are timelike. However, the fact that spacetime itself can be dynamic, owing to gravitational waves and wavelike disturbances, allows a form of travel and communication which might be described as 'faster than light' in a certain sense.

Consider a gravitational wave propagating in the z direction and incident on the two masses shown in Fig. 7.4, resulting in eqn (7.17). In a time of order π/ω the proper distance between the masses changes by approximately $h_+ d$, therefore it changes at a rate of order $h_+\nu d$ where $\nu = \omega/2\pi$ is the wave frequency. There is no fundamental reason why this rate of change of proper distance cannot exceed the speed of light. For example, if $h_+ = 0.2$ and $d = 50\lambda$ where λ is the wavelength of the gravitational wave, then $h_+\nu d = 10c$. In this example the 'warp drive' provided by the gravitational wave has driven spacetime so as to manipulate the distance between the two masses at this high speed.

The effect we are contemplating is the same sort of process as goes on all the time at the largest scales—the cosmic process which we call the expansion of the universe. In that case the proper distances between galaxies are caused to grow by the dynamic nature of spacetime, and these distances can in principle grow at any rate. We shall discuss this phenomenon in Chapters 22–25.

Returning now to more local processes, the wave effect just noted does not itself amount to either travel or communication, but one can imagine a scenario such as a gravitational disturbance in the form of a pulse with a short leading edge and a longer duration thereafter, such that two locations initially at proper separation d are quickly brought to some smaller separation, say $d/10$, and then for a while travel is possible between them in a time $d/(10c)$ as registered by clocks at either end-point. A disturbance of this magnitude is well outside the approximations of linearized GR, but the full theory allows it in principle. If, after some extraordinary event, a gravitational wave of period two hours and strain amplitude 10 swept over the solar system, with a wavefront aligned along the line between Earth and Neptune, then while the disturbance was present it would be possible to send information between Earth and a space probe near the planet Neptune in about half an hour, whereas normally the travel time for signals to Neptune is 4 hours. (There may also be large effects on proper time intervals at each location, which we shall not try to estimate.)

Can one imagine exploiting such ideas for the purpose of fast space-travel in some futuristic technology? The theory does not allow for a spaceship which, at the push of a button, causes distant places to be brought near, because it would take time to set up the required gravitational disturbance. To arrange this for a given destination would require a time at least as long as the ordinary light-speed-limited travel time to that destination. However, the theory does allow for something more like a public transportation system: a set of pre-configured dynamic links, like commercial airline routes. But it is hard to conceive of any feasible and safe way to bring about the necessary dynamic structures in spacetime.

7.3   Sources of gravitational waves

In order to understand how gravitational waves are produced by binary stars and other sources, consider the general solution to the linearized field equations, (5.44), which we repeat here, adopting the symbol $\mathbf{z}$ for locations at the source (i.e. $\mathbf{z} \equiv \mathbf{x}_{\mathrm{s}}$):

$$\bar{h}_{ab}(t, \mathbf{x}) = \frac{4G}{c^4} \int \frac{T_{ab}(t - r_{\mathrm{sf}}/c, \mathbf{z})}{|\mathbf{x} - \mathbf{z}|} \mathrm{d}^3 z \qquad (7.24)$$

where $r_{\mathrm{sf}} = |\mathbf{x} - \mathbf{z}|$ is the distance from each source event to the field event. This integral is troublesome to evaluate as it stands, owing to the denominator, to the multiple terms in the stress tensor, and to the fact that one is integrating, in effect, over the past light cone of the field event, and therefore one has to track a different source time for each source position. However, by a wonderful piece of physical and mathematical conjuring, one can convert this formula into a much simpler form for the leading order term in a multipole expansion.

The multipole expansion is a standard expansion which is often useful in considering the Poisson equation and the wave equation (whether in GR or in other areas of physics). We place the origin of coordinates somewhere at or near the source, and we suppose the field point is sufficiently far from the matter in the source that $z \ll r$ in the integral, where $r = |\mathbf{x}|$ is the distance from the origin to the field point. For $z \ll r$ the Taylor expansion of $1/|\mathbf{x} - \mathbf{z}|$ is

$$\frac{1}{|\mathbf{x} - \mathbf{z}|} = \frac{1}{r} - z^i \frac{\partial}{\partial x^i} \left(\frac{1}{|\mathbf{x} - \mathbf{z}|} \right) + \frac{1}{2!} z^i z^j \frac{\partial}{\partial x^i} \frac{\partial}{\partial x^j} \left(\frac{1}{|\mathbf{x} - \mathbf{z}|} \right) + \cdots \qquad (7.25)$$

$$= \sum_{l=0}^{\infty} \frac{(-1)^l}{l!} z^i z^j \cdots z^k \, \partial_i \partial_j \cdots \partial_k \left(\frac{1}{|\mathbf{x} - \mathbf{z}|} \right) \bigg|_{\mathbf{z}=0} \qquad (7.26)$$

where it us understood that the letters $i, j, \cdots k$ indicate a sequence of l different index letters for the term of order l, and the derivatives are to be evaluated at $\mathbf{z} = 0$. The first few derivatives are, for $l = 0, 1, 2, \ldots$,

$$\frac{1}{r}, \quad \frac{-x_i}{r^3}, \quad \frac{3x_i x_j - r^2 \delta_{ij}}{r^5}, \quad \cdots \qquad (7.27)$$

Putting all this into (7.24) we have

$$\bar{h}_{ab}(t,\mathbf{x}) = \frac{4G}{c^4} \sum_{l=0}^{\infty} \frac{(-1)^l}{l!} M_{ab}^{ij\cdots k} \, \partial_i \partial_j \cdots \partial_k \left(\frac{1}{|\mathbf{x} - \mathbf{z}|} \right) \Bigg|_{\mathbf{z}=0} \tag{7.28}$$

where

$$M_{ab}^{ij\cdots k} = \int T_{ab} \left(t - |\mathbf{x} - \mathbf{z}|/c, \, \mathbf{z} \right) z^i z^j \cdots z^k \, \mathrm{d}^3 z \tag{7.29}$$

The various $M_{ab}^{ij\cdots k}$ for different $ij\cdots k$ are called the *multipole moments* of the source distribution. Note that so far we have only tackled the $1/|\mathbf{x} - \mathbf{z}|$ factor in the original integral; we have not yet done anything to reduce the difficulty of integrating all the stresses in T_{ab}, and we still require integration over a light cone.

7.3.1   The compact source approximation

The term of order l in (7.28) falls with distance from the source as $1/r^{(l+1)}$. One of the terms is the expected $\sim 1/r$ gravitational potential that would arise in a static spherical problem. The others are owing to radiation and to the non-spherical shape in general. If we are interested in tidal effects then the leading term is not the static field but the gravitational waves, because the static field only produces tidal effects at order $1/r^3$, whereas the wave amplitude falls as $1/r$, just like electromagnetic waves from an oscillating source, as we will see. To find out about the waves in the far field, we may therefore take the leading, $l = 0$, term in (7.28):

$$\bar{h}_{ab}(t,\mathbf{x}) = \frac{4G}{c^4 r} \int T_{ab} \left(t - |\mathbf{x} - \mathbf{z}|/c, \, \mathbf{z} \right) \mathrm{d}^3 z. \tag{7.30}$$

We shall now make a further, and different approximation, which is to take the case where the various parts of the source have speeds small compared to c. In this case the quantity $|\mathbf{x} - \mathbf{z}|/c$ evaluates to the same value r/c for all parts of the source, to good approximation, so we have removed the need to integrate over a light cone; now we integrate over the source at a single time $t - r/c$:

$$\bar{h}_{ab}(t,\mathbf{x}) = \frac{4G}{c^4 r} \int T_{ab} \left(t - r/c, \, \mathbf{z} \right) \mathrm{d}^3 z. \tag{7.31}$$

So far we have employed standard techniques in multipole expansion, and we have in (7.31) a useful simplification, but it still turns out to be a rather difficult integral until we have transformed it into a more useful form. The rest of the argument is the clever part, and it will involve no further approximation other than the linearized theory itself.

First we note that the 00 term of the integral gives the total energy of the source, and evaluates to Mc^2 where M is its rest mass. We can easily prove that the latter is constant by using the energy conservation formula $\partial_\lambda T^{\lambda b} = 0$ which applies in the linearized theory. In fact of course the source is losing energy through the very radiation that we are calculating, but this appears at second order in h_{ab} so can be neglected at the level of approximation we have adopted.[2]

---

[2] And note that the calculation will remain very precise for all but the most extreme astrophysical sources.

The $0i$ and $i0$ terms in T^{ab} represent the total momentum of the source. By choosing the frame in which the source has zero total momentum, we can send these to zero. It remains then to calculate the stress terms, T^{ij}:

$$\bar{h}^{ij}(t, \mathbf{x}) = \frac{4G}{c^4 r} \int T^{ij}(t - r/c, \mathbf{z}) \, \mathrm{d}^3 z \tag{7.32}$$

where we have raised the indices for convenience in what follows. This is an integral which requires us to figure out the stresses in the source, which is itself a non-trivial job. However, if we first appeal to energy-momentum conservation, then we can convert this into a statement about time derivatives of energy density in the source, which are usually easier to find. To this end, note that the conservation formula $\partial_\lambda T^{\lambda b} = 0$ applied to the source is

$$\partial_0 T^{00} + \partial_k T^{0k} = 0, \tag{7.33}$$

$$\partial_0 T^{i0} + \partial_k T^{ik} = 0, \tag{7.34}$$

where it is understood that ∂_a now refers to $\partial/\partial z^a$ not $\partial/\partial x^a$.

We first present the flavour of the calculation to follow, using a loose notation, and then we present the exact calculation.

The idea is that we want $\int T \mathrm{d}z$ and we have $\partial_t T = -\partial_z T$ (here we are not bothering to track all the indices, we are just giving the general idea). Now the product rule tells us that $\partial_z(Tz) = (\partial_z T)z + T$ so we can write

$$\int T \mathrm{d}z = \int \partial_z(Tz) \, \mathrm{d}z - \int (\partial_z T)z \, \mathrm{d}z$$

The first term on the right-hand side is a divergence in the full argument, and it can be converted into a surface integral and thus set equal to zero. The second term can be converted into $\int (\partial_t T)z \, \mathrm{d}z$ by using momentum conservation. This completes the first step. Then we apply the same reasoning to this integral, finding $\int (\partial_t T)z \mathrm{d}z = \int \partial_t (\partial_t T)z^2 \, \mathrm{d}z$. Hence the original integral has been converted into the second time derivative of $\int T^{00} z^2 \mathrm{d}z$ because in each step a spatial part of T got replaced by a temporal part through the conservation equations (7.33), (7.34).

Now we present the exact treatment. We use the product rule and $\partial_k z^j = \delta_k^j$ to find that

$$\partial_k \left(T^{ik} z^j \right) = \left(\partial_k T^{ik} \right) z^j + T^{ij} \tag{7.35}$$

and therefore

$$\int T^{ij} \mathrm{d}^3 z = \int \partial_k \left(T^{ik} z^j \right) \mathrm{d}^3 z - \int \left(\partial_k T^{ik} \right) z^j \mathrm{d}^3 z. \tag{7.36}$$

The first term on the right-hand side is a volume integral of a divergence. It can be converted into a surface integral of a flux, and by employing a surface outside the source we conclude that it must be zero. To treat the second term, we employ (7.34) and thus obtain

$$\int T^{ij} \mathrm{d}^3 z = \int \left(\partial_0 T^{i0} \right) z^j \mathrm{d}^3 z = \frac{1}{c} \frac{\partial}{\partial t_s} \int T^{i0} z^j \mathrm{d}^3 z. \tag{7.37}$$

Now let us apply the same reasoning to the final integral, $\int T^{i0} z^j \mathrm{d}^3 z$. The product rule tells us

$$\partial_k \left(T^{0k} z^i z^j \right) = \left(\partial_k T^{0k} \right) z^i z^j + T^{0i} z^j + T^{0j} z^i. \tag{7.38}$$

Upon integrating over volume, we have on the left a total flux term which vanishes, and we can use (7.33) to replace $\partial_k T^{0k}$ on the right by $-\partial_0 T^{00}$. Hence

$$\int T^{0i} z^j + T^{0j} z^i \mathrm{d}^3 z = \int (\partial_0 T^{00}) z^i z^j \, \mathrm{d}^3 z. \tag{7.39}$$

Now look again at (7.37). Using that T^{ij} is symmetric, we see that the right-hand side must also be symmetric in i and j, so we can set the total equal to half the sum of two versions with i and j swapped, which is what we have found in (7.39). So, putting it all together, we have

$$\int T^{ij} \, \mathrm{d}^3 z = \frac{1}{2c^2} \frac{\mathrm{d}^2}{\mathrm{d}t_s^2} \int T^{00} z^i z^j \mathrm{d}^3 z. \tag{7.40}$$

This is called the Laue theorem, or von Laue's theorem, in Special Relativity. Note that the time derivative can be taken outside the integral because we are integrating over z at one time at the source, and then once the integral has been done we have a function of one source variable, namely time, so the partial derivative becomes a total derivative. The integral here is the expression for the *quadrupole moment of the energy distribution* so we have the

---

quadrupole formula

$$\bar{h}^{ij}(t, \mathbf{x}) = \frac{2G}{c^6 r} \frac{\mathrm{d}^2 I^{ij}}{\mathrm{d}t_s^2} \tag{7.41}$$

---

where

$$I^{ij} = \int T^{00} z^i z^j \, \mathrm{d}^3 z. \tag{7.42}$$

This is the quadrupole-moment tensor of the energy density of the source. It should not be forgotten that for a given field event this is to be evaluated at the time of the source event, i.e. the so-called retarded time $t_s = t - r/c$, and if radiation is happening then I^{ij} is a function of time.

Notice that the strength of the waves is to do with the quadrupole moment, not the dipole moment, of the source. This is because, in contrast to electromagnetism, the radiating body has to move its energy-momentum, not some other sort of charge, in order to generate the waves, but energy and momentum movements are subject to conservation laws. A massive body can possess a static dipole moment, but *if the body is isolated* and tries to change the size or direction of its dipole moment, by oscillating or rotating in some way, then energy-momentum conservation insists that some other part of the body has to compensate. It is essentially the same problem as trying to displace one's centre of mass by internal adjustments. It cannot be done so the dipole term does not appear in the gravitational radiation from an isolated source. For a non-isolated body this argument would not apply (and nor would (7.41)).

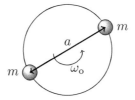

Fig. 7.6 Two stars of equal mass m and separation a on a circular orbit.

In the next section we will employ (7.41) to calculate accurately an example of gravitational radiation. Before we embark on that more precise calculation, let us first use the quadrupole formula to gain an order-of-magnitude estimate of some examples. Take a pair of 1 kilogram masses attached to one another by a string of length 10 cm, and spun around at 10 revolutions per second. We have $I^{ij} \sim 2 \times (1\,\text{kg}) \times (0.05\,\text{m})^2 \cos 2\omega t \simeq (0.005\,\text{kg}\,\text{m}^2) \cos 2\omega t$ with $\omega = 2\pi \times 10 \simeq 63\,\text{s}^{-1}$. (Notice the factor 2 in the $\cos 2\omega t$ term which is owing to the fact that the mass distribution returns to its initial form twice per revolution). At 1 metre from such a source, one finds $\bar{h}^{ij} \sim 1.6 \times 10^{-59}$. This is the fractional amount by which distances between freely floating particles will be changing owing to the gravitational waves. It is entirely negligible compared to any physical effect we might detect, and it shows that there is no hope of generating directly detectable waves using man-made devices in the near future. However, as we have already remarked, waves from astrophysical sources can be detected. For an astrophysical example, consider a pair of neutron stars with masses of order one solar mass each, orbiting at one solar radius from their centre of mass with a 7-hour period, at a distance from Earth of 100 parsec.[3] We obtain $\bar{h} \simeq 10^{-21}$.

7.3.2 Binary star/black hole source

The quadrupole formula is easily applied to the case of two point masses orbiting their centre of mass at velocities small compared to c, and in the weak field limit. This is a simple model of a binary star system. It can also model a pair of black holes moving similarly, as long as they are far enough apart that neither produces large tidal stresses at the other.[4]

For the arrangement of stars indicated in Fig. 7.6, we take the Newtonian result for the rate of rotation of the stars about their centre of mass (treating them as point masses, each of mass m):

$$\omega_{\text{o}} = \left(\frac{2Gm}{a^3} \right)^{1/2} \tag{7.43}$$

where the subscript o stands for 'orbit'. The particles are on orbits given by $\mathbf{x}_A = (a \cos \omega_{\text{o}} t, a \sin \omega_{\text{o}} t, 0)$ and $\mathbf{x}_B = -\mathbf{x}_A$. Hence the mass density function is

[3] $M_\odot = 1.99 \times 10^{30}$ kg, $R_\odot = 6.96 \times 10^8$ m, 1 parsec $= 3.0856 \times 10^{16}$ m.

[4] This second example is not obvious, but may be loosely supported from the observation that effects far from a black hole are no different from those of other massive bodies, and GR is a local theory.

$$\rho = m\delta(z)\left[\delta(x - \frac{a}{2}\cos\omega_o t)\delta(y - \frac{a}{2}\sin\omega_o t) + \delta(x + \frac{a}{2}\cos\omega_o t)\delta(y + \frac{a}{2}\sin\omega_o t)\right].$$

Substituting this into the formula (7.42) for the quadrupole moment, we have

$$[I^{ij}(t)] = \frac{mc^2 a^2}{4}\begin{pmatrix} 1 + \cos 2\omega_o t & \sin 2\omega_o t & 0 \\ \sin 2\omega_o t & 1 - \cos 2\omega_o t & 0 \\ 0 & 0 & 0 \end{pmatrix} \tag{7.44}$$

and this is to be substituted into the quadrupole formula (7.41) for $\bar{h}^{ij}$. This gives the spatial part of $\bar{h}^{ab}$. There is also a term $\bar{h}^{00}$ arising from the total mass of the source, which is constant—recall the discussion after (7.31) (this term represents the usual gravitational potential for a static source). We can separate the full $\bar{h}^{ab}$ into the constant (i.e. time-independent) part and the radiative part $\bar{h}^{ab}_{\text{rad}}$, and thus obtain

$$[\bar{h}^{ab}_{\text{rad}}(t, \mathbf{x})] = -\frac{2Gma^2\omega_o^2}{c^4 r}\begin{pmatrix} 0 & 0 & 0 & 0 \\ 0 & \cos 2\omega_o(t - r/c) & \sin 2\omega_o(t - r/c) & 0 \\ 0 & \sin 2\omega_o(t - r/c) & -\cos 2\omega_o(t - r/c) & 0 \\ 0 & 0 & 0 & 0 \end{pmatrix}. \tag{7.45}$$

This has the form of a spherical wave decaying as $1/r$, with frequency equal to $2\omega_o$ (because the quadrupole repeats itself twice per orbit of either star).

Note that the waves are indeed propagating in all directions, not just down the z axis, and they will be transverse in every case. For an observer on the z axis it is straightforward to determine from (7.45) that the waves are circularly polarized with the sense expected from the rotation of the stars. For an observer on the x axis, the form (7.45) is *not* in the transverse traceless (TT) gauge and in order to interpret it we do well to change gauge. To change to the TT gauge we adopt the simple prescription given at the end of Section 7.1.1: first set to zero the non-transverse elements (i.e. the x row and column in this example), and then subtract the same amount from all the diagonal elements so as to make the result traceless. Here the terms that remain are on the diagonal in the y, z slots. Therefore they represent a wave in the '+' polarization state relative to our choice of axes, again as one might expect, and with an amplitude $Gma^2\omega_o^2/c^4 r$. This is half the amplitude of the waves in the z direction, and furthermore there is only the + polarization, not a combination of both + and ×; in consequence the energy flux in the x direction is smaller than that in the z direction by a factor 8.

7.4 Energy flux in gravitational waves

In order to determine the energy density and flux of gravitational waves, we use eqn (5.55). We take the case of Lorenz gauge, and a traceless h_{ab}, and vacuum, so the only non-zero contribution to t_{ab} is A_{ab} given by (5.56):

Energy tensor of field in vacuum; traceless Lorenz gauge

$$t_{ab} = \frac{c^4}{32\pi G}\langle(\partial_a h_{\mu\nu})(\partial_b h^{\mu\nu})\rangle \tag{7.46}$$

Before we treat waves propagating in all directions, consider first the example of a wave propagating in the z direction (treated in TT gauge, so $\bar{h}_{ab} = h_{ab}$), with the $+$ and $\times$ components oscillating in phase:

$$
h_{ab} = \begin{pmatrix} 0 & 0 & 0 & 0 \\ 0 & h_+ & h_\times & 0 \\ 0 & h_\times & -h_+ & 0 \\ 0 & 0 & 0 & 0 \end{pmatrix} \cos(kz - \omega t).
\tag{7.47}
$$

The frequency ω here is the frequency of the wave itself. Substituting this into (7.46) and then raising indices we obtain

$$
t^{ab} = \frac{c^2}{32\pi G} \omega^2 (h_+^2 + h_\times^2) \begin{pmatrix} 1 & 0 & 0 & 1 \\ 0 & 0 & 0 & 0 \\ 0 & 0 & 0 & 0 \\ 1 & 0 & 0 & 1 \end{pmatrix}
\tag{7.48}
$$

where we used that $\langle \sin^2(kz - \omega t) \rangle = 1/2$. The 00 term of t^{ab} is the energy density of such a wave; the rest of the first row or column is the energy flux (and therefore also momentum density) up to factors of c or c^2; the 33 (bottom right) term represents a form of radiation pressure. Applied to the case of the binary star system presented in Section 7.3.2 we find that the flux in the z direction, at distance r from the source, is

$$
F_z = \frac{c^3}{32\pi G} (2\omega_{\rm o})^2 \left(\frac{2Gma^2\omega_{\rm o}^2}{c^4 r} \right)^2 2 = \frac{G}{\pi c^5} \frac{m^2 a^4 \omega_{\rm o}^6}{r^2}
\tag{7.49}
$$

where the final factor 2 arises from the circular polarization of the wave: in the case of (7.45) one has $\langle 2\cos^2 \phi + 2\sin^2 \phi \rangle = 2$ where $\phi = 2\omega_{\rm o}(t - r/c)$.

Next we generalize to arbitrary directions of propagation.

For a plane wave with wave-vector $\mathbf{k}$, we have $h_{\rm TT}^{ab} = A_{\rm TT}^{ab} \cos k_\mu x^\mu$ where $A_{\rm TT}^{ab}$ is an amplitude tensor, and we are writing only the radiative part which has no 00 term. Hence, using (7.46) and then raising indices,

$$
t^{ab} = \frac{c^4}{64\pi G} \left(A_{\rm TT}^{\mu\nu} A_{\mu\nu}^{\rm TT} \right) k^a k^b
\tag{7.50}
$$

The flux in the $\mathbf{k}$ direction is

$$
F_{\mathbf{k}} = ct^{0j}\hat{k}_j = \frac{c^5}{64\pi G} \left(A_{\rm TT}^{\mu\nu} A_{\mu\nu}^{\rm TT} \right) k^0 k^j \hat{k}_j
\tag{7.51}
$$

Now $k^j \hat{k}_j$ is equal to the size of the 3-vector $\vec{k}$, and since the 4-vector is null, this is also equal to k^0, so we have $k^0 k^j \hat{k}_j = k^0 k^0$, and therefore

$$
F_{\mathbf{k}} = c\, t^{00}.
\tag{7.52}
$$

This says that the flux is equal to the energy density of the plane wave, multiplied by its speed, which is just what one would expect.

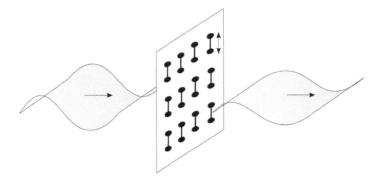

Fig. 7.7 A gravitational wave being partially absorbed by a plane of oscillators.

7.4.1 Deriving the flux another way

We will now derive (7.51) by a different approach,[5] in which we will not need the rather abstract argument given in appendix E, which led to eqns (5.55) and (5.56). This helps both to understand the result and to confirm that our ideas about gravitational wave energy are consistent.

The idea is to consider a wave that is incident on a plane of oscillators which can be driven by the wave and absorb energy from it—Fig. 7.7. The oscillators are driven to oscillate in phase with the wave, and their motion in turn generates a further wave which interferes with the incident wave. In the reflected direction the resulting change is negligible to first approximation, because at any one plane the incoming and reflected parts oscillate with respect to one another. In the transmitted direction, on the other hand, there is a constant phase relationship leading to destructive interference and thus a lower wave amplitude, consistent with the fact that the plane of oscillators has absorbed some energy. We assume that all the energy acquired by the oscillators has been lost by the wave, and thus we will be able to calibrate the amount of energy flux that is associated with a given loss of amplitude.

Let us model each oscillator as a linear pair of masses as in Fig. 7.5. In a Newtonian calculation, the rate of doing work by a force $\mathbf{f}$ is $\mathbf{f} \cdot \mathbf{v}$ where $\mathbf{v}$ is the velocity of the body to which the force is applied. Applied to a body undergoing driven harmonic motion, the terms in the force which are in phase with the position give no net work done over a cycle, but the term which is in phase with the velocity does a net work when averaged over a cycle. This is the damping term and the work is dissipated as friction. We have two bodies moving with velocity v and each subject to the damping force $f = -\nu v = -2m\gamma v$, where $v = \dot{s}/2 = (\omega R/2)\sin(\omega t + \phi)$, in which R and ϕ are given by (7.22), (7.23). Hence the total rate of energy dissipation per oscillator, averaged over a cycle, is

$$\left\langle \frac{\mathrm{d}E}{\mathrm{d}t} \right\rangle = 2(2m\gamma)(\omega R/2)^2 \langle \sin^2(\omega t + \phi) \rangle = \tfrac{1}{2} m\gamma\omega^2 R^2. \tag{7.53}$$

[5] Following Schutz.

If the number of such oscillators per unit area of the plane is σ, then the net rate of dissipation of energy per unit area is

$$\Delta F = -\sigma \langle \dot{E} \rangle = -\tfrac{1}{2}\sigma m \gamma \omega^2 R^2. \tag{7.54}$$

We have called this quantity ΔF since we are arguing that this is equal to a loss of flux from the gravitational wave.

Next consider the effect of the motion of the oscillators. Each one of them is itself a source of gravitational radiation, having quadrupole moment

$$
\begin{aligned}
I_{xx} &= mc^2 x_A^2 + mc^2 x_B^2 \\
&= mc^2 \left[(-\tfrac{1}{2}d - \tfrac{1}{2}R\cos(\omega t + \phi))^2 + (\tfrac{1}{2}d + \tfrac{1}{2}R\cos(\omega t + \phi))^2 \right] \\
&= mc^2 \left[\text{const} + dR\cos(\omega t + \phi) + \tfrac{1}{2}R^2\cos^2(\omega t + \phi) \right].
\end{aligned}
$$

We consider the case $R \ll d$ (small oscillation amplitude), then

$$\ddot{I}_{xx} = -mc^2\omega^2 dR\cos(\omega t + \phi). \tag{7.55}$$

The gravitational wave sourced by one such oscillator has an amplitude given by substituting this expression into the quadrupole formula (7.41):

$$h_{xx} = -\frac{s}{r}\cos(\omega(t - r/c) + \phi) \qquad \text{where} \quad s = 2Gm\omega^2 dR c^{-4}. \tag{7.56}$$

Here we are using the symbol s to represent the strength of one small source in the plane of oscillators. The net effect of a uniform plane array of such oscillators can be seen to be a plane wave by considering the Huygen's wavelet construction. The amplitude of this plane wave, obtained by integration over the plane—see exercise 7.4—is

$$h_{xx}^{\text{tot}} = -2\pi\sigma s c\omega^{-1}\sin(\omega(t - z/c) + \phi). \tag{7.57}$$

This is expressed in the coordinate system employed to describe the oscillators, which is not in TT gauge because it has an xx but not a yy term. After conversion to TT gauge the amplitude is divided by 2. Upon then adding this to the incident wave, we obtain that the net gravitational radiation after the plane of oscillators is a plane wave with

$$
\begin{aligned}
h_{xx}^{\text{net}} &= h_+\cos(\omega(t - z/r)) - \pi\sigma s c\omega^{-1}\sin(\omega(t - z/c) + \phi) \\
&\simeq \left(h_+ - \pi\sigma s c\omega^{-1}\sin\phi \right)\cos(\omega(t - z/r) + \alpha)
\end{aligned} \tag{7.58}
$$

where $\tan\alpha \simeq \pi\sigma s c\cos\phi/(\omega h_+)$ and we have neglected terms of order s^2. Hence the amplitude of the wave has been reduced by

$$\Delta h_+ = -\pi\sigma s c\sin(\phi)/\omega = -2\pi\sigma Gm\omega c^{-3}dR\sin\phi. \tag{7.59}$$

By combining this with (7.54) and then substituting for R and ϕ from (7.22), (7.23) we obtain

$$\frac{\Delta F}{\Delta h_+} = \frac{\gamma\omega R c^3}{4\pi Gd\sin\phi} = \frac{c^3}{16\pi G}h_+\omega^2. \tag{7.60}$$

In the final result, all the properties of the oscillators have dropped out! We have obtained a statement about the wave purely in terms of its own properties. We now argue that in the limit of small quantities, we can regard this ratio as a derivative, and then integrate it to obtain

$$F = \frac{c^3}{32\pi G} h_+^2 \omega^2. \tag{7.61}$$

This agrees with, and therefore confirms (or if you prefer, is confirmed by) eqn (7.51) since for the wave in question, $A_{\mathrm{TT}}^{\mu\nu} A_{\mu\nu}^{\mathrm{TT}} = 2h_+^2$.

This calculation can be viewed as a vindication of the approach we described at the end of Chapter 5 to treat the energy associated with the gravitational field in the linearized theory. It also illustrates very clearly what we mean by the energy flux in a gravitational wave: we mean *that which quantifies the amount of energy passed to a system which absorbs and thus diminishes the gravitational radiation.*

7.4.2   Total emitted power

Returning now to eqn (7.51), by integrating the flux over a sphere in the far field, we obtain the total rate of emission of energy by some given source of gravitational radiation:

$$L_{\mathrm{GW}} = \oint_\Omega F_{\mathbf{n}}\, r^2 \sin\theta\, \mathrm{d}\theta \mathrm{d}\phi. \tag{7.62}$$

In order to calculate this for an arbitrary source (in the weak field limit), we need to return to (7.46) so that we can connect the amplitude of the waves to the behaviour of the source. The flux in a direction $\mathbf{n}$ is

$$F_{\mathbf{n}} = c\, t^{0j} n_j \;=\; -c\, t_{0j} n^j \;=\; \frac{-c^5}{32\pi G} \left\langle \left(\frac{1}{c}\frac{\partial}{\partial t} h_{\mu\nu}^{\mathrm{TT}}\right)\left(\frac{\partial}{\partial x^j} h_{\mathrm{TT}}^{\mu\nu}\right)\right\rangle n^j \tag{7.63}$$

Taking $\vec{n}$ in the radial direction, this is

$$F_r = \frac{-c^4}{32\pi G} \left\langle \frac{\partial h_{\mu\nu}^{\mathrm{TT}}}{\partial t} \frac{\partial h_{\mathrm{TT}}^{\mu\nu}}{\partial r}\right\rangle \tag{7.64}$$

and since $h_{\mathrm{TT}}^{0b} = 0$ for the radiative part, we only need to retain the spatial components. From the quadrupole formula (7.41), we have

$$\bar{h}_{\mathrm{TT}}^{ij} = \frac{2G}{c^6 r} \frac{\mathrm{d}^2 I_{\mathrm{TT}}^{ij}}{\mathrm{d}t_s^2} \tag{7.65}$$

in which the transverse traceless part of I^{ij} can be obtained by projection, as explained in Section 7.1.1. The algebraic manipulations to follow are made simpler if one works in terms of the *reduced quadrupole-moment tensor*, defined by[6]

---

[6] J^{ij} may itself be called simply 'the quadrupole moment'; c.f. (7.27).

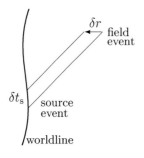

δr field event

δt_s source event

worldline

Fig. 7.8 Spacetime diagram showing the relation of displacements at the field event to changes at the source event.

$$J^{ij} = I^{ij} - \tfrac{1}{3}\delta^{ij} I^k_k. \tag{7.66}$$

This definition makes J^{ij} traceless, so $I^{ij}_{\rm TT} = J^{ij}_{\rm TT}$. Hence, using $t_s = t - r/c$ (the retardation of the source time), we have (c.f. Fig. 7.8)

$$\frac{\partial h^{\rm TT}_{\mu\nu}}{\partial t} = \frac{2G}{c^6 r}\frac{{\rm d}^3 J^{ij}_{\rm TT}}{{\rm d}t_s^3},$$
$$\frac{\partial h^{\mu\nu}_{\rm TT}}{\partial r} = -\frac{2G}{c^6 r^2}\frac{{\rm d}^2 J^{ij}_{\rm TT}}{{\rm d}t_s^2} - \frac{2G}{c^6 r}\frac{{\rm d}^3 J^{ij}_{\rm TT}}{{\rm d}t_s^3}\frac{1}{c} \simeq -\frac{2G}{c^7 r}\frac{{\rm d}^3 J^{ij}_{\rm TT}}{{\rm d}t_s^3} \tag{7.67}$$

where in the final step we neglected the $1/r^2$ term which is small compared to the other terms in the far field limit. Substituting these expressions into (7.64) gives

$$F_r = \frac{G}{8\pi c^9 r^2}\left\langle \dddot{J}^{\rm TT}_{ij}\,\dddot{J}^{ij}_{\rm TT}\right\rangle. \tag{7.68}$$

This is the flux in the radial direction for a general compact source. In order to apply this formula, it is useful first to express it in terms of J_{ij}, by using the projection described in eqn (7.11):

$$J^{\rm TT}_{ij} = \left(P^m_i P^n_j - \tfrac{1}{2}P_{ij}P^{mn}\right) J_{mn}. \tag{7.69}$$

where $P_{ij} = \delta_{ij} - n_i n_j$ is the spatial projection tensor which projects onto the direction $\vec{n}$, which we take to be the radial direction. Using this in (7.68) and expanding the brackets gives four terms, which can be simplified using

$$P^i_m P^m_j = P^i_j, \qquad P^m_m = 2 \tag{7.70}$$

(which you should confirm), giving

$$\left(P^m_i P^n_j - \tfrac{1}{2}P_{ij}P^{mn}\right)\left(P^i_k P^j_l - \tfrac{1}{2}P^{ij}P_{kl}\right) = P^m_k P^n_l - \tfrac{1}{2}P^{mn}P_{kl}. \tag{7.71}$$

After expanding these in terms of δ_{ij} and $\vec{n}$, and using $J^m_m = 0$, one finds

$$J^{\rm TT}_{ij} J^{ij}_{\rm TT} = J_{ij}J^{ij} - 2J_{im}J^{mj}n_i n_j + \tfrac{1}{2}J^{ij}J^{mn}n_i n_j n_m n_n, \tag{7.72}$$

and therefore

$$L_{\rm GW} = \frac{G}{8\pi c^9}\oint_\Omega \left\langle \dddot{J}_{ij}\dddot{J}^{ij} - 2\dddot{J}_{im}\dddot{J}^{mj}n^i n_j + \tfrac{1}{2}\dddot{J}^{ij}\dddot{J}^{mn}n_i n_j n_m n_n\right\rangle \sin\theta {\rm d}\theta {\rm d}\phi. \tag{7.73}$$

In this integral, the only factors depending on the angles θ, ϕ are the components of $\vec{n}$ (since the tensor J_{ij} is simply a property of the source), so the $\dddot{J}$ factors can be taken outside the integral. It remains to evalutate the three integrals

$$\oint d\Omega = 4\pi, \tag{7.74}$$

$$\oint n^i n_j d\Omega = (4\pi/3)\delta^i_j, \tag{7.75}$$

$$\oint n_i n_j n_m n_n d\Omega = (4\pi/15)(\delta_{ij}\delta_{mn} + \delta_{im}\delta_{jn} + \delta_{in}\delta_{jm}). \tag{7.76}$$

These are easily obtained by keeping in mind that $\vec{n}$ is a unit vector so $n^i n_i = 1$. Substituting these results into (7.73), one obtains:

Total power radiated by a compact source, in weak field approximation

$$L_{GW} = \frac{G}{5c^9} \left\langle \dddot{J}_{ij} \dddot{J}^{ij} \right\rangle \tag{7.77}$$

For example, if you shake your fist rapidly to and fro, then the oscillating part of the quadrupole moment of your body has a size of order $0.1\,\mathrm{kg\,m^2}$, oscillating at a frequency approximately $4\,\mathrm{Hz}$. The above formula then asserts that the total power in the gravitational waves you are thus emitting is 10^{-47} watts. Applied to the binary star system considered in Section 7.3.2, we find from (7.44),

$$\left[J^{ij}\right] = \frac{mc^2 a^2}{4} \begin{pmatrix} \frac{1}{3} + \cos 2\omega_\mathrm{o}t & \sin 2\omega_\mathrm{o}t & 0 \\ \sin 2\omega_\mathrm{o}t & \frac{1}{3} - \cos 2\omega_\mathrm{o}t & 0 \\ 0 & 0 & -\frac{2}{3} \end{pmatrix},$$

hence

$$\dddot{J}_{ij}\dddot{J}^{ij} = (2\omega_\mathrm{o}^2 mc^2 a^2)^2 \left(2\cos^2(2\omega_\mathrm{o}(t - r/c)) + 2\sin^2(2\omega_\mathrm{o}(t - r/c))\right) \tag{7.78}$$

and

$$L_{GW} = \frac{8G}{5c^5} m^2 a^4 \omega_\mathrm{o}^6. \tag{7.79}$$

For a gravitationally bound pair on a circular orbit, we can express ω_o in terms of a and thus find

$$L_{GW} = \frac{64G^4}{5c^5} \frac{m^5}{a^5} \tag{7.80}$$

which gives 2.001×10^{25} watts for $m = M_\odot$ and $a = R_\odot$ (one solar radius).

7.5 The detection of gravitational radiation

7.5.1 Spin-up of binary pulsar PSR B1913+16

In the year 1974 and the following years, Hulse and Taylor discovered and followed the evolution of a remarkable binary system, consisting of a pair of dense stars orbiting close to one another

Table 7.1 Orbital parameters of Hulse–Taylor binary. Values above the line are as reported in (Weisberg, Nice and Taylor, 2010). Values below the line are either defined or derived from other data or derived from all the above. Relativistic ('post Newtonian') effects are of order $GM/ac^2 \simeq 10^{-6}$.

period	P	7.75193877386(10) hours
eccentricity	e	0.6171334(5)
total mass	$M = m_1 + m_2$	2.828378(7) $M_\odot$
	m_1	1.4398(2) $M_\odot$
	m_2	1.3886(2) $M_\odot$
Julian year	yr	86400×365.25 s
	$GM_\odot/c^3$	$4.925490947 \times 10^{-6}$ s
gravitational constant	G	$6.67408(31) \times 10^{-11}$ m^{3}kg^{-1}s^{-2}
semi-major axis	a	$1.949119(2) \times 10^9$ m
	$f(e)$	11.85677(6)

with a 7.75 hour period, one of which was pulsar PSR 1913+16. The pulsar provided exquisitely accurate timing and Doppler information, allowing essentially all the orbital parameters to be obtained, some of them to many digits of precision—see Table 7.1. Hulse and Taylor realized that by observing the gradual change of the orbit over many years, they could deduce at what rate the system was losing energy, and compare this with the prediction for energy loss by gravitational radiation provided by GR.

In order to make this comparison it is important to allow for the elliptical shape of the orbit: it has a dramatic effect on the outcome. The quadrupole moment for a binary star following elliptical Keplerian orbits is described in detail in appendix A. We treat a pair of stars of masses m_1, m_2 orbiting their centre of mass such that the vector between them describes an ellipse of eccentricity e and semi-major axis a. By substituting (A.13) into (7.66) and (7.77) one finds that the power emitted, as a function of all the parameters including azimuthal angle is

$$L_{GW} = \frac{32G^4}{5c^5} \frac{m_1^2 m_2^2 (m_1 + m_2)}{(1 - e^2)^5 a^5} (1 + e\cos\phi)^4 \left[(1 + e\cos\phi)^2 + \frac{e^2}{12}\sin^2\phi \right] \tag{7.81}$$

We then take the time average of this quantity over an orbit, obtaining:

Radiated power for binary star system

$$\langle L_{GW} \rangle = \frac{32G^4}{5c^5} \frac{m_1^2 m_2^2 (m_1 + m_2)}{a^5} f(e)$$

$$f(e) = \frac{1 + (73/24)e^2 + (37/96)e^4}{(1 - e^2)^{7/2}} \tag{7.82}$$

The total Newtonian energy (gravitational plus kinetic) of the system is $E = -Gm_1 m_2/2a$, and the stars move at speed $v \sim \omega a$, so the power given by this formula is of order $L_{GW} \sim E\omega(v/c)^5$. In this sense we are studying a relativistic effect at 5th order in v/c.

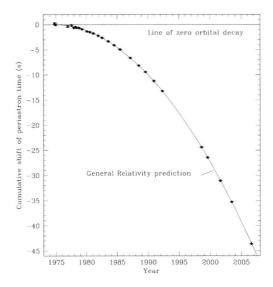

Fig. 7.9 Data on the orbital motion of the Hulse–Taylor binary (points) and the behaviour predicted by GR based on the system's loss of energy by gravitational radiation (line). [Fig. 2 from (Weisberg, Nice and Taylor, 2010)]

Putting in the data from the observed orbit timing, one finds the prediction from (7.82) for the Hulse–Taylor binary is

$$\langle L_{GW} \rangle = 7.7682(5) \times 10^{24} \text{ watts.} \qquad (7.83)$$

where the leading error is owing to uncertainty in G. This error can be avoided by looking directly at the change in the period but at present the experimental precision is limited by other factors. The observations, as reported at the time of writing, are shown in Fig. 7.9. The accumulated shift in periastron time is 46 seconds in 32.7 years. Let δt be the change in periastron time in one orbit. Then, as long as the orbit period has not changed overall very much, then the accumulated shift after N orbits is $\sum_{n=1}^{N} n\delta t = N(N+1)\delta t/2$. In 32.7 years there are $N = 36978$ orbits, so we find $\delta t = 67.28$ ns, and therefore the change in the period during this time is $N\delta t = 2.488$ milliseconds. It follows that the energy of the system has fallen by 8.045×10^{33} joules in 32.7 years, which gives that the measured rate of energy loss is 7.80×10^{24} watts—in agreement with the prediction (7.83) at 0.5% precision. A more precise analysis of the data yields that the observed rate of change of the period is 0.997 ± 0.002 of that predicted by GR (Weisberg, Nice and Taylor, 2010).

Even at their first announcement at more modest precision, these observations provided a superb test of GR, giving strong evidence of gravitational radiation, and also providing further tests as the data became more precise, and as further binary pulsars were discovered. The time and trouble taken by theoreticians in arriving at eqns (7.77)–(7.82) long before observations were available was not wasted. Through these observations it has been confirmed that the waves propagate at the speed of light, and also that they are generated via the quadrupole not the dipole moment of the source. This rules out a class of variations on GR. Furthermore, such binary systems have enabled study of the strong field regime of gravity, near to the surface of

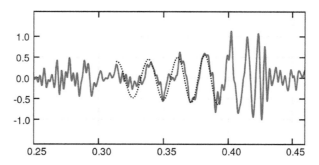

Fig. 7.10 Signal from LIGO gravitational wave detector at Hanford, USA. The dotted curve is a simple fit to part of the data. (Courtesy Caltech/MIT/LIGO Laboratory)

neutron stars, where departures from the Minkowski metric can be as large as 0.4. The Shapiro time delay and other such phenomena have been measured and GR subjected to multifold quantitative scrutiny.

7.5.2   GW150914: the first direct observation of gravity waves

At 09.51 UTC on 14 September 2015, a 40 kg mirror hanging by a thread at the end of a 50 kW laser beam in Livingston, Louisiana, USA tremored a little. Other mirrors approximately 4 km away tremored in step. 6.9 ms later a similar set of mirrors tremored, also in step, at Hanford, Washington, USA. The amplitude of the disturbance was approximately 4×10^{-18} m. This distance is approximately 200 times smaller than the radius of a proton. It is also 250,000 times smaller than the compressive effect of a flea landing on the head of a pin. But thanks to the lengthy, patient and high calibre work done by thousands of scientists in several countries, and funded by public collaboration through taxes, these tremors were detected and, having ruled out other less likely explanations, it can be deduced that this was the detection of a burst of gravitational radiation washing over the solar system and continuing onwards on its journey through space. (LIGO Scientific Collaboration and Virgo Collaboration, 2016)

This event was, in my view, the most exhilarating scientific event of the early twenty-first century. There have been other developments with greater significance to health and environment, but none so bold and patient in their journey into the unknown. It was a tremendous vindication of the 20-year-long effort to build and upgrade the facilities, and of Einstein's insight over a century before. The first published signal is shown in Fig. 7.10. The figure shows the data after a considerable amount of noise suppression by bandpass filtering and other techniques, and it is remarkable for its clarity, considering that this was the first detected such event.

To interpret the data, we note that the amplitude and frequency of the periodic oscillations first rises, gradually and then more steeply, and then there is a 'ringing' phase of rapid oscillation as the amplitude dies away. This matches what is expected for the orbit and then merger of two dense objects, when Einstein's field equation is solved by numerical analysis. The numerical methods are themselves demanding and only developed in recent years. The initial part of the signal is produced as the two objects orbit their joint centre of mass at decreasing radius as energy is radiated away; the final part is the vibration of the merged object that results.

To get further quantitative information, we employ eqn (7.45). Although this equation will model the first part of the data only approximately, and the last part not at all, it can give some correct estimates and good physical insight. We can then turn to numerical modelling to get more precise information.

The amplitude of the wave described by eqn (7.45) can usefully be written

$$\left(\frac{a\omega_o}{c}\right)^2 \left(\frac{r_s}{r}\right) \frac{m}{M_\odot} \tag{7.84}$$

where $r_s = 2GM_\odot/c^2 \simeq 2954\,\mathrm{m}$ is the Schwarzschild radius of the Sun. The factor $a\omega_o/c$ can be at most of order 1, and the Newtonian formula (7.43) relates m/a^3 to the frequency ω_o. The latter is directly observed in the signal (after a factor of 2 and a modest effect of redshift). It follows that only one further piece of information is needed in order to learn both a and m. Such a piece of information is the rate of change of frequency (the *chirp*) which can be easily obtained from the data. Therefore in principle everything about the situation at the source can be deduced and one can use the measured amplitude of the strain to determine the distance r to the source.

In exercise 7.8 the formula (7.87) is obtained for the rate of change of the period of the orbiting bodies, using a Newtonian model for the orbit and GR to predict the energy loss. By then writing a in terms of this period, and using that the period of the gravity wave is half that of the orbit, one finds that the rate of change of the period of the gravity wave is given by

$$\dot{P}_{GW} = -\frac{96}{5}\pi^{8/3} \left(\frac{GM_c}{c^3 P_{GW}}\right)^{5/3} \tag{7.85}$$

where M_c, called the *chirp mass*, is given by

$$M_c = \frac{(m_1 m_2)^{3/5}}{M^{1/5}}. \tag{7.86}$$

In Fig. 7.10 we have shown a curve of linearly increasing frequency crudely matched to the data in order to estimate the parameters. We thus find $\dot{P}_{GW} \simeq -0.07$ and $P_{GW} \simeq 0.0308\,\mathrm{s}$, giving $M_c \simeq 34\,M_\odot$. The figure quoted by Abbott *et al.* after fitting the data more fully, and allowing for the varying sensitivity of the detector, is $M_c \simeq 30\,M_\odot$. This implies the total mass in the system is greater than $70\,M_\odot$. Using this, and the fact that the orbital frequency approaches 75 Hz, one finds that the orbiting objects must have been both compact and close to one another. Their mass exceeds by an order of magnitude the upper limit believed to be possible for neutron stars, and if one were a black hole and the other not, then the black hole would have to have a large mass and consequently a large Schwarzschild radius, with the result that the coalescence would have occurred at larger separation and lower frequency. It emerges, then, that the only plausible candidate, capable of presenting both the observed chirp mass and the observed frequency before coalescence, is a pair of black holes.

After a more sophisticated analysis of the data, Abbott *et al.* report that the system comprised two black holes of masses $36 \pm 4 M_{\odot}$ and $29 \pm 4 M_{\odot}$ which coalesced to leave a final black hole of mass $62 \pm 4 M_{\odot}$. Thus an energy equal to three times the rest energy of the Sun was radiated away in gravitational waves in a fraction of a second! The power of that short burst was brighter than the sum total of the electromagnetic radiation of all the stars in the sky. By then noting the amplitude of the observed strain at the detector, the luminosity distance (defined in (22.36)) was deduced to be $410 \pm 170 \, \mathrm{Mpc}$. By deducing the orbit through the best fit to the data, the proper frequency was estimated; by comparing this with the observed frequency a redshift $z = 0.09 \pm 0.03$ was deduced.

Gravitational wave astronomy. At the time of writing, both LIGO and the European VIRGO gravity-wave detectors are operational and have detected ten binary black hole mergers and one neutron star merger. Before 2015 it was not even known whether there were any binary black holes at all. The neutron star merger has modified estimates of the proportion of heavy elements that are formed in mergers as opposed to supernovae events. The signals have also explored a regime of behaviour in which spacetime is flexed and contorted on extreme scales of curvature and at speeds approaching the speed of light. The dynamical content of strong field General Relativity has been repeatedly exhibited and shown to be correct. Thus these observations have both tested GR in extreme conditions, and also provided an important new astronomical tool.

Electromagnetic radiation typically comes from many small emitters such as atoms and electrons distributed across the source, and has a wavelength small compared to the diameter of the source. Gravitational waves, by contrast, are produced (in detectable amounts) typically by bulk movements on the scale of the whole source, and have a wavelength on this scale or larger. Consequently they cannot be used to form a detailed image of the source. In this respect they are comparable to sound. Equally, a detector of electromagnetic radiation is typically highly directional, whereas gravitational wave detectors have a sensitivity that varies only moderately with direction; they are sensitive to signals from across the sky but do not pin down the source direction precisely—again, this is analogous to hearing compared with sight.

Owing to the weakness of gravitational interactions, gravitational waves are very penetrating: they propagate through solid matter essentially undisturbed and consequently can carry information from regions that are otherwise hidden, such as stellar interiors or the early universe. The situation for observational astronomy is, then, as though the human race has acquired a new sense-organ with which to attend to our physical surroundings. One might say that via electromagnetic detection we can see the universe, and now by gravitational wave detection we are beginning to 'hear' it.

Exercises

7.1 Propose the solution $\bar{h}_{ab} = \bar{\epsilon}_{ab}\exp(ik_\mu x^\mu)$ to the wave equation, with $k^a = (k, 0, 0, k)$ and $\bar{\epsilon}_{ab}$ a constant tensor to be discovered.
(i) Show that the Lorenz gauge condition (5.39) implies $k^\lambda \bar{\epsilon}_{\lambda b} = 0$ and hence that $\bar{\epsilon}_{ab}$ must have the form

$$\bar{\epsilon}_{ab} = \begin{pmatrix} A & B & C & A \\ B & D & E & B \\ C & E & F & C \\ A & B & C & A \end{pmatrix}$$

(ii) Introduce a gauge change using $\xi^a = \rho^a \exp(ik_\mu x^\mu)$ where ρ^a is a constant 4-vector to be discovered. Obtain $\partial_a \xi_b + \partial_b \xi_a$ and $\partial^\lambda \xi_\lambda$ and hence show that the metric change given by (5.35) leads to a new polarization tensor with elements given by

$$\bar{\epsilon}'_{00} = A - 2ik\rho_0 - ik(\rho_3 - \rho_0)$$
$$\bar{\epsilon}'_{01} = B - ik\rho_1$$
$$\bar{\epsilon}'_{02} = A - ik\rho_2$$
$$\bar{\epsilon}'_{11} = D + ik(\rho_3 - \rho_0)$$
$$\bar{\epsilon}'_{12} = E$$
$$\bar{\epsilon}'_{22} = F + ik(\rho_3 - \rho_0)$$

Hence show that a solution for ρ_a exists such that the polarization tensor is transverse and traceless, giving the form (7.7).

7.2 (i) Show that for a plane wave of the form $h_{ab} = \epsilon_{ab}\exp(ik_\mu x^\mu)$ the linearized curvature and Ricci tensors are

$$R^a_{bcd} = \tfrac{1}{2}(-k_b k_c h^a_d + k_b k_d h^a_c - k^a k_d h_{bc} + k^a k_c h_{db})$$
$$R_{bd} = \tfrac{1}{2}(-k_b w_d - k_d w_b + k_b k_d h + k^\nu k_\nu h_{bd})$$

where $w_a \equiv k^\nu h_{\nu a}$. (You may like to start from (5.29)).
(ii) Hence show that if $k_\nu k^\nu \neq 0$ (i.e. a non-null wavevector), then the vacuum field equation $R_{ab} = 0$ implies $R_{abcd} = 0$. Hence there is no curvature; such a case is not a physical wave but merely an oscillation of the coordinate system.

(iii) If $k_\nu k^\nu = 0$ show that for a physical wave one requires $w_a = \tfrac{1}{2}hk_a$ and this implies $R_{abc\mu}k^\mu = 0$. Thus k^a is an eigenvector of the Riemann tensor.

7.3 A plane wave (whether gravitational or electromagnetic or other) ought to give the same effect anywhere on any given wavefront, but the pictures of the squeezing of a ring of particles appear to suggest that the wavefront has a central location. By introducing a displacement of the coordinate system, or otherwise, show that there is not really any such centre: all parts of the wavefront behave similarly.

7.4 (i) Show that

$$\int_{-\infty}^{\infty}\int_{-\infty}^{\infty} \sigma \frac{s}{r}\cos(\omega(t - r/c) + \phi)\,\mathrm{d}x\mathrm{d}y$$
$$= 2\pi \int_z^{\infty} \sigma \cos(\omega(t - r/c) + \phi)\,\mathrm{d}r$$

where r is the distance between $(0, 0, z)$ and an arbitrary point in the xy plane (this is the integral we require after eqn (7.56); c.f. Fig. 7.7).
(ii) The integration is straightforward for constant σ, but the value is undefined at $r \to \infty$. To regularize the integral, use $\sigma = \sigma_0 \exp(-\epsilon r)$ and then let $\epsilon \to 0$ after integrating. Hence obtain (7.57).

7.5 Let $(0, \vec{s})$ be a spacelike 4-vector giving the coordinate separation between two points which are close to one another, but we do not require them to be infinitesimally close. Let $\vec{\zeta}$ be a 3-vector with components given by
$$\zeta^i = s^i + \tfrac{1}{2}h^i_k s^k$$
(i) Show that the ruler distance between the points is given, to first order in h_{ab}, by

$$l^2 = \delta_{ij}\zeta^i\zeta^j.$$

In other words, the ordinary Euclidean scalar product of $\vec{\zeta}$ with itself furnishes the ruler distance.

(ii) A gravitational plane wave propagating in the z direction is incident on freely floating particles lying in a plane at some given z. Show that the effect of a $+$ polarized and a $\times$ polarized such wave, respectively, is such that the particles' ruler separations are given by

$$\vec{\zeta} = (s_x, s_y, 0) - \frac{h_+}{2}(s_x, -s_y, 0)\cos(kz - \omega t)$$

$$\vec{\zeta} = (s_x, s_y, 0) - \frac{h_\times}{2}(s_y, \; s_x, 0)\cos(kz - \omega t)$$

7.6 What will happen to Fig. 7.2 if the masses of all the blobs are doubled?

7.7 Show that a system of four stars situated at the corners of a square and rigidly rotating as a group in the plane of the square about its centre will not emit quadrupole gravitational radiation.

7.8 Starting from (7.82), show that the rate of change of the orbit period P (the time between successive periastrons) for a binary star system owing to gravitational radiation is

$$\frac{dP}{dt} = -\frac{192\pi}{5}\frac{m_1 m_2}{M^2}\left[\frac{GM}{ac^2}\right]^{5/2}f(e) \tag{7.87}$$

where $M = m_1 + m_2$. (Use a Newtonian model to relate the energy loss to the orbit parameters.)

7.9 Show that if P_{ab} is the projection tensor defined in (7.10), then for any A_{ab}, the combination

$$\left(P_a^\mu P_b^\nu - \tfrac{1}{2}P_{ab}P^{\mu\nu}\right)A_{\mu\nu}$$

is orthogonal to n_a and traceless. [Hint: first treat $P_a^\mu P_b^\nu A_{\mu\nu}n^a$; alternatively, use the fact that any tensor can be expressed as a sum of outer products of vectors.]

7.10 Verify (7.75) and (7.76) and hence obtain (7.77).

Part II

8

Manifolds

The approach taken in Volume 1 and the present text has been to learn General Relativity by the following method.

In Volume 1 we began with the principle of equivalence, and we saw that, purely on the basis of the strong equivalence principle, it was possible to derive such effects as the radius of curvature of a free-fall trajectory, and the gravitational time dilation and redshift. We then used the constantly accelerating reference in flat spacetime to introduce some further concepts, such as the metric and the use of judiciously chosen coordinate systems. Then we introduced the notion of warped space, and how intrinsic curvature can be measured by geodesic deviation and related effects. Finally, we quoted the Schwarzschild metric and used it to explore gravitational effects associated with a static spherical body, and with the simplest type of black hole.

In the present volume we have first quoted the field equations and the geodesic equation, and given some preliminary discussion of their practical use and physical meaning. Then we treated the weak-field limit using the linearized theory. This required the use of tensor notation in flat spacetime, but not the full mathematical apparatus which deals with curved spacetime. This offered the advantage of building up familiarity with gravitational phenomena and their accurate calculation, without requiring too great a degree of mathematical abstraction at the outset.

From now on we embark on the project of understanding the theory in full. We shall seek to acquire the mathematical tools needed to measure spacetime fully and precisely, and which will enable us to express how solids, fluids and electromagnetic fields inhabit and mould spacetime. We shall follow in Einstein's footsteps and see that the gravitational field equations are in some sense simple, which is to say, as simple as they can be while respecting the strong equivalence principle. We shall also find out how to derive things which were quoted without proof in Volume 1, such as the Schwarzschild metric.

8.1 The manifold

One of the axioms of General Relativity is:

Relativity Made Relatively Easy: General Relativity and Cosmology. Volume 2. Andrew M. Steane, Oxford University Press. © Andrew M. Steane 2021. DOI: 10.1093/oso/9780192895646.003.0008

Axiom: *spacetime is a manifold*. Which is to say, spacetime can be precisely modelled by the mathematical notion of *differentiable manifold*. Spacetime is, or can be precisely understood as, a physical embodiment of this mathematical notion.

The notion of *differentiable manifold* immediately brings with it the notion of *events* as points in the spacetime manifold. The word 'manifold' is the mathematical term for what physicists more commonly call a 'space'. In general a space is a set of points which can be referenced by some variables. Examples could include the state-space of a thermodynamic system, referenced by variables such as pressure and temperature, or the position-momentum space called phase space in classical mechanics. When the set of points making up the space is continuous, then we say we have a *differentiable manifold*. To be precise, for a differentiable manifold, for every point there are other points in the neighbourhood whose coordinates differ from that of the first by an infinitesimal amount, and it is possible to attach a scalar quantity to each point such that we then have a scalar field that can be differentiated to all orders. Most spaces invoked in physics are of this type, and they can include abstract spaces such as the space of all orthogonal matrices of a given order. A *discrete* abstract space such as the *Hamming space* invoked in the discussion of binary bit strings is not a differentiable manifold, and nor is a fractal set such as the Mandlebrot set. According to General Relativity, spacetime is not like Hamming space.[1]

A manifold is said to be N-dimensional if N independent real numbers are required to specify each point uniquely. Such numbers are called coordinates. The coordinates can be specified almost arbitrarily, with the one restriction that we insist that neighbouring points get neighbouring coordinates. One can imagine these coordinates as sets of numbers written on little labelled pins planted at each point: see Fig. 8.1. You can also imagine coordinate lines running through the manifold, joining the pins whose labels only differ in one coordinate. Mathematicians often take an interest in manifolds having weird and wonderful shapes, and it often arises that when one tries to set out a coordinate system that usefully maps the points in the manifold in some region, one finds that elsewhere the coordinates do not succeed, because for example one gets places where there is *degeneracy*. This is the case where many different coordinate values all refer to the same point. It occurs, for example, in a spherical polar coordinate system (r, θ, ϕ) at the locations $\theta = 0$ or π, where all values of ϕ refer to the same point. In order to map a complete manifold while avoiding degeneracy, one may use a set of *coordinate patches*. That is, one divides the complete manifold up into overlapping regions, and one assigns coordinates to each such region. The coordinate patches are called *charts*. The set of all coordinate patches that covers the whole manifold is called an *atlas*. The terms *coordinate system* and *chart* are synonymous. The term *reference frame* has a more varied usage. It is often used loosely as a synonym for *coordinate system*, and sometimes it is used to mean a set of basis vectors, which is a different concept (and arguably a better usage). In physics a good way to imagine a frame of reference is to imagine it as a physical thing: a collection of atoms with their worldlines. One speaks of observations 'in' or 'relative to' one reference frame or another; this is to indicate that

---

[1] Manifolds remain a highly non-trivial set of mathematical objects. Markov showed that the topological classification of four-dimensional manifolds can be reduced to a form of Turing halting problem, and therefore it is undecidable: no general algorithm exists for distinguishing two arbitrary manifolds with four or more dimensions.

Fig. 8.1 A coordinate system: just labels on pins, but assigned in a smoothly varying manner. The diagram shows a case where the manifold also has a fourth dimension (e.g. time) and we have shown the situation at one value of the associated coordinate.

statements are being made about how the phenomena in question relate to times and distances defined via the atoms of such a frame.

Within any given manifold there are submanifolds such as *curves* and *surfaces*. A curve is a one-dimensional manifold; we can think of it as a line wandering through the 'space' we are thinking about. It is often convenient to specify a curve parametrically. That is, we introduce a single parameter u and then state how the curve is traced out as u varies by furnishing N equations:

$$x^a = x^a(u), \qquad (a = 0, 1, 2, \ldots, (N-1)) \tag{8.1}$$

where we have chosen to count the coordinates from 0 to $(N-1)$ rather than 1 to N because this is how it is commonly done in GR, and the notation means that $x^0(u)$, $x^1(u), \ldots$ are functions of the common parameter u. In GR, for example, it is common to parametrize a worldline using the proper time τ, and then one has four functions $t(\tau)$, $x(\tau)$, $y(\tau)$, $z(\tau)$.

More generally, a submanifold of M dimensions $(M < N)$ can be specified by invoking M parameters, and giving the location of the points in the submanifold accordingly, through functions

$$x^a = x^a(u^1, u^2, \ldots u^M), \qquad (a = 0, 1, 2, \ldots, (N-1)) \tag{8.2}$$

A manifold of two dimensions is called a *surface*, and more generally the word 'surface' is often used for any submanifold of dimension above 1 and below N. In the physics community the term *hypersurface* is widely adopted for the case $M = N-1$. For example, in a static situation, at any given time, space is a hypersurface of spacetime. The particular set of events alluded to by the word 'space' at any given time can depend on reference frame.

A manifold of N dimensions can be regarded as a continuous set of $(N-1)$-dimensional manifolds, all smoothly joined together. Think of pages in a book, for example. This idea is called *foliation*; we say the pages provide a foliation of the three-dimensional space. The important feature is that the complete manifold is formed smoothly from one page to the next without any gaps or intersections or dislocations. In GR it is often useful to regard spacetime as foliated by a continuous set of spacelike hypersurfaces. Here spacetime is four-dimensional and each hypersurface is three-dimensional. The hypersurfaces do not need to be flat (and usually are not).

8.1.1   Coordinate transformation and its inverse

Let x^a be a set of coordinates mapping some region of a manifold, with $a = 0, 1, 2, \ldots (N-1)$. Let f^a be a set of single-valued functions of N variables. Then we can define a second set of coordinates x'^a using

$$
\begin{aligned}
x'^0 &= f^0(x^0, x^1, \ldots x^{N-1}) \\
x'^1 &= f^1(x^0, x^1, \ldots x^{N-1}) \\
&\cdots \\
x'^{N-1} &= f^{N-1}(x^0, x^1, \ldots x^{N-1})
\end{aligned}
\tag{8.3}
$$

The above equations are ordinarily stated in the more succinct notation

$$
x'^a = x'^a(x^0, x^1, \ldots x^{N-1}). \tag{8.4}
$$

The second set of coordinates can be used equally well as the first to map the manifold. The standard properties of partial derivatives apply, so we can write

$$
\mathrm{d}x'^a = \mathrm{d}f^a = \frac{\partial f^a}{\partial x^0}\mathrm{d}x^0 + \frac{\partial f^a}{\partial x^1}\mathrm{d}x^1 + \ldots + \frac{\partial f^a}{\partial x^{N-1}}\mathrm{d}x^{N-1}. \tag{8.5}
$$

Using the summation convention, this is written

$$
\mathrm{d}x'^a = \frac{\partial f^a}{\partial x^\mu}\mathrm{d}x^\mu, \tag{8.6}
$$

and since f^a is simply another name for x'^a we may as well write:

Coordinate transformation

$$
\mathrm{d}x'^a = \frac{\partial x'^a}{\partial x^\mu}\mathrm{d}x^\mu \tag{8.7}
$$

In view of this, the set of partial derivatives $(\partial x'^a/\partial x^b)$ is often referred to as itself the coordinate transformation.

By applying the above argument to x^a we can also obtain

$$
\mathrm{d}x^a = \frac{\partial x^a}{\partial x'^\mu}\mathrm{d}x'^\mu, \tag{8.8}
$$

and by combining (8.7) and (8.8) we have

$$\mathrm{d}x^a = \frac{\partial x^a}{\partial x'^\mu}\frac{\partial x'^\mu}{\partial x^\nu}\mathrm{d}x^\nu. \tag{8.9}$$

It follows that $(\partial x^a/\partial x'^b)$ is the inverse transformation to $(\partial x'^a/\partial x^b)$. In the context of Special Relativity, if $(\partial x'^a/\partial x^b)$ is a Lorentz transformation, then $(\partial x^a/\partial x'^b)$ is the inverse Lorentz transformation. Eqns (8.4)–(8.9) are able to handle any coordinate transformation.

When we write the partial derivative $(\partial f/\partial x^a)$, for any given f, it is understood that the variables being held constant are the other coordinates x^b, with $b \neq a$. It follows that if the function under consideration is one of those very coordinates, then we shall find

$$\frac{\partial x^a}{\partial x^b} = \delta^a_b. \tag{8.10}$$

(For example, for coordinates $x^a = \{t, r, \theta, \phi\}$, we have $(\partial t/\partial t) = 1$, $(\partial t/\partial r) = 0$, etc.) Now divide (8.8) by $\mathrm{d}x^b$ while holding the other variables constant, to obtain

$$\frac{\partial x^a}{\partial x^b} = \delta^a_b = \frac{\partial x^a}{\partial x'^\mu}\frac{\partial x'^\mu}{\partial x^b}. \tag{8.11}$$

This equation asserts that if we regard $(\partial x^a/\partial x'^\mu)$ as a matrix then $(\partial x'^\mu/\partial x^b)$ is the inverse matrix. (One may also deduce this from (8.9).)

An important formula related to (8.5) and (8.7) is:

Chain rule

$$\frac{\partial}{\partial x'^a} = \frac{\partial x^\lambda}{\partial x'^a}\frac{\partial}{\partial x^\lambda} \tag{8.12}$$

This follows from the rules of partial differentiation and therefore applies in arbitrary coordinate systems. Using the ∂_a notation it reads

$$\partial'_a = \left(\partial'_a x^\lambda\right)\partial_\lambda. \tag{8.13}$$

8.2   Riemannian manifolds

Mathematically, a manifold is specified in the first instance as simply an amorphous collection of points. In order to introduce the idea that these points bear some relationship with one another, such that geometric ideas such as distance and angle can be applied, we associate with the manifold a *metric*. The metric is a statement of how far apart any given pair of neighbouring points is, where we insist that the quantity 'how far apart' is a property of the manifold itself, not of the coordinates being used to map it. Thus, to introduce the concept of *metric* is by

> **Terminology.** A function that maps between two manifolds, is differentiable to all
> orders and has an inverse which is also differentiable to all orders is called a *diffeomor-*
> *phism.*
>
> A diffeomorphism between a manifold and itself is the same thing as a smooth coordinate
> transformation.
>
> When coordinate transformations are employed in GR, it is understood that the metric
> is adjusted accordingly, so that distances are preserved, and the net result is no change
> in the physical phenomena. This is the way the term 'diffeomorphism' has come to be
> employed by physicists. However, in the study of manifolds more generally, a map from
> one manifold to another may be kept distinct from any change in the metric, and for
> this reason the term 'diffeomorphism' is employed to mean strictly the definition above
> in the mathematical community, and this is different from the practice widespread in
> physics. You have been warned!
>
> *Diffeomorphism invariance* is another name for general covariance. It is the claim that
> given a differentiable mapping between spacetime and itself, the laws of physics under
> this transformation must remain unchanged.

definition also to introduce the concept of *scalar invariant*. For any given pair of neighbouring
points in the manifold, the distance specified by the metric is a scalar quantity that is invariant
under a change of coordinate system.

One can also have a manifold without a metric. An example is the state-space often employed
in thermodynamics, where the equilibrium states of a system described by pressure and volume
lie on a surface in p, V, T space where T is temperature. This surface is a manifold, but we do
not propose any metric on it for the purpose of thermodynamic study, because it does not make
physical sense to imagine rotations in which pressure becomes temperature, for example.

Coming now to GR, the manifold is spacetime and it has a metric. The invariant distance
specified by the metric is called the **interval**. It was mentioned in Chapter 2 that the metric
assumed in GR is of a particular kind: it is quadratic in the coordinate differences, as indicated
in eqn (2.22), which we repeat here:

$$ds^2 = g_{\mu\nu}dx^\mu dx^\nu. \tag{8.14}$$

Manifolds having this type of metric are called *Riemannian*. In strict terminology, the term
Riemannian is used when ds^2 is everywhere positive; if ds^2 can be of either sign (or zero) then
the term *pseudo-Riemannian* is used.[2] According to GR, spacetime is pseudo-Riemannian.

Let g_{ab} be the metric as given in some system of coordinates x^a, and let g'_{ab} be the metric as
given in some other system of coordinates. Then, by the invariance of the interval, we must
have

---

[2]Some treatments employ the unadorned term *Riemannian* irrespective of the signs.

> **What are the physical dimensions of dx?** One might expect a coordinate difference such as dx^a to have the physical dimensions of a length, and it often does, but it does not have to. One might be interested in using angular coordinates, for example, in which $d\theta$ and $d\phi$ are dimensionless. In this book we shall make no attempt to force all components of any given vector to have the same physical dimensions. This does not cause any problems in practice, because the mathematical formalism will take care of it. For example, in the line element eqn (8.14) the components of g_{ab} will have whatever physical dimensions are required to make the result come out as a length-squared. When calculating an inner product such as $p_\mu v^\mu$ (Chapter 10) the metric will again do the work of making the dimensions consistent.

$$g'_{\mu\nu} dx'^\mu dx'^\nu = g_{\mu\nu} dx^\mu dx^\nu \tag{8.15}$$

when both metrics are being evaluated at the same event. Divide both sides by $dx'^a dx'^b$, holding other primed variables constant, to obtain

$$g'_{\mu\nu} \frac{\partial x'^\mu}{\partial x'^a} \frac{\partial x'^\nu}{\partial x'^b} = g_{\mu\nu} \frac{\partial x^\mu}{\partial x'^a} \frac{\partial x^\nu}{\partial x'^b} \tag{8.16}$$

On the left-hand side the partial derivatives are given by Kronecker deltas (see eqn (8.10)) and thus we have:

Transformation of metric tensor

$$g'_{ab} = g_{\mu\nu} \frac{\partial x^\mu}{\partial x'^a} \frac{\partial x^\nu}{\partial x'^b} \tag{8.17}$$

We have introduced the term 'tensor' here in anticipation of a more thorough discussion of that terminology in Chapter 12.

Now look carefully at the placements of primes and indices in eqns (8.8) and (8.17). We have that *when the coordinates transform a certain way, the metric transforms the inverse way.* Invoking the terminology of 'contravariant, covariant' we say that the metric is *naturally covariant.*

It is useful to note that owing to its definition in (8.14), the metric tensor can always be taken to be symmetric. This is proved as follows. First observe that we can always write

$$g_{ab} = \frac{1}{2}\left(g_{ab} + g_{ba}\right) + \frac{1}{2}\left(g_{ab} - g_{ba}\right). \tag{8.18}$$

The first bracket here is symmetric in the indices, the second is antisymmetric. The contribution to ds^2 from the antisymmetric part vanishes exactly (as the reader should confirm), so if we replace g_{ab} by only its symmetric part $\left(g_{ab} + g_{ba}\right)/2$, then we shall get the same ds^2. We shall take it for granted in the rest of this book that symmetric metric tensors are being used.

In two and three dimensions there always exists a choice of coordinates that makes the metric tensor diagonal throughout the space, not just at one point (see box for comments on the

---

Orthogonal coordinates in two dimensions.

Suppose we would like to build a coordinate system for a two-dimensional manifold such that the metric is diagonal. This means we want our coordinate lines to cross everywhere at right angles. Observe that at any point in a two dimensional space we can identify a pair of vectors **u** and **v**, the eigenvectors of the 2×2 metric tensor **g**, such that **u** is perpendicular to **v**, and by a rotation of coordinates we can diagonalize **g** at that point. Imagine marking the directions **u**, **v** as a small cross at each point in the space. If the geometry of the space is smoothly varying then the orientation of neighbouring crosses will be very similar. We can then imagine drawing smooth curves through the crosses, thus covering the space with a curvilinear grid. These grid lines form a doubly infinite set with sensible properties—any point in the space is uniquely identifiable by specifying which pair of grid lines it lies on. We now label these grid lines by some convenient numbering (x, y). *These numbers constitute a coordinate system in which* **g** *is diagonal.* A more complete argument would have to consider the possible ambiguities that may arise if the topology gives a closed surface (e.g. the surface of a sphere).

$N = 2$ case). Such coordinates are called *orthogonal coordinates*. In higher dimensions this is not guaranteed to be possible, but in the four-dimensional spacetime of General Relativity, if there is just a modest amount of symmetry then it is often the case that orthogonal coordinates can be found.

Certain properties of the metric are independent of the coordinates. The metric **signature** is one such property. In some small enough region of a Riemannian or pseudo-Riemannian manifold we will show (below) that one can always find a coordinate system in which the metric is diagonal, and one can scale the coordinates such that the diagonal elements are either $+1$, 0 or -1. The signature is then the number of elements of each type. If it is taken for granted that there are no zeros among these diagonal elements, and the number of dimensions is known, then the word 'signature' may also be used to refer to the sum of the diagonal elements. The Minkowski metric is diagonal with elements $(-1, 1, 1, 1)$ and therefore has signature $(3, 0, 1)$ according to the first usage and signature $+2$ according to the second usage. This information can also be extracted from the eigenvalues of the metric tensor. The significance of this idea is that it is not possible, merely by coordinate transformations, to turn a metric of one signature into a metric of another signature. Thus the signature classifies what kind of metric, and consequently what kind of manifold, one has. In GR, spacetime has four dimensions and signature $+2$. One can also develop the subject using the signature -2 (see Chapter 1), but once one has made the choice one must stick to it.

8.2.1 The metric of a submanifold

Let us define an M-dimensional submanifold by using M parameters u^j ($j = 1, 2, \ldots M$) as in eqn (8.2). The u^j may themselves be considered a set of coordinates that map the submanifold. Two neighbouring events in the submanifold with parameters differing by du^j will have coordinates in the full manifold separated by

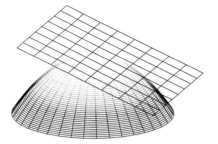

Fig. 8.2 A tangent space, tangent to a curved manifold. In this example the tangent space is two-dimensional because the manifold is two-dimensional.

$$\mathrm{d}x^a = \frac{\partial x^a}{\partial u^j}\mathrm{d}u^j \tag{8.19}$$

and therefore the invariant interval between the events is

$$\mathrm{d}s^2 = g_{\mu\nu}\frac{\partial x^\mu}{\partial u^j}\frac{\partial x^\nu}{\partial u^k}\mathrm{d}u^j\mathrm{d}u^k. \tag{8.20}$$

Hence the metric on the submanifold is

$$g_{jk}^{(\mathrm{sub})} = g_{\mu\nu}\frac{\partial x^\mu}{\partial u^j}\frac{\partial x^\nu}{\partial u^k}. \tag{8.21}$$

8.3   Local flatness and Riemann normal coordinates

We now present a very important property of Riemannian and pseudo-Riemannian manifolds: they are *locally flat*. That is to say, in the vicinity of any point the local geometry is Euclidean or pseudo-Euclidean. A space is Euclidean if there exists a coordinate choice in which the metric is the identity matrix (i.e. diagonal, with all diagonal elements equal to 1.) A space is pseudo-Euclidean if there exists a coordinate choice in which the metric is diagonal, with the elements on the diagonal equal to either 1 or -1. In this terminology, the Minkowski metric is pseudo-Euclidean. We shall use the word 'flat' as synonymous with 'Euclidean or pseudo-Euclidean'.

Let R be a Riemannian or pseudo-Riemannian space of N dimensions. A *tangent space* (or 'tangent plane') is an N-dimensional flat space *not* usually lying in R (the exception is when R is itself flat), but touching it tangentially at a point and having the same metric at that point. Think for example of a flat plane resting against a curved surface (Fig. 8.2). In General Relativity a tangent space corresponds to a local inertial frame. The notion of 'touching tangentially' is clarified in the following; in GR it refers to the fact that there exist coordinates for the manifold in which the metric takes the form (8.22) near the event in question. The metric of the tangent space is then η_{ab}.

The existence of a tangent space leads to, or is equivalent to, the statement that for any chosen point P there exists a coordinate choice that makes the metric g_{ab} Euclidean or pseudo-Euclidean at P, *and* have zero derivatives there. In GR, this is the statement that at every event, there exists a coordinate choice x'^a such that the metric tensor g'_{ab} evaluates to the Minkowski metric η_{ab} at that event, and is stationary with respect to all the coordinates, i.e.

$$g'_{ab} \overset{\text{LIF}}{=} \eta_{ab} + \mathcal{O}\left((x' - x'_P)^2\right). \tag{8.22}$$

We shall prove this by obtaining such coordinates by two different methods. Both methods employ some ideas which we shall develop fully only in the next few chapters. But the concept of local flatness is simple enough, and important enough, to warrant being presented before all those tools are studied in full. The reader is advised to skim through the following at first reading, and then return to it after completing Chapter 12.

Constructing LIF coordinates: first method.   For our first method we shall start with arbitrary coordinates and obtain LIF coordinates through a sequence of three coordinate transformations.

Suppose we start with coordinates x^a, with some arbitrary g_{ab}. First define the coordinates x'^a given by

$$x'^a = x^a - x_P^a + \frac{1}{2}\Gamma^a_{\mu\nu}(P)\left(x^\mu - x_P^\mu\right)\left(x^\nu - x_P^\nu\right). \tag{8.23}$$

where $\Gamma^a_{\mu\nu}(P)$ means the Christoffel symbol evaluated at P. We shall prove in Section 12.2.1 that in these coordinates, the metric near P takes the form

$$g'_{ab} = \text{const}_{ab} + \mathcal{O}\left((x' - x'_P)^2\right). \tag{8.24}$$

This means that so far we have eliminated the linear terms, but the metric is not yet Minkowskian at P.

Now consider a further coordinate system y^a. Let G' be the matrix having elements g'_{ab}, and let K be the matrix having elements $(\partial x'^a / \partial y^b)$, evaluated at event P. Note, here we choose that the elements of K are *constants*; they have no dependence on any coordinates (consequently they only express the coordinate transformation at one event, P). Under this assumption the coordinate change represented by K is a linear transformation.

In terms of these matrices, the transformation of the metric tensor, eqn (8.17), is written

$$\mathsf{G} = \mathsf{K}^T \mathsf{G}' \mathsf{K} \tag{8.25}$$

where G is the metric tensor in the y^a coordinates. Now let us choose for K the matrix whose columns are the normalized eigenvectors of G'. We are guaranteed to be able to find orthogonal eigenvectors because G' is symmetric. In this case we shall find that G is diagonal, with diagonal elements equal to the eigenvalues λ_a. By now rescaling the new coordinates according to $y^a \to y^a/\sqrt{|\lambda_a|}$ (with no implied summation), we shall obtain $g_{ab} = \eta_{ab}$ in the y^a coordinates at P. Furthermore, since K involves only constants, not functions of x'^a, we have not introduced any linear terms, so now we have the form as claimed in (8.22).

In GR, we shall eventually claim that the whole effect of gravity can be accounted for as a modification of the metric of spacetime, and spacetime is pseudo-Riemannian. *The geometric property of local flatness then results in the physical outcome called the strong equivalence principle.*

Constructing LIF coordinates: second method. The construction used in the second method is called *Riemann normal coordinates* or *geodesic coordinates*. Imagine yourself in a space whose curvature may be of any type, and may vary from point to point. Pick a point P in the space and trace geodesics going out in all directions from P. For Riemannian and pseudo-Riemannian manifolds these geodesics will not intersect in a region sufficiently close to P, except at P. We assign coordinates in such a way that the coordinate direction to each event is equal to the direction of the geodesic setting out from P to that event. In this way, all the events along each geodesic will lie along a straight line in the new coordinate system.

Now we shall make this idea precise.

To construct the desired coordinates, first transform to orthonormal coordinates y^a at P, as explained in the first method. Next, construct a new coordinate system x^a by assigning to each point in the neighbourhood of P a direction u^a and a distance s:

$$x^a = su^a. \tag{8.26}$$

For a given point Q, the direction is chosen by finding the unique geodesic connecting P to Q, and setting u^a equal to its tangent vector at P. It follows that if s increases smoothly along the geodesic then our coordinate system assigns coordinates in a sensible manner to all the points along each geodesic. It remains to make sure that the coordinates also vary smoothly as we pass from each geodesic to its neighbours. This is achieved by drawing a set of concentric *coordinate hyperspheres* around P in the y^a system, i.e. the locus of points satisfying $\sum (y^a)^2 = \epsilon^2$, and assigning the distance $s = \epsilon$ to all points on a given hypersphere. (We picked orthonormal y^a coordinates at the outset merely to make the equation of a coordinate hypersphere easy to write.) This construction guarantees that neighbouring events in the y^a system get neighbouring coordinates in the x^a system.

Note, this is a rectangular not a polar coordinate system: s is not itself a coordinate. A simple example is furnished by coordinates in a flat plane where we might have $u^a = (\cos\theta, \sin\theta)$ and $s = r$, giving the familiar $x_1 = r\cos\theta$, $x_2 = r\sin\theta$.

We now have a coordinate system with a useful property: any geodesic originating at P is described parametrically by a curve $x^a(s)$ which is linear in the parameter s at small s. Therefore, in the geodesic equation (2.28) we have, at the point P,

$$\ddot{x}^a = 0 \quad \Rightarrow \Gamma^a_{\mu\nu}\dot{x}^\mu\dot{x}^\nu = 0$$

where the dot signifies $\mathrm{d}/\mathrm{d}s$, so $\dot{x}^\mu = u^\mu$. Since the result holds at P for every geodesic there, it must be that $\Gamma^a_{bc} = 0$.

This establishes the central result: for any space, there exists a coordinatization such that all the Christoffel symbols vanish at some pre-assigned event P. If all the Christoffel symbols vanish then so do the first derivatives of the metric; see eqns (2.8) and (10.23). P is called the *pole* of such *geodesic coordinates* or *Riemann normal coordinates*. The argument applies to any Riemannian or pseudo-Riemannian space. Applied to GR, the geodesic coordinates are the mathematical expression of the physical concept of the local inertial frame, or freely falling cabin. Therefore they may also be called *inertial coordinates*.

8.4   Measuring length, area and volume

The length of a curve in a Riemannian or pseudo-Riemannian manifold is given by

$$s = \int \mathrm{d}s = \int \left| g_{\mu\nu} \mathrm{d}x^\mu \mathrm{d}x^\nu \right|^{1/2} = \int \left| g_{\mu\nu} \frac{\mathrm{d}x^\mu}{\mathrm{d}u} \frac{\mathrm{d}x^\nu}{\mathrm{d}u} \right|^{1/2} \mathrm{d}u \tag{8.27}$$

where the integral is taken along the path specified by the curve.

To calculate an area or a volume, first let us consider the case where the metric is diagonal, and we are interested in a region specified by some range of two or more coordinates. A diagonal metric implies that the coordinates are orthogonal, since the contributions to $\mathrm{d}s^2$ coming from the various coordinate changes add up just like in Pythagoras's theorem:

$$\mathrm{d}s^2 = g_{00} \left(\mathrm{d}x^0 \right)^2 + g_{11} \left(\mathrm{d}x^1 \right)^2 + g_{22} \left(\mathrm{d}x^2 \right)^2 + g_{33} \left(\mathrm{d}x^3 \right)^2 \tag{8.28}$$

In this situation one can see that the expression for an element of area is

$$\mathrm{d}A = \sqrt{|g_{11}g_{22}|}\,\mathrm{d}x^1 \mathrm{d}x^2 \tag{8.29}$$

where we took an illustrative case involving the coordinates indexed 1 and 2. An element of volume is

$$\mathrm{d}V = \sqrt{|g_{11}g_{22}g_{33}|}\,\mathrm{d}x^1 \mathrm{d}x^2 \mathrm{d}x^3, \tag{8.30}$$

and an element of 4-volume is

$$\mathrm{d}^4 V = \sqrt{|g_{00}g_{11}g_{22}g_{33}|}\,\mathrm{d}x^0 \mathrm{d}x^1 \mathrm{d}x^2 \mathrm{d}x^3 \tag{8.31}$$

(where, to repeat, we have restricted to diagonal metrics for the moment). For a diagonal metric, the determinant takes the form

$$g \equiv \det([g_{ab}]) \overset{\mathrm{diag}\ g}{=} g_{00}g_{11}g_{22}g_{33} \tag{8.32}$$

so the above result for the 4-volume element may be written

$$\mathrm{d}^4 V = \sqrt{|g|}\,\mathrm{d}x^0 \mathrm{d}x^1 \mathrm{d}x^2 \mathrm{d}x^3 \tag{8.33}$$

(and the generalization to N dimensions is obvious). We shall now show that this expression involving the determinant is also correct for any metric, not just diagonal metrics.

To prove this, introduce the matrix of coefficients which describe a coordinate transformation:

$$\left[\frac{\partial x'^a}{\partial x^b} \right] = \begin{pmatrix} \frac{\partial x'^0}{\partial x^0} & \frac{\partial x'^0}{\partial x^1} & \frac{\partial x'^0}{\partial x^2} & \frac{\partial x'^0}{\partial x^3} \\ \frac{\partial x'^1}{\partial x^0} & \frac{\partial x'^1}{\partial x^1} & \frac{\partial x'^1}{\partial x^2} & \frac{\partial x'^1}{\partial x^3} \\ \frac{\partial x'^2}{\partial x^0} & \frac{\partial x'^2}{\partial x^1} & \frac{\partial x'^2}{\partial x^2} & \frac{\partial x'^2}{\partial x^3} \\ \frac{\partial x'^3}{\partial x^0} & \frac{\partial x'^3}{\partial x^1} & \frac{\partial x'^3}{\partial x^2} & \frac{\partial x'^3}{\partial x^3} \end{pmatrix} \tag{8.34}$$

The determinant of this matrix is called the *Jacobian*:

$$J = \det \left[\frac{\partial x'^a}{\partial x^b} \right] \tag{8.35}$$

Let us choose for the coordinates x'^a that system of coordinates which results in a locally Minkowskian metric. Then we know that the 4-volume element in these coordinates is

$$\mathrm{d}^4 V = \mathrm{d}x'^0 \mathrm{d}x'^1 \mathrm{d}x'^2 \mathrm{d}x'^3 \tag{8.36}$$

and furthermore, using a well-known property of Jacobians,

$$\mathrm{d}x'^0 \mathrm{d}x'^1 \mathrm{d}x'^2 \mathrm{d}x'^3 = J \mathrm{d}x^0 \mathrm{d}x^1 \mathrm{d}x^2 \mathrm{d}x^3. \tag{8.37}$$

The metric in the primed coordinates is related to that in unprimed coordinates by

$$\mathsf{G}' = \mathsf{K}^T \mathsf{G} \mathsf{K}, \tag{8.38}$$

where $\mathsf{G} \equiv [g_{ab}]$ and $\mathsf{K} = [(\partial x^a / \partial x'^b)]$ is the inverse of the matrix displayed in (8.34), so $\det \mathsf{K} = J^{-1}$. Eqn (8.38) gives

$$g' \equiv \det \mathsf{G}' = (\det K)^2 g = J^{-2} g \tag{8.39}$$

and since G' is Minkowskian, $g' = -1$. Hence we have the useful result that

$$g = -J^2 \tag{8.40}$$

(and in the Euclidean case, where $g' = 1$, one has $g = +J^2$.) Substituting this into (8.37) and using (8.36) we have

$$\mathrm{d}^4 V = \sqrt{|g|} \, \mathrm{d}x^0 \mathrm{d}x^1 \mathrm{d}x^2 \mathrm{d}x^3 \tag{8.41}$$

where now we have that the result holds for any metric.

Notice that the 4-volume element here is a property of the manifold itself. Using a coordinate system, we propose a given 'box' of sides $\mathrm{d}x^0$, $\mathrm{d}x^1$, $\mathrm{d}x^2$, $\mathrm{d}x^3$. The 4-volume element then tells us how much volume there is in our box. The generalisation to any number of dimensions is obvious.

The above also shows us how to write 3-volumes and areas for general metrics. A region with fewer than N dimensions is a submanifold which can be specified parametrically as in eqn (8.2). The metric of the submanifold is then given by (8.21) and the element of M-dimensional 'area' is

$$\mathrm{d}^M V = \sqrt{|g^{(\mathrm{sub})}|} \, \mathrm{d}u^1 \dots \mathrm{d}u^M. \tag{8.42}$$

The reader is encouraged at this point to read appendix B, which illustrates many of the methods of this chapter by applying them to the 2-sphere and the 3-sphere.

Exercises

8.1 Write down the metric of three-dimensional Euclidean space in cylindrical polar coordinates x^a and spherical polar coordinates x'^a and show that the expressions are related as (8.17).

8.2 From the definition (5.9) prove that for a manifold of dimension N, one must have $g_{\mu\nu}g^{\mu\nu} = N$.

8.3 Derive all the parts of appendix B which do not depend on connection coefficients.

8.4 Sketch the set of parabolas $y = a + x^2$ on a (Euclidean) graph, for the values $a = 0, 1, 2, 3$. Now introduce coordinates $x' = y - x^2$, $y' = y$. Using (8.17) show that the metric in primed coordinates is

$$[g'_{ab}] = \frac{1}{4(y' - x')} \begin{pmatrix} 1 & -1 \\ -1 & 1 + 4(y' - x') \end{pmatrix}.$$

Express $(\mathrm{d}x', \mathrm{d}y')$ in terms of $x, y, \mathrm{d}x, \mathrm{d}y$ and hence show explicitly that

$$(\mathrm{d}x', \mathrm{d}y')[g'_{ab}](\mathrm{d}x', \mathrm{d}y')^T = \mathrm{d}x^2 + \mathrm{d}y^2$$

where T denotes the transpose (to form a column vector). Explain why the primed coordinates fail when $y' = x'$.

8.5 Consider a LIF in a Minkowski spacetime restricted to one spatial dimension z, such that the line element is $\mathrm{d}s^2 = -\mathrm{d}t^2 + \mathrm{d}z^2$ (take $c = 1$). Let the primed frame be accelerating with respect to this LIF with constant acceleration a (for some finite amount of time), such that primed coordinates are related to the LIF coordinates by $t' = t$ and $z' = z - \frac{1}{2}at^2$. Find the metric tensor in the primed system g'_{ab} and its inverse g'^{ab}.

$$\textit{Ans.} \quad [g'_{ab}] = \begin{pmatrix} -1 + a^2 t'^2 & at' \\ at' & 1 \end{pmatrix}, \quad (8.43)$$

$$[g'^{ab}] = \begin{pmatrix} -1 & at' \\ at' & 1 - a^2 t'^2 \end{pmatrix}.$$

8.6 Write down the coordinate transformation matrix for the example considered in exercise 8.5, and confirm that the metric transforms as (8.17).

$$\textit{Ans.} \quad [g'_{ab}] = \begin{pmatrix} 1 & at \\ 0 & 1 \end{pmatrix} \begin{pmatrix} -1 & 0 \\ 0 & 1 \end{pmatrix} \begin{pmatrix} 1 & 0 \\ at & 1 \end{pmatrix}.$$

8.7 Suppose we map Earth's surface with ordinary spherical polar coordinates θ, ϕ. The line element is

$$\mathrm{d}s^2 = R^2 \mathrm{d}\theta^2 + R^2 \sin^2\theta \, \mathrm{d}\phi^2 \qquad (8.44)$$

where R is Earth's radius. We wish to produce a map on a plane rectangular surface in coordinates $x = x(\theta, \phi)$ and $y = y(\theta, \phi)$.
(i) Show that the angle between two lines crossing on the sphere's surface will be equal to the angle between corresponding lines on the map if the line element can be written in the form

$$\mathrm{d}s^2 = \Omega(x, y)(\mathrm{d}x^2 + \mathrm{d}y^2) \qquad (8.45)$$

for some function $\Omega(x, t)$ [Hint: similar triangles].
(ii) The *Mercator projection* adopts

$$x = w\frac{\phi}{2\pi}, \qquad y = h\frac{1}{2\pi} \ln\left[\tan((\pi - \theta)/2)\right]$$

where w and h are constants. Show that this map satisfies the condition (8.45) when $w = h$. [*Ans.* $\Omega = (2\pi R/h)^2 \sin^2\theta$.]
(iii) Suggest an *equal area* projection, i.e. one that preserves ratios of areas.

8.8 Show that the 3-surface of a 4-sphere with radius a has a volume $2\pi^2 a^3$.

9

Vectors on manifolds

In this chapter we introduce the notion of a *vector* and the *vector field*, in the context of Riemannian manifolds. A vector is a geometric object that can be thought of as a directed line segment. There is another, more abstract, definition in terms of the *directional derivative* which we briefly outline in appendix C; we will not need it here.

In physics we use vectors for many purposes, such as for describing momentum and force, flow, electric and magnetic fields, and so on. When discussing vectors in the context of non-flat manifolds it is best to begin with the purely geometric notion of a displacement from one place to a nearby place, and then to define vectorial quantities more generally by saying that their direction and length behaves like that of such a displacement. Throughout the present chapter we will use the word 'vector' to mean such a quantity in an N-dimensional space. For example, in the context of spacetime which has $N = 4$, the word 'vector' will always mean '4-vector'. We indicate vectors by using the bold font, as in $\mathbf{v}$, $\mathbf{e}$.

A local shop for local people. In the humorous BBC television series *The League of Gentleman*, there is a memorable scene in which a visitor to a rural village enters a shop, and the people running the shop are disconcerted, because the 'outsider' is touching 'the precious things of the shop'. 'This is a local shop for local people,' they announce.

When approaching the mathematics of manifolds we have to be careful to define and discuss things in a *local* manner, in the first instance, in order to make sure we are making well-defined statements. It is a mistake, for example, to talk about the sum or the difference of two displacement vectors, if the vectors are located at different places in the manifold. Each vector has not only its own properties, but also a location: it is defined at a point in the manifold, and at that point it does not lie in the manifold, but rather in the tangent space at that point—see Fig. 8.2. A vector at some other point is located in the tangent space at its location, and since this is a different tangent space the two vectors cannot be added or compared in any way. It follows that, in GR, we shall have to take care if we wish to define concepts such as the total 4-momentum of an extended entity: one cannot simply add the 4-momenta of the parts.

Relativity Made Relatively Easy: General Relativity and Cosmology. Volume 2. Andrew M. Steane,
Oxford University Press. © Andrew M. Steane 2021. DOI: 10.1093/oso/9780192895646.003.0009

9.1   Basis vectors and the inner product

Two or more vectors defined at the same point, on the other hand, occupy the same tangent space and can be summed and compared. In particular, when the manifold has N dimensions, so does the tangent space, and therefore there can be up to N linearly independent vectors at any point.[1] Let us denote a set of N linearly independent vectors at a point by the notation

$$\mathbf{e}_{(a)}, \qquad a = 0,\, 2,\, \ldots,\, N-1. \tag{9.1}$$

Note that this expression does not indicate components of a vector, it indicates a complete vector, drawn from a set of N vectors. This is why the subscript is shown in a bracket and the symbol is shown in bold font. These vectors need not be mutually orthogonal nor normalized in the first instance, though later we will be interested in orthonormal sets. Indeed, at this stage we have not yet defined any notion of orthogonality.

We shall refer to the $\mathbf{e}_{(a)}$ as *basis vectors*. For any given manifold, there are an infinite number of ways of choosing a set of basis vectors, and they need bear no particular relation to any one coordinate system (though later we will find it useful to choose basis vectors in relation to a convenient coordinate system).

Any other vector in the same tangent space can be expressed as a linear combination of the basis vectors:

$$\mathbf{v} = \sum_{a=0}^{N-1} v^a \mathbf{e}_{(a)}. \tag{9.2}$$

Note, this is not a contraction; it is a sum of N vectors. The coefficients v^a in the sum are called the *contravariant components* of the vector $\mathbf{v}$ in the basis $\{\mathbf{e}_{(a)}\}$.

Note that no vector is itself either contravariant or covariant. A vector is simply a vector. Only its components can be contravariant or covariant. Whenever someone speaks somewhat loosely of 'a contravariant 4-vector' they either mean 'the set of contravariant components of a 4-vector' or you should press them to explain what they mean.

The basis vectors can themselves be written in terms of their contravariant components: $\mathbf{e}_{(b)} = \sum_a e^a_{(b)} \mathbf{e}_{(a)}$. One finds (and really this follows immediately from the definition of what we mean by a basis):

$$e^a_{(b)} = \delta^a_b. \tag{9.3}$$

9.1.1   The dual basis and the inner product

The inner product or scalar product is a scalar invariant quantity obtained from a combination of a vector and another vector-like object. In Euclidean geometry, for example, it can be obtained

---

[1]A set of vectors is linearly independent when no vector in the set can be expressed as a linear combination of the others.

by multiplying corresponding components of a pair of vectors, and summing the results. In Riemannian geometry we shall define the inner product in two steps. First we define its effect when combining basis vectors with another type of vector-like object (to be specified). Then we define a more general statement in such a way that everything is consistent.

Let us now introduce a new set of mathematical objects, which behave in many respects like vectors, but you should not think of them simply as vectors. Let us call them *dual objects*. These dual objects can be linearly combined, just as vectors can, and in particular we can form a basis called the *dual basis*, and express other dual objects as linear combinations of the members of such a basis. The dual objects in this basis are written $\tilde{\boldsymbol{\theta}}^{(a)}$. Just as for the basis vectors, there are N mathematical entities in the set, and none of them are either contravariant or covariant; they are just dual objects.[2] As we introduce these dual basis objects, we also define an inner product, by asserting that each dual basis object has zero inner product with all but one of the basis vectors, and its inner product with the remaining basis vector is equal to 1. This assertion is expressed:

$$\tilde{\boldsymbol{\theta}}^{(a)} \cdot \mathbf{e}_{(b)} = \delta^a_b \tag{9.4}$$

where the dot notation signifies the inner product which we are thus defining. We also assert by definition the linear property:

$$\tilde{\boldsymbol{\theta}}^{(a)} \cdot (\alpha \mathbf{e}_{(b)} + \beta \mathbf{e}_{(c)}) = \alpha \tilde{\boldsymbol{\theta}}^{(a)} \cdot \mathbf{e}_{(b)} + \beta \tilde{\boldsymbol{\theta}}^{(a)} \cdot \mathbf{e}_{(c)} \tag{9.5}$$

for any scalar coefficients α, β.

By forming linear combinations of members of the dual basis, we can make further dual objects such as

$$\tilde{\mathbf{v}} = \sum_{a=0}^{N-1} v_a \tilde{\boldsymbol{\theta}}^{(a)}. \tag{9.6}$$

The coefficients v_a in this linear combination are called the *covariant components* of the dual object $\tilde{\mathbf{v}}$. Note, there is no need to put a tilde on these coefficients since the placement of the index already distinguishes v_a from v^a. In fact, at this stage no relationship has been introduced between v_a and v^a; in $\mathbf{v}$ and $\tilde{\mathbf{v}}$ we have two entirely unrelated objects. However, in the next section we will introduce a way to relate them via the metric tensor.

From a formal point of view, one may say that the dual objects lie in a dual space to the tangent space. They do not lie in the tangent space where the vectors live, so it is utterly meaningless to try to conceive of a sum of a dual object and a vector, such as $\tilde{\mathbf{v}} + \mathbf{w}$, or to assert that a vector is equal to a dual object. It would be the equivalent of asserting, in quantum mechanics, $\langle c| = |a\rangle + |b\rangle$, or something like that: the statement has no meaning and the rules of the notation prevent one from writing it. Nevertheless, some authors prefer to drop this distinction; see box for further information.

---

[2]Since we shall agree to place the bracketed counter on the basis vectors always down for the basis and up for the dual basis, one can use the same letter $\mathbf{e}$ for both, since the notation would already be unambiguous. But we shall not adopt that practice.

One-form, dual, covector, bra

It is a universal feature of analysis using vectors that one is interested in linear functions which take a vector as input and produce a scalar as output. The inner product is such a function. This leads to the concept of 'dual'; one says that one has both vectors and other things dual to them, and in order to get a scalar one should combine a vector with a dual object. In quantum mechanics this distinction is indicated explicitly in the very useful Dirac notation, where a vector is called a *ket* and written $|a\rangle$, and a dual object is called a *bra* and written $\langle a|$. In the analysis of manifolds and differential geometry, the term *vector* is retained, and the dual object is called either a *one-form* or a *covector*.

The basic inner product involves a one-form and a vector. This book uses the term 'one-form' sparingly because such terminology can be distracting. The word 'covector' can also cause confusion because it seems to suggest that the terms 'covariant' and 'contravariant' apply to covectors and vectors, respectively. That is a mistake because vectors are neither covariant nor contravariant, and the same can be said of covectors. Rather, their sets of components can be covariant or contravariant.

A good way to think of a dual object is to think of it as very much like a gradient. It has a sense of direction, and if, in a given coordinate change, the components of a vector get larger, then the components of a one-form/covector get smaller. The latter behaviour is like the behaviour of a gradient—see Fig. 9.2. The name 'one-form' comes from the phrase 'differential form'; a gradient is a differential form, and this is generalized to 'one-form', 'two-form', etc. The terminology is:

Definition 9.1 *A p-form on a manifold M is an antisymmetric $(0, p)$ tensor field on M.*

In this terminology a 0-form is an invariant scalar function, a 1-form is a covector field, a 2-form is an antisymmetric second-rank tensor field of type $(0, 2)$, etc. where the notation (q, p) refers to the way the components behave; this will be clarified in Chapter 12.

If you know quantum mechanics, then think of the dual objects as 'bras' to the vectors' 'kets'. They inhabit another mathematical space; they are not made of kets. If $T_P(M)$ is the tangent space at location P of manifold M then the dual space is denoted $T_P^*(M)$ and called the *cotangent space* at P. The covectors or one-forms are elements of T_P^*.

In GR it is mostly harmless to ignore the distinction between a tangent space and its dual, so one can blur this distinction. I think one should not do this at the outset so I have not done it. One can get away with blurring the distinction between vectors and duals as long as one sticks to the rules for combining basis vectors with dual basis members when it comes to evaluating inner products.

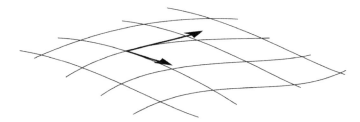

Fig. 9.1 The coordinate basis vectors are tangent to the coordinate curves.

The dual objects are commonly called *one-forms* (see box), but you can call them dual vectors if you like. We can now use contravariant and covariant components to write the inner product between any one-form and any vector, since

$$\tilde{\mathbf{v}} \cdot \mathbf{w} = \sum_{\mu} v_{\mu} \tilde{\boldsymbol{\theta}}^{(\mu)} \cdot \sum_{\nu} w^{\nu} \mathbf{e}_{(\nu)} = \sum_{\mu} \sum_{\nu} v_{\mu} w^{\nu} \tilde{\boldsymbol{\theta}}^{(\mu)} \cdot \mathbf{e}_{(\nu)}$$

$$= \sum_{\mu} \sum_{\nu} v_{\mu} w^{\nu} \delta^{\mu}_{\nu} = \sum_{\mu} v_{\mu} w^{\mu} = v_{\mu} w^{\mu} \tag{9.7}$$

where the last version adopts the summation convention.

By using (9.2) and (9.4) we have

$$\tilde{\boldsymbol{\theta}}^{(a)} \cdot \mathbf{v} = \sum_{\lambda} \tilde{\boldsymbol{\theta}}^{(a)} \cdot (v^{\lambda} \mathbf{e}_{(\lambda)}) = v^{\lambda} \delta^{a}_{\lambda} = v^{a}, \tag{9.8}$$

and similarly one finds

$$v_a = \tilde{\mathbf{v}} \cdot \mathbf{e}_{(a)}. \tag{9.9}$$

We can now write (9.2)) as

$$\mathbf{v} = \sum_{\lambda} \left(\tilde{\boldsymbol{\theta}}^{(\lambda)} \cdot \mathbf{v} \right) \mathbf{e}_{(\lambda)}. \tag{9.10}$$

(The corresponding idea in the treatment of Hilbert space would be written $|v\rangle = \sum_{\lambda} \langle \lambda | v \rangle | \lambda \rangle$.)

9.1.2 The coordinate basis

A convenient choice of basis is offered by the *coordinate basis vectors* given by

$$\mathbf{e}_{(a)} \equiv \lim_{\delta x^a \to 0} \frac{\delta \mathbf{s}(a)}{\delta x^a} \tag{9.11}$$

where $\delta \mathbf{s}(a)$ is the vector displacement between a point P and the nearby point Q differing from P in only one coordinate (x^a), by an amount δx^a. Defined this way, one finds that each basis vector is tangent to one of the coordinate curves at P (see Fig. 9.1).

Using these coordinate basis vectors, a general infinitesimal displacement (in any direction in the tangent space) can be written

$$\mathrm{d}\mathbf{s} = \sum_a \mathbf{e}_{(a)}\mathrm{d}x^a. \tag{9.12}$$

We will now introduce a very natural extension of the dot notation by defining a notion of inner product between vectors (rather than between dual objects and vectors which is the only inner product we have defined up till now). We wish to define what is meant by a notation such as $\mathbf{v} \cdot \mathbf{w}$. To this end, let us introduce

$$\begin{aligned}\mathrm{d}\mathbf{s} \cdot \mathrm{d}\mathbf{s} &= \sum_\mu \sum_\nu \mathrm{d}x^\mu \mathbf{e}_{(\mu)} \cdot \mathrm{d}x^\nu \mathbf{e}_{(\nu)} \\ &= \sum_\mu \sum_\nu \left(\mathbf{e}_{(\mu)} \cdot \mathbf{e}_{(\nu)}\right) \mathrm{d}x^\mu \mathrm{d}x^\nu.\end{aligned} \tag{9.13}$$

where the expressions follow by linearity of the inner product being introduced. We now assert that $\mathrm{d}\mathbf{s} \cdot \mathrm{d}\mathbf{s}$ given by (9.13) is that same invariant distance measure that is given by the metric through $\mathrm{d}s^2 = g_{\mu\nu}\mathrm{d}x^\mu\mathrm{d}x^\nu$. To see that this is mathematically consistent, one must convince oneself first that an infinitesimal distance in the tangent space will be equal to the corresponding distance in the manifold under consideration, and second that such an infinitesimal distance is indeed given by (9.13). The former follows from the definition of a tangent space, and the latter is arranged by defining the inner product between vectors so as to guarantee it. That is to say, we define the inner product between vectors (as opposed to the inner product between a vector and a dual object) such that

$$\sum_\mu \sum_\nu \left(\mathbf{e}_{(\mu)} \cdot \mathbf{e}_{(\nu)}\right) \mathrm{d}x^\mu \mathrm{d}x^\nu = g_{\mu\nu}\mathrm{d}x^\mu\mathrm{d}x^\nu. \tag{9.14}$$

Since this is true for any displacements $\mathrm{d}x^\mu$, $\mathrm{d}x^\nu$, we have:

Inner product of coordinate basis vectors

$$\mathbf{e}_{(a)} \cdot \mathbf{e}_{(b)} = g_{ab}. \tag{9.15}$$

It should not have escaped the reader's notice that on the right-hand side of (9.14) we have employed the summation convention, while on the left-hand side we have not. Equally, one may judge that it is not necessary to put the labels on the basis vectors in a bracket, so that one may write

$$\left(\mathbf{e}_\mu \cdot \mathbf{e}_\nu\right)\mathrm{d}x^\mu\mathrm{d}x^\nu = g_{\mu\nu}\mathrm{d}x^\mu\mathrm{d}x^\nu. \tag{9.16}$$

This less cluttered notation is perfectly consistent (and adopted by some authors); we have not adopted it in order to be clear about the meaning of symbols such as $\mathbf{e}_{(a)}$, thus avoiding confusion between a set of vectors and the components of any one vector.

We can now find the inner product between any pair of vectors by expanding each one in the coordinate basis. We thus find

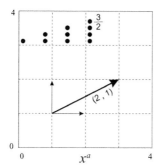 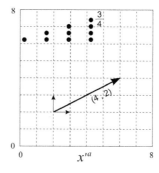

Fig. 9.2 Basis vectors, components, and change of basis. The figure shows a situation where there is a vector (the large arrow) and a scalar quantity which increases from left to right (the columns of dots). The two diagrams shows the *same* situation with two different coordinate frames (the dashed lines), with $x'^a = 2x^a$. In a change from one coordinate system to the other, any given vector does not change, but its components do: in this example they double. The exception is the *coordinate basis* vectors, which by definition extend along the coordinate directions with unit coordinate length, so they shrink or expand with the coordinate frame. The rate of growth in the scalar function per unit distance (its gradient) does not change, but the rate *per unit coordinate distance* does change; in this example it halves.

$$\mathbf{v} \cdot \mathbf{w} = \sum_{\mu} \sum_{\nu} \left(v^{\mu}\mathbf{e}_{(\mu)}\right) \cdot \left(w^{\nu}\mathbf{e}_{(\nu)}\right) = g_{\mu\nu}v^{\mu}w^{\nu}. \tag{9.17}$$

where we have adopted the summation convention in the final version. Note that owing to the symmetry of g_{ab} one will find $\mathbf{v} \cdot \mathbf{w} = \mathbf{w} \cdot \mathbf{v}$. This is different from the corresponding result in quantum physics, where one is dealing with vectors having complex-number-valued components.

In the discussion so far the vectors and the dual objects have been separate collections of mathematical objects, and therefore we have not yet defined any way in which the quantity $\tilde{\mathbf{v}} \cdot \mathbf{w}$ may be related to the quantity $\mathbf{v} \cdot \mathbf{w}$. The next step is to introduce a definition whereby these two quantities shall be equal.

Compare (9.7) with (9.17), which we display here together (after relabelling dummy indices to bring out the comparison):

$$\tilde{\mathbf{v}} \cdot \mathbf{w} = v_{\mu}w^{\mu},$$
$$\mathbf{v} \cdot \mathbf{w} = g_{\lambda\mu}v^{\lambda}w^{\mu}.$$

We now agree to make a unique association between each vector $\mathbf{v}$ and a dual object $\tilde{\mathbf{v}}$, in such a way that these two expressions give the same answer for all $\mathbf{w}$. This is achieved by picking $\tilde{\mathbf{v}}$, for any given $\mathbf{v}$, such that

$$v_{\mu} = g_{\lambda\mu}v^{\lambda} \tag{9.18}$$

We thus arrange that the metric tensor performs the job of index lowering. By this choice of $\tilde{\mathbf{v}}$ we have ensured that $\mathbf{v} \cdot \mathbf{w} = \tilde{\mathbf{v}} \cdot \mathbf{w}$. One can then rapidly prove that we also have

$$\mathbf{v} \cdot \mathbf{w} = \mathbf{w} \cdot \mathbf{v} = \tilde{\mathbf{w}} \cdot \mathbf{v}. \tag{9.19}$$

In other words, the inner product between vectors is commutative, and one can use the dual of either to evaluate it.

Eqns (9.8), (9.9) can now be written

$$v^a = \mathbf{v} \cdot \tilde{\boldsymbol{\theta}}^{(a)}, \qquad v_a = \mathbf{v} \cdot \mathbf{e}_{(a)}. \tag{9.20}$$

If one repeats the argument in terms of the dual basis to the coordinate basis then one finds

$$\tilde{\boldsymbol{\theta}}^{(a)} \cdot \tilde{\boldsymbol{\theta}}^{(b)} = g^{ab} \tag{9.21}$$

where $[g^{ab}]$ is the matrix inverse of $[g_{ab}]$. It follows that

$$\mathbf{v} \cdot \mathbf{w} = g^{\mu\nu} v_\mu w_\nu. \tag{9.22}$$

Eqns (9.7), (9.17) and (9.22) give three different ways of writing the same quantity. Since each way is valid for general vectors $\mathbf{v}$, $\mathbf{w}$ one can use these expressions to relate the contravariant and covariant components directly to one another via the metric:

$$v_a = g_{a\lambda} v^\lambda, \qquad v^a = g^{a\lambda} v_\lambda. \tag{9.23}$$

Thus g_{ab} acts to lower an index, and g^{ab} acts to raise an index, as we claimed at the beginning of Chapter 5.

The **norm** or length of a vector is defined as

$$\sqrt{|\mathbf{v} \cdot \mathbf{v}|} = \sqrt{|v_\lambda v^\lambda|}. \tag{9.24}$$

However, in relativity often we do not want to lose the information about the sign of $\mathbf{v} \cdot \mathbf{v}$, and therefore the word 'norm' or 'magnitude' is commonly used interchangeably for either of $v_\lambda v^\lambda$ or $|v_\lambda v^\lambda|^{1/2}$. The angle θ between one vector and another is defined through

$$\mathbf{v} \cdot \mathbf{w} = |v_\mu v^\mu|^{1/2} |w_\nu w^\nu|^{1/2} \cos\theta. \tag{9.25}$$

Using the transformation properties (next section), or otherwise, one can readily prove that the inner product, length and angle are all scalar invariants, that is, independent of the coordinate system that may have been employed to calculate them (exercise 9.1).

(A parenthetical comment: Any given basis vector such as $\mathbf{e}_{(a)}$ does itself have a dual which could be written $\tilde{\mathbf{e}}_{(a)}$ (this should be read as 'the dual of the vector $\mathbf{e}_{(a)}$') but be careful: this is not necessarily itself a member of the dual basis! Rather, it can be expressed as a linear combination of the members of the dual basis. In other words, $\tilde{\mathbf{e}}_{(a)} \neq \tilde{\boldsymbol{\theta}}^{(a)}$. In practice it is never necessary, and always confusing, to consider $\tilde{\mathbf{e}}_{(a)}$, so that is the last time we will ever write or refer to that quantity in this book[3].)

9.1.3 Coordinate transformation for vectors

Suppose we have two different coordinate systems x^a and x'^a, each with their associated sets of coordinate basis vectors $\mathbf{e}_{(a)}$, $\mathbf{e}'_{(a)}$. A given infinitesimal displacement $d\mathbf{s}$ can be written in either of two ways:

$$d\mathbf{s} = dx^\lambda \mathbf{e}_{(\lambda)} = dx'^\lambda \mathbf{e}'_{(\lambda)}. \tag{9.26}$$

By dividing this equation by one of the dx'^a while holding the others constant one finds

---

Transformation of coordinate basis vectors

$$\mathbf{e}'_{(a)} = \frac{\partial x^\lambda}{\partial x'^a} \mathbf{e}_{(\lambda)} \tag{9.27}$$

We also have, for some arbitrary vector $\mathbf{v}$,

$$\mathbf{v} \cdot \mathbf{e}'_{(a)} = v'_a. \tag{9.28}$$

Therefore

$$v'_a = \mathbf{v} \cdot \mathbf{e}'_{(a)} = \mathbf{v} \cdot \left(\frac{\partial x^\lambda}{\partial x'^a} \mathbf{e}_{(\lambda)} \right) = \frac{\partial x^\lambda}{\partial x'^a} \mathbf{v} \cdot \mathbf{e}_{(\lambda)} = \frac{\partial x^\lambda}{\partial x'^a} v_\lambda \tag{9.29}$$

and after making a similar argument for v'^a one finds:

Transformation of vector components

$$v'^a = \frac{\partial x'^a}{\partial x^\lambda} v^\lambda, \qquad v'_a = \frac{\partial x^\lambda}{\partial x'^a} v_\lambda. \tag{9.30}$$

Example 9.1 Suppose coordinates are changed from x^a to $x'^a = 2x^a$. Assuming the coordinate basis has been adopted, how do the two types of vector components change?
Solution.
We have $\partial x'^a / \partial x^b = 2\delta^a_b$ and $\partial x^a / \partial x'^b = (1/2)\delta^a_b$. Therefore $v'^a = 2v^a$ and $v'_a = (1/2)v_a$—c.f. Fig. 9.2.

The way to remember equations (9.30) is to notice that the prime on the right-hand side nestles alongside the free index, not the dummy index, and after a little thought one realizes that it must be so. The rules of index notation take care of the placement of the other parts of the equation.

---

[3]Well, if you really want to know: $\tilde{\mathbf{e}}_{(a)} = g_{a\lambda}\tilde{\boldsymbol{\theta}}^\lambda$, but this is not quite the same as index lowering. Rather, it says 'the components of $\tilde{\mathbf{e}}_{(a)}$ in the dual basis are given by the numbers in the a'th row (or column) of g_{ab}'. (Meanwhile $\tilde{\boldsymbol{\theta}}^{(a)} = \delta^a_\lambda \tilde{\boldsymbol{\theta}}^{(\lambda)}$ so the components of $\tilde{\boldsymbol{\theta}}^{(a)}$ in the dual basis are δ^a_b, not g_{ab}, thus $\tilde{\mathbf{e}}_{(a)} \neq \tilde{\boldsymbol{\theta}}^{(a)}$.)

> **Where does the prime go?**
> In both Special and General Relativity one is interested in the effect of a change of
> reference frame. In the case of Lorentz transformation, in Volume 1 the notation $\mathsf{P}' = \Lambda\mathsf{P}$
> was adopted, as part of a general strategy to avoid the use of indices by making use
> of the simple and widely known methods of matrix algebra. Such a statement should
> be understood as an assertion about the components of a given 4-vector, written in an
> index-free notation. In the present volume we use the bold font for vectors and we only
> ever refer to components via indices. To describe the effect of a Lorentz transformation
> we write $p'^a = \Lambda^a{}_\mu p^\mu$. In this notation the effect on the vector itself is written $\mathbf{p} = \mathbf{p}$;
> that is, the vector does not change at all. The two notations are consistent if one notes
> that P refers not directly to $\mathbf{p}$ but rather to the list of components p^a. See Chapter 1
> for further information.

9.2 An example: plane polar coordinates

We shall now illustrate many of the ideas of this and the previous chapter by presenting the
example of the flat plane treated in two coordinate systems: Cartesian coordinates and plane
polar coordinates, see Fig. 9.3.

The Cartesian coordinates are x, y; the plane polar coordinates are r, ϕ. The two systems of
coordinates are related by

$$x = r\cos\phi, \quad y = r\sin\phi \tag{9.31}$$

and

$$r = (x^2 + y^2)^{1/2}, \quad \phi = \tan^{-1}(y/x). \tag{9.32}$$

The transformation matrices are therefore

$$\begin{pmatrix} \frac{\partial r}{\partial x} & \frac{\partial r}{\partial y} \\ \frac{\partial \phi}{\partial x} & \frac{\partial \phi}{\partial y} \end{pmatrix} = \begin{pmatrix} \cos\phi & \sin\phi \\ (-\sin\phi)/r & (\cos\phi)/r \end{pmatrix} \quad \text{and} \quad \begin{pmatrix} \frac{\partial x}{\partial r} & \frac{\partial x}{\partial \phi} \\ \frac{\partial y}{\partial r} & \frac{\partial y}{\partial \phi} \end{pmatrix} = \begin{pmatrix} \cos\phi & -r\sin\phi \\ \sin\phi & r\cos\phi \end{pmatrix}. \tag{9.33}$$

Let $\mathbf{e}_x$, $\mathbf{e}_y$ be the coordinate basis vectors in the Cartesian system. We would like to obtain the
coordinate basis vectors in the polar system in terms of these, not by thinking about Fig. 9.3
but by blindly applying the formulae of this chapter and turning the crank. So we first write
down a general small displacement:

$$\mathrm{d}\mathbf{s} = \mathrm{d}x\,\mathbf{e}_x + \mathrm{d}y\,\mathbf{e}_y \tag{9.34}$$

and then we use the transformation matrices to write $\mathrm{d}x$, $\mathrm{d}y$ in the polar system:

$$\begin{pmatrix} \mathrm{d}x \\ \mathrm{d}y \end{pmatrix} = \begin{pmatrix} \cos\phi & -r\sin\phi \\ \sin\phi & r\cos\phi \end{pmatrix}\begin{pmatrix} \mathrm{d}r \\ \mathrm{d}\phi \end{pmatrix} = \begin{pmatrix} \cos\phi\,\mathrm{d}r - r\sin\phi\,\mathrm{d}\phi \\ \sin\phi\,\mathrm{d}r + r\cos\phi\,\mathrm{d}\phi \end{pmatrix}$$

$$\Rightarrow \quad \mathrm{d}\mathbf{s} = (\cos\phi\,\mathrm{d}r - r\sin\phi\,\mathrm{d}\phi)\mathbf{e}_x + (\sin\phi\,\mathrm{d}r + r\cos\phi\,\mathrm{d}\phi)\mathbf{e}_y. \tag{9.35}$$

Now the basis vectors in the polar system can be found by applying the definition (9.11):

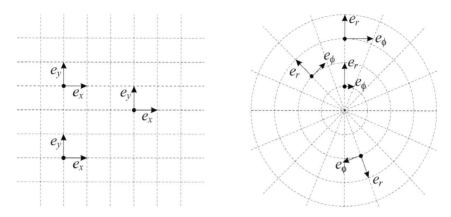

Fig. 9.3 Two coordinate systems and their basis vectors.

$$\mathbf{e}_r \equiv \frac{\partial \mathbf{s}}{\partial r} = \cos \phi \, \mathbf{e}_x + \sin \phi \, \mathbf{e}_y, \tag{9.36}$$

$$\mathbf{e}_\phi \equiv \frac{\partial \mathbf{s}}{\partial \phi} = -r \sin \phi \, \mathbf{e}_x + r \cos \phi \, \mathbf{e}_y. \tag{9.37}$$

Note that these need not be, and in this example are not, normalized. *They are* not *the same vectors as the ones often used when dealing with polar coordinates in elementary physics.*[4] You can confirm that these results also satisfy eqn (9.27) by using the transpose of (9.33).

By using $g_{ab} = \mathbf{e}_a \cdot \mathbf{e}_b$, we find that the metric tensors for the two coordinate systems are

$$[g'_{ab}] = \begin{pmatrix} 1 & 0 \\ 0 & 1 \end{pmatrix}, \qquad\qquad [g_{ab}] = \begin{pmatrix} 1 & 0 \\ 0 & r^2 \end{pmatrix}. \tag{9.38}$$

where the prime distinguishes the Cartesian from the polar coordinates. From this we find

$$\mathrm{d}s^2 = g'_{ab}\mathrm{d}x'^a\mathrm{d}x'^b = \mathrm{d}x^2 + \mathrm{d}y^2 \qquad \text{and} \qquad \mathrm{d}s^2 = g_{ab}\mathrm{d}x^a\mathrm{d}x^b = \mathrm{d}r^2 + r^2\mathrm{d}\phi^2. \tag{9.39}$$

Equally, one could use (9.35) to find these expressions, and from them deduce the components of the metric tensor.

By finding the inverses of the matrices displayed in (9.38) we obtain

$$[g'^{ab}] = \begin{pmatrix} 1 & 0 \\ 0 & 1 \end{pmatrix}, \qquad\qquad [g^{ab}] = \begin{pmatrix} 1 & 0 \\ 0 & 1/r^2 \end{pmatrix}. \tag{9.40}$$

The partial derivative operators with lower index are, by definition, simply derivatives with respect to one coordinate or another:

---

[4]The vector $\hat{\mathbf{e}}_\phi \equiv -\sin \phi \, \mathbf{e}_x + \cos \phi \, \mathbf{e}_y$ cannot be expressed as $(\partial \mathbf{s}/\partial v)$ for any coordinate v; vectors of this type can be used as basis vectors but then one has a *non-coordinate basis* or *non-holonomic basis* and many of the results applicable to coordinate bases do not apply.

$$\partial_r = \frac{\partial}{\partial r}, \qquad \partial_\phi = \frac{\partial}{\partial \phi}. \tag{9.41}$$

This is not so for the contravariant versions, which involve the metric:

$$\partial^r \equiv g^{r\lambda}\partial_\lambda = \frac{\partial}{\partial r}, \qquad \partial^\phi \equiv g^{\phi\lambda}\partial_\lambda = \frac{1}{r^2}\frac{\partial}{\partial \phi}. \tag{9.42}$$

It is commonly stated that 'partial derivatives commute', and this is true, but it does not necessarily imply that ∂^a operators commute, and usually they do not. Since $g^{a\mu}\partial_\mu f \neq \partial_\mu(g^{a\mu}f)$ for most metrics, one should avoid writing ∂^a if there is any danger of ambiguity.

Exercises

9.1 Use (9.30) to show that for any pair of vectors $\mathbf{v}$, $\mathbf{w}$, their inner $\mathbf{v} \cdot \mathbf{w}$ product is invariant (the same in all frames).

9.2 Find the coordinate basis vectors of the primed coordinate system introduced in exercise (8.4) of the previous chapter, and use (9.15) to obtain the metric tensor in the primed basis.

9.3 Two different manifolds $\mathcal{M}^A$, $\mathcal{M}^B$ are mapped by the same coordinates x^a. That is, for each point x^a in $\mathcal{M}^A$ there is a point in $\mathcal{M}^B$ having those same coordinates. If the two metrics are related as $g^{(A)}_{ab} = \Omega^2(x)g^{(B)}_{ab}$ (a conformal transformation) for some function Ω then prove that angles between coordinate lines agree at corresponding points in the two manifolds (c.f. exercise 8.7).

9.4 Verify that in the present chapter at no stage has a vector defined at a given point P in a manifold been added to, compared with, or in any way related to a vector defined at some other point in the manifold.

10

The affine connection

The business of the previous chapter has been to show that familiar properties of vectors, such as length, inner product and angle, carry over reasonably straightforwardly to general Riemannian geometry. We have also introduced the dual role of the metric, which both measures the manifold, and also relates contravariant and covariant components to one another. However, up till now each vector combination has involved vectors located at the same point in a manifold. We do not yet know how to relate a vector at one point to a vector at another point. In the present chapter we shall explore how vectors are related to nearby vectors, and from this one can develop the ideas of parallel transport and geodesics. The mathematical object which, again and again, will do the work is the set of *connection coefficients* Γ^a_{bc} which was briefly introduced in Chapter 2. In the present chapter we will define these coefficients as a statement about coordinate basis vectors, eqn (10.3), and then derive eqn (2.8).

10.1 Connection coefficients and covariant derivative

The question before us is, given some vector $\mathbf{u}$ at point P, and some vector $\mathbf{v}$ at nearby point Q, in what sense can we define a difference vector—that is, the amount that would have to be added to $\mathbf{u}$ in order to make $\mathbf{v}$? If both vectors were at the same point, then this would be easy: the difference would then be $\mathbf{v} - \mathbf{u}$. But since they are not at the same point, that is not the answer to our question. We answer our question by first finding a vector $\mathbf{w}$ at Q, which we arrange to be in some sense 'equal to' $\mathbf{u}$ (see Fig. 10.1), even though $\mathbf{u}$ is at P. Then the difference vector between $\mathbf{v}$ and $\mathbf{u}$ is $\mathbf{v} - \mathbf{w}$ (both vectors at Q).

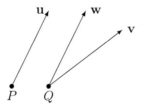

Fig. 10.1 Vectors at neighbouring points.

Relativity Made Relatively Easy: General Relativity and Cosmology. Volume 2. Andrew M. Steane, Oxford University Press. © Andrew M. Steane 2021. DOI: 10.1093/oso/9780192895646.003.0010

In order to make this precise, it is helpful to think about a *vector field*. In the previous paragraph we used different letters **u**, **v** to refer to vectors at different points P, Q. But now we want to discuss the notion of a smooth set of vectors distributed over the manifold, one at each point. For a physical example, you could think of the current density 4-vector of a flow, or the 4-vector potential of electromagnetism. We use the notation $\mathbf{v}(x)$ to signify such a vector field, with x indicating that the vector is a function of all the coordinates. Thus we have

$$\mathbf{v}(x) = v^\lambda \mathbf{e}_{(\lambda)} = v^\lambda(x)\mathbf{e}_{(\lambda)}(x) \tag{10.1}$$

where the second version makes explicit the coordinate dependence. Using the ordinary rules of differentiation, we then expect to find

$$\frac{\partial \mathbf{v}}{\partial x^c} = \frac{\partial v^\lambda}{\partial x^c}\mathbf{e}_{(\lambda)} + v^\lambda \frac{\partial \mathbf{e}_{(\lambda)}}{\partial x^c}. \tag{10.2}$$

It is the second term on the right-hand side that we now need to explore, because it expresses the notion of a change in a vector as one moves from one place to another. What we need is a way to write that change in terms of the basis vectors at the point where the derivative is being evaluated. In principle any vector can be written as a linear combination of basis vectors, so we can write:

The coefficients of the affine connection

$$\frac{\partial \mathbf{e}_{(b)}}{\partial x^c} = \Gamma^\mu_{bc}\mathbf{e}_{(\mu)} \tag{10.3}$$

This is one way to define the connection coefficients Γ^a_{bc}—they are the coefficients in the linear expansion of $\partial_c\mathbf{e}_{(b)}$ in terms of $\mathbf{e}_{(a)}$. However, the astute reader will notice that at this stage we are using them merely as a convenient way to postpone taking a definitive decision on what is meant by $\partial_c\mathbf{e}_{(b)}$.

Using the connection coefficients, (10.2) is written

$$\partial_c\mathbf{v} = (\partial_c v^\lambda)\mathbf{e}_{(\lambda)} + v^\lambda\Gamma^\mu_{\lambda c}\mathbf{e}_{(\mu)}. \tag{10.4}$$

In the last term on the right-hand side, both λ and μ are dummy indices, so we can swap them, with the aim of making it possible to factor out $\mathbf{e}_{(\lambda)}$:

$$\partial_c\mathbf{v} = (\partial_c v^\lambda)\mathbf{e}_{(\lambda)} + v^\mu\Gamma^\lambda_{\mu c}\mathbf{e}_{(\lambda)} = \left(\partial_c v^\lambda + v^\mu\Gamma^\lambda_{\mu c}\right)\mathbf{e}_{(\lambda)}. \tag{10.5}$$

We now have an expression for the derivative of the vector field **v** *as a linear combination of the basis vectors* $\mathbf{e}_{(a)}$. We can therefore extract the term in the brackets and announce that it tells us what happens to the covariant components of a vector when we take the gradient of the vector. This term is called the *covariant derivative* of the vector components, and is written:

Covariant derivative

$$\nabla_c v^a \equiv \partial_c v^a + \Gamma^a_{\mu c}v^\mu. \tag{10.6}$$

The reader should look long and hard at this expression and the expression (10.5). The first gives the partial derivative of a vector field, the second gives the covariant derivative of the contravariant components of a vector field. They are two entirely equivalent ways of giving the same mathematical information:

$$\partial_c \mathbf{v} \equiv (\nabla_c v^\lambda) \mathbf{e}_{(\lambda)}. \tag{10.7}$$

In practice, (10.6) enters into calculations much more often than (10.3), and you can, if you like, think of (10.6) as the prior idea, and then (10.3) is derived. In this way of proceeding, one begins by asserting (10.7). This equation introduces ∇_c as a way of expressing the notion of a derivative of a vector, but still, at this stage, without fully specifying what $\partial_c \mathbf{v}$ means. But the equation has content because it asserts linearity: it says that $\partial_c \mathbf{v}$ is defined to be a quantity which can be expressed as a sum of $\mathbf{e}_{(\lambda)}$, with the terms in the sum obtainable from some sort of operator applied to the components. Eqn (10.6) then does some further work by stating that this operation consists of a partial derivative added to a linear combination of the components. These definitions suffice to determine, in terms of the symbols, how differential operators act on other things such as sums of vectors, and now (10.3) can be derived.

By differentiating (9.4) we obtain

$$(\partial_c \tilde{\boldsymbol{\theta}}^{(a)}) \cdot \mathbf{e}_{(b)} = -\tilde{\boldsymbol{\theta}}^{(a)} \cdot (\partial_c \mathbf{e}_{(b)}) = -\tilde{\boldsymbol{\theta}}^{(a)} \cdot \Gamma^\mu_{bc} \mathbf{e}_\mu = -\Gamma^\mu_{bc} \delta^a_\mu = -\Gamma^a_{bc}, \tag{10.8}$$

and by starting from $\tilde{\mathbf{v}}(x) = v_\lambda \tilde{\boldsymbol{\theta}}^{(\lambda)}$ one then finds an expression for the components of the gradient of a dual vector (one-form) field, which can be written

$$\boxed{\nabla_c v_a \equiv \partial_b v_a - \Gamma^\lambda_{ac} v_\lambda. \tag{10.9}}$$

To help remember the important expressions (10.6) and (10.9) note that in each case the index with respect to which the derivative is being taken is the last index on the Christoffel symbol (c in this case), and the case with the superscript index gets the **p**ositive sign.

Notice that the difference between the covariant derivative and the partial derivative of the components arises from the second term in (10.2), which is non-zero because the coordinate basis vectors may themselves change from point to point if curvilinear coordinates are being used. The expression (10.6) for the covariant derivative has fallen out quite naturally, because we defined Γ^a_{bc} directly in terms of the change in the basis vectors, eqn (10.3). We have not yet defined what we mean by the change in the basis vectors. This is our next task.

10.1.1 Relating Γ^a_{bc} to g_{ab}

To complete the definition of what we mean by a change in a basis vector between one point and a neighbouring point, all we need to do is take the derivative of $\mathbf{e}_{(a)} \cdot \mathbf{e}_{(b)} = g_{ab}$ (eqn (9.15)) and reason cleverly:

An insight into the covariant derivative. We will present an argument in one dimension in order to get insight into the overall form of (10.6). We shall treat a vector field **v** that may have more than one dimension, and study its behaviour as a function of just one coordinate: x. To have a concrete example in mind, suppose the vector field has a size the same everywhere, and a direction along the x direction at each point. How will this constant field **v** be expressed in components v^a? It will depend on the coordinate system. The x coordinate is like a ruler. If the marks on the ruler are equally spaced then we shall find v^a is independent of x, so $\mathrm{d}v^a/\mathrm{d}x = 0$. If some other coordinate system x' is so defined that those same ruler marks are not equally spaced in terms of x', then it must be that the components v'^a *do* depend on x', so $\mathrm{d}v'^a/\mathrm{d}x \neq 0$. This illustrates the fact that the derivative of the components does not, on its own, tell us what is happening to the vector field itself.

In seeking a covariant derivative, what we seek is a way of writing the change in v^a such that it makes allowance for the changing scale of one coordinate system or another. That scale is indicated by the metric. So let us apply the above argument to the metric G (blurring the distinction, for a moment, between vector and tensor). We want to find a type of derivative such that when it is applied to the metric itself, it will say 'here is the change in the metric, after allowing for the change in the metric'. In other words, *the covariant derivative of the metric should be zero.* Thus the derivative we seek has the form (in a loose notation)

$$\nabla \approx \frac{\mathrm{d}}{\mathrm{d}x} - G^{-1}\frac{\mathrm{d}G}{\mathrm{d}x}$$

since this operator when applied to G gives zero. Eqns (10.9) and (10.15) are the more precise statement of this idea, and we will confirm in the next chapter that the covariant derivative of the metric is indeed zero.

$$
\begin{aligned}
\partial_c g_{ab} &= (\partial_c \mathbf{e}_{(a)}) \cdot \mathbf{e}_{(b)} + \mathbf{e}_{(a)} \cdot (\partial_c \mathbf{e}_{(b)}) \\
&= \Gamma^\lambda_{ac}\mathbf{e}_{(\lambda)} \cdot \mathbf{e}_{(b)} + \mathbf{e}_{(a)} \cdot \Gamma^\lambda_{bc}\mathbf{e}_{(\lambda)} \\
&= \Gamma^\lambda_{ac}\, g_{\lambda b} + \Gamma^\lambda_{bc}\, g_{a\lambda}.
\end{aligned}
\tag{10.11}
$$

Now the clever part is first to cyclically permute the indices a, b, c so as to obtain the equivalent expressions:

$$
\begin{aligned}
\partial_a g_{bc} &= \Gamma^\lambda_{ba}\, g_{\lambda c} + \Gamma^\lambda_{ca}\, g_{b\lambda}, \\
\partial_b g_{ca} &= \Gamma^\lambda_{cb}\, g_{\lambda a} + \Gamma^\lambda_{ab}\, g_{c\lambda}
\end{aligned}
$$

and then form the combination

$$\partial_c g_{ab} + \partial_b g_{ca} - \partial_a g_{bc} = \Gamma^\lambda_{ac}\, g_{\lambda b} + \Gamma^\lambda_{bc}\, g_{a\lambda} + \Gamma^\lambda_{cb}\, g_{\lambda a} + \Gamma^\lambda_{ab}\, g_{c\lambda} - \Gamma^\lambda_{ba}\, g_{\lambda c} - \Gamma^\lambda_{ca}\, g_{b\lambda}. \tag{10.12}$$

Looking carefully at the indices, we observe that the terms in $g_{a\lambda}$ add, and the other terms cancel, if the connection is symmetric in its lower indices (also using the fact that the metric is

The concept of *covariant derivative* also arises in other areas of physics. In quantum mechanics, we wish to treat a particle of charge q that interacts with an electromagnetic field. If we introduce a gauge change then the electromagnetic vector potential changes as $\tilde{A}_a = A_a + \partial_a \chi$ and the wavefunction ψ picks up a phase factor, $\tilde{\psi} = e^{i\alpha}\psi$ where $\alpha = -q\chi/\hbar$. This is consistent with the idea that no observable effects should result if we consider effects dependent only on $|\psi|^2$. However, when considering the gradient of ψ, which appears in momentum calculations for example, then we need to investigate further. We have

$$\partial_a \tilde{\psi} = e^{i\alpha}\partial_a \psi + (i\partial_a \alpha)e^{i\alpha}\psi$$

and $\partial_a \alpha = -(q/\hbar)\partial_a \chi = q(A_a - \tilde{A}_a)/\hbar$, therefore

$$\partial_a \tilde{\psi} = e^{i\alpha}\left(\partial_a \psi + (iq/\hbar)(A_a - \tilde{A}_a)\right)\psi.$$

This is not the required behaviour. Now introduce the covariant derivative

$$\nabla_a \equiv \partial_a + iqA_a/\hbar, \qquad \tilde{\nabla}_a = \partial_a + iq\tilde{A}_a/\hbar. \qquad (10.10)$$

We have

$$\tilde{\nabla}_a \tilde{\psi} = e^{i\alpha}\left((i\partial_a \alpha)\psi + \partial_a \psi + iq\tilde{A}_a\psi/\hbar\right) = e^{i\alpha}\left(\partial_a + iqA_a/\hbar\right)\psi$$
$$= e^{i\alpha}\nabla_a \psi.$$

This is the required behaviour. For example, if ψ is an eigenstate of ∇_a with eigenvalue λ then $\nabla_a \psi = \lambda\psi$ and we find $\tilde{\nabla}_a \tilde{\psi} = \lambda e^{i\alpha}\psi = \lambda\tilde{\psi}$, so the state after the gauge change is also an eigenstate and with the same eigenvalue. This means that ∇_a not ∂_a should be recognized as the operator which relates to such physical observables as momentum.

always symmetric, (8.18) and following). But is the connection symmetric in its lower indices? One can define a quantity

$$\mathcal{T}^a_{bc} \equiv \Gamma^a_{bc} - \Gamma^a_{cb}. \qquad (10.13)$$

It is shown in exercise 15.2 of Chapter 15 that this is a tensor (of rank 3) and it is called the *torsion tensor*. When one wishes to explore manifolds and connections in general, one will wish to include the case where $\mathcal{T} \neq 0$. However, in the present instance we are engaged in a mathematical task which involves elements of derivation and elements of definition at the same time. We derived (10.12) from our earlier definitions, and we are now free to choose whether or not the connection shall be symmetric in its lower indices. Both choices are consistent with what has been assumed so far. If one now makes the choice that $\Gamma^a_{bc} = \Gamma^a_{cb}$ then the connection is said to be *torsionless* or *torsion-free*, and it is called the *Levi-Civita connection*. It is this connection which is assumed in the standard treatment of GR.[1] Having made this choice, Γ^a_{bc} is symmetric in its lower indices, and using this fact in (10.12) we obtain

---

[1] One can attempt more general GR-like theories by introducing a connection with torsion; however, no great generalization is attained that way, because one can also adopt a torsionless connection and then account for physical effects related to torsion by saying that there is a further field which couples to spin and mass like a torsion tensor.

> **Which is more basic, the connection or the metric?** When we think about spacetime for the purpose of studying GR and physics, it is natural to regard the metric as the primary property of spacetime in which to take an interest, and then to develop the connection and the relationship between them. However, when studying manifolds in general, it turns out that the connection is the more general idea, in that one can define a connection without the need to have a well-defined metric. The connection can be defined, for example, through the concept of parallel transport—see Chapter 13.
>
> In the text we show that, given a metric, there is a unique metric-compatible torsion-free connection. Given a torsion-free connection, there may or may not exist a metric that gives rise to it. But if there does exist such a metric, then it is unique up to a multiplicative constant, except in exceptional cases. One such exceptional case is the case of zero curvature.

$$\partial_c g_{ab} + \partial_b g_{ca} - \partial_a g_{bc} = 2\Gamma^\lambda_{bc}\, g_{a\lambda}. \tag{10.14}$$

Hence, after multiplying by g^{ad} to make δ^d_λ on the right, and then relabelling indices,

$$\Gamma^a_{bc} = \frac{1}{2}g^{a\lambda}(\partial_b g_{c\lambda} + \partial_c g_{\lambda b} - \partial_\lambda g_{bc}). \tag{10.15}$$

This is eqn (2.8). It is an important formula which will feature in many of the arguments to come. Notice that the use of a metric, and the decision to adopt a connection without torsion, have together resulted in a unique expression for Γ^a_{bc} in terms of derivatives of g_{ab}. This proves that for a given metric there is no choice about how to define the affine connection if we want it to be torsion-free: the expression for the coefficients is determined with no free parameters, and therefore the connection (of which these are the coefficients) is unique.

We have now completed the set of ideas and definitions needed to discuss gradients of vector fields, which was our goal. Mathematicians also take an interest in treating differential geometry in the absence of a metric. One can then show that many of the ideas still apply, especially the relationship between covariant derivative and connection coefficient, but now the connection coefficients can be defined in more than one way. The version we have adopted (and which is always adopted in GR unless it is explicitly stated otherwise) is called the *Levi-Civita connection*, as we already noted. Our derivation amounts to a demonstration that the Levi-Civita connection (also called Christoffel connection) is the unique connection that is both torsion-free and **metric compatible**, which means that it satisfies

$$\nabla_c\, g_{ab} = 0 \tag{10.16}$$

(a result we shall prove in Chapter 12). The word *connection* is sometimes used somewhat loosely to refer to the covariant derivative itself (see box). The coefficients Γ^a_{bc} of the Levi-Civita connection are also called *Christoffel symbols*. Throughout this book, 'Christoffel symbol' and 'connection coefficient' will be exact synonyms (except in the two places where, as here, we are explicitly addressing the fact that other mathematical connections can be defined).

Connection terminology

  The mathematical concept called *connection* is so named because it is a way of saying how one tangent space can be related or connected to another, so that we can define things like differentiation with respect to position in a consistent manner.

The term 'affine' in 'affine connection' could (perhaps more clearly) be replaced by the word 'linear'. A linear connection is one which relates tangent spaces at different points along a curve in a linear way; this is what affine connections do. The word 'affine' is from the Latin 'affinis', ('connected with'), like 'affinity'. In mathematics more generally, a transformation is called *affine* if it has certain properties. A parameter is called affine when it varies linearly with a suitable measure of distance along a curve.

Sometimes the word *connection* is used as a name for the symbol ∇ which is an index-free way to refer to a covariant derivative, and sometimes it is used to refer to the collection of coefficients Γ^a_{bc}. Either usage is a slight abuse of terminology.

To summarize, the standard terminology runs as follows.
Connection = mathematical rule whereby vectors in different tangent spaces can be related to one another
Affine connection = a connection having the properties

$$\nabla_{\phi\mathbf{x}}\mathbf{y} = \phi\nabla_{\mathbf{x}}\mathbf{y} \tag{10.17}$$
$$\nabla_{\mathbf{x}}(\phi\mathbf{y}) = (\nabla_{\mathbf{x}}\phi)\,\mathbf{y} + \phi\nabla_{\mathbf{x}}\mathbf{y} \tag{10.18}$$

where ϕ is a scalar field and $\mathbf{x}$, $\mathbf{y}$ are vector fields, and we are using the notation introduced in (10.33).
Levi-Civita connection = the torsion-free connection which preserves the metric, i.e. $\nabla\mathbf{g} = 0$ (such a connection can be found for any manifold which has a metric, and is unique for such a manifold) (may also be called Christoffel connection).
Christoffel symbols, also called *connection coefficients* = coefficients which, together with partial derivative operators, allow the Levi-Civita connection to be expressed in terms of vector components relative to some given coordinate frame.

By contracting (10.15) over a, b we have

$$\Gamma^\mu_{\mu c} = \tfrac{1}{2}g^{\mu\lambda}(\partial_\mu g_{c\lambda} + \partial_c g_{\lambda\mu} - \partial_\lambda g_{\mu c}) = \tfrac{1}{2}(\partial^\lambda g_{c\lambda} + g^{\mu\lambda}\partial_c g_{\lambda\mu} - \partial^\mu g_{c\mu})$$
$$= \tfrac{1}{2}g^{\lambda\mu}\partial_c g_{\lambda\mu}. \tag{10.19}$$

and we present a further simplification in the next chapter, eqn (11.37).

Another useful result is, for any vector $\mathbf{w}$ (exercise 10.3),

$$\Gamma^\mu_{\nu c}w_\mu w^\nu = \tfrac{1}{2}(\partial_c g_{\mu\nu})w^\mu w^\nu. \tag{10.20}$$

The quantities

$$\Gamma_{abc} \equiv g_{a\lambda}\Gamma^{\lambda}_{bc} \tag{10.21}$$

are called the *Christoffel symbols of the first kind*. Using (10.15) we have

$$\Gamma_{abc} = \frac{1}{2}(\partial_b g_{ca} + \partial_c g_{ab} - \partial_a g_{bc}) \tag{10.22}$$

and therefore

$$\partial_c g_{ab} = \Gamma_{abc} + \Gamma_{bac}. \tag{10.23}$$

This makes it easy to see that if the connection coefficients vanish then so do all the first derivatives of the metric.

Christoffel symbols for diagonal metric.   The case of a diagonal metric often arises because we usually seek coordinates where this happens, if they are available. In this case only the term where $\lambda = a$ is non-zero in (10.15) so we have

$$\Gamma^{a}_{bc} \overset{\overset{\text{diag}}{g}}{=} \frac{1}{2g_{aa}}(\partial_b g_{ca} + \partial_c g_{ab} - \partial_a g_{bc})$$

where g_{aa} is not a sum, but merely the a'th diagonal element. One then deduces that the only non-zero cases are

$$
\begin{aligned}
\Gamma^{a}_{bb} &\overset{\overset{\text{diag}}{g}}{=} -\frac{1}{2g_{aa}}\partial_a g_{bb} && a \neq b, \text{ no sum} \\
\Gamma^{a}_{ac} = \Gamma^{a}_{ca} &\overset{\overset{\text{diag}}{g}}{=} \frac{1}{2g_{aa}}\partial_c g_{aa} && \text{no sum, } c \neq a \text{ or } c = a
\end{aligned}
\tag{10.24}
$$

10.2   Differentiation along a curve

Suppose we have a vector **v** that has well-defined values at all points along a curve $x^a(u)$, where u is a parameter. For example, **v** could be the energy-momentum of a particle or the 4-spin of a particle, and the curve is the worldline. Or **v** could be the electromagnetic 4-vector potential, and we are interested in its value at events along some particular line. In this context one is commonly interested in the derivative of the vector along the curve. Such a derivative appears in the Euler–Lagrange equations, for example, and in the study of fluid flow. It is also, by definition, this kind of derivative that relates 4-acceleration to 4-velocity (discussed in the next chapter).

At any given point on the curve the vector can be written

$$\mathbf{v}(u) = v^a(u)\mathbf{e}_{(a)}(u) \tag{10.25}$$

where the $v^a(u)$ are the components at the point associated with the parameter value u, and $\mathbf{e}_{(a)}(u)$ are the coordinate basis vectors at that point. Differentiating along the curve, we obtain

$$\frac{d\mathbf{v}}{du} = \frac{dv^a}{du}\mathbf{e}_{(a)} + v^a\frac{d\mathbf{e}_{(a)}}{du} = \frac{dv^a}{du}\mathbf{e}_{(a)} + v^a\frac{\partial\mathbf{e}_{(a)}}{\partial x^\nu}\frac{dx^c}{du} \qquad (10.26)$$

where we used the chain rule. Using the definition of the affine connection (10.3) this can be written

$$\frac{d\mathbf{v}}{du} = \frac{dv^a}{du}\mathbf{e}_{(a)} + v^a\Gamma^\lambda_{ac}\mathbf{e}_{(\lambda)}\frac{dx^c}{du}. \qquad (10.27)$$

Now, just as in the argument for the covariant derivative, we may bring out a factor of $\mathbf{e}_{(a)}$ by swapping the dummy indices a and λ in the last term, obtaining

$$\frac{d\mathbf{v}}{du} = \left(\frac{dv^a}{du} + v^\lambda\Gamma^a_{\lambda c}\frac{dx^c}{du}\right)\mathbf{e}_{(a)} \qquad (10.28)$$

The term in the bracket is the set of contravariant components of the result, and it is called the *absolute derivative* or the *intrinsic derivative* along the curve:

Intrinsic derivative along curve $x^c(u)$

$$\frac{Dv^a}{du} \equiv \frac{dv^a}{du} + v^\lambda\Gamma^a_{\lambda c}\frac{dx^c}{du} \qquad (10.29)$$

(The notation Dv^a/Du may also be used.) By similar reasoning, you can show that the covariant components are given by

$$\frac{Dv_a}{du} \equiv \frac{dv_a}{du} - v_\lambda\Gamma^\lambda_{ac}\frac{dx^c}{du}. \qquad (10.30)$$

These results are thus easily obtained, but one should pause to reflect on eqn (10.29). The first term on the right-hand side is itself a rate of change along the curve, and it can be used to form the vector $(dv^a/du)\mathbf{e}_{(a)}$, but beware! that vector is not $d\mathbf{v}/du$: see (10.28).

By invoking the chain rule to rewrite the first term in the intrinsic derivative, one finds that (10.29) can also be written

$$\frac{Dv^a}{du} = (\nabla_c v^a)\frac{dx^c}{du} \qquad (10.31)$$

which leads to the convenient form

$$\frac{D}{du} = T^c\nabla_c \qquad (10.32)$$

---

where $T^c = dx^c/du$ is the tangent vector along the curve. Note that this version is only strictly valid when the vector field being operated on is well-defined off the curve as well as on it, since only then do its partial derivatives with respect to all the coordinates make sense. This will not be the case for particle properties such as spin, but it may be the case for a vector field such as the 4-vector potential or the flux of a continuous fluid.[2]

[2]The use of (10.32) can in principle be generalized to vector fields restricted to a subspace if one claims that, for a given vector field defined only on a subspace, a vector field can be defined throughout the larger space which matches the given field in the subspace.

Index-free notation.   We can now proceed to a natural abstraction: since (10.32) applies to a curve in any direction T^a, we introduce the notation

$$\nabla_{\mathbf{T}} \equiv T^c \nabla_c \tag{10.33}$$

and thus we have a notion of derivative which is liberated from the use of the coordinate basis or any particular basis. This notion is used sufficiently often in the subject that the symbol ∇ is often called 'the connection', and the covariant derivative can be regarded as the form under which the connection appears in one setting or another. In a similar way, the notation

$$\nabla \mathbf{t} \tag{10.34}$$

signifies *that tensor whose components are given by* $\nabla_b t^a$ *or* $\nabla_b t_a$ *or* $\nabla_c t^{ab}$ *etc., as the case may be* (we are here anticipating the notion of a tensor of any rank, to be discussed in Chapter 12). In order to be unambiguous, this notation relies on the fact that index raising and lowering commutes with covariant differentiation—which it does, because both the metric tensor and its inverse behave as constants under the action of ∇ (eqns (10.16) and (12.63)).

10.3   Extending the example: plane polar coordinates

We continue the example which was begun in Section 9.2. The example is a simple one in which the manifold has no curvature, but the methods will apply equally well to arbitrary Riemannian or pseudo-Riemannian manifolds.

Previously we obtained the coordinate basis vectors and the metric in the two coordinate systems illustrated in Fig. 9.3. Next we find the connection coefficients from the metric. There are 8 coefficients in all. For the rectangular system, they are all zero. For the polar system, using (10.24) we find (no implied summation)

$$\Gamma^r_{\phi\phi} = -\frac{1}{2}\partial_r r^2 = -r, \qquad \Gamma^r_{r\phi} = \Gamma^r_{\phi r} = 0,$$
$$\Gamma^\phi_{rr} = 0, \qquad\qquad \Gamma^\phi_{\phi r} = \Gamma^\phi_{r\phi} = \frac{1}{2r^2}\partial_r r^2 = \frac{1}{r} \tag{10.35}$$

and the others are all zero.

The covariant derivative is given by (10.6): $\nabla_b v^a = \partial_b v^a + \Gamma^a_{\lambda b} v^\lambda$. In the polar system we find

$$\nabla_r v^r = \partial_r v^r + \Gamma^r_{\lambda r} v^\lambda = \partial_r v^r \tag{10.36}$$
$$\nabla_r v^\phi = \partial_r v^\phi + \Gamma^\phi_{\lambda r} v^\lambda = \partial_r v^\phi + \frac{1}{r}v^\phi \tag{10.37}$$
$$\nabla_\phi v^r = \partial_\phi v^r + \Gamma^r_{\lambda\phi} v^\lambda = \partial_\phi v^r - r v^\phi \tag{10.38}$$
$$\nabla_\phi v^\phi = \partial_\phi v^\phi + \Gamma^\phi_{\lambda\phi} v^\lambda = \partial_\phi v^\phi + \frac{1}{r}v^r. \tag{10.39}$$

For example, the divergence of a vector field is given by

$$\boldsymbol{\nabla} \cdot \mathbf{v} = \nabla_r v^r + \nabla_\phi v^\phi = \partial_r v^r + \partial_\phi v^\phi + \frac{1}{r} v^r \;\; = \frac{1}{r} \frac{\partial}{\partial r} \left(r v^r \right) + \frac{\partial v^\phi}{\partial \phi}. \qquad (10.40)$$

If we apply this result to a vector field which is itself the gradient of a scalar field f, then we shall find the Laplacian operator. The covariant components of such a field are given by

$$v_r = \frac{\partial f}{\partial r}, \qquad v_\phi = \frac{\partial f}{\partial \phi} \qquad (10.41)$$

and therefore, using (9.40), the contravariant components are

$$v^r = \frac{\partial f}{\partial r}, \qquad v^\phi = \frac{1}{r^2} \frac{\partial f}{\partial \phi}, \qquad (10.42)$$

where we used the metric to raise the index. Hence, using (10.40) we obtain

$$\begin{aligned} \boldsymbol{\nabla} \cdot \boldsymbol{\nabla} f &= \frac{1}{r} \frac{\partial}{\partial r} \left(r \frac{\partial f}{\partial r} \right) + \frac{\partial}{\partial \phi} \left(\frac{1}{r^2} \frac{\partial f}{\partial \phi} \right) \\ &= \frac{1}{r} \frac{\partial}{\partial r} \left(r \frac{\partial f}{\partial r} \right) + \frac{1}{r^2} \frac{\partial^2 f}{\partial \phi^2}. \end{aligned} \qquad (10.43)$$

This gives the Laplacian operator in plane polar coordinates. It can also be derived by repeated use of the chain rule, but that turns out to involve a surprisingly lengthy sequence of steps. The connection coefficients simplify the calculation considerably, and this is even more so in three or more dimensions.

Exercises

10.1 Starting from (10.3), obtain the useful

$$\Gamma^a_{bc} = \tilde{\boldsymbol{\theta}}^{(a)} \cdot \partial_c \mathbf{e}_{(b)} = -\mathbf{e}_{(b)} \cdot \partial_c \tilde{\boldsymbol{\theta}}^{(a)} \quad (10.44)$$

$$\text{and } \partial_c \tilde{\boldsymbol{\theta}}^{(a)} = -\Gamma^a_{\lambda c} \tilde{\boldsymbol{\theta}}^{(\lambda)}. \quad (10.45)$$

10.2 Obtain (10.15) by developing $\partial_c g_{ab}$.

10.3 Prove (10.20) [hint: dummy indices]

10.4 Use (10.24) to obtain all the Christoffel symbols in the case of a metric of the form (6.6) with $W_i = 0$, $\Phi = \Phi(z)$.

10.5 Prove the following:

$$\partial_c g_{ab} = \Gamma_{abc} + \Gamma_{bac}$$

$$\partial_c g^{ab} = -\Gamma^a_{\lambda c} g^{\lambda b} - \Gamma^b_{\lambda c} g^{a\lambda}$$

$$g_{a\mu} \partial_c g^{\mu b} = -g^{\mu b} \partial_c g_{a\mu} \qquad (10.46)$$

10.6 For the case of a metric $ds^2 = g_{tt} dt^2 + g_{zz} dz^2 + dx^2 + dy^2$, where g_{tt} and g_{zz} depend only on z, show that the only non-zero Christoffel symbols are

$$\Gamma^z_{tt} = -\frac{1}{2 g_{zz}} \partial_z g_{tt}, \;\; \Gamma^z_{zz} = \frac{1}{2 g_{zz}} \partial_z g_{zz},$$

$$\Gamma^t_{tz} = \Gamma^t_{zt} = \frac{1}{2 g_{tt}} \partial_z g_{tt}. \qquad (10.47)$$

10.7 Consider the surface of a cylinder in Euclidean space. Verify that there exists a choice of coordinates in which all the connection coefficients vanish everywhere. (Method: guess coordinates; write down the metric; use (10.15).)

10.8 Work through appendix B, obtaining or verifying all the results.

10.9 Suppose a vector field has contravariant components everywhere equal to $(0, 1, 0, 0)$ in some coordinate system. Is the covariant derivative of such a vector field guaranteed to be zero? If not, then why not?

10.10 Obtain (10.24).

10.11 Obtain all the connection coefficients for the metric shown in (8.43). [*Ans.* $\Gamma^1_{00} = a$; others zero.]

10.12 Using (10.35), verify explicitly that $\nabla_r g_{ab} = 0$ and $\nabla_r g^{ab} = 0$ for the flat plane treated in polar coordinates.

10.13 Confirm that $\nabla_{\mathbf{e}_{(a)}} = \nabla_a$ (where the first version adopts the notation (10.33)).

10.14 Express $u^\mu w^\nu v^\lambda \nabla_\lambda T_{\mu\nu}$ and $(\nabla^\mu w^a)(\nabla_\lambda v_\mu) u^\lambda$ in index-free notation. [*Ans.* $\mathbf{u} \cdot (\nabla_{\mathbf{v}} \mathbf{T}) \cdot \mathbf{w}$; $\nabla_{(\nabla_{\mathbf{u}} \mathbf{v})} \mathbf{w}$]

10.15 Show that if the metrics of two different manifolds can be related by $\tilde{g}_{ab}(x) = \Omega^2(x) g_{ab}(x)$ (a conformal transformation) where Ω is some function and x are coordinates, then the connection coefficients are related by

$$\tilde{\Gamma}^a_{bc} = \Gamma^a_{bc} + (\delta^a_c \partial_b \Omega + \delta^a_b \partial_c \Omega - g_{bc} g^{a\lambda} \partial_\lambda \Omega)/\Omega. \tag{10.48}$$

11

Further useful ideas

In this chapter we introduce some physical ideas which will begin to ground the subject after the wealth of abstract ideas introduced in the previous three chapters. We define some 4-vectors useful to GR, and discover some of their relationships. Then we discuss the *tetrad*, and its use in connecting general formulae to physical observations in a LIF. We then define GR equivalents to such familiar friends as div, grad, curl.

11.1 Some physics related to 4-velocity

We are now able to define some standard concepts related to particle motion. If a massive particle of rest mass m has worldline $x^a(\tau)$, then the 4-velocity, 4-momentum and 4-acceleration are defined as

$$u^a \equiv \frac{\mathrm{d}x^a}{\mathrm{d}\tau} \tag{11.1}$$

$$p^a \equiv mu^a \qquad \text{but see also (13.36)} \tag{11.2}$$

$$a^a \equiv \frac{\mathrm{D}u^a}{\mathrm{d}\tau} = \frac{\mathrm{d}u^a}{\mathrm{d}\tau} + u^\mu \Gamma^a_{\mu\nu} \frac{\mathrm{d}x^\nu}{\mathrm{d}\tau} = \frac{\mathrm{d}u^a}{\mathrm{d}\tau} + \Gamma^a_{\mu\nu} u^\mu u^\nu \tag{11.3}$$

Notice that in this set of equations, the ordinary derivative (along the curve) of x^a is used to define the 4-velocity, but the intrinsic derivative of u^a is used to define the 4-acceleration. The reason is that the quantity we choose to call velocity is one whose direction is *along the worldline*; we say it is *tangential* to the worldline, which means that at any event the 4-vector u^a is in the same spacetime direction as $\mathrm{d}x^a$ between neighbouring events on the worldline. The quantity $\mathrm{D}x^a/\mathrm{d}\tau$, by contrast, has little or no physical meaning and amounts to a misuse of the mathematical ideas. It is a misuse because at any given event P on the worldline, the quantities x^a are the set of coordinates of the event, but they do not amount to components of a 4-vector defined at that event.[1] This is in contrast to $\mathrm{d}x^a$ which do form a 4-vector at P.

Coming now to 4-acceleration, it is the intrinsic derivative we want, because we wish to define 4-acceleration $\mathbf{a}$ such that $\mathbf{a} = \mathrm{d}\mathbf{u}/\mathrm{d}\tau$ where $\mathbf{u}$ is 4-velocity.

---

[1] Of course one can always define a quantity $\mathbf{x} \equiv x^a \mathbf{e}_{(a)}$ and thus announce that we have defined a 4-vector in the mathematical sense, but this 4-vector has no relevance to physics and only by an arbitrary assertion can it be said to be located at P.

Relativity Made Relatively Easy: General Relativity and Cosmology. Volume 2. Andrew M. Steane, Oxford University Press. © Andrew M. Steane 2021. DOI: 10.1093/oso/9780192895646.003.0011

Now let us evaluate the inner product between **a** and **u**:

$$u_\lambda a^\lambda = u_\lambda \frac{\mathrm{d}u^\lambda}{\mathrm{d}\tau} + u_\lambda \Gamma^\lambda_{\mu\nu} u^\mu u^\nu \tag{11.4}$$

It might seem as if we have our work cut out to simplify this equation, but not so! For, if we adopt a LIF for the purpose of evaluating this inner product, then we shall find both that the Christoffel symbols are zero and that the first term is zero, just as in Special Relativity. To make it really easy, do not adopt just any old LIF; adopt the instantaneous rest LIF, where u^a is purely temporal and a^a is purely spatial. Hence we find that

$$u_\lambda a^\lambda \stackrel{\mathrm{LIF}}{=} 0. \tag{11.5}$$

This is so in the chosen LIF, but since the equation is tensorial we must therefore get the same answer in any frame (i.e. in any coordinatization). Hence

$$u_\lambda a^\lambda = 0. \tag{11.6}$$

Note the logic here: both sides of this equation are scalar invariants, therefore it does indeed suffice to find the answer in one frame in order to find it in all frames. Such is the beauty of tensor methods!

Using the previous formula we have

$$\frac{\mathrm{d}}{\mathrm{d}\tau}(u_\lambda u^\lambda) = 2u_\lambda \frac{\mathrm{d}u^\lambda}{\mathrm{d}\tau} = 0 \tag{11.7}$$

where it is taken as understood that the derivative is along the worldline. We deduce that the length of **u** is constant, no matter what the worldline may be. You should be able to see that this is guaranteed from the way **u** is defined in terms of displacement along the worldline and the proper time. For the proper time is by definition the spacetime distance between the events whose separation is $\mathrm{d}x^a$. It is a useful fact to remember: *the 4-velocity has constant size in General Relativity, just like in Special Relativity; always and everywhere one finds*

$$u^\lambda u_\lambda = -c^2. \tag{11.8}$$

Another useful observation is: *the 4-acceleration is zero for a particle in free fall.* You can deduce this by comparing (11.3) with (2.28).

$$\frac{\mathrm{d}\mathbf{u}}{\mathrm{d}\tau} = 0 \qquad \text{free fall} \tag{11.9}$$

This can seem a strange sort of statement at first. 'In free fall we have acceleration, right?' No: wrong. *Free fall is inertial motion.* 'But how can my 4-velocity be unchanging as I fall? What does it even mean?' This is a legitimate question, and at the heart of it is the question of how we can compare a vector at one location with a vector at another location—that is, at one event or another in GR. The best way to make such a comparison, says differential geometry (and perhaps the only way that really makes sense), is to say that a vector changes if and only if it becomes different from a parallel-transported version of itself. But the equation of free-fall motion in GR is the equation for that type of motion in which the tangent vector is parallel-transported. Hence **a** = 0. This idea is presented more fully in Chapter 13.

11.2   Tetrads

Suppose that a particle of 4-momentum **p** is observed by an observer whose 4-velocity is **u**. If you draw on your familiarity with Special Relativity, you should be able to see that the energy of the particle, as observed by the observer, is given by

$$E = -\mathbf{p} \cdot \mathbf{u}. \tag{11.10}$$

This familiar special relativistic result carries over to GR because we can evaluate the right-hand side in any frame and get the same result, and in particular we can adopt a LIF in which the observer is momentarily at rest, and in this LIF Special Relativity applies and the result does indeed give the energy that the observer detects.

This idea is generalized by the use of *tetrads*.

A set of N orthornormal vectors in an N-dimensional manifold is called a *vielbein* (from the German for 'many legs') and when $N = 4$ this is called a *tetrad*. Tetrads are a very useful labour-saving device for figuring how things in spacetime appear to one observer or another.

For, suppose an observer F has some worldline, such that the observer's 4-velocity is **u**. We can construct a tetrad such that one of the vectors is along **u**, so it is timelike, and the others are orthogonal to this and to each other. It follows that they are spacelike and furthermore they are all in the direction of the plane which observer F considers to be the plane of simultaneity for events near to him in his own instantaneous rest frame. In short, such a tetrad is the coordinate basis of a LIF, and in particular we have constructed the tetrad for an instantaneous rest LIF of F. (We say 'an' not 'the' because there are a set of such frames, related to one another by spatial rotations.) Note that the tetrad is made of vectors not one-forms.

At any given event, the members of any given pair of tetrads are related to one another by Lorentz transformations.

Let $\hat{\mathbf{e}}_{(a)}$ be the members of such a tetrad for a given observer. If we wish to calculate what the observer finds in his LIF, we do not need to carry out coordinate transformations. The covariant components of some 4-vector **w**, as observed by F, will be

$$\left(\mathbf{w} \cdot \hat{\mathbf{e}}_{(0)}, \ \ \mathbf{w} \cdot \hat{\mathbf{e}}_{(1)}, \ \ \mathbf{w} \cdot \hat{\mathbf{e}}_{(2)}, \ \ \mathbf{w} \cdot \hat{\mathbf{e}}_{(3)} \right) \tag{11.11}$$

(c.f. eqn (9.20)). The first of these is calculated using

$$\mathbf{w} \cdot \hat{\mathbf{e}}_{(0)} = w_\lambda \hat{e}^\lambda_{(0)} \tag{11.12}$$

and similarly for the others. For example, if a particle of 4-momentum **p** is observed by an observer whose tetrad is $\{\hat{\mathbf{e}}_{(a)}\}$ then the observed 3-momentum is

$$p'_i = \mathbf{p} \cdot \hat{\mathbf{e}}_{(i)}$$

where the prime is there to indicate that these are the momentum components as observed in the primed frame, which we are taking as the frame of the observer.

Each member of the tetrad is a 4-vector and there is also a dual basis related to these vectors in the ordinary way:

$$\hat{\boldsymbol{\theta}}^{(a)} \cdot \hat{\mathbf{e}}_{(b)} = \delta^a_b \tag{11.13}$$

(c.f. eqn (9.4)). By construction, we have

$$\hat{\mathbf{e}}_{(a)} \cdot \hat{\mathbf{e}}_{(b)} = \eta_{ab} \tag{11.14}$$

which makes the set of one-forms of the dual basis easy to find: they are

$$\hat{\boldsymbol{\theta}}^{(a)} \; '=' \; \eta^{a\lambda} \hat{\mathbf{e}}_{(\lambda)} \tag{11.15}$$

since then

$$\hat{\boldsymbol{\theta}}^{(a)} \cdot \hat{\mathbf{e}}_{(b)} = \eta^{a\lambda} \hat{\mathbf{e}}_{(\lambda)} \cdot \hat{\mathbf{e}}_{(b)} = \eta^{a\lambda} \eta_{\lambda b} = \eta^a_b = \delta^a_b. \tag{11.16}$$

Strictly, one ought not to write (11.15) since it asserts that a one-form is equal to a sum of vectors, which abuses the distinction between a vector space and its dual. Equation (11.15) should be understood as a convenient shorthand or mnemonic for a result which can be more carefully expressed by using components:

$$\hat{\theta}^{(a)}_{\mu} = \eta^{a\lambda} \hat{e}_{(\lambda)\mu} = \eta^{a\lambda} g_{\mu\nu} \hat{e}^{\nu}_{(\lambda)}. \tag{11.17}$$

The main point is that if we use a prime to indicate components relative to the tetrad, then we have, for any $\mathbf{v}$:

The use of a tetrad

$$v'_a = \hat{e}^{\mu}_{(a)} v_{\mu},$$
$$v'^a = \hat{\theta}^{(a)}_{\mu} v^{\mu} = \eta^{a\lambda} \hat{e}_{(\lambda)\mu} v^{\mu} = \eta^{a\lambda} \hat{e}^{\mu}_{(\lambda)} v_{\mu} = \eta^{a\lambda} v'_{\lambda} \tag{11.18}$$

---

where in the second line the μ index performed a see-saw.

Be careful here: the members of the dual tetrad are not to be confused with either covariant or contravariant components of the tetrad vectors in the starting coordinate system. Those components should be related to one another, if one wanted to relate them, in the standard way using g_{ab} and g^{ab}.

Furnished with $\hat{\mathbf{e}}_{(a)}$ and $\hat{\boldsymbol{\theta}}^{(a)}$, we are now able to quickly identify how the covariant and contravariant components of any vector $\mathbf{w}$ will come out when calculated by our friend F: simply take inner products with $\hat{\mathbf{e}}_{(a)}$ for the former, and take inner products with with $\hat{\boldsymbol{\theta}}^{(a)}$ for the latter. In eqn (11.10), $\mathbf{u}$ is itself serving as the timelike vector in a tetrad, and the equation gives the zeroth component of the 4-momentum if it were to be evaluated in the rest frame of the detector whose 4-velocity is $\mathbf{u}$. To be precise, in this example we have $\mathbf{u} = c\hat{\mathbf{e}}_{(0)}$ and

$$E = p'^0 c = \mathbf{p} \cdot \hat{\boldsymbol{\theta}}^{(0)} c = -\mathbf{p} \cdot \hat{\mathbf{e}}_{(0)} c = -\mathbf{p} \cdot \mathbf{u}. \tag{11.19}$$

Let us try applying this procedure to the metric itself (anticipating a little the general notion of a tensor which is introduced in the next chapter). Suppose we have been using a coordinate system in which the metric g_{ab} is not equal to η_{ab} but suppose, for the sake of simplicity, that it so happens that g_{ab} is diagonal. Let us pick as our tetrad a set of vectors along the axes of the coordinate system (since a diagonal metric implies that these are mutually orthogonal). In order that the tetrad vectors are normalized, they must be (up to overall signs)

$$
\begin{aligned}
\hat{e}^a_{(0)} &= (1/\sqrt{-g_{00}},\ 0,\ 0,\ 0) \\
\hat{e}^a_{(1)} &= (0,\ 1/\sqrt{g_{11}},\ 0,\ 0) \\
\hat{e}^a_{(2)} &= (0,\ 0,\ 1/\sqrt{g_{22}},\ 0) \\
\hat{e}^a_{(3)} &= (0,\ 0,\ 0,\ 1/\sqrt{g_{33}})
\end{aligned}
\tag{11.20}
$$

where we have specified them by listing their contravariant components. Let us check the normalization:

$$
\hat{\mathbf{e}}_{(0)} \cdot \hat{\mathbf{e}}_{(0)} = \hat{e}^\mu_{(0)} g_{\mu\nu} \hat{e}^\nu_{(0)} = (-g_{00})^{-1/2} g_{00} (-g_{00})^{-1/2} = -1,
$$
$$
\hat{\mathbf{e}}_{(1)} \cdot \hat{\mathbf{e}}_{(1)} = \hat{e}^\mu_{(1)} g_{\mu\nu} \hat{e}^\nu_{(1)} = (g_{11})^{-1/2} g_{11} (g_{11})^{-1/2} = 1,
$$

and similarly for the others. Now, in the spirit of what has been said about vectors and their components in this chapter, let us take a look at how the metric will come out when projected onto the tetrad. Writing g'_{ab} for such a projection, we find:

$$
g'_{ab} = g_{\mu\nu} \hat{e}^\mu_{(a)} \hat{e}^\nu_{(b)}
\tag{11.21}
$$

and by carrying through the calculation you will find (exercise) that[2]

$$
g'_{ab} = \eta_{ab}.
\tag{11.22}
$$

This is precisely what we should have expected: it illustrates how the tetrad can be used as an alternative to coordinate transformations, and it illustrates the general idea of identifying how a given geometric quantity (here, the metric) 'looks' to one observer or another.

By facilitating a move from any coordinates to Minkowski coordinates and back, tetrads (and more generally vielbeins) play a significant role in more complex manipulations in differential geometry, but we shall not need to explore this usage. We will use tetrads mostly as we have introduced them in this section: as a way to answer questions about what will be observed using instruments whose behaviour is understood, in the first instance, in terms of physics as it applies in an instantaneous rest LIF of the instrument in question.

11.3 Vector operators

Next we develop some useful results related to the curved space generalization of div, grad, curl.

---

[2]In a text that has adopted a notation where Roman indices refer to one frame, and Greek indices to another, this result would be written without a prime, but we have not adopted such a notation.

First note that the gradient of a scalar field ϕ has components

$$\nabla_a \phi = \partial_a \phi. \tag{11.23}$$

Noting that this expression yields a set of covariant components, one observes that the gradient of a scalar field yields a one-form field not a vector field. That is,

$$(\text{gradient of } \phi) = (\partial_\lambda \phi)\, \tilde{\boldsymbol{\theta}}^{(\lambda)}. \tag{11.24}$$

A suitable index-free notation for this quantity could be $\tilde{\boldsymbol{\nabla}}\phi$, but another notation is widely employed in differential geometry: the notation $(\text{gradient of } \phi) = \mathrm{d}\phi$. This is done in the interests of uncluttered notation, but it can be confusing when the letter d is widely used for infinitesimals, so we shall not adopt it.

The statement

$$(\boldsymbol{\nabla}\phi)_a = \partial_a \phi \tag{11.25}$$

asserts that $(\boldsymbol{\nabla}\phi)$ is a field whose a'th covariant component is given by $\partial_a \phi$. To find the contravariant components, raise the index using the metric:

$$(\boldsymbol{\nabla}\phi)^a \equiv g^{a\lambda}(\boldsymbol{\nabla}\phi)_\lambda = g^{a\lambda}\partial_\lambda \phi \tag{11.26}$$

which we shall call $\nabla^a \phi$. It is important to note the placement of $g^{a\lambda}$ here; the result is not in general equal to $\partial_\lambda(g^{a\lambda}\phi)$. But it is equal to $\nabla_\lambda(g^{a\lambda}\phi)$.

Proceeding now to operators on vector fields, note the useful fact that

$$\nabla_a v_b - \nabla_b v_a = \partial_a v_b - \partial_b v_a \tag{11.27}$$

because the connection coefficients cancel in this expression (which you should confirm).[3] It makes sense to call this combination a curl, so we have

$$(\text{curl } \mathbf{v})_{ab} = (\boldsymbol{\nabla}\times\mathbf{v})_{ab} \equiv \nabla_a v_b - \nabla_b v_a = \partial_a v_b - \partial_b v_a. \tag{11.28}$$

The divergence can be defined

$$\boldsymbol{\nabla}\cdot\mathbf{v} \equiv \nabla_\lambda v^\lambda = \partial_\lambda v^\lambda + \Gamma^\lambda_{\mu\lambda} v^\mu \tag{11.29}$$

and this expression can be simplified by developing a simpler expression for $\Gamma^\lambda_{\lambda b}$. To this end, first let us recall two facts about determinants and matrix inverses:

$$g \equiv \det(g_{ab}) = \sum_\nu g_{a\nu}\Delta^{a\nu}, \qquad (\text{no sum over } a) \tag{11.30}$$

$$g^{ab} = \frac{1}{g}\Delta^{ba} \tag{11.31}$$

where Δ^{ab} is the cofactor of the element a, b of g_{ab}; it is equal to $(-1)^{a+b}$ times the determinant of the matrix obtained by deleting row a and column b from the metric. In the first equation,

---

[3]This can be generalized to all ranks by defining an antisymmetric combination of partial derivatives known as the *exterior derivative*, given by $(p+1)\partial_{[\mu_0}A_{\mu_1\mu_2\cdots\mu_p]}$.

the sum is explicit because we are *not* using implied summation in this equation; there is no sum over a. For a one simply picks any one of the possible values. The second equation then states how the inverse matrix is obtained from the determinant and the matrix of cofactors.

We can regard g as a function of the elements of g_{ab}. Let us now consider the partial derivative of g with respect to one of the elements in row a, while the others remain constant. Using (11.30) one finds

$$\frac{\partial g}{\partial g_{ad}} = \sum_\nu \left(\frac{\partial g_{a\nu}}{\partial g_{ad}} \Delta^{a\nu} + g_{a\nu} \frac{\partial \Delta^{a\nu}}{\partial g_{ad}} \right) = \sum_\nu \delta_\nu^d \Delta^{a\nu} = \Delta^{ad} \tag{11.32}$$

where we used the fact that since each cofactor $\Delta^{a\nu}$ is obtained by deleting the row and column a, ν, it must be independent of all the elements of row a. Now, again regarding g as a function of the elements of g_{ab}, we have

$$dg = \frac{\partial g}{\partial g_{\mu\nu}} dg_{\mu\nu} = \Delta^{\mu\nu} dg_{\mu\nu} = g g^{\mu\nu} dg_{\mu\nu} \tag{11.33}$$

where the summation is assumed and we used (11.32) and (11.31) and the fact that the metric is symmetric. It follows that

$$\frac{\partial g}{\partial x^c} = g g^{\mu\nu} \frac{\partial g_{\mu\nu}}{\partial x^c} \tag{11.34}$$

Now $\partial_c g_{\mu\nu} = \Gamma_{\mu\nu c} + \Gamma_{\nu\mu c}$ (eqn (10.23)) so we have

$$\partial_c g = g g^{\mu\nu} \left(\Gamma_{\mu\nu c} + \Gamma_{\nu\mu c} \right) = g \left(\Gamma_{\nu c}^\nu + \Gamma_{\mu c}^\mu \right) = 2g \Gamma_{\mu c}^\mu \tag{11.35}$$

hence

$$\Gamma_{\mu c}^\mu = \frac{\partial_c g}{2g} = \frac{1}{2} \frac{\partial \ln |g|}{\partial x^c}. \tag{11.36}$$

By bringing the factor $(1/2)$ into the log function as a square root, this can also be written

$$\Gamma_{\mu c}^\mu = \frac{1}{\sqrt{|g|}} \frac{\partial \sqrt{|g|}}{\partial x^c} \tag{11.37}$$

Using this in (11.29) we obtain:

Divergence of vector

$$\boldsymbol{\nabla} \cdot \mathbf{v} = \partial_a v^a + \frac{\partial_\mu \sqrt{|g|}}{\sqrt{|g|}} v^\mu = \frac{1}{\sqrt{|g|}} \partial_a \left(v^a \sqrt{|g|} \right) \tag{11.38}$$

This result can be read 'take the divergence you might have guessed, but fix it up by factors of $\sqrt{|g|}$.' The expression can also be written

$$\partial_a \left(\sqrt{|g|}\, v^a \right) = \sqrt{|g|}\, \nabla_a v^a. \tag{11.39}$$

In order to gather results together, we will now anticipate the next chapter a little and quote the associated result for the divergence of a second-rank tensor. Using (12.60) and an argument essentially the same as the above, one finds

Divergence of tensor

$$\nabla_\mu T^{\mu b} = \frac{1}{\sqrt{|g|}} \partial_a \left(\sqrt{|g|}\, T^{\mu b} \right) + \Gamma^b_{\lambda\mu} T^{\mu\lambda} \tag{11.40}$$

In the case of an *antisymmetric* tensor, the last term vanishes because it is the complete contraction of a symmetric with an antisymmetric object, and then one obtains

Divergence of antisymmetric tensor

$$\nabla_\mu F^{\mu b} = \frac{1}{\sqrt{|g|}} \partial_a \left(\sqrt{|g|}\, F^{\mu b} \right) \tag{11.41}$$

The next useful operator is div grad. Its effect on a scalar field can be written

$$\nabla_a \nabla^a \phi = \frac{1}{\sqrt{|g|}} \partial_a \left(\sqrt{|g|} g^{a\lambda} \partial_\lambda \phi \right) \tag{11.42}$$

and for tensors of higher rank a more complicated expression would be obtained. This is the operator which is commonly called 'Laplacian' in three dimensions, and 'd'Alembertian' in four dimensions. In the latter case it is often written $\Box^2$ or $\Box$ but we shall not do that because we are using the symbol $\Box^2$ for the operator $\partial_a \partial^a$. The covariant d'Alembertian $\nabla_a \nabla^a \phi$ is **not** equal to $\partial_a \partial^a$ except in special circumstances such as constant g_{ab}. (Another possible notation is ∇^2.)

11.4   Gauss' divergence theorem

Gauss' divergence theorem is one of the most useful theorems in differential geometry. It can be obtained as a special case of a generalized Stokes theorem using advanced methods. Here we will obtain it using simpler methods.

By thinking carefully about the fact that scalars at different places can be added and subtracted with impunity, one can prove that the divergence theorem extends to curved manifolds without needing to be adapted in the first instance, but one must keep clear the distinction between coordinate volume and metric or proper volume. For example, in four dimensions if coordinates x^a are orthogonal at P, then the region bounded by $x^a(P)$, $x^a(P) + dx^a$ has coordinate volume $dx^0 dx^1 dx^2 dx^3$ and proper volume $\sqrt{|g|} dx^0 dx^1 dx^2 dx^3$ (see (8.41)).

Let $v^a(x)$ be a vector field on some manifold $\mathcal{M}$. We can if we wish take an interest in an integral such as

$$\int_{\mathcal{R}} \partial_\mu v^\mu \, d^4 x \tag{11.43}$$

We want to relate this to an integral over the surface $\partial \mathcal{R}$ of the region $\mathcal{R}$, and we can regard it as a problem in integral calculus. Here the metric of $\mathcal{M}$ plays no role. We treat the coordinates

x^a as a set of variables; we are simply carrying out a multi-dimensional integral in the usual way, and for the purpose of our first result we can regard these variables as coordinates in an abstract space with Euclidean properties. It follows that the usual proofs of Gauss' theorem apply, and we shall find

$$\int_{\mathcal{R}} \partial_\mu v^\mu \, \mathrm{d}^4 x = \oint_{\partial \mathcal{R}} v^\mu n_\mu \mathrm{d}S \tag{11.44}$$

where n^a is the unit outward normal to the boundary $\partial \mathcal{R}$ in the abstract Euclidean space adopted for the purpose of this calculation and the notation $\mathrm{d}S$ signifies an element of 'area' of this hypersurface (that is, a three-dimensional region in this example) in the abstract Euclidean space.

To be clear, (11.44) applies to a vector field v^a defined on any manifold, whether curved or flat, but one must be careful to understand what the result does and does not assert, especially on the right-hand side of the equation.

The result can be immediately generalized to other numbers of dimensions, and tensors of higher rank (by noting that at higher rank the further indices 'come along for the ride').

The expression (11.44) is not manifestly covariant. This does not make it incorrect, but a covariant result is sometimes more useful; this is our next goal.

By using (11.39) and (11.44) we find that for any vector field v^a and region $\mathcal{R}$,

$$\int_{\mathcal{R}} (\nabla_\mu v^\mu) \sqrt{|g|} \, \mathrm{d}^4 x = \oint_{\partial \mathcal{R}} v^\mu n_\mu \sqrt{|g|} \, \mathrm{d}S \tag{11.45}$$

where we have treated a four-dimensional manifold for illustration and we will generalize later. The left-hand side of this equation is manifestly covariant, but the right-hand side is not as it stands. It takes a little thought to discern precisely what has been assumed concerning n_a and $\mathrm{d}S$ when we use (11.44) and thus obtain (11.45). The essential point is that *if* the coordinates at the hypersurface $\partial \mathcal{R}$ were orthonormal, with one coordinate (say x^0) perpendicular to the surface and the others in the surface, *then* $\mathrm{d}S$ could be written $\mathrm{d}S = \mathrm{d}x^1 \mathrm{d}x^2 \mathrm{d}x^3$.

We shall now develop such a coordinate system, and this will show us how to write the right-hand side of our equation in a manifestly covariant form. The coordinates we seek are called *Gaussian normal coordinates*.

A surface is said to be null, timelike or spacelike according as its normal is null, spacelike or timelike. We shall restrict to non-null surfaces, but we allow that a given surface may be timelike in some regions and spacelike in others.

The hypersurface $\partial \mathcal{R}$ in which we are interested is a submanifold of the four-dimensional manifold $\mathcal{M}$ under discussion. Let u^1, u^2, u^3 be coordinates in this submanifold. Define coordinates x^a throughout $\mathcal{M}$ as follows. For each point (u^1, u^2, u^3) on $\partial \mathcal{R}$ find the geodesic through that point which starts out orthogonal to $\partial \mathcal{R}$. For non-null $\partial \mathcal{R}$, every point in $\mathcal{M}$ (and not too far from the hypersurface) is on a unique such geodesic; let the x^i coordinates of the point be

$x^i = u^i$ and let x^0 be equal to the value of an affine parameter[4] along that geodesic, a parameter scaled such that the tangent vector to the geodesic has length ± 1 at $\partial \mathcal{R}$. With these choices we must have that the line element has the form

$$ds^2 = \pm (dx^0)^2 + \gamma_{ij} du^i du^j \tag{11.46}$$

(with the sign depending on whether the hypersurface is timelike or spacelike at any given point), where γ_{ij} is the induced metric on the submanifold. Hence

$$\det([g_{ab}]) = \pm \det([\gamma_{ij}]) \tag{11.47}$$

therefore $|g| = |\gamma|$ where $\gamma \equiv \det([\gamma_{ij}])$.

If the coordinates u^i were orthonormal then we would have $dS = du^1 du^2 du^3$ and therefore (11.45) could be written

$$\int_{\mathcal{R}} (\nabla_\mu v^\mu) \sqrt{|g|}\, d^4x = \oint_{\partial \mathcal{R}} v^\mu n_\mu \sqrt{|\gamma|}\, d^3u. \tag{11.48}$$

The right-hand side is now manifestly covariant, so now we can assert that it still holds for any coordinates, whether for the full manifold on the left, or for the submanifold on the right, or both. Hence, after also generalizing to any number of dimensions:

Gauss' theorem; manifestly covariant form, non-null $\partial \mathcal{R}$

$$\int_{\mathcal{R}} (\nabla_\mu v^\mu) \sqrt{|g|}\, d^Nx = \oint_{\partial \mathcal{R}} v^\mu n_\mu \sqrt{|\gamma|}\, d^{N-1}u \tag{11.49}$$

---

in which u^j is a set of coordinates on the submanifold $\partial \mathcal{R}$, γ_{ij} is the induced metric on the submanifold, $\gamma = \det([\gamma_{ij}])$, and $d^{N-1}u \equiv du^1 du^2 \cdots du^{N-1}$.

Exercises

11.1 Show that if two particles have 4-velocities **u**, **v** as they pass close to one another then the relative speed (that is, the speed of either particle when observed in the instantaneous rest frame of the other) is $c\sqrt{1 - \gamma^{-2}}$ where

$$\gamma = -\mathbf{u} \cdot \mathbf{v}/c^2. \tag{11.50}$$

11.2 In stationary conditions the metric is independent of t, where $t = x^0$ is a timelike coordinate. Let the locations at fixed spatial coordinates x^i be called lattice points. Explain why the worldline $x^a(\tau)$ of a lattice point satisfies $dx^i/d\tau = 0$. If the metric is in the canonical form (4.14), show also that

---

[4]This is a parameter which increases by equal amounts for equal distances along the geodesic; c.f. Section 13.1.2.

$dx^0/d\tau = \exp(-\Phi/c^2) = \text{const}$. Show that the 4-acceleration a^a of a lattice point is

$$a^a = e^{2\Phi/c^2} \Gamma^a_{00} \qquad (11.51)$$

and hence obtain (4.15) and (4.16).

11.3 A binary star system consists of a pair of neutron stars of mass $2M_\odot$ each, orbiting on a common circular orbit with a period of one minute. An astronaut is moving, without any rocket pack, in the vacuum near to these stars. What can be determined concerning the astronaut's 4-acceleration? Explain qualitatively what forces the astronaut experiences, and use a Newtonian calculation to estimate their size.

11.4 Write down the contravariant and covariant components of the 4-velocity and 4-acceleration of yourself, expressed first in a LIF momentarily at rest relative to yourself, then in a frame S' permanently fixed to yourself. For the second case, obtain the result first by coordinate transformation using (9.30), and then from (11.3) by using the connection coefficients evaluated in exercise 10.11 of Chapter 10. Why are the components of your 4-acceleration not zero in the frame S'?

11.5 Having noted the method, obtain (11.35) and hence (11.38) without referring back to the text.

11.6 Show that if n^a is a timelike unit vector, then in a coordinate system where only one of its components is non-zero,

$$n_0 = \pm\sqrt{|g_{00}|}. \qquad (11.52)$$

12

Tensors

In Chapter 9 we defined vectors as geometric objects having the same character as directed line segments. One can also define a vector as *that whose components transform, under a change of coordinates, as shown in (9.30)*. In the present chapter we consider a more general object, also having a geometric character, called *tensor*. In Volume 1 we introduced the second-rank tensor as *that which can act upon a vector so as to give a vector result*, and also as *that which behaves under coordinate transformations like the outer product of two vectors*. We will show in the present chapter that these two definitions can be extended to GR. The two definitions are equivalent, and both are useful.

We have distinguished between a vector $\mathbf{v}$, and the set of its components v^a. For clear thinking, it is important to make this distinction, and to continue to make it for tensors of any rank. One should keep clear before one's mind the fact that a tensor 'sits' at some point in a manifold, and it is what it is, independent of how that manifold may be being mapped by one coordinate system or another. When we change reference frame, tensors do not change, but their components may.

Once we have defined vectors, then a tensor of any rank can be defined in either of the following ways:

1. A tensor of rank k is a mathematical calculating machine, which, upon being furnished with k vectors as input, will report a single scalar as output, such that this output value will not depend on the choice of coordinate system which may have been adopted in order to describe the vector inputs, and the action of this function is linear in each of its arguments. (In more formal language, one says that a tensor is a linear function mapping a list of k vectors to an invariant scalar.)
2. A tensor of rank k in an N-dimensional manifold is completely specified by a set of N^k components. These components can be furnished in any given frame, and as a set they transform from one frame to another in the same way as the components of an outer product of k vectors.

In order to understand how these definitions work, let us begin by considering a first-rank tensor: that is, a vector. If we have a vector $\mathbf{v}$ then we can furnish another vector $\mathbf{w}$, and supply it to the first as 'input', such that a mathematical operation will be performed, yielding

Relativity Made Relatively Easy: General Relativity and Cosmology. Volume 2. Andrew M. Steane, Oxford University Press. © Andrew M. Steane 2021. DOI: 10.1093/oso/9780192895646.003.0012

an invariant scalar as the result of the operation. The operation in question is the one we call *inner product*, so the statement is summarized as:

$$\mathbf{v}(\mathbf{w}) = \mathbf{v} \cdot \mathbf{w}. \tag{12.1}$$

On the left-hand side of this expression, we are regarding the $\mathbf{v}$ as a scalar-valued function on the set of all vectors. The linearity requirement stated in the definition is

$$\mathbf{v}(\alpha\mathbf{u} + \beta\mathbf{w}) = \alpha\mathbf{v}(\mathbf{u}) + \beta\mathbf{v}(\mathbf{w}) \tag{12.2}$$

and indeed the inner product does have this property.

Now let us proceed to a second-rank tensor $\mathbf{t}$. According to the first definition given above, we can furnish $\mathbf{t}$ with two vectors and get a scalar output. We write this as:

$$\mathbf{t}(\mathbf{v}, \mathbf{w}). \tag{12.3}$$

Be careful to read this correctly. It makes reference to a second-rank tensor $\mathbf{t}$, which we can tell because we see two arguments, but the overall quantity on display is a scalar.

A useful example of this, and an important second-rank tensor, is the metric tensor, whose value, for any given pair of vectors, is their inner product:

$$\mathbf{g}(\mathbf{v}, \mathbf{w}) = \mathbf{v} \cdot \mathbf{w}. \tag{12.4}$$

The linearity requirement for second-rank tensors is

$$\begin{aligned}\mathbf{t}(\alpha\mathbf{u} + \beta\mathbf{v}, \gamma\mathbf{w} + \epsilon\mathbf{x}) &= \alpha\mathbf{t}(\mathbf{u}, \gamma\mathbf{w} + \epsilon\mathbf{x}) + \beta\mathbf{t}(\mathbf{v}, \gamma\mathbf{w} + \epsilon\mathbf{x}) \\ &= \alpha\gamma\mathbf{t}(\mathbf{u}, \mathbf{w}) + \alpha\epsilon\mathbf{t}(\mathbf{u}, \mathbf{x}) + \beta\gamma\mathbf{t}(\mathbf{v}, \mathbf{w}) + \beta\epsilon\mathbf{t}(\mathbf{v}, \mathbf{x}), \end{aligned} \tag{12.5}$$

and similar results apply at any rank.

Getting a tensor of lower rank. Consider the quantity $\mathbf{t}(\mathbf{v}, \cdot, \cdot)$. This is a quantity which started off with three 'slots' or function inputs, but one of the slots has been filled by the given vector $\mathbf{v}$, and the other two are empty. This means that $\mathbf{t}(\mathbf{v}, \cdot, \cdot)$ is waiting for **two** vectors to come along as input, and upon receiving such a pair it will generate its scalar output. Therefore, $\mathbf{t}(\mathbf{v}, \cdot, \cdot)$ is a second-rank tensor.

We can generalize this argument to any rank, and assert that a tensor of rank k may be considered to be an object that, upon being furnished with a single vector, gives as a result of its operation a tensor of rank $k - 1$.

12.1 The components of a tensor

Notice what happens when we furnish the coordinate basis vectors as 'inputs' in the case of a first-rank tensor. We have

$$\mathbf{v}(\mathbf{e}_{(a)}) \equiv \mathbf{v} \cdot \mathbf{e}_{(a)} = (v^\lambda \mathbf{e}_{(\lambda)}) \cdot \mathbf{e}_{(a)} = v^\lambda g_{\lambda a} = v_a \tag{12.6}$$

by using (9.15). Similarly,

$$\mathbf{v}(\tilde{\boldsymbol{\theta}}^{(a)}) = v^a. \tag{12.7}$$

In the case of a second- or higher-rank tensor, we *define* the components as the outcomes when the tensor is evaluated on coordinate basis vectors:

$$
\begin{aligned}
t_{ab} &\equiv \mathbf{t}(\mathbf{e}_{(a)}, \mathbf{e}_{(b)}), & t^{ab} &\equiv \mathbf{t}(\tilde{\boldsymbol{\theta}}^{(a)}, \tilde{\boldsymbol{\theta}}^{(b)}), \\
t_a{}^b &\equiv \mathbf{t}(\mathbf{e}_{(a)}, \tilde{\boldsymbol{\theta}}^{(b)}), & t^a{}_b &\equiv \mathbf{t}(\tilde{\boldsymbol{\theta}}^{(a)}, \mathbf{e}_{(b)}).
\end{aligned}
\tag{12.8}
$$

You should confirm that this works for the metric tensor (12.4), for example. These sets of components are called covariant, contravariant, mixed and mixed, respectively. In a slight abuse of terminology, one may also refer to a 'contravariant tensor' or 'covariant tensor' or 'mixed tensor' (this is an example of blurring the distinction between a tensor and its components). A useful terminology is:

Definition 12.1 *A (p, q) tensor is a tensorial object whose components have p upper and q lower indices. (Such a tensor has rank $p + q$, and the pair of values (p, q) is called the valence.)*

Thus g_{ab} is a $(0, 2)$ tensor, $s^{ab}{}^c_{d\,e}$ is a $(3, 2)$ tensor, and so on.

Now we can use the linearity property to deduce that

$$\mathbf{t}(\mathbf{v}, \mathbf{w}) = \mathbf{t}(v^\mu \mathbf{e}_{(\mu)}, w^\nu \mathbf{e}_{(\nu)}) = v^\mu w^\nu \mathbf{t}(\mathbf{e}_{(\mu)}, \mathbf{e}_{(\nu)}) = v^\mu w^\nu t_{\mu\nu}. \tag{12.9}$$

This shows that we can use the components themselves to discover the action of $\mathbf{t}$ on any given vectors, in terms of their components.

By similar reasonsing one may also show that

$$\mathbf{t}(\mathbf{v}, \mathbf{w}) = v^\mu w^\nu t_{\mu\nu} = v_\mu w_\nu t^{\mu\nu} = v^\mu w_\nu t_\mu{}^\nu = v_\mu w^\nu t^\mu{}_\nu. \tag{12.10}$$

This, together with similar results at higher rank, establishes the 'see-saw' rule, namely that for a dummy index, the superscript and subscript positions can be interchanged without affecting the result.

We have shown that the components of a tensor themselves suffice to determine how the tensor acts on any given set of vectors. If follows that one can determine everything about a tensor by finding out about its components. This is how we often work with tensors in practice. (And this is why the distinction between a tensor and its components can in practice be blurred for the purposes of many calculations.)

By supplying coordinate basis vectors in the standard way which we have described, one finds that, for given vector $\mathbf{v}$, the contravariant components of the tensor $\mathbf{t}(\mathbf{v}, \cdot, \cdot)$ are

$$v_\lambda t^{\lambda ab}. \tag{12.11}$$

Symmetric and antisymmetric combinations. Notice that tensors do not necessarily have to have any symmetry with respect to their inputs. That is, it is not necessarily the case that $\mathbf{t}(\mathbf{v}, \mathbf{w})$ has any particular relationship to $\mathbf{t}(\mathbf{w}, \mathbf{v})$. But if, for every pair of vectors, a second-rank tensor $\mathbf{t}$ gives $\mathbf{t}(\mathbf{v}, \mathbf{w}) = \mathbf{t}(\mathbf{w}, \mathbf{v})$ then $\mathbf{t}$ is said to be *symmetric*, and if $\mathbf{t}(\mathbf{v}, \mathbf{w}) = -\mathbf{t}(\mathbf{w}, \mathbf{v})$ then $\mathbf{t}$ is said to be *antisymmetric*. By using basis vectors as inputs, it is easy to show that for a symmetric tensor, $t_{ab} = t_{ba}$, $t^{ab} = t^{ba}$ and $t_a{}^b = t^b{}_a$. For an antisymmetric tensor, $t_{ab} = -t_{ba}$, $t^{ab} = -t^{ba}$ and $t_a{}^b = -t^b{}_a$.

Given any second-rank tensor $\mathbf{t}$, one can obtain from it symmetric and antisymmetric tensors $\mathbf{S}$, $\mathbf{A}$ with components given by

$$S_{ab} = \tfrac{1}{2}(t_{ab} + t_{ba}), \qquad A_{ab} = \tfrac{1}{2}(t_{ab} - t_{ba}) \tag{12.12}$$

and then one has $\mathbf{t} = \mathbf{S} + \mathbf{A}$. A commonly used notation is

$$t_{(ab)} \equiv \tfrac{1}{2}(t_{ab} + t_{ba}), \qquad t_{[ab]} \equiv \tfrac{1}{2}(t_{ab} - t_{ba}). \tag{12.13}$$

This is, a pair of indices placed in a parenthesis indicates the symmetric combination of the sets of components, and a pair of indices placed in a square bracket indicates the antisymmetric combination. A bracket around a larger number of indices means the totally symmetric or totally antisymmetric combination. For example,

$$t_{[abc]} \equiv \frac{1}{6}\left(t_{abc} - t_{acb} + t_{cab} - t_{cba} + t_{bca} - t_{bac}\right). \tag{12.14}$$

One can also place these brackets across a pair of tensors, as for example:

$$g_{a[b}\, t_{c]d} \equiv \tfrac{1}{2}(g_{ab}t_{cd} - g_{ac}t_{bd}). \tag{12.15}$$

Raising and lowering indices. Consider some arbitrary tensor $\mathbf{t}$, which for the sake of having a concrete example we shall take to be of third rank. By definition, we have

$$\mathbf{t}(\mathbf{e}_{(a)}, \mathbf{e}_{(b)}, \mathbf{e}_{(c)}) = t_{abc}, \qquad \mathbf{t}(\mathbf{e}_{(a)}, \mathbf{e}_{(b)}, \tilde{\boldsymbol{\theta}}^{(c)}) = t_{ab}{}^c \tag{12.16}$$

and by using linearity,

$$\mathbf{t}(\mathbf{e}_{(a)}, \mathbf{e}_{(b)}, \tilde{\boldsymbol{\theta}}^{(c)}) = \mathbf{t}(\mathbf{e}_{(a)}, \mathbf{e}_{(b)}, g^{c\lambda}\mathbf{e}_{(\lambda)}) = g^{c\lambda}\mathbf{t}(\mathbf{e}_{(a)}, \mathbf{e}_{(b)}, \mathbf{e}_{(\lambda)}) = g^{c\lambda}t_{ab\lambda}. \tag{12.17}$$

Therefore

$$g^{c\lambda}t_{ab\lambda} = t_{ab}{}^c. \tag{12.18}$$

By generalizing this, one sees that g^{ab} can be used for raising any chosen index on any chosen tensor, and similarly one finds that g_{ab} can be used for lowering any chosen index on any chosen tensor.

12.1.1 Coordinate transformation for tensors

The argument of Section 9.1.3 can readily be extended to tensors of any rank. At rank two one obtains

$$t'_{ab} = \frac{\partial x^\mu}{\partial x'^a} \frac{\partial x^\nu}{\partial x'^b} t_{\mu\nu}, \tag{12.19}$$

$$t'^{ab} = \frac{\partial x'^a}{\partial x^\mu} \frac{\partial x'^b}{\partial x^\nu} t^{\mu\nu}, \tag{12.20}$$

$$t'^{\ b}_a = \frac{\partial x^\mu}{\partial x'^a} \frac{\partial x'^b}{\partial x^\nu} t^{\ \nu}_\mu, \tag{12.21}$$

and the general result is that, for a tensor of arbitrary rank, each lower index is transformed via the coefficients $(\partial x^\mu / \partial x'^a)$ and each upper index is transformed via the coefficients $(\partial x'^a / \partial x^\mu)$. If we define

$$K^a_{\ b} \equiv \frac{\partial x'^a}{\partial x^b}, \qquad \bar{K}^{\ b}_a \equiv \frac{\partial x^b}{\partial x'^a} \tag{12.22}$$

and let T', T, K denote the matrices whose components are given by t'^{ab}, t^{ab}, $K^a_{\ b}$ respectively, then (12.20) can be written

$$\mathsf{T}' = \mathsf{K}\mathsf{T}\mathsf{K}^T \tag{12.23}$$

where ^{T} is the matrix transpose operation.

12.2 Transformation of Γ and relation to geodesic coordinates

We have already warned the reader that the symbol Γ^a_{bc} is not itself tensorial. One way to prove this is to note that at any point one can find coordinates in which $\Gamma^a_{bc} = 0$, whereas we know that in general it will not be zero in other coordinate systems. But transformation of a tensor can never give a non-zero tensor in the new coordinates starting from a zero tensor in the old coordinates; it follows that Γ^a_{bc} does not transform as a tensor. Let us see how it does transform:[1]

$$
\begin{aligned}
\Gamma'^a_{bc} &= \tilde{\boldsymbol{\theta}}'^a \cdot \frac{\partial \mathbf{e}'_b}{\partial x'^c} && \text{[from (10.44)]} \\
&= \left(K^a_{\ \mu} \tilde{\boldsymbol{\theta}}^\mu \right) \cdot \frac{\partial}{\partial x'^c} \left(\bar{K}^{\ \nu}_b \mathbf{e}_\nu \right) && \text{[using (9.27), (12.22)]} \\
&= \left(K^a_{\ \mu} \tilde{\boldsymbol{\theta}}^\mu \right) \cdot \left(\frac{\partial \bar{K}^{\ \nu}_b}{\partial x'^c} \mathbf{e}_\nu + \bar{K}^{\ \nu}_b \frac{\partial \mathbf{e}_\nu}{\partial x'^c} \right) \\
&= K^a_{\ \mu} \bar{K}^{\ \nu}_b \tilde{\boldsymbol{\theta}}^\mu \cdot \frac{\partial \mathbf{e}_\nu}{\partial x'^c} + K^a_{\ \mu} \frac{\partial \bar{K}^{\ \nu}_b}{\partial x'^c} \delta^\mu_\nu.
\end{aligned}
\tag{12.24}
$$

Now using rules of partial differentiation (the chain rule), $(\partial \mathbf{e}_\nu / \partial x'^c) = (\partial x^\lambda / \partial x'^c)(\partial \mathbf{e}_\nu / \partial x^\lambda)$ and therefore we have

---

[1]To reduce clutter, we here drop the use of brackets on labels of basis vectors.

$$\Gamma'^a_{bc} = K^a_{\ \mu} \bar{K}_b^{\ \nu} \bar{K}_c^{\ \lambda} \Gamma^\mu_{\ \nu\lambda} + K^a_{\ \mu} \frac{\partial \bar{K}_b^{\ \mu}}{\partial x'^c}. \tag{12.25}$$

In this result, the first term is what one would obtain if Γ^a_{bc} were a $(1,2)$ tensor; the second term shows that Γ^a_{bc} is not transforming that way.

Now suppose that for the primed coordinates in this calculation we take any coordinatization in which we might be interested, and for the unprimed coordinates we take Riemann normal coordinates (RNC), also called geodesic coordinates or a LIF. In this case, we shall have $\Gamma^\mu_{\ \nu\lambda} = 0$ at the origin, and therefore

$$\Gamma'^a_{bc} \overset{\text{RNC}}{\underset{X}{=}} K^a_{\ \mu} \frac{\partial \bar{K}_b^{\ \mu}}{\partial x'^c} = \frac{\partial x'^a}{\partial X^\mu} \frac{\partial^2 X^\mu}{\partial x'^c \partial x'^b} \tag{12.26}$$

where X^a are Riemann normal coordinates. This shows us another way of understanding what the connection coefficients are: they constitute a statement about the relation of our coordinate system to Riemann normal (or LIF) coordinates.

If in the above derivation of Γ'^a_{bc} one starts from (10.8) then one obtains

$$\Gamma'^a_{bc} = K^a_{\ \mu} \bar{K}_b^{\ \nu} \bar{K}_c^{\ \lambda} \Gamma^\mu_{\ \nu\lambda} - \frac{\partial x^\mu}{\partial x'^b} \frac{\partial^2 x'^a}{\partial x'^c \partial x^\mu}. \tag{12.27}$$

The second term in this expression can also be written

$$-\frac{\partial x^\mu}{\partial x'^b} \frac{\partial^2 x'^a}{\partial x^\lambda \partial x^\mu} \frac{\partial x^\lambda}{\partial x'^c}, \tag{12.28}$$

hence by taking $\Gamma^a_{bc} = 0$ we find

$$\Gamma'^a_{bc} \overset{\text{RNC}}{\underset{X}{=}} -\frac{\partial X^\mu}{\partial x'^b} \frac{\partial^2 x'^a}{\partial X^\mu \partial x'^c} = -\frac{\partial X^\mu}{\partial x'^b} \frac{\partial X^\lambda}{\partial x'^c} \frac{\partial^2 x'^a}{\partial X^\mu \partial X^\lambda}. \tag{12.29}$$

This will be used in a study of the covariant derivative in Section 12.4.

12.2.1 Finding Riemann normal coordinates

We will now provide the proof of the claim made in Section 8.3 with regard to eqn (8.23), which showed how to find coordinates such that the metric tensor has no linear dependence on the coordinates near to some chosen point P. The equation defining the primed coordinates reads

$$x'^a = x^a - x^a_P + \frac{1}{2}\Gamma^a_{\mu\nu}(P)\left(x^\mu - x^\mu_P\right)\left(x^\nu - x^\nu_P\right) \tag{12.30}$$

where $\Gamma^a_{\mu\nu}(P)$ is the connection coefficient evaluated at P (so it is not a function of x). Taking the derivative with respect to x^b gives

$$\frac{\partial x'^a}{\partial x^b} = \delta^a_b + \Gamma^a_{b\mu}(P)(x^\mu - x^\mu_p) \tag{12.31}$$

and therefore

$$\frac{\partial^2 x'^a}{\partial x^c x^b} = \Gamma^a_{b\mu}(P)\delta^\mu_c \;=\; \Gamma^a_{bc}(P). \tag{12.32}$$

Evaluating the first of these results at the point P gives

$$\frac{\partial x'^a}{\partial x^b}(P) = \delta^a_b, \qquad \frac{\partial x^a}{\partial x'^b}(P) \;=\; \delta^a_b, \tag{12.33}$$

where the second follows since it is the inverse of the first. Now we substitute all the above information into the equation for the transformation of Γ^a_{bc}, eqn (12.27), making use of (12.28):

$$\Gamma'^a_{bc}(P) = \delta^a_\mu \delta^\nu_b \delta^\lambda_c \Gamma^\mu_{\nu\lambda} + \delta^\mu_b \Gamma^a_{\lambda\mu}\delta^\lambda_c \;=\; \Gamma^a_{bc} - \Gamma^a_{bc} \;=\; 0. \tag{12.34}$$

Thus we have found coordinates in which all the connection coefficients vanish at the chosen point P. Equation (8.24) follows. QED.

12.3   Tensor algebra

The legal operations of tensor algebra are:

1. Multiplication by a scalar
2. Summing tensors of the same rank and sets of tensor components of the same valence
3. Outer product
4. Contraction (and hence inner product)
5. Index permutation (applied to all terms in a sum)
6. Index raising and lowering

These were discussed in Volume 1 in the context of Special Relativity. Using the tools developed in this chapter it is easy to show that they carry over immediately to Riemannian manifolds in general and hence to GR.

The important and useful **quotient theorem** was also discussed in Volume 1. It states that if we have some quantity X, which, when contracted with any tensor always gives a tensor result, then X must itself be a tensor. See exercise 12.3 for a proof.

12.4   Covariant derivative of tensors

12.4.1   The covariant derivative is itself a tensor

We introduced the covariant derivative of a vector in the previous chapter through eqn (10.5). Its most important properties are:

1. The covariant derivative is not the same as the partial derivative in general.
2. The covariant derivative of a tensor is itself a tensor (of one higher rank).
3. If at some point (i.e. in GR, at some event) one finds $\Gamma^a_{bc} = 0$ in the coordinate system which has been adopted, then at that point the first-order covariant derivative and the first-order partial derivative agree.

We have defined the covariant derivative in such a way that all these properties hold, so they are not in any doubt. But it will be instructive to prove the second property by using the third. The proof will use the fact that at every point it is possible to find a coordinate system such that $\Gamma^a_{bc} = 0$ at that point (Chapter 1 and Sections 8.3, 12.2.1.)

The argument from (10.3) (connection coefficients) to (10.6) (covariant derivative) shows that the quantity

$$\nabla_a v^b \tag{12.35}$$

is tensorial; to be precise, it is the set of components of a tensor of type (1,1). However, suppose we did not know this. Suppose, for example, that we did not know how to relate Γ^a_{bc} to the coordinate basis vectors. Then we should want to prove that $\nabla_a v^b$ is tensorial by some other argument. We shall do it by a method based on coordinate transformation.

We explore the differentiation of a vector by starting out from local Riemann normal coordinates X^a. An arbitrary vector **v** has components v^a in this coordinate system, with partial derivatives

$$\frac{\partial v^a}{\partial X^c}. \tag{12.36}$$

We begin the argument by asserting that this set of quantities is indeed the set of components of a second-rank tensor, because it can be calculated in the tangent space and in that space this is precisely the gradient operator with which we are familiar in Special Relativity. (The argument hinges on the fact that in Riemann normal coordinates not only is the metric Cartesian at the pole of the coordinate system, but also it has zero first derivatives there; Section 8.3). Let us call these components $t^a_{\ c}$. By now changing to any other coordinate system, we deduce that the set of quantities

$$t'^a_{\ c} = \frac{\partial x'^a}{\partial X^\lambda}\frac{\partial X^\mu}{\partial x'^c}t^\lambda_{\ \mu} \stackrel{\text{RNC}}{=} \frac{\partial x'^a}{\partial X^\lambda}\frac{\partial X^\mu}{\partial x'^c}\frac{\partial v^\lambda}{\partial X^\mu} \tag{12.37}$$

are the components of a second-rank tensor. Note the logic: the left-hand side of this equation is tensorial because it has been obtained by the correct transformation rule applied to a known tensor. The right-hand side is therefore also tensorial, but it can only be expressed by this particular combination of partial derivatives when the X^a coordinates are Riemann normal coordinates.

Now let us write down the components v'^a and take their partial derivatives. We have

$$\frac{\partial v'^a}{\partial x'^c} = \frac{\partial}{\partial x'^c}\left(\frac{\partial x'^a}{\partial X^\lambda}v^\lambda\right) = \frac{\partial^2 x'^a}{\partial x'^c \partial X^\lambda}v^\lambda + \frac{\partial x'^a}{\partial X^\lambda}\frac{\partial v^\lambda}{\partial x'^c}$$

$$= \frac{\partial^2 x'^a}{\partial x'^c \partial X^\lambda}v^\lambda + t'^a_{\ c} \tag{12.38}$$

where we used the chain rule and then (12.37). The presence of the first term on the right-hand side shows that the overall result is not tensorial. It also invites us to consider the quantity obtained by moving this term to the left-hand side:

$$\frac{\partial v'^a}{\partial x'^c} - \frac{\partial^2 x'^a}{\partial x'^c \partial X^\lambda} v^\lambda = t'^a{}_c. \tag{12.39}$$

Now examine eqn (12.29), which we repeat here for convenience:

$$\Gamma'^a_{bc} \overset{\text{RNC}}{\underset{X}{=}} -\frac{\partial X^\mu}{\partial x'^b} \frac{\partial^2 x'^a}{\partial X^\mu \partial x'^c} \tag{12.40}$$

It follows that

$$\Gamma'^a_{\nu c} v'^\nu = -\frac{\partial X^\mu}{\partial x'^\nu} \frac{\partial^2 x'^a}{\partial X^\mu \partial x'^c} \left(\frac{\partial x'^\nu}{\partial X^\lambda} v^\lambda \right)$$

$$= -\delta^\mu_\lambda \frac{\partial^2 x'^a}{\partial X^\mu \partial x'^c} v^\lambda = -\frac{\partial^2 x'^a}{\partial X^\lambda \partial x'^c} v^\lambda. \tag{12.41}$$

By substituting this into (12.39) we obtain

$$\frac{\partial v'^a}{\partial x'^c} + \Gamma'^a_{\nu c} v'^\nu = t'^a{}_c. \tag{12.42}$$

The left-hand side of this equation is the covariant derivative of v'^a. We have thus proved that this derivative gives a tensorial result. QED.

The extension to tensors of any rank can be made by using that every tensor of rank above zero can be written as a sum of outer products of vectors, and the Leibniz rule applies to the covariant differentiation of such products.

12.4.2 Partial differentiation and covariant differentiation

In order to proceed confidently in GR, it is crucial to be clear about the roles played by both partial differentiation and covariant differentiation. It is not true to say that covariant differentiation merely replaces partial differentiation. In fact both are needed and both play important roles.

We will now take some trouble to spell out these roles in detail.

If we have a manifold $\mathcal{M}$ then in principle there can be invariant scalar fields, and vector fields and tensor fields defined on this manifold. There can also be coordinates, written x^a, and there can be components of vector and tensor fields, and there can be basis vectors. All of these quantities may vary from one place to another in the manifold. If v^a are the contravariant components of a vector field, then each v^a (for any one particular value of the index a) is itself a single number which can be a function of position in the manifold, so in this sense it too is a field.

Let P be a point in $\mathcal{M}$, and let Q be a nearby point.

If ϕ is an invariant scalar field, then the symbol $\mathrm{d}\phi$ refers to a small change in ϕ. The change we have in mind is the difference between $\phi(Q)$ and $\phi(P)$:

$$\mathrm{d}\phi \equiv \phi(Q) - \phi(P) \tag{12.43}$$

and it is understood that eventually this small quantity will be compared to some other small quantity and a limit taken. So far this is all very straightforward: we begin in this simple way in order to build towards some more subtle things.

Using the definition of partial differentiation, we must now find

$$\mathrm{d}\phi = \frac{\partial \phi}{\partial x^{\mu}}\mathrm{d}x^{\mu} \tag{12.44}$$

where

$$\mathrm{d}x^{\mu} \equiv x^{\mu}(Q) - x^{\mu}(P). \tag{12.45}$$

That is, $\mathrm{d}x^{\mu}$ is a set of four quantities that are given by the differences between the coordinate values at Q and P. The question now arises, are these quantities the contravariant components of a vector, or does the notation mislead us? Should we adopt a different notation? By looking at the coordinate transformation (8.7), we satisfy ourselves that the $\mathrm{d}x^{\mu}$ do indeed transform the right way to earn the name 'vector'. Using the quotient rule, we can then deduce from (12.44) that $\partial_{\mu}\phi$ are the components of a dual vector (equally, one may say they are covariant components of a 4-vector.) In fact there are a number of ways to establish these basic facts, and mathematicians often prefer to start by announcing that $\partial_{\mu}\phi$ are the components of a dual vector (see appendix C), but let us move on.

If we want to know how some other (i.e. not invariant) type of scalar quantity f varies from place to place, then we can write

$$\mathrm{d}f \equiv f(Q) - f(P) = \frac{\partial f}{\partial x^{\mu}}\mathrm{d}x^{\mu} \tag{12.46}$$

and this is true no matter what kind of quantity f is: it need not be rank-zero tensor; it need not be tensorial at all. It could be just one component of a vector, for example, or it could be one 'component' (a, b, c) of a Christoffel symbol Γ^{a}_{bc}. In all cases, if one wants to find out by how much a single number f has changed between P and Q, then one can use the partial derivatives and add them up via the chain rule. The main difference between (12.44) and (12.46) is that *whereas $\partial_{a}\phi$ is tensorial* (to be precise, it is the set of covariant components of a vector), $\partial_{a}f$ *is not (unless f is itself an invariant)*.

Next we would like to introduce the idea of a change in a vector field. To this end, define the following notation

$$\mathbf{v}(P) \equiv \text{the value of the vector field } \mathbf{v} \text{ at } P$$
$$\mathbf{v}(Q) \equiv \text{the value of the vector field } \mathbf{v} \text{ at } Q$$
$$T_{\|}(\mathbf{v}, P, Q) \equiv \begin{array}{l} \text{the result of taking the vector } \mathbf{v}(P) \text{ and} \\ \text{moving it from } P \text{ to } Q \text{ by parallel transport} \end{array}$$

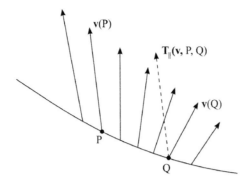

Fig. 12.1 The full arrows show part of a vector field. Each vector is in the tangent space of the point at its foot. The dashed vector is defined in the text. The curve is a geodesic.

(see Fig. 12.1). Here the term 'parallel transport' means that as the vector is moved along any given geodesic, the angle it makes with that geodesic does not change, and the size of the vector does not change. For Q sufficiently close to P, we can always find a unique geodesic connecting P to Q, so this definition is all we need in the present context. We can now define

$$\mathrm{d}\mathbf{v} \equiv \mathbf{v}(Q) - T_{\parallel}\left(\mathbf{v}(P), P, Q\right) \tag{12.47}$$

and

$$\frac{\partial \mathbf{v}}{\partial x^b} \equiv \lim_{\delta \to 0} \frac{\mathbf{v}(P + \delta x^b) - T_{\parallel}\left(\mathbf{v}(P), P, P + \delta x^b\right)}{\delta} \tag{12.48}$$

where the notation $P + \delta x^b$ indicates the location which differs from P by a change δ in the b'th coordinate, other coordinates remaining constant. This is essentially a standard notion of partial differentiation, but we have been careful to say exactly what we mean by a change in a vector field. You should now pause to convince yourself that this definition agrees with the one reported in the discussion of (10.2) and (10.6), as long as the covariant derivative gives zero when $\mathbf{v}(Q)$ is equal to the parallel-transported version of $\mathbf{v}(P)$. That point will be discussed more fully in Chapter 13 (indeed it falls out immediately if one defines parallel transport using eqn (13.2)).

With the definition of $\partial_b \mathbf{v}$ established, we can now define

$$\nabla_b v^a \equiv \text{the set of contravariant components of } \frac{\partial \mathbf{v}}{\partial x^b}, \text{ for each } b,$$
$$\text{if } \mathrm{d}\mathbf{v} \text{ is expressed using the basis vectors at } P \tag{12.49}$$

This is precisely what we already asserted in eqn (10.7).

The central point can now be expressed:

$$v^a(Q) - v^a(P) = \frac{\partial v^a}{\partial x^\mu}\mathrm{d}x^\mu \qquad \text{BUT } \frac{\partial v^a}{\partial x^\mu} \text{ IS NOT A TENSOR} \tag{12.50}$$

$$\nabla_b v^a = \frac{\partial v^a}{\partial x^b} + \Gamma^a_{\mu b} x^\mu \qquad \text{and } \nabla_b v^a \text{ IS a tensor} \tag{12.51}$$

where we have allowed the shorthand of saying 'is a tensor' for 'is the set of components of a tensor'.

In view of the above, it is perfectly correct to write

$$\mathrm{d}v^a = \frac{\partial v^a}{\partial x^\mu}\mathrm{d}x^\mu \tag{12.52}$$

and to avoid getting in a muddle one simply needs to remember that

$$\mathrm{d}\mathbf{v} \neq \mathrm{d}v^a \mathbf{e}_{(a)} \tag{12.53}$$

because

$$\mathrm{d}\mathbf{v} = \mathrm{d}v^a \mathbf{e}_{(a)} + v^a \mathrm{d}\mathbf{e}_{(a)}. \tag{12.54}$$

To find the contravariant components of the change in $\mathbf{v}$ for some small movement along a curve $x^a(u)$, use

$$(\nabla_\mu v^a)\mathrm{d}x^\mu \tag{12.55}$$

—the intrinsic derivative, eqn (10.31). Thus one finds

$$\mathrm{d}\mathbf{v} = (\nabla_\mu v^a)\mathrm{d}x^\mu \mathbf{e}_{(a)}, \qquad \text{along curve } x^a(u) \tag{12.56}$$

whereas

$$\mathrm{d}\mathbf{v} \neq \frac{\partial v^a}{\partial x^\mu}\mathrm{d}x^\mu \mathbf{e}_{(a)}. \tag{12.57}$$

12.4.3 Derivatives of higher-rank tensors

A very nice property of covariant differentiation is that it obeys the rule of partial differentiation applied to products, called the Leibniz rule. For, in Riemann normal coordinates the two types of differentiation are identical, so when such coordinates are adopted we may write

$$\nabla_c(u^a v^b) \overset{\text{RNC}}{=} \partial_c(u^a v^b) = (\partial_c u^a)v^b + u^a(\partial_c v^b) \overset{\text{RNC}}{=} (\nabla_c u^a)v^b + u^a(\nabla_c v^b). \tag{12.58}$$

But thus we find

$$\nabla_c(u^a v^b) = (\nabla_c u^a)v^b + u^a(\nabla_c v^b) \tag{12.59}$$

and now the whole equation is tensorial, and being a tensor equation, it must be true in all coordinates.

Consider now a second-rank tensor $\mathbf{t}$ that can be expressed as the outer product of two first-rank tensors $\mathbf{u}$, $\mathbf{v}$. The components will then take the form $t^{ab} = u^a v^b$ and using the above result we find

$$\begin{aligned}
\nabla_c t^{ab} &= (\nabla_c u^a)v^b + u^a(\nabla_c v^b) \\
&= \left(\partial_c u^a + \Gamma^a_{\lambda c}u^\lambda\right)v^b + u^a\left(\partial_c v^b + \Gamma^b_{\lambda c}v^\lambda\right) \\
&= \partial_c t^{ab} + \Gamma^a_{\lambda c}t^{\lambda b} + \Gamma^b_{\lambda c}t^{a\lambda}.
\end{aligned} \tag{12.60}$$

By linearity, the same result applies to all second-rank tensors, and thus we find the general rule for covariant differentiation of second-rank tensors: one uses the partial derivative plus further factors involving the connection, one for each index.

Using (10.9) you can now quickly show that

$$\nabla_c t_{ab} = \partial_c t_{ab} - \Gamma^\lambda_{ac} t_{\lambda b} - \Gamma^\lambda_{bc} t_{a\lambda} \tag{12.61}$$

$$\nabla_c t^a{}_b = \partial_c t^a{}_b + \Gamma^a_{\lambda c} t^\lambda{}_b - \Gamma^\lambda_{bc} t^a{}_\lambda \tag{12.62}$$

and the above argument from outer products is readily extended to all higher ranks. A covariant derivative of a tensor of rank k is given by a partial derivative added to a sum of k terms involving Γ. The rule for the terms involving Γ is that one has a **plus** sign when summing over an **up** index on t, and a minus sign when summing over a down index on t.

The covariant derivative of the metric tensor.   A property of the metric tensor which proves to be very important for GR is that its covariant derivative is zero:

$$\nabla_c g_{ab} = 0, \qquad \nabla_c g^{ab} = 0 \tag{12.63}$$

and therefore raising and lowering commutes with covariant differentiation:

$$\nabla_b v_a \equiv \nabla_b(g_{a\mu} v^\mu) \;=\; g_{a\mu} \nabla_b v^\mu. \tag{12.64}$$

The metric is said to be *covariantly constant*. Equally, the connection is said to be *metric-compatible*. To derive either of the above results one can start by choosing local Riemann normal coordinates, at some given point P, and then both results follow from the fact that in these coordinates g_{ab} has no *first-order* dependence on the coordinates. One then argues that we have a tensor equation, and a coordinate transformation can never give a non-zero tensor from a zero tensor, so the result must be zero in all frames. Furthermore, this argument can be applied at any point P.

Note that the above argument cannot be applied at higher order, and indeed in general $\nabla_c \nabla_d g_{ab}$ need not be zero, because even when Γ^a_{bc} is zero at one point, its derivatives there need not be.

It is a useful exercise to obtain the same answer by other routes, for example working in arbitrary coordinates and employing (10.6) and (10.15) (the definition of ∇_c and the relation of Γ to g_{ab}).

12.5   Tensors of rank zero

We have, throughout the chapter, discussed tensors of rank 1 or above. It remains to assert that the whole framework continues to make sense if we assert that an invariant scalar is a tensor of rank zero. An example of an invariant scalar is the inner product of a pair of vectors.

For scalar fields, the covariant derivative and the partial derivative agree:

$$\nabla_a \phi \equiv \partial_a \phi \tag{12.65}$$

and the intrinsic derivative along a curve is

$$\frac{\mathrm{D}\phi}{\mathrm{d}u} \equiv \frac{\mathrm{d}\phi}{\mathrm{d}u} = \frac{\partial\phi}{\partial x^\lambda}\frac{\mathrm{d}x^\lambda}{\mathrm{d}u}. \tag{12.66}$$

We can show that this is consistent by employing the previously established action of ∇_a on second-rank tensors, and then contracting:

$$\nabla_c(u^a v_b) = \partial_c(u^a v_b) + \Gamma^b_{\lambda c}u^\lambda v_b - \Gamma^\lambda_{bc}u^a v_\lambda$$
$$\Rightarrow \qquad \nabla_c(u^a v_a) = \partial_c(u^a v_a) + \Gamma^a_{\lambda c}u^\lambda v_a - \Gamma^\lambda_{ac}u^a v_\lambda$$
$$= \partial_c(u^a v_a) \tag{12.67}$$

where the cancellation can be seen by swapping the labels of dummy indices in one of the terms involving Γ.

Notice, for example, one simple but useful application. If we have a velocity field, such as the velocity field of a flowing fluid, then everywhere and at all times the velocity u^a of a fluid element has constant size $-c^2$, so we may write

$$\nabla_c(u^\mu u_\mu) = 0 \tag{12.68}$$

and thus obtain (exercise)

$$u^\mu \nabla_c u_\mu = 0. \qquad \text{[any fluid or other continuous medium]} \tag{12.69}$$

This equation can be read as *a velocity field is orthogonal to its own covariant gradient.*

12.6   Tensor density and the Hodge dual

The *Levi-Civita symbol* $\tilde{\epsilon}_{abcd}$ is defined

$$\tilde{\epsilon}_{abcd} \equiv \begin{cases} +1 \text{ if } abcd \text{ is an even permutation of 0123} \\ -1 \text{ if } abcd \text{ is an odd permutation of 0123} \\ 0 \text{ otherwise} \end{cases} \tag{12.70}$$

This is called a 'symbol' not a 'tensor' because it is not a tensor. It is defined to be this set of values and does not change at all when we change coordinates. However, it is in some sense 'almost a tensor' because in order to convert it into a tensor we need only multiply by a power of the metric determinant g (exercise 12.11). Quantities having this character are called *tensor densities.*[2] The *Levi-Civita tensor* is

$$\epsilon_{abcd} \equiv \sqrt{|g|}\,\tilde{\epsilon}_{abcd}. \tag{12.71}$$

We will not in fact need this tensor in this book, but one can see that it is liable to arise in the discussion of determinants, and it is involved in the definition of the *Hodge dual*, denoted by a star. The Hodge dual is a mapping from a p-form to a $(N-p)$-form in N dimensions,

---

[2]The word 'density' is here only loosely related to the notion of 'per unit volume'.

where a p-form is a completely antisymmetric tensor of type $(0,p)$. This is a p-index object which changes sign under the exchange of any two indices. It follows that there are N choose p independent components to a p-form. From basic combinatorics we have

$$\binom{N}{p} = \frac{N!}{p!(N-p)!} = \binom{N}{N-p} \tag{12.72}$$

so a p-form and an $(N-p)$-form have the same number of independent components.

In an N-dimensional manifold the Hodge dual of a p-form is an $(N-p)$-form given by

$$(*A)_{a_1 \cdots a_{N-p}} \equiv \epsilon^{\mu_1 \cdots \mu_p}{}_{a_1 \cdots a_{N-p}} A_{\mu_1 \cdots \mu_p} \frac{1}{p!}. \tag{12.73}$$

Two examples of the use of this idea (Maxwell's equations and angular momentum) were described in Volume 1. In GR we also introduce the *left-dual* and *right-dual* for an object such as the curvature tensor which is antisymmetric in some pairs of its indices:

$$^*R_{abcd} \equiv \tfrac{1}{2}\epsilon_{ab}{}^{\mu\nu} R_{\mu\nu cd}, \qquad R^*_{abcd} \equiv \tfrac{1}{2}\epsilon_{cd}{}^{\mu\nu} R_{ab\mu\nu}. \tag{12.74}$$

In linear algebra one may also introduce the *wedge product*. This is an antisymmetrized outer product; the simplest example is the wedge product of two one-forms which is $(A \wedge B)_{ab} = A_a B_b - A_b B_a$. We mention this merely for completeness; it will not be needed in this book.

Exercises

12.1 Prove that $\partial^2 \phi / \partial x^a \partial x^b$ is not a tensor.

12.2 Prove that δ^a_b is a tensor, and furthermore its components are the same in all frames.

12.3 **Quotient theorem**. Prove the quotient theorem by the following method. First consider the simple case where u^a is unknown (i.e. it is unknown whether or not it is a tensor) and for any rank one tensor v^a we are given that $u_\mu v^\mu$ is a tensor of rank zero (a scalar invariant). Therefore

$$u'_\mu v'^\mu = u_\lambda v^\lambda.$$

But since v^a is a 4-vector, we have $v^\lambda = (\partial x^\lambda / \partial x'^\mu) v'^\mu$ so

$$u'_\mu v'^\mu = u_\lambda \frac{\partial x^\lambda}{\partial x'^\mu} v'^\mu$$

$$\implies \left(u'_\mu - u_\lambda \frac{\partial x^\lambda}{\partial x'^\mu} \right) v'^\mu = 0.$$

Since this is true for all v'^μ, the bracket must be zero, therefore

$$u'_\mu = (\partial x^\lambda / \partial x'^\mu) u_\lambda$$

which proves that u^a is a rank one tensor. Your task is to generalize this to an unknown of any rank. The starting point is the assertion that the unknown, when contracted with any tensor of some given rank, gives a tensorial result. This is sufficient to prove that the unknown is also tensorial (your proof should allow the known tensor to be of any rank).

12.4 Using the symmetries listed in (2.15) show that the Bianchi identity (2.14) can be written

$$\nabla_{[e} R_{ab]cd} = 0 \qquad (12.75)$$

in the notation illustrated in (12.14), (12.15).

12.5 Show that if t_{ab} is symmetric and $v_a t_{bc} + v_b t_{ca} + v_c t_{ab} = 0$ then either $\mathbf{t} = 0$ or $\mathbf{v} = 0$. [Hint: start by considering the component $a = b = c = 0$]

12.6 Show that $B_{abc} = \partial_a A_{bc} + \partial_b A_{ca} + \partial_c A_{ab}$ is a tensor, and determine its symmetry properties.

12.7 (i) Prove that

$$\text{if} \qquad M_{\mu\nu} u^\mu u^\nu = 0 \ \forall u^a$$
$$\text{then} \qquad M_{ab} + M_{ba} = 0. \qquad (12.76)$$

(that is, **M** is antisymmetric). [Hint: first assume four dimensions and start by considering the vector $u^a = (1,0,0,0)$, then the vector $u^a = (1,1,0,0)$ and spot the pattern].

(ii) Prove that

$$\text{if} \qquad M_{\lambda\mu\nu} u^\lambda u^\mu u^\nu = 0 \ \forall u^a$$
$$\text{then} \quad M_{abc} + M_{bca} + M_{cab} = 0. \quad (12.77)$$

[Hint: like previous method, but now you will require $u^a = (1,-1,0,0)$ and $(1,1,1,0)$ as well as $(1,0,0,0)$ and $(1,1,0,0)$]

12.8 Show that for a vector field to have a vanishing covariant derivative $\nabla_b v^a$ everywhere, it must satisfy

$$\left(\partial_b \Gamma^a_{\lambda c} - \partial_c \Gamma^a_{\lambda b} - \Gamma^a_{\mu c} \Gamma^\mu_{\lambda b} + \Gamma^a_{\mu b} \Gamma^\mu_{\lambda c} \right) v^\lambda = 0. \qquad (12.78)$$

12.9 Show that covariant differentiation of a product of tensors of any rank obeys the Leibniz rule.

12.10 Show that under a coordinate transformation the determinant of the metric transforms as

$$g' = J^{-2} g \qquad (12.79)$$

where $g = \det(g_{ab})$, $g' = \det(g'_{ab})$ and $J = \det(\partial x'^a / \partial x^b)$.

12.11 The determinant of any 4×4 matrix M can be written $\det(M) = \tilde{\epsilon}_{\alpha\beta\gamma\delta} M^\alpha{}_0 M^\beta{}_1 M^\gamma{}_2 M^\delta{}_3$ and therefore

$$\tilde{\epsilon}_{abcd} \det(M) = \tilde{\epsilon}_{\alpha\beta\gamma\delta} M^\alpha{}_a M^\beta{}_b M^\gamma{}_c M^\delta{}_d.$$

Apply this to $M^a{}_b = \partial x^a / \partial x'^b$ and hence show that

$$\tilde{\epsilon}'_{abcd} = J \frac{\partial x^\alpha}{\partial x'^a} \frac{\partial x^\beta}{\partial x'^b} \frac{\partial x^\gamma}{\partial x'^c} \frac{\partial x^\delta}{\partial x'^d} \tilde{\epsilon}_{\alpha\beta\gamma\delta} \quad (12.80)$$

where $J = \det(\partial x'^a / \partial x^b)$. Using this and the previous exercise, show that ϵ_{abcd} defined in (12.71) is a tensor. (Noting the power of J in (12.79) and (12.80), the Levi-Civita symbol is said to be a tensor density of *weight* 1, and g is a (rank 0) tensor density of weight -2.)

13

Parallel transport and geodesics

13.1 Parallel transport

In Chapters 9 to 12 we have discussed vectors such that each vector is located at one point (in GR, at one event). When discussing a vector field, for example, we have in mind that each point has its own vector. However, in considering the notion called *parallel transport* we are interested in the idea of carrying or moving a vector from one place to another. Parallel transport of a vector is defined as a series of infinitesimal displacements of a vector, such that in each such displacement the vector does not change in either size or direction in the local tangent space.

To make this concept precise, suppose we start with a vector $\mathbf{w}$ at a single point on some curve, and then construct a set of further vectors at all the other points on the curve. We then have a vector field $\mathbf{w}(u)$ that is defined everywhere along a curve $x^a(u)$. We will say that this field constitutes the result of parallel transport of the vector $\mathbf{w}(0)$ if and only if:

Parallel transport

$$\frac{\mathrm{d}\mathbf{w}}{\mathrm{d}u} = 0 \qquad \left(\text{which is equivalent to} \quad \frac{Dw^a}{\mathrm{d}u} = 0 \right). \tag{13.1}$$

For the avoidance of all doubt, the quantity expressed in the first equation here is defined as explained in Chapter 10 where the connection coefficients were defined. The second version then gives the equivalent statement about the components, obtained by using the intrinsic or absolute derivative, eqn (10.29).

Using now (10.29), we find that for a vector undergoing parallel transport,

parallel transport

$$\mathrm{d}w^a = \frac{\partial w^a}{\partial x^\nu}\mathrm{d}x^\nu = -\Gamma^a_{\mu\nu}w^\mu\mathrm{d}x^\nu,$$

$$\frac{\mathrm{d}w^a}{\mathrm{d}u} = -\Gamma^a_{\mu\nu}w^\mu\frac{\mathrm{d}x^\nu}{\mathrm{d}u}. \tag{13.2}$$

Relativity Made Relatively Easy: General Relativity and Cosmology. Volume 2. Andrew M. Steane,
Oxford University Press. © Andrew M. Steane 2021. DOI: 10.1093/oso/9780192895646.003.0013

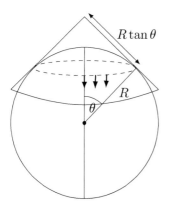 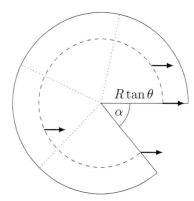

Fig. 13.1 A vector is parallel-transported around the dashed curve (a line of latitude) on a 2-sphere. The cone is constructed so as to touch the sphere along this path. When the cone is cut and unrolled to lie flat one can see the way the parallel transport of the arrow changes its direction relative to the path and to other features such as the lines of longitude. During one circuit around the path on the sphere the arrow rotates by the angle α relative to its initial direction.

Since this is a first order differential equation for the components, we see that once the vector has been given at one point on the curve, its components at all other points are uniquely fixed by integration of this equation.

Another very useful way to write the equation for parallel transport is to employ the covariant derivative. Using (10.32) in (13.1) gives

---

parallel transport again

$$\nabla_{\mathbf{T}} w^a \equiv T^\lambda \nabla_\lambda w^a \;=\; 0. \qquad \text{where} \;\; T^b = \mathrm{d}x^b/\mathrm{d}u \qquad (13.3)$$

Equation (13.3) asserts 'if the vector field w^a is the result of parallel transport, then its covariant derivative along the tangent direction is zero'. More generally, a covariant derivative of a general vector field evaluated at a point P expresses the difference (per unit distance) between the vector evaluated at $P + \mathrm{d}x^c$ and the vector obtained by parallel transport from P. If the vector field in question has itself been obtained by parallel transport from P, then clearly its covariant derivative must be zero! Indeed, one can use this property as a starting-point with which to define the covariant derivative. This is an alternative way of presenting the logic—see box.

Example 13.1 A vector is located at $(\theta, 0)$ on the surface of a 2-sphere, pointing initially in the $\mathbf{e}_{(\theta)}$ direction. It is then parallel-transported around the line of latitude at θ (i.e. a line of constant θ). What is its direction upon returning to the starting point?
Solution.
We shall answer this question first by appealing to a geometric construction, and then by using the tools of differential geometry. The geometric construction is shown in Fig. 13.1. By dropping

An alternative route to the one we have taken in Chapter 10 is to start by asserting that (13.3) is a suitable definition of parallel transport, with ∇_λ some sort of differential operator whose full definition is to be determined. Next we take some arbitrary vectors **v** and **w** at a point, and use them to create vector fields by parallel transport along some arbitrary curve with tangent vector **T**. Then we assert that the scalar product is preserved as the vectors are transported, so

$$T^\lambda \nabla_\lambda (g_{\mu\nu} v^\mu w^\nu) = 0. \tag{13.4}$$

We next assert by definition that ∇ obeys the Leibniz rule (product rule), so we have

$$T^\lambda (\nabla_\lambda g_{\mu\nu}) v^\mu w^\nu + g_{\mu\nu} T^\lambda (\nabla_\lambda v^\mu) w^\nu + v^\mu T^\lambda (\nabla_\lambda w^\nu) = 0. \tag{13.5}$$

But since **v** and **w** were created by parallel transport, they each satisfy (13.3) so the second and third terms vanish and we deduce

$$T^\lambda v^\mu w^\nu (\nabla_\lambda g_{\mu\nu}) = 0. \tag{13.6}$$

Since this is true for any $\mathbf{T}, \mathbf{v}, \mathbf{w}$ it follows that $\nabla_\lambda g_{\mu\nu} = 0$.

Next we use (13.2) as a *definition* of Γ^a_{bc}. With this in hand, take a vector A^a evaluated at x and parallel displace it through a small distance $\mathrm{d}x$, then subtract the resulting vector from A^a evaluated at $x + \mathrm{d}x$:

$$A^a(x + \mathrm{d}x) - \left[A^a(x) - \Gamma^a_{\lambda\mu} A^\lambda \mathrm{d}x^\mu \right] = \frac{\partial A^a}{\partial x^\mu} \mathrm{d}x^\mu + \Gamma^a_{\lambda\mu} A^\lambda \mathrm{d}x^\mu$$

Since this is the difference of two vectors evaluated at the same place (namely $x + \mathrm{d}x$), it must be a vector. It is a vector for all $\mathrm{d}x^\mu$, therefore, by the quotient theorem, $\partial_b A^a + \Gamma^a_{\lambda b} A^\lambda$ is a tensor—hence we have the covariant derivative. By considering the covariant derivative of the metric tensor (noting from the above that it must vanish) one can then find the expression for Γ^a_{bc} in terms of g_{ab}.

a cone of just the right opening angle onto the sphere, we can find a flat space (the surface of the cone, excluding its vertex) which is tangential to the sphere at all points along the path. After laying this flat space onto a flat Euclidean plane (the diagram on the right in Fig. 13.1) we observe that when the vector comes back to the starting point on the sphere, it reaches the line of the cut on the cone. Then, using the fact that in a parallel transport in a flat space, a vector does not rotate, we deduce that its angle relative to the line of longitude has changed by α, the opening angle of the cone.

To find α, note that the length of the path is $2\pi R \sin\theta$ and that it lies along an arc of a circle of radius $R \tan\theta$ in the flat space. Therefore $2\pi R \tan\theta = 2\pi R \sin\theta + \alpha R \tan\theta$ which yields $\alpha = 2\pi(1 - \cos\theta)$.

Now let us do the calculation by using (13.2). The vector being transported is $\mathbf{w}$. The curve $x^a(u)$ can be conveniently parameterized by ϕ, and since it is a line of constant θ, we have $\mathrm{d}x^\theta/\mathrm{d}\phi = 0$ and $\mathrm{d}x^\phi/\mathrm{d}\phi = 1$. Hence (13.2) simplifies a little, giving

$$\frac{\mathrm{d}w^a}{\mathrm{d}\phi} = -\Gamma^a_{\phi\nu}w^\nu \tag{13.7}$$

The only non-zero Christoffel symbols for the 2-sphere are $\Gamma^\theta_{\phi\phi} = -\sin\theta\cos\theta$, $\Gamma^\phi_{\theta\phi} = \Gamma^\phi_{\phi\theta} = \cot\theta$ so we have

$$\frac{\mathrm{d}w^\theta}{\mathrm{d}\phi} = -\Gamma^\theta_{\phi\nu}w^\nu = \sin\theta\cos\theta\, w^\phi, \tag{13.8}$$

$$\frac{\mathrm{d}w^\phi}{\mathrm{d}\phi} = -\Gamma^\phi_{\phi\nu}w^\nu = -\cot\theta\, w^\theta. \tag{13.9}$$

After differentiating the first and using the second, one finds

$$\frac{\mathrm{d}^2w^\theta}{\mathrm{d}\phi^2} = -(\cos^2\theta)\, w^\theta \tag{13.10}$$

whose general solution is

$$w^\theta = A\cos(\phi\cos\theta) + B\sin(\phi\cos\theta) \tag{13.11}$$

(since we are dealing with a case where θ is constant). The solution for w^ϕ is then

$$w^\phi = -\frac{A}{\sin\theta}\sin(\phi\cos\theta) + \frac{B}{\sin\theta}\cos(\phi\cos\theta). \tag{13.12}$$

The initial conditions $w^a = (1,0)$ at $\phi = 0$ give $A = 1$, $B = 0$ for the constants of integration. Hence

$$\mathbf{w} = \cos(\phi\cos\theta)\mathbf{e}_{(\theta)} - \sin(\phi\cos\theta)\frac{\mathbf{e}_{(\phi)}}{\sin\theta}. \tag{13.13}$$

This is a vector of length 1 which makes an angle $(-\phi\cos\theta)$ with the lines of longitude. When $\phi = 2\pi$ the angle is $-2\pi\cos\theta = \alpha$ modulo 2π, in agreement with the previous calculation.

The cone construction was useful in this example for added insight, but it will not be available in more general cases, whereas the algebraic method is always available.

13.1.1   The affine geodesic: a curve of no turning

The above definition of parallel transport applies to any vector and any curve. You can pick any curve you like, and find out how vectors parallel-transported along that curve behave. But parallel transport is also a useful way to single out one set of curves for special attention.

We now come to an important idea. Among all the curves in any given manifold, there are a special set that have the property of *no turning*, in the following sense: as one proceeds along the curve, the next little segment continues straight on from the one you are on, without changing direction in the local tangent space. A curve having this property is called an *affine geodesic*.

Definition 13.1 An affine geodesic is a curve having the property that there exists a tangent vector which undergoes a parallel transport as one moves along the curve.

To see how this definition works, first pick a point P and a direction. Let **t** be a vector in the direction you have picked. We shall construct the curve in such a way that **t** is its *tangent vector*. That is to say, if the curve is $x^a(u)$ then

$$t^a = \frac{\mathrm{d}x^a}{\mathrm{d}u}. \tag{13.14}$$

Note that this is an ordinary derivative, not an intrinsic derivative, because it is a statement about the direction of the curve itself, not about the changes in some other quantity defined along the curve. The property stated in the above definition is then

$$\frac{\mathrm{D}t^a}{\mathrm{d}u} = 0, \qquad \Rightarrow \qquad \frac{\mathrm{d}t^a}{\mathrm{d}u} + \Gamma^a_{\mu\nu} t^\mu \frac{\mathrm{d}x^\nu}{\mathrm{d}u} = 0 \tag{13.15}$$

which immediately gives

Geodesic equation

$$\frac{\mathrm{d}^2 x^a}{\mathrm{d}u^2} + \Gamma^a_{\mu\nu} \frac{\mathrm{d}x^\mu}{\mathrm{d}u} \frac{\mathrm{d}x^\nu}{\mathrm{d}u} = 0 \qquad \Leftrightarrow \qquad t^\lambda \nabla_\lambda t^a = 0. \tag{13.16}$$

This is the important *geodesic equation* which we first presented in Chapter 2 (eqn (2.28)).

13.1.2 Affine parameter

The definition of parallel transport is such that both the length and the direction of the vector is maintained as the vector is moved along. In defining an affine geodesic, on the other hand, while we require that the tangent vector should not change direction (in the local tangent space as the parameter u changes by $\mathrm{d}u$), we do not need to require that it should also preserve its length. This means that a geodesic can also be defined by the property

$$\frac{\mathrm{d}\mathbf{t}}{\mathrm{d}u} = f(u)\mathbf{t} \tag{13.17}$$

for any function f, leading to the equation

$$\frac{\mathrm{d}^2 x^a}{\mathrm{d}u} + \Gamma^a_{\mu\nu} \frac{\mathrm{d}x^\mu}{\mathrm{d}u} \frac{\mathrm{d}x^\nu}{\mathrm{d}u} = f(u) \frac{\mathrm{d}x^a}{\mathrm{d}u}. \tag{13.18}$$

At first sight this equation looks different from (13.16), but it is not: it is simply that (13.16) describes the case where $f(u) = 0$ for all u, and we will now show that it is always possible to ensure that $f(u) = 0$ by choosing u appropriately.

How to construct an affine parameter without using a metric

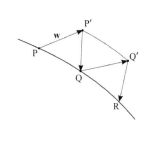

The notion of parallel transport does not require a metric tensor; it only requires a connection. However, when we move along a geodesic, we would like some notion of moving the same 'distance' each time. Also, we need such a notion for null geodesics where the metric does not directly provide it. The figure shows a geometric way to solve the problem (as an alternative to the differential equations provided in the main text). All we need is the ability to construct small parallelograms, and parallel transport is sufficient to do that. Create a small vector **w** at P, and transport it along the geodesic in question to some nearby point Q. At the other end of the vector are points P' and Q'. Now construct the vector $P'Q$ at P', and parallel-transport it to Q'. At the other end of this vector locate point R, which will also lie on the geodesic in the limit where **w** is infinitesimal (we omit the details which would be needed for a complete proof). We now construct our affine parameter u by asserting that it increases by the same amount between Q and R as it does between P and Q.

For any given curve, the direction of the tangent vector is fixed by the curve, but the length of the tangent vector given by (13.14) depends on how we choose to parameterize the curve using the parameter u. If we choose the increase of u along the curve in such a way that the vector with components dx^a/du has constant length, then we have the case $f(u) = 0$. When u has this form, then u is called an *affine parameter*. For any given curve it is always possible to find an affine parameter (exercise 13.4), and therefore we can decide to adopt that case for the purpose of writing down the equation for a geodesic. This is the choice almost always made in GR. Note, however, that once one has identified a given geodesic, by whatever method, then it is not guaranteed to satisfy (13.16) for just any old parameterization: for that simpler result (as opposed to (13.18)) it is a requirement that the parameterization be affine.

For non-null geodesics, a convenient affine parameter is the distance along the curve (as furnished by the metric). For null geodesics this does not work, because the distance along the curve is then zero for all points, but we want a function u that increases monotonically as one moves along the curve. If one has a parameter such that the geodesic obeys (13.16) (and not just (13.18)), then the parameter is called affine.

It should be reasonably apparent from the above that if, for a given curve, u is an affine parameter, then so is $s = \alpha u + \beta$ for any constant α, β, with the exception $\alpha = 0$. This is proved in exercise 13.4.

13.2   Metric geodesic: most proper time and least distance

We now approach a second way of defining geodesics, using the notion of 'shortest path' (to be precise, path of stationary length). Curves defined this way are called *metric geodesics*, and we will show that when the connection is torsion-free, the metric geodesics and the affine geodesics agree. Since GR adopts the Levi-Civita connection, which is torsion-free, we shall then assume this case and drop the distinction.

A curve in a manifold is called null if the invariant distance between all pairs of neighbouring points on the curve is zero, and non-null if the invariant distance between all pairs of neighbouring points on the curve is non-zero. We shall not need to concern ourselves with curves having both null and non-null sections, because exercise 13.1 assures us that this does not happen for geodesic curves. For null curves, the distance along the curve is always zero so it is not possible to define a notion of least (or most) distance. One may handle this by defining a null geodesic as the limiting case of nearby non-null geodesics.

The distance along a curve extending between points (in spacetime, events) P and Q is given by (8.27):

$$ s = \int_P^Q \left| g_{a\mu\nu} \frac{\mathrm{d}x^\mu}{\mathrm{d}u} \frac{\mathrm{d}x^\nu}{\mathrm{d}u} \right|^{1/2} \mathrm{d}u \tag{13.19} $$

We can find a curve giving a stationary value of this distance by using the calculus of variations, leading to Euler–Lagrange equations. It was explained in Section 14.4 of Volume 1 that the equations can be obtained by taking the 'Lagrangian' to be $\mathcal{L} = g_{\mu\nu}\dot{x}^\mu \dot{x}^\nu$, where the dot signifies differentiation with respect to an appropriate measure of distance along a curve in spacetime, such as s, and the word 'Lagrangian' is in inverted commas because it quantifies the *square* of the infinitesimal length along a given curve, not the length itself, but nevertheless when we want to find curves of stationary length, they are correctly given by Euler–Lagrange equations associated with $\mathcal{L}$ (see Volume 1 for the proof, which involves using s to parameterize the path *after* the variational procedure has been completed). In order for the method to work, one must use an affine parameter along the worldline; s is a suitable affine parameter.[1]

The path of stationary length can now be found by solving the Euler–Lagrange equation

$$ \frac{\mathrm{d}}{\mathrm{d}s} \left(\frac{\partial \mathcal{L}}{\partial \dot{x}^a} \right) = \frac{\partial \mathcal{L}}{\partial x^a}. \tag{13.20} $$

We have

$$ \frac{\partial \mathcal{L}}{\partial x^a} = (\partial_a g_{\mu\nu})\dot{x}^\mu \dot{x}^\nu, $$

$$ \frac{\partial \mathcal{L}}{\partial \dot{x}^a} = g_{a\nu}\dot{x}^\nu + g_{\mu a}\dot{x}^\mu = 2g_{a\mu}\dot{x}^\mu, $$

---

[1] If instead one uses the square root of our $\mathcal{L}$, then the condition that the parameter be affine can be dropped, but this is usually a less straightforward method.

therefore

$$\frac{\mathrm{d}}{\mathrm{d}s}\left(\frac{\partial \mathcal{L}}{\partial \dot{x}^a}\right) = \frac{\mathrm{d}}{\mathrm{d}s}\left(2g_{a\mu}\dot{x}^\mu\right) = 2(\partial_\nu g_{a\mu})\dot{x}^\mu \dot{x}^\nu + 2g_{a\mu}\ddot{x}^\mu \tag{13.21}$$

(using the chain rule for the first term on the right). Hence the Euler–Lagrange equation reads

$$2(\partial_\nu g_{a\mu})\dot{x}^\mu \dot{x}^\nu + 2g_{a\mu}\ddot{x}^\mu = (\partial_a g_{\mu\nu})\dot{x}^\mu \dot{x}^\nu. \tag{13.22}$$

Since swapping the dummy labels in the first term has no effect, we can split it into two terms and thus obtain

$$\left(\partial_\nu g_{a\mu} + \partial_\mu g_{a\nu} - \partial_a g_{\mu\nu}\right)\dot{x}^\mu \dot{x}^\nu + 2g_{a\mu}\ddot{x}^\mu = 0. \tag{13.23}$$

Now notice that the term in the bracket is two times the Christoffel symbol of the first kind, eqn (10.22). Hence after raising a we can write (13.23) in the form

$$\ddot{x}^a + \Gamma^a_{\mu\nu}\dot{x}^\mu \dot{x}^\nu = 0 \tag{13.24}$$

and we have the geodesic equation (13.16) once again.

In GR, for spacelike geodesics, the length along the curve is a minimum with respect to spatial changes, and it may be either minimal or maximal with respect to timelike changes. For timelike geodesics the length is a maximum of the proper time along the curve.

Finding geodesics. The geodesic equation is important, but do not let me find you trying to calculate geodesics using it! Rather, use the Euler–Lagrange method, and take advantage of symmetries where possible. For example, if g_{ab} does not depend on some coordinate then the associated canonical momentum is conserved along the geodesic. This observation can be converted into a statement about the tangent vector, as follows.

If $\mathcal{L}$ is independent of one specific coordinate x^z (say) then

$$\frac{1}{2}\frac{\partial \mathcal{L}}{\partial \dot{x}^z} = g_{z\mu}\dot{x}^\mu = \dot{x}_z = \text{constant}.$$

But $\dot{x}^a = t^a$ where $\mathbf{t}$ is the tangent to the worldline, so the result is $t_z = $ constant. In words, *if the metric does not depend on the coordinate x^z then the covariant component t_z of the tangent vector is a conserved quantity along an affinely parameterized geodesic.*.

The Euler–Lagrange equations are sufficient to fix a geodesic, but it is often simpler to drop one of them and employ instead the equation

$$|g_{\mu\nu}\dot{x}^\mu \dot{x}^\nu| = 1 \tag{13.25}$$

for non-null geodesics, which holds when the parameter is distance s along the curve (or c times proper time for worldlines). For null geodesics one has

$$g_{\mu\nu}\dot{x}^\mu \dot{x}^\nu = 0. \tag{13.26}$$

Also, by using (10.30) and (10.20) the geodesic equation can be written

---

Geodesic equation again

$$\ddot{x}_a = \tfrac{1}{2}\left(\partial_a g_{\mu\nu}\right)\dot{x}^\mu \dot{x}^\nu. \tag{13.27}$$

(exercise 13.5). This forms shows explicitly the conservation of $\dot{x}_z$ when g_{ab} is independent of x^z, already noted above, and is useful for rapidly obtaining suitable equations in that case.

Example 13.2 Find the equations for timelike geodesics in the uniform gravitational field described in Chapter 3.
Solution.
We treat the metric given by (3.1) which we repeat here:

$$ds^2 = -\alpha^2(x)dt^2 + dx^2 + dy^2 + dz^2. \tag{13.28}$$

where $\alpha = ce^{kx}$. Hence

$$\mathcal{L} = g_{\mu\nu}\dot{x}^\mu \dot{x}^\nu = -\alpha^2 \dot{t}^2 + \dot{x}^2 + \dot{y}^2 + \dot{z}^2. \tag{13.29}$$

We adopt proper time as the affine parameter. We have

$$\frac{\partial \mathcal{L}}{\partial t} = 0, \qquad \frac{\partial \mathcal{L}}{\partial \dot{t}} = 2\alpha^2 \dot{t}$$

so the EL equation relating to the coordinate t is

$$\frac{\mathrm{d}}{\mathrm{d}\tau}\left(2\alpha^2 \dot{t}\right) = 0 \qquad \Rightarrow \qquad \frac{\mathrm{d}^2 t}{\mathrm{d}\tau^2} = -\frac{2}{\alpha}\frac{\mathrm{d}\alpha}{\mathrm{d}x}\dot{x}\dot{t}. \tag{13.30}$$

Similarly, for coordinate x we find

$$\frac{\partial \mathcal{L}}{\partial x} = 2\alpha \frac{\mathrm{d}\alpha}{\mathrm{d}x}\dot{t}^2, \quad \frac{\partial \mathcal{L}}{\partial \dot{x}} = -2\dot{x} \quad \Rightarrow \quad \frac{\mathrm{d}^2 x}{\mathrm{d}\tau^2} = -\alpha \frac{\mathrm{d}\alpha}{\mathrm{d}x}\dot{t}^2. \tag{13.31}$$

These equations are eqns (3.10), (3.11).

13.3 Inertial motion

Any region of spacetime is locally flat, and therefore is described by the Minkowski metric η_{ab} in a suitable coordinate system, in the vicinity of one point of that coordinate system, to first order in coordinate changes. Let those locally Minkowskian coordinates be $X^a = (T, X, Y, Z)$. The equation of inertial motion in this flat region is that of no acceleration, i.e.

$$\frac{\mathrm{d}^2 X^a}{\mathrm{d}\tau^2} = 0.$$

Now introduce a change of coordinates to some arbitrary set (t, x, y, z). Since $\mathrm{d}X^a$ is a 4-vector it can be expressed in terms of the new coordinates as $\mathrm{d}X^a = (\partial X^a / \partial x^\lambda)\mathrm{d}x^\lambda$, so the equation of motion reads

$$\frac{d}{d\tau}\left(\frac{\partial X^a}{\partial x^\lambda}\frac{dx^\lambda}{d\tau}\right) = 0 \tag{13.32}$$

$$\Rightarrow \quad \frac{\partial^2 X^a}{\partial x^\mu \partial x^\lambda}\frac{dx^\mu}{d\tau}\frac{dx^\lambda}{d\tau} + \frac{\partial X^a}{\partial x^\lambda}\frac{d^2 x^\lambda}{d\tau^2} = 0 \tag{13.33}$$

Now premultiply by $\partial x^b/\partial X^a$ in order to get δ^b_λ in the second term, hence

$$\frac{\partial x^b}{\partial X^a}\frac{\partial^2 X^a}{\partial x^\mu \partial x^\lambda}\frac{dx^\mu}{d\tau}\frac{dx^\lambda}{d\tau} + \frac{d^2 x^b}{d\tau^2} = 0. \tag{13.34}$$

After relabelling this can be written

$$\frac{d^2 x^a}{d\tau^2} + \Gamma^a_{\mu\nu}\frac{dx^\mu}{d\tau}\frac{dx^\nu}{d\tau} = 0 \tag{13.35}$$

by using (12.26).

We have the geodesic equation yet again! What is going on?

Mathematically, the above shows that the geodesic equation is the equation for a curve which proceeds in a straight line in each local tangent space. So far, so unsurprising. But we presented the argument in a physical way, by talking about *inertial motion*. We said that inertial motion is that kind of motion in which there is no acceleration relative to a locally Minkowskian frame. By the strong equivalence principle, this *must* be the equation of free-fall motion. So we have now derived from the equivalence principle, and from the assertion that spacetime is a manifold, an important statement about motion under gravity: we can now assert that the equation of particle motion in free fall is the equation for a geodesic. We have thus derived one of the basic elements of GR which was asserted in Chapter 2. (Another way to proceed is to argue that the Einstein field equation is the fundamental equation, and then the equation for motion of test particles can be derived from that; we shall present this approach in Chapter 15.)

13.3.1   Inertial forces in flat spacetime

The equation of motion (13.35) applies equally well in flat spacetime as in curved spacetime. In flat spacetime the term involving the affine connection will be zero in any inertial frame, but in accelerating frames it can be non-zero, and it exactly and completely gives the forces called *inertial forces* experienced in accelerating frames. For example, in a rotating frame it describes the centrifugal and Coriolis forces; the reader is invited to confirm this (exercise).

13.4   Conservation laws and Killing vectors

We noted in (13.27) and the discussion leading up to it that if the metric is independent of one or more coordinates, then a corresponding property of a particle in free fall is conserved. In particular, if the metric is independent of time then p_0 is conserved, where $\mathbf{p} = m\mathbf{u}$ is the momentum 4-vector (11.2), so we may as well call $-p_0$ (note, not p^0) the energy of the particle. More generally, the canonical momentum

$$p_a \propto (\partial \mathcal{L}/\partial \dot{x}^a) \tag{13.36}$$

is often judged to be the best way to define and think of particle momentum, in preference to $m\mathbf{u}$, in which case one regards momentum as a one-form, not a vector, in the first instance.

Now let us consider the quantity $\mathbf{K} \cdot \mathbf{p}$ where $\mathbf{p}$ is the 4-momentum of a particle in free fall, and $\mathbf{K}(x)$ is a vector field to be discovered. We want to find a $\mathbf{K}$ such that $\mathbf{K} \cdot \mathbf{p}$ is conserved for free-fall motion. Now the change in any quantity, as one moves along some curve, is given by the derivative of that quantity along the curve. So we are seeking a $\mathbf{K}$ such that

$$\frac{\mathrm{d}}{\mathrm{d}\tau}(\mathbf{K} \cdot \mathbf{p}) = 0 \qquad \Leftrightarrow \qquad u^\lambda \nabla_\lambda (K_\mu p^\mu) = 0 \tag{13.37}$$

(c.f. (10.31)). Hence

$$p^\lambda (\nabla_\lambda K_\mu) p^\mu + p^\lambda K_\mu \nabla_\lambda p^\mu = 0 \tag{13.38}$$

where we used that $\mathbf{p}$ is itself proportional to the 4-velocity. The second term here is $K_\mu (p^\lambda \nabla_\lambda p^\mu)$ and this is zero since p^μ satisfies the geodesic equation. The first term is $p^\lambda p^\mu \nabla_\lambda K_\mu$. This is the complete contraction of a symmetric tensor $(p^\lambda p^\mu)$ with another tensor, so we deduce that it will give zero if the second tensor is antisymmetric. Hence we find that any $\mathbf{K}$ such that $\nabla_\lambda K_\mu$ is antisymmetric will imply the conservation of $\mathbf{K} \cdot \mathbf{p}$. This condition on $\mathbf{K}$ may be expressed:

Killing's equation

$$\nabla_a K_b + \nabla_b K_a = 0 \tag{13.39}$$

This equation, named after the German mathematician Wilhelm Karl Joseph Killing, offers us a partial differential equation which can be used, in principle, to find vector fields $\mathbf{K}(x)$ which capture a conservation law: the conservation of the component of $\mathbf{p}$ in the direction $\mathbf{K}$ for a test body in free fall. We will show in Chapter 15 that such fields also capture an *isometry*, which is a symmetry of spacetime itself. A vector field satisfying the equation is called a *Killing vector field*. This is usually abbreviated to *Killing vector*. The properties of Killing vectors are discussed further in Section 15.5.

13.5 Fermi–Walker transport

Let us now recall the notion of *tetrad* introduced in Section 11.2, and especially the use of a tetrad to deduce the observations made by some observer. We use the word 'observation' as a shorthand for the measurements which may be made at any given event by an inertially moving observer at that event, using apparatus that is not moving relative to that observer. We then allow for accelerating observers by using a sequence of inertial observers.

As explained in Section 11.2, observations in the LIF where such an observer is momentarily at rest can be obtained through the use of an orthonormal tetrad attached to the observer. Such a tetrad consists of four vectors $\hat{\mathbf{e}}_{(a)}$ defined such that

$$\hat{\mathbf{e}}_{(a)}(\tau) \cdot \hat{\mathbf{e}}_{(b)}(\tau) = \eta_{ab} \tag{13.40}$$

and

$$\hat{\mathbf{e}}_{(0)}(\tau) = \mathbf{u}(\tau)/c \tag{13.41}$$

where τ is proper time along the observer's worldline, and $\mathbf{u}$ is the 4-velocity of the observer. Thus one of the tetrad vectors is a unit vector along the worldline, and the others, being orthogonal to this, indicate purely spatial directions relative to the observer.

As one moves along the worldline, the tetrad evolves so as to keep track of the changing direction of the trajectory through spacetime. This evolution is partly, but not completely, specified by the above pair of equations. These equations uniquely determine one of the vectors, and only determine that the others are of unit length and orthogonal to the first and each other. In physical terms, one can imagine constructing a small cubic cabin whose sides are along the vectors $\hat{\mathbf{e}}_{(1)}$, $\hat{\mathbf{e}}_{(2)}$, $\hat{\mathbf{e}}_{(3)}$. As the cabin follows the observer's worldline it may be spinning and tumbling in any way.

We would like now to specify that form of evolution of the tetrad which corresponds to a cabin which is not spinning or tumbling. To be precise, what we require is

1. Vector $\hat{\mathbf{e}}_{(0)}$ is along $\mathbf{u}$.
2. Vectors $\hat{\mathbf{e}}_{(i)}$ are orthogonal to $\hat{\mathbf{e}}_{(0)}$ and each other.
3. All the tetrad members remain of unit length.
4. Linear combinations of $\hat{\mathbf{e}}_{(i)}$ that are orthogonal to the acceleration $\mathbf{a}$ do not evolve.

It is the last requirement which 'pins down' the cabin. The observer's acceleration $\mathbf{a}$ will be purely spatial in the LIF, and absence of rotation is defined as absence of rotation of spatial vectors that are orthogonal to $\mathbf{a}$. Combined with the rigidity (i.e. orthonormality) of the tetrad, this is sufficient to fix the evolution of the tetrad completely. The solution is that the evolution of the tetrad is given by:

Fermi–Walker transport

$$\frac{\mathrm{d}\hat{\mathbf{e}}_{(b)}}{\mathrm{d}\tau} = \frac{1}{c^2}\left((\mathbf{a} \cdot \hat{\mathbf{e}}_{(b)})\mathbf{u} - (\mathbf{u} \cdot \hat{\mathbf{e}}_{(b)})\mathbf{a}\right) \tag{13.42}$$

You can easily confirm that this equation gives $\mathrm{d}\hat{\mathbf{e}}_{(0)}/\mathrm{d}\tau = \mathbf{a}/c$, and the orthonormality is preserved (exercise 13.7). Also, vectors orthogonal to both $\mathbf{u}$ and $\mathbf{a}$ do not evolve, which is what we require.

A spin which is accelerated such that there is no torque in each instantaneous rest frame will undergo a Fermi–Walker transport. In the case of flat spacetime, such motion results in the Thomas precession.

Suppose now that an observer is located in a cabin moving along some trajectory (the cabin may or may not be in free fall). Can the observer determine whether or not the cabin is undergoing Fermi–Walker transport? The answer is yes, because the observer always has the option of noting the behaviour of isolated objects released inside the cabin (and therefore in free fall, whether or not the cabin is). Such objects will neither gain nor lose angular momentum from one moment to the next; therefore their amount of rotation relative to a non-rotating cabin will not change. The observer can check for this. In particular, the use of a gyroscope mounted in a gimbal is a convenient way to check for such effects.

13.5.1   The freely falling cabin

Since the cabin undergoing Fermi–Walker transport is not necessarily in free fall, its tetrad is not necessarily undergoing a parallel transport. However, in the case of a freely falling cabin, the velocity **u** is undergoing a parallel transport, and the acceleration **a** is zero. By substituting **a** = 0 into (13.42) we discover that the whole tetrad is then undergoing a parallel transport.

Pick one of the spatial members of such a tetrad. As time goes on, it occupies a sequence of positions, and one can imagine it laid down in spacetime at each successive position, defining a sort of railway track. The rails of the track are parallel to the worldline; the lines across the rails (called sleepers in railway terminology) are locations of a tetrad vector at a sequence of discrete moments. Now add to your imagination the other spatial parts of the tetrad, and thus form a spatial box at each moment. This box traces out a tube in spacetime. Inside the tube we have an inertial frame. It is called a freely falling frame, and it is interesting because it can be extended a long way in the temporal direction—all the way until it encounters some extreme behaviour such as a spacetime singularity. Seen from outside, the tube may appear to bend and twist its way through spacetime. It might form a helix around a massive object, for example. Experienced from inside, there is no sense of any twisting or bending; all is calm, all is right. Around the extended worldline compiled, gravity is purely tidal and mild.

13.6   Gravitational redshift

The gravitational redshift is the observation that if electromagnetic waves propagate in empty space from A to B, then the frequency at A (as determined by a standard clock at A) is not in general the same as the frequency at B (as determined by a standard clock at B), and this difference cannot always be wholly accounted for as a special relativistic Doppler shift. Specifically, in a stationary situation, the gravitational shift is predicted to be non-zero between any pair of events at different gravitational potential, for measuring devices at rest relative to the stationary lattice. If A is at a low position, close to some gravitating body, and B is at a high position, far from any gravitating body, then $\omega_B < \omega_A$—the frequency is lower at B. We therefore call it a *red* shift. We can imagine the observer at B saying: 'look, this wave came from a distant caesium atom emitting on the hyperfine transition which I know to be at 9 GHz in all the caesium atoms near me, but when I receive the wave I find its frequency to be lower, only 8 GHz; it has been red-shifted owing to gravitational redshift, because the atom which emitted it is low down in the gravitational potential well and the wave had to climb out of the well in order to reach me.'

We shall now make this qualitative statement more precise. There is a beautiful argument whereby a simple formula of great generality can be obtained.

It will be convenient to discuss the phenomenon in terms of the energy and momentum of 'a photon'. This is merely for convenience; all the statements can equally well be made about the energy and momentum of a section of a continuous stream of classical electromagnetic radiation, and about the frequency and wave-vector of such an entity. In either case, the energy is proportional to the frequency, and the momentum is proportional to the wave-vector.

First, recall eqn (11.10): if a photon has 4-momentum $\mathbf{p}$ then its energy, as observed by a detector whose 4-velocity is $\mathbf{u}$, will be

$$E = -\mathbf{p} \cdot \mathbf{u}. \tag{13.43}$$

It follows that the ratio of frequencies observed by detectors at A, B with 4-velocities $\mathbf{u}(A)$, $\mathbf{u}(B)$ is

---

redshift

$$\frac{\omega_B}{\omega_A} = \frac{p_\mu(B)u^\mu(B)}{p_\mu(A)u^\mu(A)} \tag{13.44}$$

This is the completely general formula for gravitational redshift.

Now let us specialize to detectors which are at rest relative to the local coordinate system being adopted at A and B, while allowing a completely general spacetime. In this case $u^i = 0$. But we have $u_\mu u^\mu = -c^2$ always, so $u^0 = c/\sqrt{-g_{00}}$ at either A or B (c.f. (2.34)). Substituting this into (13.44), along with $u^i = 0$, we find

---

Gravitational redshift, detectors fixed in the coordinates

$$\frac{\omega_B}{\omega_A} = \frac{p_0(B)}{p_0(A)} \sqrt{\frac{g_{00}(A)}{g_{00}(B)}} \tag{13.45}$$

This is a remarkably simple formula. It asserts that the zeroth component of the 4-momentum directly gives the frequency we are interested in, as long as we remember to make allowance for the relationship between proper time and coordinate time at either detector, furnished by the local value of the metric. If one now wishes to consider moving detectors at A or B, it suffices to apply the special-relativistic Doppler shift formula in a LIF at either location. In this sense, our result is still completely general.

Finally, let us specialize to stationary conditions:

$$\partial_0 g_{ab} = 0 \tag{13.46}$$

where we suppose that this is true all along the photon's worldline. In this case we have a conservation law for the zeroth covariant component of the tangent vector along the worldline (c.f. (13.27)), but the photon 4-momentum is proportional to that tangent vector, with a fixed constant of proportionality, so we can deduce

$$p_0(B) = p_0(A). \tag{13.47}$$

Hence

Gravitational redshift, stationary conditions

$$\frac{\omega_B}{\omega_A} = \sqrt{\frac{g_{00}(A)}{g_{00}(B)}} \tag{13.48}$$

In stationary conditions, for any photon worldline the gravitational redshift can be obtained purely from the metric at the emission and reception events. Note that the result applies to photons on complicated rosette-shaped orbits as well as other orbits, and it applies for the spacetime around rotating as well as static stars and black holes.

Example 13.3 A disc is in rigid rotation at constant angular velocity in flat spacetime. A source A on the edge of the disc emits a photon which is subsequently absorbed by a detector B at another location on the edge of the disc; see Fig. 13.2. What is the ratio between the emitted and received frequencies, as determined by a local observer on the disc at each location?
Solution.
If we were to argue in terms of the special relativistic Doppler effect then one method of calculation would be to use the Doppler effect formula twice to find two shifts, being careful to get angles and signs right, and after looking carefully at our equations we would arrive at an answer. But if instead we adopt the GR approach, we note that the conditions are stationary and we can construct a coordinate frame fixed relative to the disc, and by symmetry it is clear that we will find $g_{00}(A) = g_{00}(B)$ for locations at the same radius. Hence, using (13.48), we predict $\omega_B = \omega_A$. This simple (and correct) method illustrates the power of the geometric approach.

An interesting feature of (13.48) is that the *coordinate* period of the electromagnetic wave is the same at the two events:

$$\Delta t_A = \Delta t_B. \tag{13.49}$$

This can be deduced by noting that the observed frequency is by definition the inverse of the proper time between reception events of successive wavefronts at the detector, and this is related to the coordinate time by $\sqrt{g_{00}}$ for events at the same spatial coordinate location.

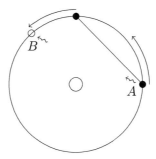

Fig. 13.2 Light sent between observers on a disc in rigid rotation in flat spacetime.

Einstein proposed three tests for General Relativity, in the interests of finding explicit departures from Newtonian gravity: the advance of the perihelion of Mercury, the bending of light rays by twice the amount one might reasonably expect on Newtonian lines, and the gravitational redshift. The first of these was confirmed by Einstein himself by calculation and comparison with existing astronomical data. The second was observed at just sufficient accuracy in 1919, and subsequently with improved accuracy. The redshift was not observed unambiguously until after Einstein's death in 1955. The first definitive observation was performed by Pound and Rebka in 1959 at Harvard University's Jefferson Tower. They measured the frequency shift of a 14.4 keV gamma ray falling through 22.6 m. Pound and Rebka were able to measure the shift in energy (a few parts in 10^{14}) by taking advantage of the Mössbauer effect, which was itself only recently discovered at the time. More recent experiments such as *Gravity Probe A*, a space-based experiment launched in 1976, have tested the prediction to a precision of some parts in 10^5.

Gravitational redshift is an important part of the astronomical toolkit. It is used to interpret the cosmic expansion, and to understand the emission spectrum of accretion disks. It plays an important role in the study of the cosmic microwave background radiation, through the Sachs–Wolfe effect.

13.7 Cause and effect

Considered as a problem in calculus, the geodesic equation is a set of N coupled second-order differential equations for N functions x^a, where N is the dimensionality of the manifold. It follows that the solution will involve two constants of integration for each function. These can be fixed, for example, by specifying a starting value and the first derivative for each function. Translated into geometric terms, this is saying that once one has fixed a point and an initial direction, one has specified a unique geodesic passing through that point in that direction. This is the way that GR incorporates the notion of *cause and effect* for the motion of test particles.

Cause and effect is itself quite a subtle notion when one tries to unpack it in philosophical or metaphysical terms. For present purposes we will not need a thorough metaphysical analysis, but it is worth noticing that the idea of 'initial conditions–motion–final conditions' with which

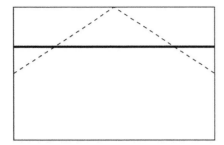 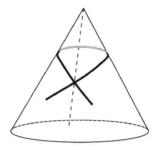

Fig. 13.3 Take a flat rectangular piece of paper and draw a pair of dashed lines from a point at the centre of one of the long edges, making equal angles with the edge, as shown. Draw also a full line right across the paper, parallel to a long edge, and intersecting the dashed lines about half way down them. Now roll your paper into a cone in such a way that the dashed lines lie on top of one another. Now look at where your geodesic (the full line) has gone. This illustrates that in a suitably curved space a geodesic can intersect itself.

we are familiar in everyday life and in Newtonian mechanics is connected to the presence, in the theory, of second-order differential equations for the position coordinates of moving entities subject to the influence of a given environment.

The uniqueness of the whole geodesic in a given direction at a given point serves to fulfil the criteria associated with the notion of cause and effect. However, there is a further subtlety. What if a geodesic winds around in a manifold and comes back to where it started, but now in a new direction? You can find such geodesics on a cone, for example (see Fig. 13.3). When this happens we may say that the two crossing lines are parts of a single geodesic. In the case of spacelike geodesics, this does not pose any special conundrum for physics, but in the case of timelike geodesics, it does. If a worldline brings a physical entity back to the very event where it started out, it suggests that information is being brought to that event from its own future, leading to various causality paradoxes. One can find examples of spacetimes which appear to be allowed by GR in which such *closed timelike curves* can exist. However, they involve rather exotic scenarios and there has been no report, to date, of any impact of such studies on observed physical phenomena.

A different type of difficulty that can arise is the presence of singularity in the functions used to describe geodesics. Singularities commonly arise in two forms in GR: *coordinate singularity* and *curvature singularity*. The former is not necessarily anything to do with the manifold, but may be merely owing to the use of poorly chosen coordinates near a given event. The curvature singularity, on the other hand, is a property of the manifold itself and at a curvature singularity the analysis breaks down. From a physical point of view, that would be a strong hint that the physical theory under consideration is itself no longer valid; this is a matter for further research in physics.

Exercises

13.1 If vectors **v** and **w** are parallel-transported along a curve, show that $\mathbf{v} \cdot \mathbf{w}$ remains constant. Hence show that if a geodesic is timelike (or null or spacelike) at some point, it is timelike (or null or spacelike) at all points.

13.2 Would you say that the marker lines on a steel measuring tape, of the type used by carpenters and builders, are parallel to one another? Obtain such a tape. Explore whether or not, by manipulating the tape, you can cause the one part of the tape to lie flat on top of another part while they cross at a non-zero angle. Explain your observations in the language of manifolds, curvature and parallel transport.

13.3 Revisit the example treated in Fig. 13.1 and equation (13.13).
(i) What happens when θ is small? In which direction is the rotation? What does the cone look like? What happens when θ is close to, but just less than $\pi/2$? What happens when $\theta = \pi/3$?
(ii) Show that the geometric construction yields the correct result for transport through any ϕ, eqn (13.13).
(iii) Also consider the following argument. If we stand at $\phi = 0$ and then walk a short way along the ϕ direction, carrying a pole whose two ends rest on wheels which are geared to rotate by equal amounts (and there is no slipping on the ground), then the two wheels are following arcs of radius $R\sin(\theta \pm L/2R)$ where L is the length of the pole. Show that for small ϕ this implies a rotation of the pole in agreement with (13.13).
(iv) How does the behaviour of such a pole relate to the geodesics?

13.4 **Affine parameter** (i). Show that if we have a geodesic parameterized such that it satisfies (13.18) for some function f, then if the parameter is changed from u to $s(u)$, so that the same curve is defined by new functions $X^a(s)$ (where $X^a(s) = x^a(u)$), then one obtains the form (13.18) but with

$$\left(f(u)\frac{\mathrm{d}s}{\mathrm{d}u} - \frac{\mathrm{d}^2 s}{\mathrm{d}u^2} \right) \left(\frac{\mathrm{d}s}{\mathrm{d}u} \right)^{-2} \frac{\mathrm{d}X^a}{\mathrm{d}s} \quad (13.50)$$

on the right-hand side. It follows that the new parameter will be affine if it is a solution of the differential equation $\ddot{s} = f(u)\dot{s}$.
(ii). Show that if u is affine then so is any s related linearly to u.

13.5 By starting out from the covariant form of the intrinsic derivative (10.30), and employing (10.20), obtain (13.27).

13.6 Obtain the Christoffel symbols for a rotating frame in flat spacetime and confirm that (13.35) then gives the centrifugal and Coriolis forces.

13.7 Confirm that if a tetrad evolves by Fermi–Walker transport (13.42) then after any small time $\mathrm{d}\tau$,

$$(\hat{\mathbf{e}}_{(b)} + \mathrm{d}\hat{\mathbf{e}}_{(b)}) \cdot (\hat{\mathbf{e}}_{(c)} + \mathrm{d}\hat{\mathbf{e}}_{(c)}) = \eta_{bc} + O(\mathrm{d}\tau^2).$$

13.8 If two vectors are Fermi–Walker transported along some worldline (not necessarily geodesic), show that their scalar product $\mathbf{v} \cdot \mathbf{w}$ is preserved.

13.9 Find the gravitational redshift of light emitted from the surface of the Sun and observed far away.

13.10 Find the ratio of the rate of a clock in circular orbit around the Earth to that of one on the ground at the north pole. [*Ans.* $1 + (GM/c^2)(1/R - 3/2r)$]

13.11 Show that in static conditions the temperature T of a body in thermal equilibrium is distributed such that $T(|g_{00}|)^{1/2}$ is uniform (the *Ehrenfest–Tolman effect*). [Hint: energy flux]

13.12 **Rindler metric**. Recall the Rindler metric from Chapter 3:

$$\mathrm{d}s^2 = -(ax)^2\mathrm{d}t^2 + \mathrm{d}x^2 + \mathrm{d}y^2 + \mathrm{d}z^2.$$

(i) Find the timelike and null geodesics, working in the (t, x, y, z) coordinates.
(ii) Introduce coordinates $T = x\sinh(at)$, $X = x\cosh(at)$, $Y = y$, $Z = z$ and show that in these coordinates all the geodesics found in part (i) are straight lines.

14

Physics in curved spacetime

In this chapter we shall examine physical effects such as electromagnetism and hydrodynamics (fluid flow) in curved spacetime. This will also serve to illustrate how physics is approached in general in the presence of gravity.

We are now contemplating a task in theoretical physics, and one approach would be to say that in the absence of any other prior knowledge, one simply makes informed guesses of what types of physical theory merit to be explored. One then elucidates what the observable phenomena would be, and proceeds to empirical tests. Equally, one may already have data ready to be elucidated, and one uses this to shape ones guesses. By 'merits to be explored', one might include, for example, the requirement that the theory respects the strong equivalence principle. Or, to put it slightly more generally and in a more cogent way, one chooses to construct ideas using purely the methods of tensor algebra. So then the main criteria are simplicity (Occam's razor) and tensorial correctness. This is the method we used in order to obtain the Maxwell equations in Chapter 13 of Volume 1. All we need to do now is use the more general notion of tensors in any Riemannian or pseudo-Riemannian manifold.

Another very fruitful method is to bring in everything one knows about physics in the absence of gravity—to be precise, in flat spacetime—and then seek to generalize from there. We can do this in two steps. First, we take equations and ideas in flat spacetime, and generalize to reference frames moving in an arbitrary way but still within flat spacetime. Then, in the second step, we propose that the equations we have thus obtained remain valid in curved spacetime. This is not guaranteed to be true, but it is a good starting point and often it turns out that it is true.

Equations that retain their form under Lorentz transformations are called *Lorentz covariant*. Equations that retain their form under general coordinate transformations are called *generally covariant*. So one might say that the subject of the present chapter is general covariance.

If one has some Lorentz covariant equations, then generally covariant versions of them can often be obtained by the following simple recipe:

1. judiciously replace partial derivatives with covariant derivatives;

Relativity Made Relatively Easy: General Relativity and Cosmology. Volume 2. Andrew M. Steane,
Oxford University Press. © Andrew M. Steane 2021. DOI: 10.1093/oso/9780192895646.003.0014

2. replace ordinary derivatives along curves with intrinsic derivatives (but not quite always; c.f. the discussion of 4-velocity in Section 10.2);

3. replace η_{ab} by g_{ab}.

4. replace $\int \mathrm{d}^4x$ by $\int \sqrt{|g|}\,\mathrm{d}^4x$

One should not follow this procedure blindly. One should pay attention at each stage to what one is doing, in case some further consideration relating to gravity might be relevant. For example, if the tidal effects of gravity are liable to be important, then the above recipe may not always do the whole work of arriving at a correct theory. For the examples treated in the present chapter, however, it is sufficient. The procedure outlined above, or the link between gravitation and other phenomena which can be expressed this way, is called *minimal coupling*.[1]

14.1 Electromagnetism

We already presented the field equations of classical electromagnetism in curved spacetime in the introductory Chapter 2; eqns (2.25), (2.26), and the electromagnetic force term was included in the equation of motion for test particles, (2.10). We repeat the field equations here:

$$\nabla_\lambda \mathbb{F}^{\lambda b} = -\mu_0 j^b \tag{14.1}$$

$$\partial_c \mathbb{F}^{ab} + \partial_a \mathbb{F}^{bc} + \partial_b \mathbb{F}^{ca} = 0. \tag{14.2}$$

It is extremely easy to obtain these equations from the versions which apply in flat spacetime; the 'recipe' indicated above is perfectly sufficient. All we need to do now is add that, although the recipe is not guaranteed to give the right theory, it does on this occasion—or, at least, at the time of writing there is no experimental evidence to suggest otherwise. As far as we know this is indeed how electromagnetic fields behave in the presence of gravity in the classical (i.e. not quantum mechanical) limit. However, even the simple recipe is not unambiguous on this occasion—see exercise 16.6 of Chapter 16.

Note that the second electromagnetic field equation employs partial derivatives not covariant derivatives. This is because we have made use of a useful fact. For any vector A^a and for any **antisymmetric** tensor F^{ab},

$$\nabla_b A^a - \nabla_a A^b = \partial_b A^a - \partial_a A^b \tag{14.3}$$

$$\nabla_c F^{ab} + \nabla_a F^{bc} + \nabla_b F^{ca} = \partial_c F^{ab} + \partial_a F^{bc} + \partial_b F^{ca}. \tag{14.4}$$

This is because for these cases the Γ terms cancel, as you are invited to confirm.[2] These facts are useful in the treatment of any physical theory in curved spacetime. In the case of electromagnetism (14.4) serves to simplify the second of the field equations. From this result we can deduce that the field tensor can be related to a vector potential through

---

[1] See also after eqn (28.18) for another usage of the same term.

[2] Eqn (14.3) can be generalized by defining the *exterior derivative* mentioned in the footnote after eqn (11.27). Such an operation gives a tensorial result when it is applied to p-forms (i.e. totally antisymmetric tensors of type $(0,p)$).

$$\mathbb{F}^{ab} = \partial_a A^b - \partial_b A^a \tag{14.5}$$

and then (14.3) assures us that the result is indeed tensorial in the generally covariant sense.

From 14.5 it follows that we can make gauge transformations of the form

$$A_a \rightarrow A_a + \nabla_a \chi = A_a + \partial_a \chi \tag{14.6}$$

without affecting the fields, where χ is any scalar function. In particular, one can choose χ so as to obtain the **Lorenz gauge** where A^a satisfies

$$\nabla_\lambda A^\lambda = 0, \tag{14.7}$$

and then (exercise for the reader), the first field equation can be written

$$\nabla^\mu \nabla_\mu A_a = -\mu_0 j_a \tag{14.8}$$

Eqn (14.8) looks like a wave equation, but if only that were so! In fact it is not; the Γ symbols in ∇_a make it a lot more complicated, and even in the absence of gravitating masses, the electromagnetic field itself carries energy-momentum and stress and consequently causes spacetime curvature.

We have not yet interpreted j_a, except to say that it is a 4-vector which acts as source of the field. Its interpretation is easy in any LIF: there it is the familiar 4-current, with components equal to the local electric charge density and flux. Using the same argument as for the special relativistic case, one can also obtain

$$\nabla_\mu j^\mu = 0 \tag{14.9}$$

by taking the divergence of the first field equation, and this is a continuity equation which expresses charge conservation.

The electromagnetic field acts on its sources, pushing them. One can obtain the equation of motion of charged particles either from energy conservation, a method will shall present in Chapter 15 (and exercise 14.6), or by making the following intelligent guess:

$$\frac{d\mathbf{p}}{d\tau} = q\mathbb{F} \cdot \mathbf{u} \tag{14.10}$$

where $\mathbf{p} = m\mathbf{u}$ is the 4-momentum of the particle. Using (10.28) and (10.29) we thus obtain

$$m\frac{Du^a}{d\tau} = q\mathbb{F}^a{}_\lambda u^\lambda \tag{14.11}$$

which is eqn (2.10). In terms of the worldline $x^a(\tau)$ this is

$$\frac{d^2 x^a}{d\tau^2} + \Gamma^a{}_{\mu\nu}\frac{dx^\mu}{d\tau}\frac{dx^\nu}{d\tau} = q\mathbb{F}^a{}_\lambda \frac{dx^\lambda}{d\tau}. \tag{14.12}$$

Thus we see that the electromagnetic force acts to push the charged particle away from the geodesic which it would otherwise follow.

14.2   Fluid flow and continuous media

The study of fluid flow is called *hydrodynamics*. Many of the concepts and results apply also to solids.

Let us start with a simple result concerning energy conservation. If we take a sufficiently small region of space, then Special Relativity is sufficient, and energy conservation for a simple compressible system is expressed by the thermodynamic result[3]

$$\mathrm{d}(\rho c^2 V) = T\mathrm{d}S - p\mathrm{d}V. \tag{14.13}$$

This is just like the non-relativistic formula, except that the total energy (including mass energy) is needed on the left. The pressure term accounts correctly for the increase of this energy by mechanical work and the entropy and temperature account for heat. It is useful to express the result also in terms of number density $n = N/V$ where N is some constant particle number. In the presence of pair creation and annihilation or other subatomic processes one uses for N a conserved quantity such as baryon number. In terms of n and $s \equiv S/N$ (14.13) becomes

$$c^2\mathrm{d}\rho = nT\mathrm{d}s + (\rho c^2 + p)\frac{\mathrm{d}n}{n}. \tag{14.14}$$

The conservation of baryon number is expressed by the continuity equation $\nabla_\mu(nu^\mu) = 0$ where u^a is the flow velocity of a continuous fluid of baryons.

Next we consider the motion of a continuous entity more generally, such as the flow of a fluid or the vibration of a solid. The final chapter of Volume 1 was devoted to the important stress-energy or energy-momentum tensor T^{ab}, which we are here calling simply the energy tensor. It was explained how and why the conservation of energy and momentum in flat spacetime is expressed by the formula

$$\partial_\mu T^{\mu b} \overset{\mathrm{LIF}}{=} 0. \tag{14.15}$$

where T^{ab} is the energy tensor of an isolated system. In the case of electromagnetism, for example, one has $\partial_\mu T^{\mu b}_{(\mathrm{em})} \neq 0$ if $T^{ab}_{(\mathrm{em})}$ is the energy tensor of the electromagnetic field alone, but

$$\partial_\mu T^{\mu b}_{(\mathrm{em})} + \partial_\mu T^{\mu b}_{(\mathrm{charged\ matter})} \overset{\mathrm{LIF}}{=} 0. \tag{14.16}$$

Staying in flat spacetime for the moment, but allowing arbitrary coordinate systems (which is the same as saying arbitrarily accelerating and wobbling reference frames, not just inertial ones), this same criterion (energy-momentum conservation) is expressed mathematically by

$$\nabla_\mu T^{\mu b} = 0. \tag{14.17}$$

Proceeding now to the case of curved spacetime, a subtlety arises. We retain (14.17) (c.f. eqn (16.3)) but its physical interpretation is subtly changed. In the absence of any gravitational effects, this equation would express conservation of energy and momentum of the matter and

---

[3]Eqn (14.13) is often referred to as the first law of thermodynamics, but in fact it implicitly invokes the second law as well as the first because it employs an entropy function.

(non-gravitational) radiation whose energy tensor is T^{ab}. In the presence of gravitational effects, on the other hand, we do not expect matter and (non-gravitational) radiation necessarily to conserve its total energy, because it can absorb and emit gravitational radiation, or, which amounts to the same thing, it can give rise to a different gravitational environment as it becomes configured differently. When we throw a ball upwards into the air, for example, its rest mass and kinetic energy contributes to T^{ab} (in any given reference frame) but that is all: T^{ab} does *not* include any notion of gravitational potential energy. But as we know, and as the geodesic equation predicts, such a ball will slow and come to a stop at the top of its trajectory. So its energy is *not conserved*. In our study of gravity, eqn (14.17) is no longer an expression of conservation of energy and momentum. Instead, it is the **equation of motion** of matter and non-gravitational radiation under the influence of gravity. If $\Gamma^a_{bc} = 0$ then we have $\partial_\mu T^{\mu b} = 0$ and the matter's energy is conserved;[4] if $\Gamma^a_{bc} \neq 0$ then we have $\partial_\mu T^{\mu b} \neq 0$ and the matter's energy is not conserved.

In order to investigate the sense in which (14.17) is an equation of motion, we can now develop the equations describing fluid flow.[5] We consider the case of an **ideal fluid**, which is defined to be a physical system having energy tensor:

Ideal or perfect fluid

$$T^{ab} = (\rho + p/c^2)u^a u^b + pg^{ab} \tag{14.18}$$

---

where u^a is the local fluid 4-velocity at any event. In this equation, T^{ab} is a second-rank tensor, and therefore in order for this to be a correct tensorial result, ρ and p must be tensors of rank zero—that is, invariant scalars. They can be related to physical measurements as follows. A *fluid element* is a small part of the fluid having a well-defined flow velocity; it may be composed of smaller things such as particles. ρc^2 is the energy density in the local rest frame of a fluid element; p is the pressure in the local rest frame of a fluid element. To be clear, ρ includes all contributions to the energy of the matter of the fluid in the rest frame of a fluid element, which is not necessarily (nor usually) the rest frame of any one particle in the fluid. In the case of a gas, for example, the pressure is owing to random motion of the molecules and this motion is present in the rest frame of the gas. ρ also includes internal energy such as rotational and vibrational energy and the energy of chemical bonds.

Substituting this T^{ab} into (14.17) gives[6]

$$u^a u^\mu \nabla_\mu(\rho + p/c^2) + (\rho + p/c^2)\nabla_\mu(u^\mu u^a) + g^{a\mu}\nabla_\mu p = 0 \tag{14.19}$$

where we used that the metric is covariantly constant (eqn (12.63)). By contracting with u_a and using $u_a u^a = -c^2$, one finds

$$-c^2 u^\mu \nabla_\mu(\rho + p/c^2) + (\rho + p/c^2)\left(-c^2\nabla_\mu u^\mu + u^\mu u_a \nabla_\mu u^a\right) + u^\mu \nabla_\mu p = 0. \tag{14.20}$$

---

[4]To avoid the lengthy expression 'matter and non-gravitational radiation' we often just say 'matter' and take it as understood that electromagnetic radiation is included. Sometimes one says 'matter and radiation' and then it is taken as understood that gravitational radiation is excluded.

[5]See Volume 1 Section 16.3 for a special relativistic introduction to these equations.

[6]In any equation, where ∇_a acts on a scalar one could equally put ∂_a, but here we retain ∇_a to make it manifestly obvious that the equation is generally covariant.

Now use that since $u_a u^a$ is constant, we have $u_a \nabla_\mu u^a = 0$ (eqn (12.69)). Hence the term in (14.20) involving this combination is zero, and after dividing through by $-c^2$ we obtain

$$\nabla_\mu(\rho u^\mu) + \frac{p}{c^2} \nabla_\mu u^\mu = 0. \tag{14.21}$$

This equation can be interpreted as a type of continuity equation for a flow whose 4-current is ρu^a. It should be 'read' as a statement about energy flow. It can also be written

$$\frac{\mathrm{D}\rho}{\mathrm{d}\tau} \equiv u^\mu \nabla_\mu \rho = -\left(\rho + p/c^2\right) \nabla_\mu u^\mu,$$

$$\text{or} \qquad \boldsymbol{\nabla_\mathbf{u}}\rho = -\left(\rho + p/c^2\right) \boldsymbol{\nabla} \cdot \mathbf{u} \tag{14.22}$$

where the second version shows the same result in index-free notation.

By substituting this result into (14.19) one can simplify the latter somewhat, obtaining

---

Equation of motion ('Euler equation') of ideal fluid

$$\left(\rho + \frac{p}{c^2}\right) u^\mu \nabla_\mu u^a = -\left(g^{a\mu} + \frac{u^a u^\mu}{c^2}\right) \nabla_\mu p,$$

$$\text{or} \qquad \left(\rho + \frac{p}{c^2}\right) \boldsymbol{\nabla_\mathbf{u}}\mathbf{u} = -\boldsymbol{\nabla} p - \frac{1}{c^2}\mathbf{u}\boldsymbol{\nabla_\mathbf{u}}p. \tag{14.23}$$

This is a very important and useful result, which we will apply to particle motion in Chapter 15, to stars in Chapter 18 and to cosmology in Chapters 23 and 24.

We shall now extract the behaviour in the weak field limit for a slow-moving fluid. From the definition of the covariant derivative, we have

$$u^\mu \nabla_\mu u^i = u^\mu \partial_\mu u^i + u^\mu \Gamma^i_{\nu\mu} u^\nu$$

$$\simeq \frac{\mathrm{d}u^i}{\mathrm{d}\tau} + \Gamma^i_{00} c^2 = \frac{\mathrm{d}u^i}{\mathrm{d}\tau} + \partial^i \Phi.$$

We take $g^{ab} = \eta^{ab} + h^{ab}$ with $|h^{ab}| \ll 1$ and a flow velocity $u^i \ll c$. Hence the spatial part of (14.23) becomes

$$(\rho + p/c^2)\left(\frac{\mathrm{d}\vec{u}}{\mathrm{d}\tau} + \vec{\nabla}\Phi\right) = -\vec{\nabla}p \tag{14.24}$$

where $\vec{\nabla}$ is the 3-gradient and $\vec{u}$ is the 3-velocity. This is the classical Euler equation for the fluid, with the inertia augmented by pressure (this is predicted by Special Relativity), and with a force term proportional to $\vec{\nabla}\Phi$. Notice how this gravitational force has made an appearance. It was not explicitly present in the starting eqn (14.23). That equation has the appearance of describing a fluid with no external force. However, gravity has had its effect by making the covariant derivative different from a partial derivative, through the Christoffel symbol—in other words, by influencing the metric.

Example 14.1 For a perfect fluid undergoing adiabatic stationary flow in a stationary gravitational field, show that along the flow lines

$$u_0(\rho + p/c^2)/n = \text{constant} \qquad \text{(Bernoulli equation)} \qquad (14.25)$$

where n is the density of a conserved particle number such as baryon number.

Solution.

Taking $c = 1$, the Euler equation (14.23) gives

$$(\rho + p)\frac{\mathrm{D}u_0}{\mathrm{d}\tau} = -\delta_0^\mu \partial_\mu p - u_0 \frac{\mathrm{D}p}{\mathrm{d}\tau}$$

where we used that ∇ acts like ∂ for a scalar, and $g_a^b = \delta_a^b$. In stationary conditions, p may change along the flow lines, but it does not change with time at any particular place, so $\partial_0 p = 0$. Hence we have

$$(\rho + p)\frac{\dot{u}_0}{u_0} = -\dot{p} = -(\dot{\rho} + \dot{p}) + \dot{\rho} = -(\dot{\rho} + \dot{p}) + (\rho + p)\dot{n}/n$$

where the dot signifies $\mathrm{D}/\mathrm{d}\tau$ and the last step used (14.14) in adiabatic conditions. Integrating, this gives

$$\int \frac{\mathrm{d}u_0}{u_0} = -\int \frac{\mathrm{d}(\rho + p)}{\rho + p} + \int \frac{\mathrm{d}n}{n}$$

and (14.25) follows.

14.2.1   Hydrostatic equilibrium

Hydrostatic equilibrium is an interesting special case of (14.23), and is important in the study of stellar structure. In such equilibrium, there is a coordinate choice in which the metric is static and the fluid's 4-velocity has no spatial part: $u^a = (u^0, 0, 0, 0)$. Using $u_\lambda u^\lambda = -c^2$ we then find $u^0 = c\sqrt{-1/g_{00}}$. Now the covariant derivative appearing on the left-hand side of (14.23) is

$$u^\mu \nabla_\mu u^a = u^0 \nabla_0 u^a = \partial_0 u^a + \Gamma_{\lambda 0}^a u^\lambda = \Gamma_{00}^a u^0 \qquad (14.26)$$

where we used that $u^i = 0$ so only one of the terms contributes, and in the last step we also used that u^0 is independent of time. Using (2.8) to obtain the connection coefficient from the metric, we have

$$\Gamma_{00}^a = \frac{1}{2} g^{a\lambda} \left(\partial_0 g_{0\lambda} + \partial_0 g_{\lambda 0} - \partial_\lambda g_{00} \right) = -\frac{1}{2} \partial^a g_{00} \qquad (14.27)$$

where we used that the time derivatives vanish for a static case.

By similarly reasoning from u^μ and the static conditions, the term $u^\mu \nabla_\mu p$ on the right-hand side of (14.23) also vanishes, and we have

$$\left(\rho + \frac{p}{c^2} \right) \frac{c^2}{g_{00}} \frac{1}{2} \partial^a g_{00} = -\partial^a p \qquad (14.28)$$

where we used that the covariant derivative of a scalar is equal to the partial derivative. After lowering the index and expressing the derivative on the left in terms of a log, we obtain

---

Equation of hydrostatic equilibrium

$$\partial_a p + \left(\rho c^2 + p\right) \partial_a \ln \sqrt{|g_{00}|} = 0 \qquad (14.29)$$

If we have an expression for p as a function of ρ (the *equation of state*) then this expression can be integrated immediately (exercises 14.7, 14.8). By introducing the gravitational potential Φ, defined such that $g_{00} = -c^2 \exp(2\Phi/c^2)$, the equation can be written

$$\partial_a p + \left(\rho + p/c^2\right) \partial_a \Phi = 0. \qquad (14.30)$$

The Newtonian equation for hydrostatic equilibrium is

$$\vec{\nabla} p + \rho \vec{\nabla} \Phi = 0. \qquad (14.31)$$

14.3 How General Relativity works

We have seen in this chapter that electromagnetic fields will change and fluids will flow differently when the gravitational environment changes. In subsequent chapters we will calculate the metric in the presence of some given matter configuration such as a star. But sometimes it is hard to tell what is the physical import of a given mathematical change—after all, some changes can be accounted for merely as coordinate relabelling. We saw in Chapter 11 how to use LIFs and tetrads to get unambiguous predictions for various types of observable phenomena. Here we will briefly reprise the ideas.

Suppose we have prepared the labelled foam or jelly depicted in Fig. 8.1, and then someone sneaks up behind our back, pulling a neutron star with them. Or suppose someone comes along with a binary star system sending out gravitational waves. What will happen?

One thing that will happen is that the timelike geodesics will change. Therefore particles in freefall will follow different trajectories from the ones they would have followed before the metric changed. This could be noticed by its effect on light signals exchanged between any given pair of particles, for example— it could be observed via gravitational redshift. Another immediately noticeable effect is the *tidal* effect: the squeezing or stretching influence which happens when the acceleration due to gravity is a function of position or time. The tidal effect can be detected by the simple instrument depicted in Fig. 14.1—or by a modern equivalent such as the laser interferometer depicted in Fig. 7.3.

Suppose we have a set of labelled pins floating freely, and near to them some diamond rods.[7] Then the tidal effect will move the pins relative to the diamond rods. This can be calculated by first solving the Einstein field equation for the metric tensor, and using it to discover the proper distance between any two pins. The equivalence principle tells us that a diamond rod in free fall does not change its proper length (unless the tidal stress is exceptionally large, and

---

[7]Diamond has been chosen merely as an example of a material which is very stiff (it has a large coefficient of elasticity or Young's modulus).

Fig. 14.1 A device for detecting the tidal effect of gravity. Two masses can slide along a rod, with a weak spring between them. A gravitational field will cause the geodesics to change in such a way that the two masses will approach or move away from one another, as the case may be. The extension or compression of the spring therefore indicates the tidal effect. [Image: Elsie Steane]

this case can also be calculated). Therefore if we have a rod of proper length one millimetre, and the metric tells us that the coordinate locations marked by a given pair of pins have proper separation two millimetres, then we have the prediction that the diamond rod will not reach between those two pins. If previously (i.e. before our friend with the neutron star came long) it did reach, then we can notice that something has changed.

We just compared the distance between neighbouring pins with the length of a rod in free fall and momentarily at rest relative to the pins. Let us call this rod (A). We might also take an interest in a diamond rod (B) that is not in freefall, but is fixed relative to some particular set of pins which are themselves not in free fall. For example, the pins could be at fixed locations in the Schwarzschild coordinates commonly used to chart spacetime around a spherically symmetric object. If rod B is fixed relative to these coordinates then it is accelerating relative to any LIF. The length of B will be affected by the electromagnetic and other non-gravitational forces which are keeping it from falling. General relativity makes no generic prediction about where such forces will act. For example, the rod may be hanging from a rope, and therefore in tension, or standing on a post, and therefore in compression—this will depend on circumstances. But the effects of such forces can in principle be calculated using materials science in a LIF which is momentarily at rest relative to B. Thus we can predict from materials science the length of a rod B fixed relative to the pins, as observed in a LIF, provided that the rod B is not itself very long.[8] We can then use this information to deduce how the length of rod B compares to the separation of some given pair of pins.

In these examples the pins illustrated the physical meaning which can be attached to a coordinate system, and the rods illustrated the physical meaning which can be attached to statements about proper distance. One could evoke quartz crystal oscillators, or more precise timekeepers such as atomic clocks, to make similar statements about intervals of proper time.

---

[8]And if it were long, then we can perform an integral to get a measure of the total length, but there is no universal prescription for such an integral.

14.4 Generally covariant physics

The term 'general covariance' was introduced by Einstein as a generalization of the Lorentz covariance of Special Relativity. The term is intended to capture the idea that predictions about physical behaviour ought not to depend on coordinate labelling, but it is not always clear how this notion should be expressed in mathematical terms. The assertion that a theory is 'generally covariant' is usually taken to mean that the fundamental equations of the theory can be written down using tensor algebra, such that all the equations are well-formed tensorial expressions. This is not the only way to capture, in a mathematical treatment, the idea of coordinate invariance, but we will adopt this meaning here.

It is an interesting question whether or not physics has to be generally covariant in this sense. Sometimes the term 'generally covariant' is taken to mean merely that the physical content of a theory should be independent of arbitrary human choices such as the sequence of values assigned to a given coordinate which is being employed to track events. If this is what the term means then it becomes self-evident that physics is generally covariant, and such an assertion becomes devoid of physical content. However, the assertion that physics can be expressed in tensor language has content, because there are many expressions that are not well-formed tensorial expressions, and by ruling them out we impose a significant and useful constraint on a theory. This is not the same thing as ruling out the use of non-tensorial expressions completely. We rule out only those that purport to be tensorial but are not properly formed.

It may surprise modern physicists to learn that Einstein's first ideas about gravity deliberately avoided general covariance, because of an argument called the 'hole argument' which seemed to suggest that a generally covariant theory would not respect causality and determinism. In fact that argument was mistaken—see appendix F.[9]

The formulation of physical theories that respect general covariance is done by adopting tensor algebra at the outset. A remarkable feature of GR is that this permits the influence of gravity to be incorporated automatically. It might seem a daunting task to work out how to generalize Maxwell's equations so as to express classical electromagnetism in the presence of gravitation, but as we have seen, it is mostly achieved simply by replacing ∂_a with ∇_a. One can adopt a similar approach to other areas of physics, including condensed matter physics and quantum physics. This is not to say that this mere replacement does all the work; it does not, because at any stage there may be further curvature-related effects that were simply omitted altogether from a theory in flat spacetime. But tensor methods remain a powerful mathematical tool in the expression and elaboration of physical theories, and at the time of writing the Standard Model of particle physics does not involve curvature-related terms directly; the effects of gravity are accounted for indirectly via the covariant derivative. This situation is called *minimal coupling*.

[9]The 'hole argument' refers to the fact that the metric inside some finite region of spacetime (the 'hole') is not uniquely determined by the metric around the boundary, nor is it determined up to coordinate transformations. This seems to suggest that what happens inside the region is not a unique continuation of what happens outside it. It was subsequently established that this is a false impression, and the freedom here is a form of gauge freedom.

Exercises

14.1 Write out the components of the energy tensor (14.18) in the rest frame of a fluid element, and interpret them physically.

14.2 Show that (14.1) can be written

$$\partial_\lambda \left(\mathbb{F}^{\lambda b} \sqrt{|g|} \right) = -\mu_0 j^b \sqrt{|g|} \qquad (14.32)$$

(Thus one can write Maxwell's equations in curved spacetime without explicitly using the covariant derivative).

14.3 Prove from the Maxwell equations that electric charge is conserved in the presence of gravity.

14.4 Let us assert that the energy tensor of the electromagnetic field is symmetric and quadratic in the field tensor; that is, it has the form

$$T^{ab}_{(\mathrm{em})} = \alpha \mathbb{F}^a{}_\lambda \mathbb{F}^{\lambda b} + \beta g^{ab} \mathbb{F}_{\mu\nu} \mathbb{F}^{\mu\nu} \qquad (14.33)$$

with constants α and β to be determined. Show that $\nabla_a T^{ab}_{(\mathrm{em})} = 0$ in the absence of charges and currents if $\beta = \alpha/4$ and find the value of α.

14.5 Charged dust consists of electrically charged particles at sufficiently low density as to have negligible pressure. If ρ is the proper mass density and σ the proper charge density, show that the equation of motion is

$$\rho u^\mu \nabla_\mu u^a = \sigma \mathbb{F}^a{}_\mu u^\mu. \qquad (14.34)$$

Hence show that

$$\nabla_\mu T^{a\mu}_{(\mathrm{d})} = \mathbb{F}^a{}_\mu j^\mu, \qquad \nabla_\mu T^{a\mu}_{(\mathrm{em})} = -\mathbb{F}^a{}_\mu j^\mu, \qquad (14.35)$$

where $T^{ab}_{(\mathrm{d})} = \rho u^a u^b$, and physically interpret the result. What can you say about the energy tensor for dust and field combined?

14.6 Obtain the charged particle equation of motion (2.10) by two methods:
(i) by using the Euler–Lagrange equation

$$\frac{\mathrm{D}}{\mathrm{d}\tau} \left(\frac{\partial \mathcal{L}}{\partial \dot{x}^a} \right) = \nabla_a \mathcal{L}$$

and the Lagrangian

$$\mathcal{L} = -mc(-g_{\mu\nu} \dot{x}^\mu \dot{x}^\nu)^{1/2} + q \dot{x}^\mu A_\mu, \qquad (14.36)$$

(or, alternatively, $\mathcal{L} = \frac{1}{2} mc g_{\mu\nu} \dot{x}^\mu \dot{x}^\nu + q \dot{x}^\mu A_\mu$)
(ii) by asserting $\nabla_\mu T^{a\mu}_{(\mathrm{em})} = -\mathbb{F}^a{}_\mu j^\mu$ where $T^{ab}_{(\mathrm{em})}$ is given by (16.15).

14.7 Derive the 'fishing rod equation' (3.15) from the equation for hydrostatic equilibrium (14.29). (The fishing line has significant tension but negligible mass).

14.8 An isotropic ball of electromagnetic radiation has the equation of state $p = \rho c^2/3$. Show that for such a ball in hydrostatic equilibrium, $\rho \propto (-g_{00})^{-2}$. Note that, assuming g_{00} is finite, it follows that ρ cannot vanish and therefore such a ball cannot be confined gravitationally in static conditions.

14.9 Show that in static conditions with a diagonal metric, the energy tensor for a perfect fluid takes the form $T^{00} = -g^{00} \rho c^2$, $T^{ij} = p g^{ij}$, $T^{i0} = T^{0i} = 0$.

14.10 Show that if the metric g_{ab} is diagonal in some given coordinate system then, for any T^{ab}, there exists a LIF in which $T'^{00} = -g_{00} T^{00}$ and $T'^{ii} = g_{ii} T^{ii}$ (no sum), where the prime indicates the LIF. [Hint: tetrad]

14.11 Two containers of the same size and shape sit side by side in a static gravitational environment described by $\mathrm{d}s^2 = A(z)\mathrm{d}t^2 + B(z)\mathrm{d}z^2 + \mathrm{d}x^2 + \mathrm{d}y^2$ where $A(z)$ is the same for each. Their energy densities match one another at corresponding points but their stresses do not. Show that they weigh the same (that is, the force, electromagnetic or other, which has to be provided at the base to prevent motion is the same for each). [This problem is not easy; a solution may be found in (Lightman, Press, Price and Teukolsky, 1975), problem 14.6.]

15

Curvature

The mathematical concept of *Gaussian curvature* was introduced in Volume 1 through easily visualized examples, and some of the mathematical methods were presented. We are now ready to develop a more thorough treatment. The goal of the present chapter is to quantify and understand the presence and degree of curvature or 'warping' in a general Riemannian or pseudo-Riemannian manifold. The main player will be the Riemann curvature tensor; we will define it and examine its role in various settings. The ideas are applicable to any Riemannian or pseudo-Riemannian manifold.

15.1 Quantifying curvature

We will develop three ways to quantify curvature:

1. Geodesic deviation: curvature is manifested by nearby geodesics diverging from or approaching one another in a non-linear way.
2. Parallel transport: curvature is manifested by the result of the parallel transport of a given vector being path-dependent.
3. Non-commutation of covariant derivatives: curvature is manifested when $\nabla_c \nabla_d \neq \nabla_d \nabla_c$.

In each case, we seek a tensorial treatment. It will emerge that all these measures of curvature amount to different ways of measuring the same property of the manifold and connection. Therefore, that property earns the name 'curvature tensor'.

We begin with the third of our methods, by defining a fourth-rank tensor R^a_{bcd} through the equation

$$\nabla_c \nabla_d v^a - \nabla_d \nabla_c v^a = R^a_{\lambda cd} v^\lambda \quad + \quad \text{torsion} \qquad (15.1)$$

where v^a is an arbitrary vector and the torsion term is zero in GR (see exercise 15.2 for more information). In order to check that R^a_{bcd} in this definition is a property of the manifold, not the vector $\mathbf{v}$, one should carry out the above covariant differentiation. The result (see box) is eqn (2.13), which we repeat here:

Relativity Made Relatively Easy: General Relativity and Cosmology. Volume 2. Andrew M. Steane,
Oxford University Press. © Andrew M. Steane 2021. DOI: 10.1093/oso/9780192895646.003.0015

Curvature tensor. Here is the derivation of (15.2) from (15.1), N.B. *without* assuming that Γ_{bc}^a is symmetric in its lower indices. We have

$$\nabla_c \nabla_d v^a = \partial_c(\nabla_d v^a) - \Gamma_{cd}^\lambda \nabla_\lambda v^a + \Gamma_{c\mu}^a \nabla_d v^\mu$$

$$= \partial_c \left(\partial_d v^a + \Gamma_{d\lambda}^a v^\lambda \right) - \Gamma_{cd}^\lambda \nabla_\lambda v^a + \Gamma_{c\mu}^a \left(\partial_d v^\mu + \Gamma_{d\lambda}^\mu v^\lambda \right)$$

$$= \partial_c \partial_d v^a + (\partial_c \Gamma_{d\lambda}^a) v^\lambda + \Gamma_{d\lambda}^a (\partial_c v^\lambda) + \Gamma_{c\mu}^a \left(\partial_d v^\mu + \Gamma_{d\lambda}^\mu v^\lambda \right) - \Gamma_{cd}^\lambda \nabla_\lambda v^a$$

and

$$\nabla_d \nabla_c v^a = \text{ditto, but with } c \leftrightarrow d$$

When we now form the combination $\nabla_c \nabla_d v^a - \nabla_d \nabla_c v^a$ three pairs of terms cancel, and three pairs survive, giving

$$\nabla_c \nabla_d v^a - \nabla_d \nabla_c v^a = \left(\partial_c \Gamma_{d\lambda}^a - \partial_d \Gamma_{c\lambda}^a + \Gamma_{c\mu}^a \Gamma_{d\lambda}^\mu - \Gamma_{d\mu}^a \Gamma_{c\lambda}^\mu \right) v^\lambda - \left(\Gamma_{cd}^\lambda - \Gamma_{dc}^\lambda \right) \nabla_\lambda v^a$$

$$= R_{\lambda cd}^a v^\lambda - \mathcal{T}_{cd}^\lambda \nabla_\lambda v^a \tag{15.3}$$

where R_{bcd}^a is given by (15.2) and $\mathcal{T}_{cd}^\lambda = \Gamma_{cd}^\lambda - \Gamma_{dc}^\lambda$ (the torsion tensor).

Riemann curvature tensor

$$R_{bcd}^a \equiv \partial_c \Gamma_{bd}^a - \partial_d \Gamma_{bc}^a + \Gamma_{c\lambda}^a \Gamma_{db}^\lambda - \Gamma_{d\lambda}^a \Gamma_{bc}^\lambda. \tag{15.2}$$

Note that we have constructed the Riemann curvature tensor purely through the connection coefficients; we did not require a metric tensor, only a definition of ∇_a.

Equation (15.1) proves that R_{bcd}^a is a tensor, through the quotient theorem, and it can be interpreted as a statement about parallel transport. For single-valued functions we have $\partial_a \partial_b f - \partial_b \partial_a f = 0$. Therefore the non-vanishing of the left-hand side of (15.1) is not owing to the functional form of the components of **v**. It must be owing to a difference in the final vector when it is parallel transported by the different routes: ($\mathrm{d}x^d$ followed by $\mathrm{d}x^c$) compared with ($\mathrm{d}x^c$ followed by $\mathrm{d}x^d$). This is the relationship between Riemann curvature and parallel transport which we will treat more thoroughly in Section 15.2.

After lowering the first index in (15.2) one can obtain the following alternative forms:

$$R_{abcd} = \partial_d \Gamma_{abc} - \partial_c \Gamma_{abd} + \Gamma_{bd}^\lambda \Gamma_{\lambda ac} - \Gamma_{bc}^\lambda \Gamma_{\lambda ad} \tag{15.4}$$

$$= \frac{1}{2} \left(\partial_b \partial_d g_{ac} + \partial_a \partial_c g_{bd} - \partial_b \partial_c g_{ad} - \partial_a \partial_d g_{bc} \right) + \Gamma_{bd}^\lambda \Gamma_{\lambda ac} - \Gamma_{bc}^\lambda \Gamma_{\lambda ad} \tag{15.5}$$

In LIF coordinates, all the undifferentiated Γ symbols vanish. This makes it easy to find the symmetries listed in (2.15), (and the result, being tensorial, remains generally valid). Note in particular that R_{bcd}^a is antisymmetric in the last two indices, which follows immediately from the definition (15.1). The Bianchi identity is also easily obtained by adopting LIF coordinates, where the first covariant derivative equals the partial derivative.

> **Are curvature and torsion properties of a manifold or a connection or both?**
> The ideas of *manifold* and *curvature* are sufficiently closely related that in GR we usually say that it is the manifold itself which is curved. However, in differential geometry more generally one must insist that since the curvature tensor depends on the connection, it will in general change if the connection does. Therefore the same manifold will display different amounts of curvature, and different amounts of torsion, depending on which connection is adopted for the purpose of relating tangent spaces to one another. In this sense, curvature and torsion are properties of the manifold and connection together, not of the manifold alone. This distinction is mostly glossed-over in GR because we assume that the manifold has a metric and that we are using the Levi-Civita connection; in this case, once the metric is specified, the curvature follows (and the torsion is zero).

One may also note the generalization of (15.1) to tensors of any rank:

$$(\nabla_c \nabla_d - \nabla_d \nabla_c) T^a{}_b = R^a{}_{\lambda cd} T^\lambda{}_b - R^\lambda{}_{bcd} T^a{}_\lambda. \tag{15.6}$$

This gives a representative example. More generally, one needs one positive term for each up index, and one negative term for each down index.

Another way to introduce the mathematics of curvature is to define an operator $\mathcal{R}$ whose action, when applied to a vector field, is given by

$$\mathcal{R}(\mathbf{u}, \mathbf{v})\mathbf{w} = \nabla_{\mathbf{u}} \nabla_{\mathbf{v}} \mathbf{w} - \nabla_{\mathbf{v}} \nabla_{\mathbf{u}} \mathbf{w} - \nabla_{[\mathbf{u}, \mathbf{v}]} \mathbf{w} \tag{15.7}$$

where we have adopted index-free notation and $[\mathbf{u}, \mathbf{v}]$ is the *Lie bracket* defined in (15.67). Upon expanding all the parts of this statement into index notation one finds that (15.3) is equivalant to the statement

$$(\mathcal{R}(\mathbf{u}, \mathbf{v})\mathbf{w})^a = -R^a{}_{\lambda\mu\nu} w^\lambda u^\mu v^\nu \tag{15.8}$$

This is an alternative way of defining $R^a{}_{bcd}$ and simultaneously proving that it is a tensor. One can also present the result in the following way. Let

$$\mathbf{R}(\cdot, \mathbf{w}, \mathbf{u}, \mathbf{v}) \equiv -\mathcal{R}(\mathbf{u}, \mathbf{v})\mathbf{w} \tag{15.9}$$

Here the symbol $\mathbf{R}(\cdot, \mathbf{w}, \mathbf{u}, \mathbf{v})$ is a fourth-rank tensor which has been provided with three vectors as inputs, and consequently gives a vector as output. Equation (15.8) is the translation of this statement into index notation.

Geodesic plane. For any given manifold $\mathcal{M}$, a *geodesic plane* is a two-dimensional space, not necessarily flat (indeed, usually it is curved), lying in $\mathcal{M}$. It is the space generated by a set of geodesics that set out from a given point in directions $\lambda\mathbf{v} + \mu\mathbf{w}$ where $\mathbf{v}$ and $\mathbf{w}$ single out a pair of fixed 'axes', and λ and μ give all their linear combinations. The two-dimensional subspace of $\mathcal{M}$ that results from this definition can be useful for defining curvature. For, geodesics of $\mathcal{M}$ that happen to lie in the subspace $\mathcal{S}$ must also be geodesics of $\mathcal{S}$ (think about it: if a path already has a shortest length in the higher space, then there are not any shorter paths to be found in reduced parts of that space). Therefore the geodesic deviation in $\mathcal{S}$ is a property that

A caution on sign conventions. The Gaussian curvature of a spacelike 2-sphere (e.g. the surface of a three-dimensional spherical ball in Euclidean space) is by convention called positive. However, the sign of R_{1212} depends on conventions that are not universally agreed. It may depend on the order of indices adopted in the definition (15.1), and it may depend on the metric signature. In the literature there are a range of conventions on the definitions of R_{abcd} and R_{ab}, some of which result, for a given metric, in a sign change compared to the choices made here. See Chapter 1 for further information.

can be legitimately associated with $\mathcal{M}$; it is a measure of curvature of $\mathcal{M}$ associated with the plane singled out by $\mathbf{v}$ and $\mathbf{w}$. It is called the *sectional curvature*.

In the case of a two-dimensional space, since there is only one intrinsic curvature at any point,[1] we can write the equation of geodesic deviation without needing to specify a direction in the space, but we shall need a sign convention if we wish to handle timelike and spacelike geodesics by a single formula. At any given point, any pair of non-null geodesics in the given geodesic plane has a separation[2] $\eta(u)$ that varies with affine parameter u according to

$$\frac{\mathrm{d}^2\eta}{\mathrm{d}u^2} = -\sigma K \eta \tag{15.10}$$

where σ is assigned the value 1 for a pair of neighbouring geodesics of positive length (these are spacelike in our convention) and -1 for a pair of neighbouring geodesics of negative length (these are timelike in our convention). In this way all such pairs end up with the same value of K.

Now let us connect this to $R^a{}_{bcd}$. Consider any given event and a pair of geodesics which intersect at that event. This pair suffice to single out a geodesic plane. Since the plane is two-dimensional, it has a curvature at the chosen point specified by a single number: the Gaussian curvature K given by (15.10). By treating the case of a 2-sphere, you can show (exercise 15.3) that the Riemann curvature tensor of the 2-sphere is completely specified by its symmetries and the single element

$$R_{1212} = gK \tag{15.11}$$

where K is the Gaussian curvature (equal to $1/a^2$ for the surface of a sphere of radius a) and $g = \det(g_{ab})$. A similar calculation for the hyperbolic 2-sphere shows that the result is valid for all types of surface in two dimensions. It follows that we can interpret individual elements of R_{abcd} as Gaussian curvature of the surface whose spacetime direction is singled out by the indices.

We mentioned in Chapter 2 that Gaussian curvature relates not only to geodesic deviation but also to angle excess per unit area (see (2.18) and accompanying text). Both these ideas may now be applied more generally by using R_{abcd}.

---

[1]This fact was explored in Volume 1 and we shall derive it in Section 15.1.1.

[2]Define distance from a given point P to any nearby point Q as metric distance along the unique short geodesic from P and Q, and define an arc as a locus of points at constant distance from P. These definitions allow η to be defined.

15.1.1 The local metric at second order

We shall next derive the following rather beautiful result. In Riemann normal coordinates near any given point, the metric tensor takes the form

$$g_{ab}(x) \stackrel{\text{RNC}}{=} g_{ab}(0) - \frac{1}{3} R_{a\mu b\nu} x^\mu x^\nu + \mathcal{O}(x^3) \tag{15.12}$$

That is, the curvature tensor exactly supplies all the second-order terms in the Taylor expansion of g_{ab}, once we have adopted coordinates in which the first-order terms vanish.

To derive this result, we proceed in steps. First we examine some counting arguments which suggest the outcome is at least feasible.

Suppose we have some general metric tensor g_{ab}, and we wish, by means of a coordinate transformation, to convert it to g'_{ab} satisfying

$$g'_{ab}|_P = \eta_{ab}, \qquad \left.\frac{\partial g'_{ab}}{\partial x'^c}\right|_P = 0 \tag{15.13}$$

at some point P. The first condition represents $N(N+1)/2$ constraints (the number of elements in a symmetric tensor); the second condition represents $N^2(N+1)/2$ constraints.

The coordinate transformation gives the old coordinates in terms of the new; it can be expressed as a Taylor series about P:

$$x^a = x^a_P + \left(\frac{\partial x^a}{\partial x'^\mu}\right)_P h'^\mu + \frac{1}{2}\left(\frac{\partial^2 x^a}{\partial x'^\mu x'^\nu}\right)_P h'^\mu h'^\nu + \frac{1}{6}\left(\frac{\partial^3 x^a}{\partial x'^\lambda x'^\mu x'^\nu}\right)_P h'^\lambda h'^\mu h'^\nu + \cdots$$

where $h'^a = (x'^a - x'^a_P)$. For an N-dimensional space, the number of independent quantities in the various partial derivatives are as follows:

$\partial x^a/\partial x'^\mu$	N^2
$\partial^2 x^a/\partial x'^\mu \partial x'^\nu$	$\frac{1}{2}N^2(N+1)$
$\partial^3 x^a/\partial x'^\lambda \partial x'^\mu \partial x'^\nu$	$\frac{1}{6}N^2(N+1)(N+2)$

Using just the linear term in the Taylor expansion, we have N^2 parameters which are sufficient to get the elements of g'_{ab} into the form we want at P. It might appear that since g'_{ab} is always symmetric we will only require $\frac{1}{2}N(N+1)$ parameters, and so we have some to spare. However, this is not so, owing to the symmetries of η_{ab} under rotation and boost. $\frac{1}{2}N(N-1)$ is the number of N-dimensional 'rotations' that leave η_{ab} unchanged so there will be this number of further relationships among the parameters and therefore we have none to spare.[3]

---

[3]For example, in GR we have $N = 4$, $N^2 = 16$ and $\frac{1}{2}N(N+1) = 10$; the group of Lorentz transformations is 6-dimensional (3 to specify rapidity, 3 to specify rotation angle). It follows that there are $16 - 6 = 10$ independent conditions on g_{ab} up to Lorentz transformation. The 6 other pieces of information available from the original coordinate transformation merely enable the result to select among the realizations of η_{ab} in one inertial frame or another and these are all equal.

Table 15.1 Number of independent components of
the curvature tensor, eqn (15.16)

N:	1	2	3	4	5
n. components:	0	1	6	20	50

Next we notice that the second derivatives in the Taylor expansion offer just sufficient parameters to allow the first derivatives of g'_{ab} to be set to zero at P, which is what we require.

Proceeding now to second derivatives of g'_{ab}, we have $N^2(N+1)(N+2)/6$ parameters from the third-order term in the Taylor expansion, but $(\frac{1}{2}N(N+1))^2$ requirements on $\partial g'_{ab}/\partial x'^c \partial x'^d$. Therefore we have $N^2(N^2-1)/12$ parameters too few, so we cannot, by a coordinate transformation, arrange for all the second derivatives of g'_{ab} to vanish at P (except in special cases, such as a flat space).

We will now show that the Riemann curvature tensor has precisely this number of degrees of freedom, which suffices to show that (15.12) is plausible, and then we will prove it.

To count the number of independent elements in R_{abcd} we need to determine the effect of all the symmetries listed in (2.15) and (2.16). We can think of the indices as a pair of pairs: $[ab][cd]$. The tensor is antisymmetric with respect to each pair, and symmetric with respect to swapping the groups. The antisymmetric combination of two things drawn from N has $m = N(N-1)/2$ possibilities; the symmetric combination of two things drawn from m has $m(m+1)/2$ possibilities, so if there were no further symmetries then the number of independent elements would be

$$\tfrac{1}{2}m(m+1) = \tfrac{1}{4}N(N-1)\left(\tfrac{1}{2}N(N-1)+1\right). \tag{15.14}$$

Now let us define

$$A_{abcd} \equiv R_{abcd} + R_{acdb} + R_{adbc}, \tag{15.15}$$

then the cyclic symmetry described by (2.16) indicates that $A_{abcd} = 0$. The pair symmetries guarantee that A_{abcd} is totally antisymmetric, so the condition $A_{abcd} = 0$ gives no new constraint unless a, b, c, d are all distinct. Hence the number of new constraints represented by (2.16) is given by the number of ways of choosing 4 distinct objects from N, which is $N!/(N-4)!4!$. Therefore (15.14) overstates the number of independent elements of R_{abcd} by this amount, and the true number is

$$\frac{1}{4}N(N-1)\left(\tfrac{1}{2}N(N-1)+1\right) - \frac{1}{24}N(N-1)(N-2)(N-3) = \frac{1}{12}N^2(N^2-1) \tag{15.16}$$

(see Table 15.1). This is precisely the number of parameters which we needed to supply the quadratic part of g'_{ab} in (15.12) after the coordinate transformation has contributed what it can. This makes (15.12) plausible. We now present the proof.

When Riemann normal coordinates are adopted, the equation of any geodesic near the origin is

$$x^a(s) \overset{\text{RNC}}{=} x^a_0 + sk^a$$

where $x^a_0 \equiv x^a(0)$, s is an affine parameter and k^a is a constant vector. Hence for these geodesics $\dot{x}^a = k^a$ and $\ddot{x}^a = 0$, not only at the origin but also for points near to the origin. By substituting these results into the geodesic equation one finds

$$\Gamma^a_{\mu\nu}\left((x^a_0 + sk^a)\right)k^\mu k^\nu \overset{\text{RNC}}{=} 0 \qquad \forall k^b \tag{15.17}$$

where the double-bracket signifies that we are considering Γ^a_{bc} as a function of position near the origin. Using (12.76) and $\Gamma^a_{bc} = \Gamma^a_{cb}$ we deduce $\Gamma^a_{bc}\left((P)\right) = 0$ where P signifies the origin (also called the pole of geodesic coordinates). This fact is already known to us from our previous study of these coordinates, but now we can go further by differentiating (15.17) with respect to s along any given geodesic:

$$\left(\frac{\mathrm{d}}{\mathrm{d}s}\Gamma^a_{\mu\nu}\left((x^a_0 + sk^a)\right)\right)k^\mu k^\nu \overset{\text{RNC}}{=} 0 \tag{15.18}$$

where we used that k^b is a constant vector. Employing the chain rule, we thus find

$$\frac{\mathrm{d}x^\lambda}{\mathrm{d}s}\left(\partial_\lambda\Gamma^a_{\mu\nu}\right)k^\mu k^\nu \overset{\text{RNC}}{=} 0 \qquad \Rightarrow \qquad (\partial_\lambda\Gamma^a_{\mu\nu})k^\lambda k^\mu k^\nu \overset{\text{RNC}}{=} 0 \tag{15.19}$$

and therefore, using (12.77),

$$\partial_d\Gamma^a_{bc} + \partial_b\Gamma^a_{cd} + \partial_c\Gamma^a_{db} \overset{\text{RNC}}{=} 0. \tag{15.20}$$

Now consider

$$R^a_{bcd} + R^a_{cbd} \overset{\text{RNC}}{=} \partial_c\Gamma^a_{bd} - \partial_d\Gamma^a_{bc} + \partial_b\Gamma^a_{cd} - \partial_d\Gamma^a_{cb} \overset{\text{RNC}}{=} -3\partial_d\Gamma^a_{bc} \tag{15.21}$$

using first (15.2) and then (15.20). Also, from $\partial_c g_{ab} = \Gamma_{abc} + \Gamma_{bac}$ (eqn (10.23)) we have

$$\partial_d\partial_c g_{ab} = \partial_d\Gamma_{abc} + \partial_d\Gamma_{bac} \overset{\text{RNC}}{=} -\tfrac{1}{3}\left(R_{abcd} + R_{acbd} + R_{bacd} + R_{bcad}\right)$$
$$= -\tfrac{1}{3}\left(R_{acbd} + R_{bcad}\right) = -\tfrac{1}{3}\left(R_{acbd} + R_{adbc}\right). \tag{15.22}$$

This quantity is the one appearing in the quadratic term in the Taylor expansion of the metric tensor as a function of the coordinates about the origin:

$$g_{ab}(x) \overset{\text{RNC}}{=} \eta_{ab} + \tfrac{1}{2}(\partial_\mu\partial_\nu g_{ab})x^\mu x^\nu + \mathcal{O}(x^3)$$
$$\overset{\text{RNC}}{=} \eta_{ab} - \tfrac{1}{6}\left(R_{a\mu b\nu} + R_{a\nu b\mu}\right)x^\mu x^\nu + \mathcal{O}(x^3)$$
$$= \eta_{ab} - \tfrac{1}{3}R_{a\mu b\nu}x^\mu x^\nu + \mathcal{O}(x^3) \tag{15.23}$$

QED.

15.2 Relating R^a_{bcd} to parallel transport

We defined parallel transport in Chapter 13. We now study the fact that the result of parallel transport can be path-dependent.

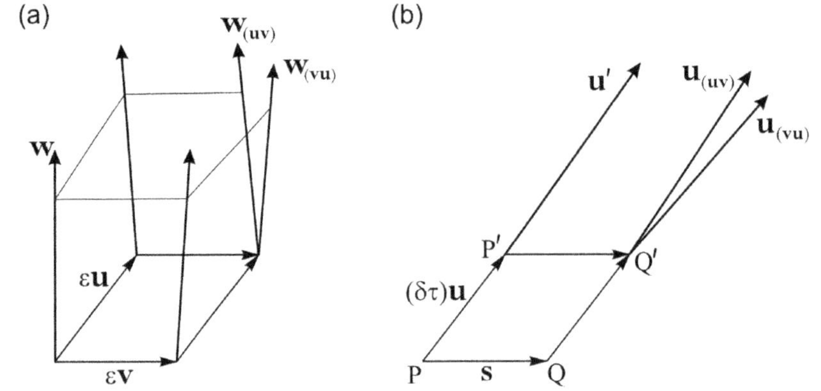

Fig. 15.1 Parallel transport. (a) An arbitrary vector **w** is carried by parallel transport from one point to a nearby point, either by the route $\epsilon\mathbf{u}$ then $\epsilon\mathbf{v}$, or by the route $\epsilon\mathbf{v}$ then $\epsilon\mathbf{u}$. The difference between the resulting vectors is a measure of curvature of the space. (b) The same argument, but now applied to parallel transport of the same vector **u** as the one used to define one side of the parallelogram. In this case the difference between the resulting vectors is a measure of geodesic deviation. The arrow marked **u**′ shows the result of the transport from P to P′, it is shown with a short length merely for convenience on the diagram: in fact this vector is much longer than $\delta\tau\mathbf{u}$. If **u** is a 4-velocity then $(\mathbf{u}_{(\mathbf{vu})} - \mathbf{u}_{(\mathbf{uv})})/\delta\tau$ is the relative 4-acceleration of the worldlines PP′ and QQ′, where $\mathbf{v} = \mathbf{s}/\epsilon$.

There is no point trying to define 'parallel' for vectors that are far apart in a curved manifold (think of two vectors on the surface of a sphere, for example). No consistent definition is possible, because if we transport a vector from A to far away B in such a way that each small displacement is a parallel transport, then when the vector arrives at B its orientation will depend on the path it was transported along. This fact is intimately connected to R^a_{bcd}, as we now show.

From the parallel transport point of view, the meaning of the curvature tensor can be stated, somewhat loosely, as:

> R^a_{bcd} expresses the change which a 4-vector w^b experiences after undergoing a parallel transport around a small loop in the x^c, x^d plane, per unit area of the loop and size of the 4-vector.

More precisely, consider a pair of vectors **u**, **v** at a point P in some manifold. Take another vector **w** and parallel transport it first by $\epsilon\mathbf{u}$ and then by $\epsilon\mathbf{v}$ (two sides of a parallelogram)—see Fig. 15.1(a). The result is some vector $\mathbf{w}_{(\mathbf{uv})}$. Alternatively, transport it first by $\epsilon\mathbf{v}$ and then by $\epsilon\mathbf{u}$. The result $\mathbf{w}_{(\mathbf{vu})}$ is in general different from $\mathbf{w}_{(\mathbf{uv})}$. This difference expresses the net change in **w** when it is parallel-transported around a small parallelogram of sides $\epsilon\mathbf{u}$, $\epsilon\mathbf{v}$, per unit area of the parallelogram. We will define this parallelogram more carefully in Section 15.2.2, and then we will show that

$$\lim_{\epsilon \to 0} \frac{\mathbf{w}_{(\mathbf{vu})} - \mathbf{w}_{(\mathbf{uv})}}{\epsilon^2} = \mathcal{R}(\mathbf{u}, \mathbf{v})\mathbf{w} \tag{15.24}$$

where $\mathcal{R}$ is the curvature operator defined in (15.7). This result can also be written

$$\mathrm{d}w^a = -R^a_{\lambda\mu\nu} w^\lambda \mathrm{d}u^\mu \mathrm{d}v^\nu \tag{15.25}$$

which can be read as a change in w^a which is proportional to an 'area' $\mathrm{d}u^\mu \mathrm{d}v^\nu$. We will derive this from (15.2) after the following clarification of a geometric issue.

15.2.1 Parallelograms close but cubes do not

Towards deriving (15.25), let us now consider some issues in the geometry of a curved space.

Suppose we try to construct a cube. This is a regular three-dimensional convex polyhedron with six faces. In Euclidean space, it is a perfectly coherent idea, and one finds that all the edges of a cube are at right angles to one another. In a general Riemannian manifold, on the other hand, things go differently.

Let us map the neighbourhood of some point P using Riemann normal coordinates, and write the metric near P as $g_{ab} = \eta_{ab} + h_{ab}$. Then we know that h_{ab} is of order ϵ^2 for points at coordinate separation ϵ from P. The proper intervals between such points will therefore be

$$\mathrm{d}s^2 = g_{ab}\mathrm{d}x^a \mathrm{d}x^b \sim (\eta_{ab} + \epsilon^2)\epsilon^2$$
$$\Rightarrow \mathrm{d}s \sim (\pm 1 + \tfrac{1}{2}\epsilon^2)\epsilon = \pm\epsilon + \tfrac{1}{2}\epsilon^3$$

where the notation is intended merely to indicate that the true values differ at **third** order in ϵ from their Euclidean approximations.

Fig. 15.2 shows two attempts to construct a cube in some manifold of dimension 3 or more. The construction begins at the corner marked P. First focus on the left-hand diagram. We will construct a cube by starting at P, and proceeding to the diagonally opposite corner.

Arm yourself with three small vectors $\mathbf{u}, \mathbf{v}, \mathbf{w}$ at P, mutually orthogonal and of the same length. Starting from P, first proceed along $\mathbf{u}$, dragging $\mathbf{v}$ and $\mathbf{w}$ with you by parallel transport. Now proceed along the transported $\mathbf{v}$, dragging $\mathbf{w}$ with you. Finally, walk along the transported $\mathbf{w}$. You have arrived at Q_1.

Next, start out from P again, but following the route $\mathbf{u}$ then $\mathbf{w}$ then $\mathbf{v}$. You arrive near to, but not quite at, Q_1.

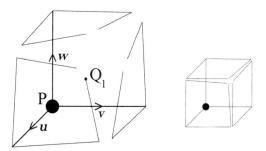

Fig. 15.2 Constructing a cube in a curved space. See text for details.

Following other routes, you can arrive at four other places, to make 6 candidates in all for the role of opposite corner to P in our 'cube'. Notice that these 6 points come in 3 pairs.

The right-hand diagram of Fig. 15.2 shows the same construction, after the length of each of the three vectors has been reduced by a factor 2. The shape is now more cube-like, and this is not merely because distances have shrunk. It is also because the curvature effects have shrunk. To be precise, in passing from the left to the right diagram, every gap in the cube's corners that was owing to curvature has shrunk by a factor not 2 but $2^3 = 8$.

Now suppose we want to make arguments concerning the *faces* of the cube. If the lengths of the edges are scaling as ϵ, and the gaps at the corners as ϵ^3, then if we artificially close the gap by changing the length and direction of one or more edges, then our argument will have imprecision of order ϵ^3, but the area of a face scales as ϵ^2. Therefore the imprecision will be negligible compared to the area, in the limit $\epsilon \to 0$. More generally, the lesson is that it is legitimate to construct parallelograms by dragging vectors around: in the absence of torsion, small enough parallelograms close.[4] This justifies the claim, implicit in eqn (15.24), that routes **uv** and **vu** take one to the same place.

When we come to the cube as a whole, on the other hand, the imprecision ϵ^3 that would be involved if we forced the cube to close scales in the same way as the volume of the cube. Therefore it will not be negligible in comparison to the volume in the limit $\epsilon \to 0$. It is precisely this imprecision that the Riemann curvature tensor quantifies.

15.2.2 Calculation of transport around a loop

Consider a vector **w** that is parallel-transported along a curve described parametrically by $x^a(u)$. For parallel transport, we have, from (13.2)

$$\frac{\mathrm{d}w^a}{\mathrm{d}u} = -\Gamma^a_{bc} w^b \frac{\mathrm{d}x^c}{\mathrm{d}u}, \tag{15.26}$$

therefore the total change in w^a is

$$\Delta w^a = -\int \Gamma^a_{bc} w^b \mathrm{d}x^c. \tag{15.27}$$

In this integral both Γ^a_{bc} and w^b are functions of x^c. We want to treat the case of transport around a small closed curve, and so we can use a Taylor expansion of these functions about some point P on the curve:

---

[4]One can show that in the presence of torsion, this closure is not guaranteed, and this property can be used to define torsion: it signifies the degree to which a loop made of parallel transported vectors is not closed.

$$\Gamma^a_{bc} = (\Gamma^a_{bc})_P + (\partial_\nu \Gamma^a_{bc})_P (x^\nu - x^\nu_P) + \cdots \tag{15.28}$$

$$w^b = w^b_P + \left(\frac{\mathrm{d}w^b}{\mathrm{d}u}\right)_P (u - u_P) + \cdots$$

$$= w^b_P - \left(\Gamma^b_{\mu\nu} w^\mu \frac{\mathrm{d}x^\nu}{\mathrm{d}u}\right)_P (u - u_P) + \cdots$$

$$= w^b_P - \left(\Gamma^b_{\mu\nu} w^\mu\right)_P (x^\nu - x^\nu_P) + \cdots \tag{15.29}$$

Hence, up to first order in $(x^\nu - x^\nu_P)$ we have

$$\Delta w^a = -\left(\Gamma^a_{bc}\right)_P w^b_P \int \mathrm{d}x^c$$

$$- \left((\partial_\nu \Gamma^a_{bc})_P w^b_P - (\Gamma^a_{bc})_P (\Gamma^b_{\mu\nu} w^\mu)_P\right) \int (x^\nu - x^\nu_P)\mathrm{d}x^c. \tag{15.30}$$

When we take the integral around a closed curve, we have $\oint \mathrm{d}x^c = 0$, so the first integral vanishes and the second simplifies. After a little dummy index relabelling in order to bring out a factor of w^b_P, we then have

$$\Delta w^a = -\left(\partial_\nu \Gamma^a_{bc} - \Gamma^a_{\mu c}\Gamma^\mu_{b\nu}\right)_P w^b_P \oint x^\nu \mathrm{d}x^c. \tag{15.31}$$

By interchanging dummy indices ν, c we can obtain a similar equation, and then write Δw^a as half the sum of the two versions. But since for any curve,

$$\int_P^Q \mathrm{d}(x^\nu x^c) = (x^\nu x^c)_Q - (x^\nu x^c)_P, \tag{15.32}$$

for a closed curved one has

$$\oint \mathrm{d}(x^\nu x^c) = 0 \qquad \Rightarrow \qquad \oint (x^\nu \mathrm{d}x^c + x^c \mathrm{d}x^\nu) = 0, \tag{15.33}$$

so $\oint x^\nu \mathrm{d}x^c = -\oint x^c \mathrm{d}x^\nu$. Therefore

$$\Delta w^a = -\frac{1}{2}\left[(\partial_\nu \Gamma^a_{bc} - \Gamma^a_{\mu c}\Gamma^\mu_{b\nu})_P - (\nu \leftrightarrow c)_P\right] w^b_P \oint x^\nu \mathrm{d}x^c. \tag{15.34}$$

The term in the square bracket is the Riemann curvature tensor $R^a_{b\nu c}$. Thus we have

---

Parallel transport around a small loop

$$\Delta w^a = -\frac{1}{2} R^a_{\lambda\mu\nu} w^\lambda \oint x^\mu \mathrm{d}x^\nu. \tag{15.35}$$

In order to connect this to (15.8), adopt, for example, a coordinate system where the x^2, x^3 coordinates are aligned with the sides of the loop (which we now take to be a coordinate

Fig. 15.3 A small loop in the yz coordinate plane.

parallelogram), and for ease of calculation suppose the loop is a coordinate rectangle of sides δy, δz, see Fig. 15.3. In this case one finds

$$\left[\oint x^\mu \mathrm{d}x^\nu\right] = \begin{pmatrix} 0 & 0 & 0 & 0 \\ 0 & 0 & 0 & 0 \\ 0 & 0 & 0 & \delta y \delta z \\ 0 & 0 & -\delta y \delta z & 0 \end{pmatrix} \tag{15.36}$$

For clarity, let us refer to the tensor with these components as $A^{\mu\nu}$. Substituting this into (15.35) gives

$$\delta w^a = -\frac{1}{2}\left(R^a_{\lambda 23}A^{23} - R^a_{\lambda 32}A^{32}\right)w^\lambda \tag{15.37}$$

since all the other contributions are zero. Using now the antisymmetry of R^a_{bcd} in its last two indices, we obtain

$$\delta w^a = -R^a_{\lambda 23}w^\lambda \delta y \delta z. \tag{15.38}$$

This is eqn (15.8), for the case we considered, and there was no essential loss of generality in the method of calculation. For the avoidance of all doubt, note that the second-rank tensor $(u^\mu v^\nu)$ on the right-hand side of (15.8) has only a single non-zero element for the case we treated, unlike the tensor $\oint x^\mu \mathrm{d}x^\nu$, so there is no hidden factor of 2 unaccounted for.

15.3   Geodesic deviation

Equation (15.8) is closely connected to geodesic deviation. For, there is always some geodesic setting off in the **u** direction from P, and $\epsilon\mathbf{v}$ is the displacement from this geodesic to a neighbouring event Q, and one can identify another geodesic at Q which starts off parallel to the first; that is, its tangent vector at Q is the one given by parallel transport of **u** from P—see Fig. 15.1(b).

For the sake of clarity, let us examine timelike geodesics, although the argument will apply to any type, and let the displacement along these geodesics be $(\delta\tau)\mathbf{u}$ and the displacement between them be $\mathbf{s} \equiv \epsilon\mathbf{v}$. Then $\delta\tau$ is some small time interval, **s** is a displacement which tends to zero as ϵ does, and **u** is the 4-velocity of a particle starting out from P, and also of another particle starting out from Q as shown in Fig. 15.1(b). Under inertial motion, the velocities are carried forward by parallel transport, so after time $\delta\tau$ the two particles have new velocities at events P', Q'. To compare these, we transport the velocity at P' over to Q'. The difference between this and the velocity of the other particle gives the relative acceleration **a** of the particles:

$$\mathbf{a} = \lim_{\delta\tau\to 0} \frac{\mathbf{u}_{(\mathbf{vu})} - \mathbf{u}_{(\mathbf{uv})}}{\delta\tau} \tag{15.39}$$

(we will return to interpret this expression more fully in a moment). Therefore, using (15.24),

$$\lim_{\epsilon \to 0} \frac{\mathbf{a}}{\epsilon} = \mathcal{R}(\mathbf{u}, \mathbf{v})\mathbf{u} = \mathcal{R}\left(\mathbf{u}, \frac{\mathbf{s}}{\epsilon}\right)\mathbf{u}. \tag{15.40}$$

This is the **geodesic deviation equation**. Using (15.8) it can be written

$$\lim_{\epsilon \to 0} \frac{a^a}{\epsilon} = R^a_{\lambda\mu\nu} u^\lambda \frac{s^\mu}{\epsilon} u^\nu. \tag{15.41}$$

The right-hand side of this equation is finite in the limit, and therefore so is the left-hand side. The acceleration a^a becomes small in the limit because it is not the acceleration of any one particle, but a measure of the relative acceleration of neighbouring particles. We can now deduce that

$$a^a = R^a_{\lambda\mu\nu} u^\lambda s^\mu u^\nu \tag{15.42}$$

and this is the way the geodesic deviation equation is usually written—once we have agreed the interpretation of $\mathbf{a}$.

So far we have not brought in the fact that $\mathbf{a}$ is related not only to $\mathbf{u}$, but also to $\mathbf{s}$. What we need to notice is that the difference velocity $\mathbf{u}_{(vu)} - \mathbf{u}_{(uv)}$ is itself a statement about $\mathbf{s}$, because the velocities in question are the tangents to the geodesics. Therefore their difference quantifies the degree to which the geodesics are now moving apart (or together) more than they were at P and Q. To be precise,

$$\mathbf{a} = \lim_{\delta\tau \to 0} \frac{\mathbf{u}_{(vu)} - \mathbf{u}_{(uv)}}{\delta\tau} = \lim_{\delta\tau \to 0} \frac{\delta\left(\frac{d\mathbf{s}}{d\tau}\right)}{\delta\tau} = \frac{d^2\mathbf{s}}{d\tau^2} \tag{15.43}$$

Hence we find

Geodesic deviation equation

$$\frac{D^2 s^a}{d\tau^2} = -R^a_{\lambda\mu\nu} u^\lambda s^\mu u^\nu = \left(R^a_{\lambda\mu\nu} u^\lambda u^\mu\right) s^\nu \tag{15.44}$$

This is the generalization of (15.10).

The above derivation is perfectly rigorous and robust, but for further elucidation we will now present a second derivation. This second approach takes as its starting point a smooth collection of neighbouring geodesics $\gamma_s(u)$ parameterized by two parameters: s and u (Fig. 15.4). The idea

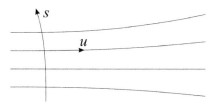

Fig. 15.4 A collection of adjacent non-crossing geodesics parameterized by two parameters.

is that u is an affine parameter along each geodesic, and s labels the geodesics in such a way that s increases continuously as one moves from each geodesic to the next; the geodesics are taken such that this is possible. For example, they do not cross one another in the (small) region under study. The tangent vector to any one geodesic $x^a(u)$ is

$$T^a = \frac{dx^a}{du} \tag{15.45}$$

and this satisfies the definition of the affine geodesic, namely that its value along the geodesic satisfies the definition of parallel transport, so we have

$$\nabla_{\mathbf{T}} T^a \equiv T^\lambda \nabla_\lambda T^a = 0. \tag{15.46}$$

We can also regard the two parameters u, s as furnishing coordinates for our small two-dimensional region. Therefore we can equally well write

$$T^a = \left(\frac{\partial x^a}{\partial u}\right)_s \tag{15.47}$$

where now x^a is a general point. We can also define a vector

$$S^a \equiv \left(\frac{\partial x^a}{\partial s}\right)_u. \tag{15.48}$$

Clearly this vector indicates something about the separation between the geodesics. Let us introduce two more vectors:

$$\mathbf{v} \equiv \nabla_{\mathbf{T}} \mathbf{S}, \qquad \mathbf{a} \equiv \nabla_{\mathbf{T}} \mathbf{v} \tag{15.49}$$

In component form, this is

$$a^a \equiv T^\mu \nabla_\mu \left(T^\nu \nabla_\nu S^a\right). \tag{15.50}$$

This is the quantity that we shall agree to call the relative acceleration between geodesics.

Now you can show (exercise) that

$$T^\nu \nabla_\nu S^a = S^\nu \nabla_\nu T^a. \tag{15.51}$$

This is an example of 'parallelograms close'; we will use it twice in the following:

$$\begin{aligned}
a^a &\equiv T^\mu \nabla_\mu \left(T^\nu \nabla_\nu S^a\right) \\
&= T^\mu \nabla_\mu \left(S^\nu \nabla_\nu T^a\right) & \text{using (15.51)} \\
&= (T^\mu \nabla_\mu S^\nu)\nabla_\nu T^a + T^\mu S^\nu \nabla_\mu \nabla_\nu T^a & \text{Leibniz (product) rule} \\
&= (S^\mu \nabla_\mu T^\nu)\nabla_\nu T^a + T^\mu S^\nu \nabla_\mu \nabla_\nu T^a. & \text{using (15.51)}
\end{aligned} \tag{15.52}$$

Next, we will reverse the pair of covariant derivatives in the second term by using the curvature (eqn (15.1)):

$$a^a = (S^\mu \nabla_\mu T^\nu)\nabla_\nu T^a + T^\mu S^\nu \left(\nabla_\nu \nabla_\mu T^a + R^a_{\lambda\mu\nu} T^\lambda\right) \tag{15.53}$$

and this can be simplified by noting that

$$0 = \nabla_\nu(T^\mu \nabla_\mu T^a) = (\nabla_\nu T^\mu)\nabla_\mu T^a + T^\mu \nabla_\nu \nabla_\mu T^a \tag{15.54}$$

(the result is zero because of (15.46)). Therefore the first two terms in (15.53) sum to zero and we are left with

Geodesic deviation again

$$a^a = R^a_{\lambda\mu\nu} T^\lambda T^\mu S^\nu. \tag{15.55}$$

This is precisely the same equation as (15.44), with the advantage that every term has now been defined explicitly using covariant derivatives.

Now, we can always adopt for our coordinate system a LIF in which $u^a = (c,0,0,0)$ and we can choose the x axis along **s**. We then find the geodesic deviation equation takes the form

$$\frac{\ddot{\eta}}{\eta} \overset{\text{LIF}}{=} R^1_{001} \tag{15.56}$$

so the Gaussian curvature of the tx 'plane' near P is $K = -R^1_{001} = R^1_{010}$. For a static metric, a similar argument for spacelike geodesics would yield $K = R^2_{121}$ in the xy plane, assuming LIF coordinates in which the x and y axes are aligned with spacelike geodesics. Similar arguments can be made for other directions and hence other components of R^a_{bcd}.

Geodesic deviation can be thought of both as a probe of spacetime, and also as a notable physical effect: it presents the **tidal effect of gravity**. The above equation shows how particles in free fall near to one another will draw apart or together as they fall. If the particles are part of a body, and the internal forces of that body act to prevent the particles changing their separation, then the body must have internal stress. To quantify this, we define:

Tidal stress tensor

$$S^a_{\ d} \equiv -R^a_{\mu\nu d} u^\mu u^\nu, \qquad \frac{D^2 s^a}{d\tau^2} = -S^a_{\ d} s^d \tag{15.57}$$

The sign of this definition is arranged in such a way that when the equation says that freely falling particles are accelerating apart, i.e. $D^2 s^a/D\tau^2 > 0$, then $S^a_{\ d}$ is negative, indicating the tension in any object whose internal forces prevent such free motion. Conversely, when the geodesic equation says $D^2 s^a/D\tau^2 < 0$ then there will be pressure.

The imposition of tidal stress is the primary physical effect of gravity, once we have accounted for the mere steering around of the timelike geodesics that tell where free-fall motion goes.

15.3.1 Parallel transport and non-locality

In eqn (15.35) we found the result of parallel transport around an infinitesimal loop. To find the result of parallel transport around a still small but somewhat larger loop, one can divide the loop up as explained in Fig. 15.5. One first identifies a surface bounded by the loop, and then one traces a meandering path which encircles every small part of this surface in the same sense. According to the parallel transport equation (15.26), the contribution to δw^a from any given path segment dx^a will be exactly cancelled if the same segment is traversed in the opposite

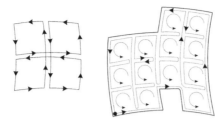

Fig. 15.5 The left-hand picture shows how a square region can be divided up into smaller squares and a single continuous path can be found that encircles all the squares in the same sense. The right-hand picture shows a more general construction that allows regions of any shape to be divided up by a single path that encircles all the smaller regions in the same sense. The internal parts which traverse the same line segment in opposite senses cancel, and therefore the line integral along the loop around the border is equal to the integral along the meandering path, where the latter may lie on *any* surface bounded by that border.

direction. Therefore the contribution of the internal parts of the path depicted in 15.5 cancel, and we find that the line integral along the meandering path is equal to the line integral around the loop which bounds the surface. Therefore, writing $(\Delta w^a)_n$ for the contribution of the n'th small loop, we have for the total effect:

$$\Delta w^a = \sum_n (\Delta w^a)_n \tag{15.58}$$

and by using (15.25) one finds that in the limit $n \to \infty$ this sum becomes an integral over the surface. This is a generalization of Stokes' theorem relating the curl to the line integral of a vector field. HOWEVER, there is a problem. Each part of such an integral, or each term in the sum in (15.58), is a vector, and these vectors are not all at the same location. How are they to be added up then? There is no fixed prescription and therefore we do not have a precise version of Stokes' theorem, but the method is useful up to some finite accuracy of the order of the cube of the diameter of the region.

The effect of parallel transport of a given vector around a large loop is a well-defined mathematical idea, and for a given loop there will be a unique and precise result. As we have just seen, this result cannot be calculated precisely by using an integral over a surface bounded by the loop. However, one *can* infer from the above, without requiring any approximation, that if the result of parallel transport around a loop is non-zero, then there must be non-zero curvature somewhere within any surface bounded by the loop. This consequence is illustrated by paths on the cone depicted in Fig. 15.6. The cone is a 2-surface which is everywhere flat except at the apex. For the purpose of the argument, let us round the apex off a little, so that there is no singular behaviour. Then the surface has $R^a_{bcd} = 0$ everywhere except in a small region near the top. It follows that the terms in (15.58) will all be zero for any loop that does not enclose the top of the cone, so one expects $\Delta w^a = 0$ for all such loops. For a loop which does enclose the top of the cone, on the other hand, it is not hard to show, by unrolling the cone, that Δw^a is then given by the angle between the cut edges of the cone when it is unrolled.[5] It follows from this

---

[5] This is a precise statement for any loop that does not enter the small curved region taken up by the rounded-off apex of the cone.

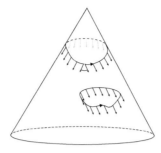

Fig. 15.6 Parallel transport around loops on the surface of a cone.

that an observer who never visits the apex, but does experiments only in the flat part of the surface, can nevertheless deduce that there is a curved region somewhere. They can deduce this by observing the non-zero value of Δw^a for some of the loops they walked around.

Before you think this is too bizarre, note that similar statements can be made about the presence of charge and current in electromagnetism in flat space. In the case of the electromagnetic field, it is sufficient to observe the flux of $\mathbf{E}$ through a surface, and the line integral of $\mathbf{B}$ around a loop, to discover whether the surface encloses some charge somewhere, and whether the loop encircles a current (or displacement current). One need never make a measurement right at the charge or current. In the case of electromagnetism, this is not a case of action at a distance or any other sort of magic; it is a case of field equations. The Maxwell equations constrain the fields so that one thing leads to another in such a way that the flux and line integrals probe the charge and current distributions in the way they do. Similar statements can be made in the case of differential geometry. It is *the very fact that we are dealing with a manifold on which parallel transport can be defined* that results in the generalized Stokes theorem and what it implies. Differentiable manifolds in general are rich and complex, but they cannot do absolutely anything. For example, they cannot stay intrinsically flat everywhere without giving zero rotation for parallel transport around all loops.

15.4   Lie derivative

The Lie derivative will not be needed in the rest of this book except in the next section, but we include it since it is a basic element of differential geometry. The idea of the Lie derivative is to enquire how a tensor field $\mathbf{A}$ (of any rank) changes under the influence of a 'flow' or movement of points described by a vector field $\mathbf{v}$. The definition is

$$\mathcal{L}_{\mathbf{v}} \mathbf{A} \equiv \lim_{\epsilon \to 0} \frac{\mathbf{A}(x^c) - \mathbf{A}'(x^c - \epsilon v^c)}{\epsilon} \tag{15.59}$$

where the prime indicates a version of the tensor field $\mathbf{A}$ which we shall define in the following by specifying its components.

Suppose event P is located at $x^a(P)$ and nearby event Q is located at $x^a(P) - \epsilon v^a$. You can think of Q as located a little 'downstream' in the flow given by $\mathbf{v}$. Let us define new coordinates

x'^a in such a way that the values of the new coordinates at Q are equal to the values of the old coordinates at P. This is achieved by

$$x'^a = x^a + \epsilon v^a \tag{15.60}$$

since then

$$
\begin{aligned}
x'^a(Q) = x^a(Q) + \epsilon v^a(Q) &= x^a(P) - \epsilon v^a(P) + \epsilon v^a(Q) \\
&= x^a(P) + O(\epsilon^2)
\end{aligned}
\tag{15.61}
$$

since $v^a(Q) - v^a(P) = O(\epsilon)$. For this coordinate change, we have

$$
\begin{aligned}
\frac{\partial x'^a}{\partial x^b} &= \delta^a_b + \epsilon \partial_b v^a, \\
\frac{\partial x^a}{\partial x'^b} &= \delta^a_b - \epsilon \partial'_b v^a = \delta^a_b - \epsilon \partial_b v^a + O(\epsilon^2).
\end{aligned}
\tag{15.62}
$$

We shall consider first the Lie derivative of a vector field $\mathbf{w}$. We have

$$
\begin{aligned}
w'^a(Q) = \frac{\partial x'^a}{\partial x^\lambda} w^\lambda(Q) &= (\delta^a_\lambda + \epsilon \partial_\lambda v^a) w^\lambda(Q) \\
&= (\delta^a_\lambda + \epsilon \partial_\lambda v^a) \left(w^\lambda(P) - \epsilon v^\mu \partial_\mu w^\lambda(P) + O(\epsilon^2) \right)
\end{aligned}
\tag{15.63}
$$

where in the second line we expressed the value of w^λ at Q by using a Taylor expansion about P, using that Q is located at $x^a(P) - \epsilon v^a$. Hence we find

$$w'^a(Q) = w^a(P) + \epsilon \left(w^\lambda \partial_\lambda v^a - v^\mu \partial_\mu w^a \right) + O(\epsilon^2). \tag{15.64}$$

This is telling us that the components of $\mathbf{w}(Q)$ for the primed basis differ from the components of $\mathbf{w}(P)$ for the unprimed basis by an amount which is itself a vector (which you should check). Keep in mind that the primed basis here is not just any old basis; it is the one defined by (15.60). The Lie derivative is, by definition, this result, expressed in the limit:

Lie derivative of vector field

$$\mathcal{L}_{\mathbf{v}} w^a = \lim_{\epsilon \to 0} \frac{w^a(P) - w'^a(Q)}{\epsilon} = v^\mu \partial_\mu w^a - w^\lambda \partial_\lambda v^a \tag{15.65}$$

In the limit we find that we do not need to appeal to the basis change; the result can be written entirely in unprimed quantities. The excursion via P and Q and the coordinate bases was merely an attempt to explain what this derivative is doing: what kind of change it is quantifying.

The notation $\mathcal{L}_{\mathbf{v}} \mathbf{w}$ is also used. This is an index-free way to write that vector field whose components are given by

$$(\mathcal{L}_{\mathbf{v}} \mathbf{w})^a \equiv \mathcal{L}_{\mathbf{v}} w^a. \tag{15.66}$$

The **Lie bracket** of any two vector fields $\mathbf{X}, \mathbf{Y}$ is written $[\mathbf{X}, \mathbf{Y}]$ and is defined to be that vector field whose components are

$$[\mathbf{X}, \ \mathbf{Y}]^a = X^\mu \partial_\mu Y^a - Y^\mu \partial_\mu X^a. \tag{15.67}$$

By comparing this with (15.65) one finds

$$\mathcal{L}_\mathbf{v} \mathbf{w} = [\mathbf{v}, \ \mathbf{w}]. \tag{15.68}$$

The result for a one-form can be obtained by a similar analysis. Owing to the different transformation law, there is a sign difference and also a different combination in the second term; one finds

$$\mathcal{L}_\mathbf{v} w_a = \lim_{\epsilon \to 0} \frac{w_a(P) - w'_a(Q)}{\epsilon} = v^\mu \partial_\mu w_a + w_\lambda \partial_a v^\lambda. \tag{15.69}$$

When applied to a scalar field, the result is

$$\mathcal{L}_\mathbf{v} \phi = v^\lambda \partial_\lambda \phi. \tag{15.70}$$

The expression on the right is also called the *directional derivative*. Some of these ideas are discussed further in appendix C.

Finally, we consider a set of covariant components of a second-rank tensor. The transformation rule is now

$$T'_{ab} = \frac{\partial x^\mu}{\partial x'^a} \frac{\partial x^\nu}{\partial x'^b} T_{\mu\nu} = (\delta^\mu_a - \epsilon \partial_a v^\mu)(\delta^\nu_b - \epsilon \partial_b v^\nu) T_{\mu\nu} \tag{15.71}$$

and after employing a Taylor expansion one finds (exercise):

Lie derivative of covariant components of second-rank tensor

$$\mathcal{L}_\mathbf{v} T_{ab} = T_{a\nu} \partial_b v^\nu + T_{\mu b} \partial_a v^\mu + v^\lambda \partial_\lambda T_{ab} \tag{15.72}$$

15.5 Symmetries of spacetime

We conclude the chapter with a presentation of some mathematical ideas related to symmetry. The material is only used in a few places later in the book, and could be omitted at first reading, but it is part of the basic vocabulary of GR and the Killing vector field proves to be a very useful concept in more advanced study.

A change which leaves spacetime unchanged is called an *isometry*. To be precise, an isometry is a mapping of spacetime to itself (or to another spacetime) which leaves all intervals between events unchanged (and the concept generalizes to metric spaces in any number of dimensions). Examples include translation and rotation in Euclidean space, Lorentz transformation in flat spacetime, and rotation about the origin in Schwarzschild spacetime. One can infer that there is such an isometry if there is a coordinate system in which the metric is independent of one or more coordinates. However, it may be that one is presented with, or has otherwise discovered, a metric which describes a space with isometries, but this is not self-evident in the coordinate

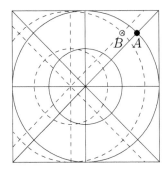

Fig. 15.7 We would like to know whether spacetime is the same at A and B. The diagram shows two coordinate systems (full and dashed) related by $x'^a = x^a + v^a \epsilon$ where $v^a = (1,0)$. If $g_{ab}(A)$ is the metric in the first coordinates evaluated at A, then the metric in the second coordinates at that same event A is $g'_{ab}(A) = (\partial x^\mu / \partial x'^a)(\partial x^\nu / \partial x'^b) g_{\mu\nu}(A)$. The metric in the second coordinates evaluated at B is then $g'_{ab}(B) = g'_{ab}(A) - \epsilon v^\lambda \partial'_\lambda g'_{ab}$ to first order in ϵ. Spacetime is the same at A and B if $g'_{ab}(B) = g_{ab}(A)$ and the corresponding result also applies at events throughout the neighbourhood of A and B. For a Killing vector field we extend the argument to all events in the manifold.

system which has been adopted. We would like a coordinate-independent way of finding and expressing isometries. This can be especially useful in the task of finding geodesics, since we already established a link between symmetry of the metric and conservation along a geodesic (eqn (13.27)). It is also useful when considering conserved quantities, and it can aid in solving further differential equations such as Maxwell's equations.

A good way to proceed is to consider applying to every event an infinitesimal shift in coordinates (the shift being chosen in a systematic way), and enquiring whether the metric changes. It is convenient to express this shift by writing it as the product of a finite vector field $\mathbf{v}$ and an infinitesimal scalar ϵ as in eqn (15.60). For example, if the Euclidean plane has been charted in rectangular coordinates (x, y), then to express a small rotation about the origin of coordinates we might use $v^a = (-y, \ x)$ and ϵ is the rotation angle.

Under the change of coordinates, the metric tensor transforms in the standard way, and we compare the metric tensor in the new coordinates (that is, g'_{ab}) evaluated at B with the metric tensor in the old coordinates (g_{ab}) evaluated at A (c.f. Fig. 15.7). This is precisely the concept employed in the definition of the Lie derivative.

The condition for the coordinate change to map spacetime back to itself with no change is

$$g'_{ab}\left(\!\left(x^c\right)\!\right) = g_{ab}\left(\!\left(x^c\right)\!\right) \tag{15.73}$$

and therefore

$$\mathcal{L}_{\mathbf{v}} g_{ab} = 0. \tag{15.74}$$

The reader should pause to reflect on why these two expressions say the same thing, and why they do indeed give the correct condition; figure (15.7) and exercise 15.17 may help. If the metric satisfies (15.73) then it is said to be *form invariant*. Using (15.72) we find that in order

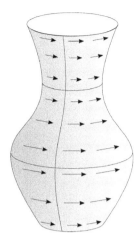

Fig. 15.8 An example Killing vector field.

for this invariance to hold, the metric and the vector field together must satisfy the differential equation

$$g_{a\nu}\partial_b v^\nu + g_{\mu b}\partial_a v^\mu + v^\lambda \partial_\lambda g_{ab} = 0. \qquad (15.75)$$

Owing to the special properties of the metric, one can show (exercise) that this combination of derivatives can also be written

$$\nabla_a v_b + \nabla_b v_a = 0. \qquad (15.76)$$

This is **Killing's equation** which we already met in (13.39). It states that *the gradient of a Killing vector field is antisymmetric.*

If a given spacetime has a timelike Killing vector field, then we can use that vector field to identify a temporal coordinate t, and thus arrive at a metric which is independent of time ($\partial_0 g_{ab} = 0$). The Killing vector field will then be tangent to the coordinate lines associated with t. Hence a spacetime is stationary (or admits a coordinate frame such that it is stationary) if and only if it possesses a timelike Killing vector.

It is not hard to show (exercise 15.20) that a Killing vector **K** must also satisfy the following:

$$\boldsymbol{\nabla} \cdot \mathbf{K} = 0, \qquad (15.77)$$

$$\nabla_\mu \nabla^\mu K^a = -R^a_\lambda v^\lambda, \qquad (15.78)$$

$$\nabla_c \nabla_b K_a = R_{abc\lambda} K^\lambda, \qquad (15.79)$$

$$\nabla_\mu \nabla_b K^\mu = R_{b\lambda} K^\lambda, \qquad (15.80)$$

$$K^\lambda \nabla_\lambda R = 0, \qquad (\nabla_\mathbf{K} R = 0). \qquad (15.81)$$

The last result states that the derivative of the Ricci scalar along a Killing vector must vanish; this illustrates the fact that the geometry of the manifold is not changing as one moves along a Killing vector (rotational symmetries and stationary spacetimes furnish examples). We will use these results in the discussion of gravitational energy in the next chapter.

Eqns (15.77) and (15.78) make it easy to show that a Killing vector $\mathbf{A}$ is a possible vector potential for an electromagnetic field in vacuum. For, the first shows that the Lorenz gauge condition is satisfied, and the second shows that if $R_{ab} = 0$ then $\mathbf{A}$ satisfies the electromagnetic field equation in Lorenz gauge.

15.5.1   Curvature invariants

The **Weyl tensor**, also called the **conformal curvature tensor**, is defined for $N \geq 3$ by

$$C_{abcd} = R_{abcd} - \frac{2}{N-2}\left(g_{a[c}R_{d]b} - g_{b[c}R_{d]a}\right) + \frac{2}{(N-1)(N-2)}g_{a[c}g_{d]b}R \qquad (15.82)$$

This tensor is completely traceless, in the sense that taking the contraction over any pair of indices gives zero; it has been defined in such a way as to bring this about. The Ricci tensor and scalar capture all information about traces of the curvature tensor; the Weyl tensor captures the trace-free parts. Together they specify the curvature.

The Weyl tensor has the remarkable property that it is unchanged under conformal transformations of the metric. That is, if one replaces g_{ab} by $\Omega(x)g_{ab}$ for any function $\Omega(x)$ of the coordinates, then C_{abcd} does not change. Note that such a transformation is *not* a mere coordinate change; it can change the curvature tensor—but not the Weyl tensor. Regions of spacetime in which the Weyl tensor vanishes contain no gravitational radiation and are conformally flat.[6] We shall not need to explore these facts; we have introduced this tensor mainly in order to make a few general remarks in Section 16.1, and to help us list the principal invariants that are associated with curvature. These are

$$
\begin{array}{ll}
\text{Ricci scalar} & R \equiv R^{\mu\nu}{}_{\mu\nu} \\[4pt]
\text{Kretschmann scalar} & K_1 \equiv R_{abcd}R^{abcd} \\[4pt]
\text{Chern–Pontryagin scalar} & K_2 \equiv {}^*R_{abcd}R^{abcd} \\[4pt]
\text{Euler scalar} & K_3 \equiv {}^*R^*_{abcd}R^{abcd} \\[4pt]
& I_1 \equiv C_{abcd}C^{abcd} \\[4pt]
& I_2 \equiv {}^*C_{abcd}C^{abcd}
\end{array}
\qquad (15.83)
$$

where the star indicates the Hodge dual (Section 12.6). K_i are called the principal invariants of the Riemann tensor; I_i are called the principal invariants of the Weyl tensor. They are related through

$$K_1 = I_1 + 2R_{ab}R^{ab} - \frac{1}{3}R^2, \qquad (15.84)$$

$$K_3 = -I_1 + 2R_{ab}R^{ab} - \frac{2}{3}R^2. \qquad (15.85)$$

---

[6]To be precise: vanishing of the Weyl tensor everywhere is both a necessary and sufficient condition for conformal flatness, and a metric for a spacetime without gravitational waves cannot be changed into a metric for a spacetime with gravitational waves by a conformal transformation.

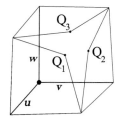

Fig. 15.9 Visual demonstration of first Bianchi identity; see exercise 15.4.

Curvature invariants are useful for characterizing manifolds, up to a point. If a curvature invariant is non-zero, then so is the curvature; if a curvature invariant diverges, then so does the curvature. However, one cannot argue in the other direction: there can be curved spacetimes, and even singular behaviour, with all curvature invariants evaluating to zero. Similar statements arise in electromagnetism, where the main invariants are $\mathbb{F}_{\mu\nu}\mathbb{F}^{\mu\nu} = 2(E^2 - B^2/c^2)$ and $(*\mathbb{F})_{\mu\nu}\mathbb{F}^{\mu\nu} = 4\mathbf{E}\cdot\mathbf{B}/c$. Both of these invariants are zero for electromagnetic waves—but that does not imply the fields themselves are zero.

Exercises

15.1 Reproduce the derivation of (15.2) from (15.1).

15.2 **Torsion**. Prove (e.g. by checking its transformation properties) that $\mathcal{T}_{cd}^a$ is indeed a tensor. [Hint: (12.28)] It follows that the torsion term in (15.3) is tensorial. The left-hand side is manifestly tensorial. It follows that the curvature term is also tensorial, and therefore, by the quotient theorem, so is R_{bcd}^a.

15.3 Adopting the metric $ds^2 = dr^2 + \alpha^2 d\phi^2$ for the 2-sphere, with $\alpha(r) = R\sin(r/R)$, show that the non-zero Christoffel symbols are $\Gamma_{22}^1 = -\alpha d\alpha/dr$, $\Gamma_{12}^2 = \Gamma_{21}^2 = \alpha^{-1}d\alpha/dr$, and hence obtain eqn (15.11).

15.4 Fig. 15.9 shows a cube constructed as in Fig. 15.2, but drawn with each face closed since each face is a parallelogram.
(i) Explain why the directed line segment Q_1Q_2 is equal to $\mathcal{R}(\mathbf{u}, \mathbf{v})\mathbf{w}$
(ii) After arguing similarly for Q_2Q_3 and Q_3Q_1, deduce (2.16).

15.5 Use the simple geodesic deviation equation (15.10) and the measured acceleration due to gravity to obtain the Gaussian curvature of the (t, r) geodesic plane near the surface of the Earth.

15.6 Show that a manifold with line element $ds^2 = ydx^2 + xdy^2$ is curved, but one with line element $ds^2 = y^2dx^2 + x^2dy^2$ is not.

15.7 Show that for $ds^2 = dr^2 + f^2(r)d\theta^2$ the Gaussian curvature is $K = -f''/f$.

15.8 A rigidly rotating disc in flat spacetime is treated using the coordinates t, r, z, ϕ defined in (4.13) leading to (4.12),(4.13). A notion of simultaneity for observers fixed to the disc can be defined *locally*, for any small part of the disc, by finding the direction in spacetime that is orthogonal to the world-lines of those observers (this notion cannot be extended globally, for the whole disc at once, owing to ambiguities when ϕ increments by 2π). Such an approach leads to the spatial metric

$$dl^2 = dr^2 + \frac{r^2}{1 - r^2\omega^2/c^2}d\phi^2 \qquad (15.86)$$

Use this to reproduce the results of exercise 4.1 of Chapter 4, and find the Gaussian curvature.

15.9 Obtain (15.4) and (15.5) and hence the symmetries of R_{abcd}. [Hint: (10.23)]

15.10 Show that, for any second tank tensor, $(\nabla_\mu\nabla_\nu - \nabla_\nu\nabla_\mu)T^{\mu\nu} = 0$.

15.11 Show that for a two-dimensional manifold one has $R_{abcd} = K(x)(g_{ac}g_{bd} - g_{ad}g_{bc})$ [Hint: how many components do you need to evaluate?]. Using the Bianchi identity, show that if the curvature of a manifold of higher dimension can be expressed this way, then K must be constant (Schur's lemma).

15.12 A **maximally symmetric space** is defined to be one which is both homogeneous and isotropic. In an N-dimensional space there are N linearly independent translations and $N(N-1)/2$ linearly independent rotations; therefore the maximum possible number of isometries is $N + N(N-1)/2 = N(N+1)/2$. It can be shown that in a maximally symmetric space there are $N(N+1)/2$ linearly independent Killing vectors and the curvature tensor can be written in the form

$$R_{abcd} = K(g_{ac}g_{bd} - g_{ad}g_{bc}) \qquad (15.87)$$

where K is a constant (exercise (15.11)). Show that for such a space,

$$R_{ab} = (N-1)Kg_{ab} = (R/N)g_{ab},$$

and that the 3-sphere has this property.

15.13 Obtain (15.6) [Hint: outer product]

15.14 Show that along a line of constant ϕ on the surface of a sphere the geodesic deviation is given by $D^2s^\theta/d\tau^2 = 0$, $D^2s^\phi/d\tau^2 = -s^\phi\dot\theta^2$. Hence find, as a function of θ, the separation between two neighbouring such geodesics both starting from $\theta = 0$.

15.15 Obtain the Newtonian equation of geodesic deviation from the weak field limit of GR.

15.16 (i) Show that in three-dimensional Euclidean space, the following are Killing vectors: $(1,0,0)$; $(0,1,0)$; $(0,0,1)$; $(-y,x,0)$; $(z,0,-x)$; $(0,-z,y)$ and the following are not: $(y,0,0)$, $(0,x,0)$, where the components are given in Cartesian coordinates. Interpret these results.
(ii) Show that in Minkowski spacetime $K^a = (x, ct, 0, 0)$ is a Killing vector.

15.17 **Killing's equation example.** Consider the Euclidean plane charted by polar coordinates (r, θ). The metric is $g_{ab} = \text{diag}(1, r^2)$. Introduce primed coordinates related to the unprimed coordinates by an infinitesimal shift in the x direction (i.e. the direction of the axis from which θ is measured).
(i) Show that the relationship between the coordinates is $x'^a = x^a + v^a dx$ where $v^a = (\cos\theta, -\sin\theta/r)$
(ii) Obtain $g_{a\nu}\partial_b v^\nu$ and hence show that $g_{a\nu}\partial_b v^\nu + g_{\mu b}\partial_a v^\mu = \text{diag}(0, -2r\cos\theta)$.
(iii) Show that $v^\lambda\partial_\lambda g_{ab} = \text{diag}(0, 2r\cos\theta)$, and hence that v^a, g_{ab} satisfy (15.75).
(iv) Show by coordinate transformation that, to first order, $g'_{ab} = \text{diag}(1, r^2 - 2r\cos\theta dx) = \text{diag}(1, r'^2)$. Explain whether or not the metric tensor is form invariant under the given coordinate transformation.

15.18 Confirm that the condition (15.73) implies (15.75) and hence Killing's equation (13.39).

15.19 Show that any Killing vector K^a satisfies the geodesic deviation equation $D^2 K^a/du^2 = R^a_{\lambda\mu\nu}v^\lambda v^\mu K^\nu$ where $v^a = dx^a/du$ and $x^a(u)$ is a geodesic.

15.20 Derive eqns (15.77)–(15.81). [Hint: the first is found by contracting Killing's equation; the second by starting with Riemann curvature then using Killing's equation in one term and contracting; the third by first using the cyclic symmetry of Riemann, followed by the use of Killing's equation to reduce six terms to three. The fourth then follows. For the last one, start with the covariant derivative of the Einstein tensor to obtain $K^\mu\nabla_\mu R = 2K^\mu\nabla_\nu R^\nu_\mu$. Then take the divergence of (15.80) and spot terms involving contractions of symmetric with antisymmetric tensors.]

16

The Einstein field equation

In the previous chapter we followed in the footsteps of Gauss, Riemann, Cartan and Poincaré. Now we pick up Einstein's trail. We shall find the field equation of General Relativity, and use it to obtain general insights into the mutual influence of matter and spacetime on one another.

16.1 Derivation of the field equation

We are ready to 'derive' the Einstein field equations—in the same sense in which Maxwell's equations were 'derived' in Volume 1: we seek the simplest possible theory consistent with the demands of covariance. We lay claim to the Equivalence Principle, adopting Einstein's wonderful intuition that gravitational forces are none other that inertial forces in disguise. We find that the equation of motion in general coordinates (i.e. in an accelerating frame) (13.35), looks just like the equation for a geodesic in a curved space, (13.16). We dimly see that something very special must be going on to allow almost arbitrarily defined coordinate systems to nevertheless agree on physical predictions. We need to understand two things: how to express the constraint on the coordinate systems that brings about this wonderful conspiracy (this is the field equation in free space), and how to express the influence of matter so that the geodesics reflect gravitational influences (this is the general field equation).

These reflections lead us to propose that spacetime may be curved, and its Riemann curvature tensor is related to a suitable tensorial expression of mass. After trying the Poynting vector we make little progress, so we turn to the stress-energy tensor. We have a fourth-rank curvature tensor and a second-rank stress-energy tensor. To relate them we can either contract the former or expand the latter. Either job can be done using the only other generally available tensor, namely the metric tensor. In the interest of simplicity, we contract the curvature tensor. Owing to its symmetries, many contractions give zero. One that does not is over the first and third index, yielding the Ricci tensor (2.2). So we try setting the Ricci tensor equal to the stress-energy tensor:

$$R_{ab} \overset{?}{=} k T_{ab}$$

with a proportionality constant k to be discovered by comparison with Newtonian theory. This is the Einstein field equation in empty space (where $T_{ab} = 0$). In the presence of matter it is not quite what we want, because the covariant derivative of R_{ab} is not always zero, whereas we

Relativity Made Relatively Easy: General Relativity and Cosmology. Volume 2. Andrew M. Steane,
Oxford University Press. © Andrew M. Steane 2021. DOI: 10.1093/oso/9780192895646.003.0016

would like to assert conservation of energy and momentum, which means we expect a suitable derivative (and the covariant derivative is the only one to hand) of T_{ab} will always be zero:

$$\nabla_\lambda T^{\lambda b} = 0. \tag{16.1}$$

What is the covariant divergence of R_{ab} then? By contracting the Bianchi identity twice (exercise 16.1), we find it is

$$\nabla_\lambda R_b^\lambda = \frac{1}{2}\nabla_b R \tag{16.2}$$

where R is the Ricci scalar. This crucial insight allows us to construct a tensor defined in terms of the spacetime properties alone (i.e. curvature) whose covariant divergence is guaranteed to be zero. It is the **Einstein tensor**:

$$G_{ab} \equiv R_{ab} - \frac{1}{2}g_{ab}R, \qquad \nabla_\lambda G_b^\lambda = 0. \tag{16.3}$$

We can now propose the equation $G_{ab} = kT_{ab}$, and obtain the constant k by comparison with Newtonian theory in the weak field limit. This is the Einstein field equation! QED.

The above sequence of steps went directly to the required result, which makes it all seem rather easy and obvious, but this is only possible with the benefit of a century of hindsight. The derivation of the divergence-free tensor G_{ab} was itself a remarkable development (Einstein did not have the benefit of familiarity with the Bianchi identities), and one must perform further analysis to check that the field equation makes physically reasonable predictions. We have already seen that it reproduces Newtonian gravity in the weak-field limit, which lends considerable encouragement.

In the course of the last one hundred years, various other approaches to GR have been worked out. Some of them yield the same theory but expressed in terms of different sets of equations, like the difference between Lagrangian and Hamiltonian mechanics. One of the most important is the Hamiltonian formulation worked out by R. Arnowitt, S. Deser and C. W. Misner and commonly called the **ADM formalism**. This formalism foliates spacetime with spacelike surfaces and treats the metric of each such surface as a dynamical variable. By carefully defining suitable conjugate momenta and constructing a Hamiltonian, one arrives at a classical field theory whose predictions match those of GR. This approach has proved to be especially valuable in numerical calculations and in some attempts to develop a quantum theory of gravity. However, we shall not say more about it here.

The Weyl tensor or conformal tensor was introduced in (15.82). By taking its divergence and invoking the Einstein field equation, one finds that it satisfies

$$\nabla_\mu C^\mu_{bcd} = 8\pi G \left(\nabla_{[c}T_{d]b} + \tfrac{1}{3}g_{b[c}\nabla_{d]}T \right). \tag{16.4}$$

Whereas R_{ab} is related to T_{ab} by an algebraic (not differential) equation, the Weyl tensor is related to T_{ab} through this first order differential equation. This leaves some room for a variety of solutions for any given T_{ab}. In other words, the distribution of matter does not uniquely

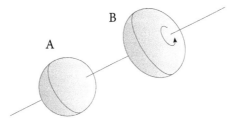

Fig. 16.1 A thought-experiment involving two planets of similar composition, in an otherwise empty universe. B is rotating about the common axis whereas A is not, and consequently B has an equatorial bulge and A does not. But should such a situation be symmetric between A and B? GR says it is not symmetric, and the bulge will happen.

specify the metric and the curvature. This is as it should be. In the complete absence of matter, for example, one can have either flat spacetime or gravitational waves. The distribution of matter combined with a boundary condition, on the other hand, does uniquely specify the spacetime except in extraordinary cases associated with singularities and closed timelike curves—see Appendix F. (The matter distribution combined with some symmetry can also suffice).

16.1.1 Inertial motion and the distant universe

> Like the great pendulum in its rotunda scribing through the long day movements
> of the universe of which you may say it knows nothing and yet know it must.
>
> From *The Road*, Cormac McCarthy

A thought-experiment that troubled Einstein is shown in Fig. 16.1. Suppose that there were a universe containing just two planets, of equal size and composition, in a state of relative rotation about their common axis. If we suppose for a moment that planet A is not rotating and planet B is, then planet B will show an equatorial bulge, and other rotation effects, while planet A will not. But, before developing his theory, Einstein felt that the situation here ought to be symmetric, so that neither planet bulged more than the other. In this he was motivated by ideas due to Mach. But according to GR—in other words, according to the theory that Einstein himself developed in due course—there is no need for there to be a symmetry here. It is perfectly possible that one planet should bulge and the other not, even without anything else in the universe. We can say this by basing our statement on the equations of GR and the rest of physics. It turns out that these equations do *not* respect Mach's principle, if that principle is formulated so as to imply symmetry in the two-planets thought-experiment.[1]

It will be useful to press this point home. Let us imagine a slightly different scenario: two aluminium rings floating in space, with radii a few tens of metres, separate but centred on the same axis, in a state of rigid rotation relative to one another about that axis. We make the rings light enough that their gravitation can be neglected. Are there any simple experiments

---

[1] *Mach's principle* is not one which can be formulated in a pithy statement; it is broadly the idea that there is a link between the matter distribution on the largest scales of the universe and the definition of which type of motion is inertial in any given region.

which we can perform in order to find out which if any of these rings is not rotating relative to a LIF? Newtonian physics provides some easy things to do. For example, one might simply place a small inert (electrically neutral and non-magnetic) item, such as a ceramic ball, near the rings (not on the axis), and release it such that initially it is not moving relative to some nearby part of the ring under investigation. If the item stays at constant distance from every part (e.g. every atom) of the ring, then the latter is not rotating. If atoms fixed in the ring undergo circular motion in the rest frame of the item, then the ring is rotating. The main point for our present discussion is that *GR makes almost no change in this experiment*. The only change is to note that the experiment determines whether or not the ring is rotating relative to a local inertial frame (LIF). GR does allow that there can be LIFs which rotate relative to the distant stars (owing to fast rotation of some massive nearby object).

Now let us add to Fig. 16.1 the matter distribution of our universe, taking it that the two planets are far from other objects. There will be almost no difference to any experiments local to A and B, for times short compared to billions of years, because it so happens that the cosmic matter distribution causes almost zero spatial curvature on average. Now suppose that in the first instance the distant stars are not in orbit about A. Then an observer fixed on B sees them rise and set. Once they understand physics well enough, the B observer will not say 'the stars are really orbiting me.' They will say 'I can if I like adopt a frame in which the stars orbit me, but that would not be a local inertial frame.' Now suppose that instead the whole large-scale matter distribution of the universe is somehow set in rotational motion, matching that of B. The observer on B will not say, 'oh look: now my planet is not rotating, but planet A is.' Well they might say that, but they should not. They should say 'although there is no relative rotation between my planet and the distant stars, nevertheless the frame in which my planet is non-rotating is not a LIF, whereas the frame in which A is non-rotating is a LIF.'[2] Reflecting on this, the B observer might well add, 'How strange that the universe should thus be rotating about the very axis that my planet also rotates about.' To be clear: the large-scale structure of our universe does *not* show evidence of such an overall rotation. The point is, if it were rotating then we would be able to tell—according to GR.[3]

16.2 Stability and energy conditions

In Newtonian gravity we are familiar with the idea that mass is always positive, and in Special Relativity we have the idea that rest energy is always positive. This notion can be expressed as the assertion that physically possible configurations of matter will always give a non-negative value of the energy density component of the energy tensor. This observation is captured by

---

[2]To very good approximation: we are ignoring the slight frame dragging.

[3]If GR is wrong (and after all, all theories of physics are probably wrong at some level) then the comments in this paragraph may have to be modified. However, it is unlikely that the main point here is wrong. A modified theory of gravity was proposed by Brans and Dicke, with a view to constructing a theory in which the distant universe is more influential so that the theory is more Machian. Experiments to date rule out this way of modifying GR at the level which would be required to significantly alter our story of the planets A and B.

Definition 16.1 Weak energy condition.

$$T_{\mu\nu}u^{\mu}u^{\nu} \geq 0 \ \text{ for all timelike } u^a. \tag{16.5}$$

This quantity expresses the energy density that would be observed by a local inertial observer whose 4-velocity is u^a. Investigation of the Einstein field equation suggests that it is required that the universe should satisfy some such condition as this in order that the vacuum solutions can be stable. Otherwise some region might acquire positive energy-density while another acquires negative energy-density and such a fluctuation can grow. Note, this observation does not constitute a derivation; the weak energy condition is a reasonable conjecture that is not known to fail, and the evidence suggests it holds since otherwise such instability would, presumably, have made its presence felt.

It is also natural to suppose that it will not be possible to find mass-energy flowing at rates above the speed of light. This leads one to propose

Definition 16.2 Dominant energy condition. *The weak energy condition holds, and for every future-pointing null or timelike u^a, the vector $-T^a_{\mu}u^{\mu}$ is future pointing null or timelike.*

This is believed to hold as far as classical physics is concerned, but in quantum physics the uncertainty principle may allow some limited exceptions.

Another condition, which asserts that energy density exceeds pressure in such a way that gravitation is always attractive, is called the **strong energy condition**. This is obeyed by ordinary matter but the evidence for a non-zero cosmological constant suggests that the strong energy condition does not hold in general. Finally, we have the **null energy condition** which is like the weak energy condition, but concerns null rather than timelike u^a. Most evidence suggests that this condition is satisfied, with the possible exception of certain exotic cases in quantum field theory.

For an ideal fluid and $c = 1$ the weak energy condition equates to $\{\rho \geq 0, \ \rho + p \geq 0\}$ and the dominant energy condition equates to $\rho \geq |p|$—the energy density 'wins' against the pressure or tension.

16.3 Field equation for a small region

We shall next derive two simple versions of the Einstein field equation:

$$K = \frac{8\pi G}{3c^2}\rho, \tag{16.6}$$

$$\frac{\mathrm{d}^2 V}{\mathrm{d}t^2} = -4\pi G\left(\rho + 3p/c^2\right) V \tag{16.7}$$

These are equations (10.17) and (10.19) of Volume 1. The first gives the Gaussian curvature of any spatial plane in static, spatially isotropic conditions, in terms of the proper[4] mass density ρ. The second gives the rate of change of the volume of a small spherical shell of test particles in the same conditions, the test particles being initially at rest, where p is the pressure in the local matter and ρ is its proper density. Note that p contributes to the second equation but not the first; this is not a mistake.

The first of these results can be obtained by considering the properties of maximally symmetric spaces mentioned after (15.87). But here we shall derive it using more elementary methods. To this end, observe that in four dimensions it is always true that

$$R_{00} = 0 \quad\ +R^1_{010} +R^2_{020} +R^3_{030},$$
$$R_{11} = R^0_{101} +0 \quad\ +R^2_{121} +R^3_{131},$$
$$R_{22} = R^0_{202} +R^1_{212} +0 \quad\ +R^3_{232},$$
$$R_{33} = R^0_{303} +R^1_{313} +R^2_{323} +0,$$

since the symmetry conditions require that the diagonal elements in this set of equations are zero. In a spatially isotropic situation in LIF coordinates we have $R^1_{010} = R^2_{020} = R^3_{030}$ (by cyclicly permuting the axis labels) and $R^1_{212} = R^2_{323} = R^3_{131}$; $R^0_{101} = R^0_{202} = R^0_{303}$; $R^2_{121} = R^3_{232} = R^1_{313}$. The sum of the diagonal elements of the Ricci tensor is then

$$R_{00} + R_{11} + R_{22} + R_{33} = 3(R^1_{010} + R^0_{101}) + 3(R^1_{212} + R^2_{121}). \tag{16.8}$$

N.B. this sum is not a contraction and therefore does not yield a Lorentz scalar. However, it is useful because for LIF coordinates we know that the metric tensor is $\mathrm{diag}(-1,1,1,1)$ and therefore $R^1_{010} = R_{1010} = R_{0101}$ whereas $R^0_{101} = -R_{0101}$. Therefore the terms in the first bracket in (16.8) cancel, while those in the second add, and we have

$$R + 2R_{00} = 6K \tag{16.9}$$

where K is the Gaussian curvature of any spatial plane, and we used $K = R^2_{121}$—see the observation made after eqn (15.56). Using the Einstein field equation (2.1) this gives

$$K = \frac{8\pi G}{3c^4} T_{00}. \tag{16.10}$$

This applies at any event where spacetime is static and spatially isotropic, as long as T_{00} is specified in LIF coordinates. But, in such coordinates, $T_{00} = \rho c^2$ where ρ is the proper density, so we obtain eqn (16.6).

Next consider the tidal effects. Consider a collection of test particles, initially all at rest relative to one another and situated on the surface of a small spherical shell of radius ϵ (these statements are well-defined as long as ϵ is small enough). We can adopt coordinates such that the particles are initially at rest. Changes in the size and shape of the shell are then all second-order in coordinate time. Also, for a non-rotating sphere any small enough change either leaves the shape spherical or changes it to an ellipsoid, therefore we can align coordinate axes with the

---

[4]Proper mass density is mass density in the rest frame of the lump of matter under discussion.

principle axes of the ellipsoid and the motion for small times is completely characterized by the changes in the three radii:

$$r_i(t) = \epsilon + \tfrac{1}{2}a_i t^2 + O(t^3).$$

Using eqn (15.56) we find, in the limit $\epsilon \to 0$, that $a_1 = -R^1_{010}c^2$ and similarly for a_2, a_3. Therefore

$$\lim_{V \to 0} \frac{\ddot{V}}{V} = -R^\lambda_{0\lambda 0}c^2 = -R_{00}c^2 \tag{16.11}$$

where V is the volume of the shell and all four terms can be included in the sum since $R^0_{000} = 0$.

For the case of a perfect fluid we have the energy tensor (14.18):

$$T^{ab} = (\rho + p/c^2)u^a u^b + pg^{ab} \qquad \Rightarrow \qquad T^\lambda_\lambda = -\rho c^2 + 3p \tag{16.12}$$

(using $u^\lambda u_\lambda = -c^2$ and $g^\lambda_\lambda = 4$). This fluid should be understood to be the gravitating body, not the set of test particles. If we now adopt LIF coordinates in the local rest frame of the gravitating body, then $u^a = (c, 0, 0, 0)$ and $g_{00} = -1$ so the field equation (2.7) gives

$$R_{00} = \frac{4\pi G}{c^4}\left(\rho c^2 + 3p\right).$$

Substituting this into (16.11) gives eqn (16.7). In fact it is not necessary to assume that the pressure is isotropic; allowing that the pressures in different directions can disagree gives the slightly more general form

$$\lim_{V \to 0} \frac{\ddot{V}}{V} = -\frac{4\pi G}{c^2}\left(\rho c^2 + p_x + p_y + p_z\right). \tag{16.13}$$

This result is useful because it has not assumed the weak field limit and it gives a good general insight into the effects of gravity on the motion of test particles. One sees that gravity is attractive ($\ddot{V} < 0$) when the pressure is zero or positive, and gravity remains attractive at negative pressure (i.e. tension) as long as $\sum_i |p_i| < \rho c^2$. Ordinary matter has $|p_i| \ll \rho c^2$ and electromagnetic fields have $T^\lambda_\lambda = 0$.

Equation (16.11) can be seen as an example of a more general insight into the geometric meaning of the Ricci tensor. The Ricci tensor expresses the change in volume with geodesic 'flow' in any direction. That is, if one moves the boundaries of a three-dimensional region along a set of neighbouring initially parallel geodesics in whatever direction, then the relevant component of the Ricci tensor tells how the 3-volume changes as one moves along. In a space of N dimensions, the same can be said of subspaces of dimension $(N-1)$—see Fig. 16.2.

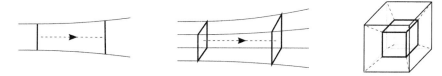

Fig. 16.2 Examples of eqn (16.11) in two, three and four dimensions (in the last case the fourth dimension may be imagined as time).

16.4 Motion of matter from the field equation

In the previous section we obtained a statement about motion of test particles from the equation for geodesic deviation. The argument assumed that such particles follow geodesics, and up till now we have asserted that claim as an axiom of GR.

It is a remarkable feature of GR that the claim that test particles[5] follow geodesics does not need to be introduced as an axiom. It is not, and cannot be, separated from the field equation, but rather follows from it. For, examine the (exact) eqn (14.23) for the motion of an ideal fluid, and take the case of zero pressure, $p = 0$. We obtain

$$\rho u^\mu \nabla_\mu u^a = 0. \tag{16.14}$$

But this is none other than ρ multiplied by the formula for a geodesic. For $\rho \neq 0$ it is the geodesic equation!

What we need to do is think carefully about the calculation we have just presented. First note that a fluid with no pressure must be moving inertially. It is like a collection of particles of dust which float along without bumping into one another, and without exerting any other force on one another. You could, if you like, set ρ to zero everywhere except in the vicinity of one timelike curve (it must be timelike in order to obey conservation of particles), and then the equation tells you that that curve has to be a geodesic.

Next, note that the derivation of fluid motion started out from $\nabla_\mu T^{\mu b} = 0$. Once we made that claim, the rest followed by the definition of the covariant derivative and the form of T^{ab}. But what gave us the right to claim $\nabla_\mu T^{\mu b} = 0$ in the first place? It was the Einstein field equation! For, the left-hand side of the field equation says, 'Here are some derivatives and other combinations involving the metric tensor; these are all about the manifold we call spacetime, and that is all. I make no mention of particles or anything like that. But look: all these derivatives and so forth end up making a combination which has zero covariant divergence: $\nabla_\mu G^{\mu b} = 0$. So you guys over on the right-hand side of this equation had better have zero covariant divergence too. So distribute yourselves in spacetime accordingly!' The effect of this instruction to the matter providing T^{ab} is that it must move along a geodesic if it is not subject to forces causing pressure or stress.

Here is another way of looking at it. A small,[6] not very massive piece of matter creates a small bump in the gravitational field. The field equation gets to grips with this bump and tells it how to move.

Comparison with electromagnetism. This property of gravity is often contrasted with other field theories such as electromagnetism, as if if it a completely new feature. That claim is

---

[5] Test particles are particles that do not themselves distort the sources causing whatever is the local gravitational field.

[6] To avoid infinite gradients make it small but not infinitesimal.

somewhat of an exaggeration, however. Similar reasoning can be applied to Maxwell's equations in flat spacetime, as follows.

First, we develop the Maxwell equations somehow or other. One could follow the route outlined in Volume 1, or one could start from a Lagrangian. One then takes an interest in energy and one seeks an energy tensor that can be written in terms of the field tensor. One finds

$$T_{ab} = \epsilon_0 c^2 \left(-\mathbb{F}_{a\mu}\mathbb{F}^{\mu}{}_b - \tfrac{1}{4}g_{ab}\mathbb{F}_{\mu\nu}\mathbb{F}^{\mu\nu} \right) \tag{16.15}$$

and one notes that this tensor satisfies $\nabla_{\mu}T^{\mu b} = 0$ in vacuum. So far, so good: it looks like energy is conserved by our field. But then one notices that $\nabla_{\mu}T^{\mu b} \neq 0$ at any spatial location where the sources of the field are present and moving in certain ways. In dismay, one starts to abandon the theory, but then one realizes in a flash of inspiration that there is no need to do that. All we need to do is realize that $\nabla_{\mu}T^{\mu b}$ must be the rate, per unit volume, at which the field is transferring energy onto the charged matter it is interacting with. Our field equations tell us that $\nabla_{\mu}T^{\mu b} = -\mathbb{F}^{a\nu}j_{\nu}$. It follows that $\mathbb{F}^{a\nu}j_{\nu}$ must be the force (per unit volume) on the matter, because if this is so then we will get energy conservation overall. Thus in classical electromagnetism we can *derive* the Lorentz force equation from the field equations plus energy conservation.

The argument in the case of General Relativity does not need to bring in a further principle (energy conservation) in addition to what is already there in the field equation, and this is a notable difference, but in other respects the two cases are similar.

16.5   The cosmological constant

When Einstein first proposed his field equation, he was aware of the possibility of including a term Λg_{ab} in addition to the Ricci tensor, when describing the field. Because the metric is guaranteed to be covariantly constant, one will find $\nabla_{\lambda}\tilde{G}^{\lambda}_b = 0$ when $\tilde{G}_{ab} = R_{ab} - \tfrac{1}{2}g_{ab}R + \Lambda g_{ab}$ for any constant Λ. He decided to include such a term, on the argument that it would allow for static solutions for the cosmos on the largest scale. When it subsequently emerged that the cosmos is not static, he is said to have referred to this as his 'greatest blunder'. (Nussbaumer, 2014) No-one else has agreed with Einstein's self-assessment here however; to include Λ was never a blunder but an important move which has to be taken seriously.[7]

Λ is called the cosmological constant, and when the term Λg_{ab} is regarded as being added to the Einstein tensor, then it is referred to as the cosmological constant term. However, as we saw in Chapter 2, one can equally well move this term to the right-hand side of the field equation, and regard it as a contribution to the energy tensor. Since Λ here really is constant, in both space and time, this contribution is present whether or not there is any ordinary matter present. In particular, it is present in that part of spacetime which we ordinarily call vacuum. Astronomical observations suggest that a term either exactly of this form, or behaving like it in some respects,

---

[7]Perhaps what Einstein regretted was that he did not take the hint from his equation and announce that the cosmos was not static.

is indeed present and influences the large scale shape and dynamics of the universe. This will be discussed in Chapter 25. It is called 'dark energy' in order to keep open the possibility that it may not be exactly of the cosmological constant form; for example it may evolve over time. The word 'dark' alludes to the fact that it does not have electromagnetic interactions; 'energy' says it contributes to the energy tensor.

By examining the metric in any LIF, one can see that $(-\Lambda g_{ab})$, when interpreted as a form of energy density (after multiplying by $c^4/8\pi G$), has both energy density and tension, indicated by the diagonal elements, and furthermore the amount of tension far exceeds what would be possible for any ordinary matter of the same energy density. We can regard this tension as a way of speaking about an aspect of spacetime, or as a property of a form of exotic matter. However we interpret it, its gravitational effects are repulsive: see eqn (16.13) for example (recalling that tension is negative pressure). So if $\Lambda > 0$, as cosmological studies suggest, then we have a form of anti-gravity, in the sense that every part of spacetime everywhere is in a state of gravitational repulsion from other parts, except where this repulsion is overcome by the presence of ordinary gravitating matter. The evidence for, and implications of, the cosmological constant will be explored in Chapters 22–25. We shall ignore it up until then since its numerical value is sufficiently small that its influence is negligible at scales smaller than that of galaxy clusters.

16.6   Energy and momentum

We will discuss the total energy and momentum that can be said to be associated with a source of gravity, including whatever energy momentum is assigned to spacetime itself. Our treatment is limited to stationary systems (those possessing a timelike Killing vector). It is thus somewhat limited, by very beautiful.

In the case of a spacetime with a timelike Killing vector $\mathbf{K}$, define a current[8]

$$J^a \equiv K_\mu R^{a\mu} \;=\; 8\pi G K_\mu (T^{\mu a} - \tfrac{1}{2} T g^{\mu a}). \tag{16.16}$$

The second version shows that we can interpret this current as something to do with the flow of energy and momentum in the direction picked out by $\mathbf{K}$. For example, this could be the direction called time in a static universe, so J^a is something like an energy density (the zeroth component) and an energy flux (the other components). The remarkable thing about this current is that its divergence vanishes:

$$\nabla_\nu J^\nu = (\nabla_\nu K_\mu) R^{\nu\mu} + K_\mu \nabla_\nu R^{\nu\mu} \;=\; 0 \tag{16.17}$$

where we used that the first term vanishes since it contracts an antisymmetric with a symmetric tensor (recall that Killing's equation (13.39) asserts that $\nabla_\nu K_\mu$ is antisymmetric), and the second term vanishes by writing it in terms of R using $\nabla_\nu G^{\nu\mu} = 0$ and then using (15.81). From the vanishing of this divergence we can deduce that $\mathbf{J}$ describes the density and flow of a conserved quantity!

---

[8]One may also present the argument without the use of Killing vectors, by adopting coordinates as one ordinarily would for a stationary spacetime. Then $J^a = 4\pi G(2T^{0a} - T g^{0a})$.

The energy whose conservation we have inferred is

$$E = \frac{1}{4\pi G} \int_V n_\mu J^\mu \sqrt{|\gamma|} \mathrm{d}^3 x \tag{16.18}$$

where γ is the determinant of the induced metric on the submanifold (see (8.42)); the submanifold in this case being three-dimensional—it is the whole of space at some instant of time. More formally, it is any spacelike hypersurface and n^a is a unit timelike vector everywhere perpendicular to this hypersurface. This is our expression for the total energy of a stationary spacetime, or, if you prefer, of the matter content of a given stationary spacetime, including the influence of gravitational binding energy. See (18.8) for an example.

Now we employ (15.80): $\nabla_\nu \nabla_b K^\nu = R_{b\lambda} K^\lambda$, hence

$$J^a = \nabla_\nu (\nabla^a K^\nu). \tag{16.19}$$

This is significant because it shows that J^a is itself a divergence, and therefore we can use Gauss' theorem (11.49) to convert its volume integral into a surface integral. The statement of the theorem in the form we need it here is

$$\int_V n_\mu (\nabla_\nu A^{\mu\nu}) \sqrt{|\gamma|} \, \mathrm{d}^3 x = \int_S n_\mu \sigma_\nu A^{\mu\nu} \sqrt{|\gamma^{(2)}|} \, \mathrm{d}^2 x \tag{16.20}$$

where $A^{\mu\nu}$ is any antisymmetric second-rank tensor, σ_ν is an outward-pointing unit normal to the surface, and $\gamma^{(2)}$ is the determinant of the induced metric on the surface. In order to obtain this from (11.49), first adopt coordinates such that $n_\mu = (1,0,0,0)$ and argue that it is correct in those coordinates, then generalize by using that the expression is tensorial. Applied to our case, where $A^{\mu\nu} = \nabla^\mu K^\nu$ we obtain

$$E = \frac{1}{4\pi G} \int_S n_\mu \sigma_\nu (\nabla^\mu K^\nu) \sqrt{|\gamma^{(2)}|} \, \mathrm{d}^2 x. \tag{16.21}$$

This is the **Komar integral** associated with the timelike Killing vector **K**. It contains the remarkable result that the total energy throughout some volume can be obtained by an integral over the surface, which can be taken in the asymptotic region, far away from the gravitating bodies. It means that the energy is encoded in the way spacetime behaves 'at infinity'. The result is comparable to the way an electric field at any closed surface encodes the information about the total charge in the region enclosed. But note that the energy we are dealing with here is that of the spacetime and source together.

A further insight can be obtained by arguing that far from gravitating bodies spacetime will tend to the flat (Minkowski) form. Such a region is called the *asymptotic limit*. Note, in order to obtain a well-defined asymptotic limit we need a region which consists largely of vacuum, and this is not necessarily a good approximation when considering very large-scale structure in the universe—the subject of cosmology. But outside any given solar system or galaxy it can be a very good approximation. In such a region a timelike Killing vector will tend to a constant in the limit, and it picks out a direction in spacetime. The energy we have calculated can be considered to be the rest energy of the collection of gravitating bodies and the spacetime around them, and we can use it to construct a complete energy-momentum 4-vector

$$P^a = EK^a. \tag{16.22}$$

The Killing vector provides the 4-velocity. In the asymptotic limit the quantity thus constructed will transform between inertial frames by Lorentz transformations and it captures the physical notion of 'the 4-momentum of the solar system' or 'the 4-momentum of the galaxy', etc., as the case may be, with gravitational binding energy correctly accounted for.

Similar reasoning applies to angular momentum. Now the Killing vector will correspond to rotation in the asymptotic limit; it might for example be the ϕ direction of a polar coordinate system. This is a spacelike direction but the argument leading to (16.17) still applies so we have a conserved quantity again. The conserved angular momentum is given by the right-hand side of (16.21), with K^a the appropriate Killing vector, multiplied by a suitable scalar whose value will depend on the size of K^a.

The above line of reasoning is precisely consistent with the idea that we define the mass of a gravitating body by examining its gravitational influence in the asymptotic limit. Indeed, mass cannot be uniquely defined in GR except by appeal to an asymptotic limit, because only in such a limit does the Komar integral become independent of further increases in the size of the region of integration. You have to get away from something, to climb out of its gravitational potential well, in order to find out how much mass it presents. But one cannot get away from the whole physical universe. It follows that the notion of a 'mass of the universe' is undefined; there is no way we can assign it a value.

Exercises

16.1 **Einstein tensor**. Obtain (16.2) by the following steps. First write down the Bianchi identity (2.14) and obtain

$$\nabla_e R_{bd} + \nabla_a R^a_{bde} + \nabla_d R^a_{bea} = 0. \tag{16.23}$$

Now use (anti)symmetries of the Riemann tensor to deduce $\nabla_d R^a_{bea} = -\nabla_d R_{be}$, and hence

$$\nabla_b R^b_d + \nabla_a R^{ab}_{db} - \nabla_d R = 0. \tag{16.24}$$

Finally, show that $\nabla_a R^{ab}_{db} = \nabla_a R^a_d$ and note that this is equal to the first term. Equation (16.2) follows.

16.2 **Lorenz gauge**.
(i) Let $\Gamma^a \equiv g^{\mu\nu}\Gamma^a_{\mu\nu}$. Show that for coordinates x^a, $\nabla^\lambda \nabla_\lambda x^a = -\Gamma^a$.
(ii) Show that $\nabla_\lambda g^{\lambda b} = 0$ (use a LIF) and by using this together with (11.37), obtain

$$\Gamma^b = \frac{1}{\sqrt{|g|}}\partial_\lambda\left(\sqrt{|g|}\,g^{\lambda b}\right). \tag{16.25}$$

Hence obtain the Lorenz gauge condition (F.4).

16.3 Show that in the case of an ideal fluid, the weak energy condition (16.5) requires both $\rho \geq 0$ and $p \geq -\rho c^2$.

16.4 Estimate the Gaussian curvature of space-time (i) inside a lump of metal (after averaging over the atomic spacing) (ii) inside an atomic nucleus. What diameter would an entity of mass M need to have in order that curvature effects are significant?

16.5 **An exact plane wave**. Consider a metric of the form

$$ds^2 = -c^2 dt^2 + dx^2 + f^2(u)dy^2 + g^2(u)dz^2 \tag{16.26}$$

where $u = ct - x$. Obtain the Ricci tensor, and hence show that this is a solution to the vacuum field equation if

$$g\ddot{f} + f\ddot{g} = 0 \tag{16.27}$$

where the dot signifies d/du. Show also that if $\ddot{f} = \ddot{g} = 0$ then $R_{abcd} = 0$. The effect of such a wave passing over a collection of dust is quite subtle. In the flat spacetime after the wave has passed, it can place the dust on a set of trajectories corresponding to a type of collapse described by Milne.

16.6 *Alternative equations for electromagnetism.* In flat spacetime we have $\partial_\lambda F^{\lambda b} = \partial_\lambda \partial^\lambda A^b - \partial_\lambda \partial^b A^\lambda$ and since partial derivatives commute this is also $\partial_\lambda \partial^\lambda A^b - \partial^b \partial_\lambda A^\lambda$. Show that, upon replacing partial derivatives by covariant derivatives, the field equation now becomes

$$\nabla_\lambda \mathbb{F}^{\lambda b} + R^b_\lambda A^\lambda = -\mu_0 j^b. \tag{16.28}$$

This equation makes different predictions to the one ordinarily assumed (14.1), therefore at most of them is correct.

16.7 **Komar energy example**. Show that the energy given by the Komar integral for Schwarzschild spacetime is equal to Mc^2 where M is the mass parameter in (2.33). [Hint: taking $K^a = (1,0,0,0)$ in Schwarzschild coordinates you should find that $\nabla^0 K^1 = g^{00}\Gamma^1_{0\lambda}K^\lambda = -GMc^2/r^2$.]

Part III

17

Schwarzschild–Droste solution

We are now ready to embark on the exact calculation of the curvature of spacetime by solving the field equations. Generally speaking it is extremely hard to find the exact metric for any given physical scenario. One usually needs a degree of symmetry in the problem, a judicious choice of coordinates, and a bit of luck. If the conditions are sufficiently simple the problem becomes tractable. An important such case is that of a *static* field in *empty space* with *spherical symmetry* in the spatial directions. This yields the *Schwarzschild–Droste metric*, the first and still the most important non-flat solution to the vacuum field equation. It is sobering to note that Karl Schwarzschild worked out this solution in 1916 while serving as an artillery officer in the first world war; he became a casualty of that war soon after. Johannes Droste discovered the same result independently at the same time, and published a more complete analysis.[1] The solution applies to the space outside a spherically symmetric distribution of matter. We shall first derive the metric, and then discuss some of the associated physics, such as orbits, gravitational redshift, and tidal effects.

17.1 Obtaining the metric

The technique to find simple solutions to Einstein's field equation is much like that employed to find solutions of differential equations more generally. We guess the overall form, for example by specifying a function with the expected symmetries, and then 'plug it in' to the equations. If the equations are all satisfied simultaneously, placing mutually consistent requirements on the function we guessed, then we are assured that we have a valid solution.

Proceeding in this spirit, we first guess that in simple cases it will probably be possible to find a coordinate system such that the metric tensor is diagonal. We shall assume a diagonal form anyway, and the field equation will tell us our mistake if it is one. One may motivate this

---

[1] The result is often called simply the *Schwarzschild* metric; we shall employ this name interchangeably with *Schwarzschild–Droste* metric; this is simply another name for precisely the same result, and makes the terminology more even-handed and historically informed. Droste's contribution was as great as Schwarzschild's, and prior to other early work of David Hilbert and Hermann Weyl. Arguably the 'Schwarzschild coordinates' should be called 'Droste coordinates' but we shall employ the former term nonetheless.

assumption a little further as follows. In two and three dimensions it is always possible to find orthogonal coordinates, so in four dimensions it may also be possible if the problem has even a modicum of symmetry. All we need for present purposes is the assurance that a diagonal metric is worth trying.

For a static field it should furthermore be possible to find coordinates such that the metric is independent of the time coordinate —so let us assume this. By a 'time coordinate' we mean one (call it t) such that $g_{tt} < 0$.

So far we have reduced the problem to the general form

$$ds^2 = -A(x, y, z)dt^2 + B(x, y, z)dx^2 + C(x, y, z)dy^2 + D(x, y, z)dz^2$$

where the functions A, B, C, D are all positive (to make sure dt gives a timelike and dx, dy, dz a spacelike contribution)—but we relax that assumption later.

Next we assert spherical symmetry. This means that if we choose a polar coordinate system and look at displacements with $dr = 0$, $dt = 0$ we must get a two-dimensional space whose curvature is everywhere the same. This means it must be a 2-sphere, and the metric must be of the form

$$ds^2 = -A(r)dt^2 + B(r)dr^2 + C(r)(d\theta^2 + \sin^2\theta d\phi^2).$$

One of the pair of functions $B(r), C(r)$ is redundant, because (except at singular behaviour) we can always rescale the r coordinate to achieve either $B = 1$ or $C = 1$ or some other dependence. Therefore let us choose r such that $C(r) = r^2$. Then the spherical surface at given r will have circumference $2\pi r$ and area $4\pi r^2$. This choice of radial coordinate amounts to *defining* r *not* as 'distance from the origin' but as 'circumference/2π' of a circle around the origin. We now have the metric

$$ds^2 = -A(r)dt^2 + B(r)dr^2 + r^2(d\theta^2 + \sin^2\theta\, d\phi^2), \tag{17.1}$$

i.e.

$$g_{ab} = \begin{pmatrix} -A(r) & & & \\ & B(r) & & \\ & & r^2 & \\ & & & r^2 \sin^2\theta \end{pmatrix}. \tag{17.2}$$

We have reduced the problem to that of finding just two functions. We now have to substitute g_{ab} into the vacuum field equations $G_{ab} = 0$, which are also $R_{ab} = 0$ since $R = 0$ in vacuum (eqn (2.5)). Evaluating all the derivatives is laborious. A labour-saving device (the Euler–Lagrange method) will be described shortly (in Section 17.1.1) but that still leaves some work which you can do by checking the results supplied in Table 17.1. One then has to substitute these results in eqn (2.2) for the Ricci tensor. Setting $c = 1$, one finds

Table 17.1 Connection coefficients for the general spherically symmetric metric. First we write the Lagrangian. Then we write the four Euler–Lagrange equations. Each such equation then gives those connection coefficients which can be extracted by comparing the equation to the geodesic equation. Since no other velocity terms appeared in any Euler–Lagrange equation, the other connection coefficients must be zero.

$$\mathcal{L} = -A(r)\dot{t}^2 + B(r)\dot{r}^2 + r^2\dot{\theta}^2 + r^2\sin^2(\theta)\dot{\phi}^2, \qquad \frac{d}{dt}\frac{\partial\mathcal{L}}{\partial\dot{x}^a} = \frac{\partial\mathcal{L}}{\partial x^a}$$

① $\qquad -2A\ddot{t} - 2A'\dot{r}\dot{t} = 0 \qquad\qquad\qquad\qquad\qquad$ (17.3)

② $\qquad 2B\ddot{r} + 2B'\dot{r}^2 = -A'\dot{t}^2 + B'\dot{r}^2 + 2r(\dot{\theta}^2 + \sin^2(\theta)\dot{\phi}^2) \qquad$ (17.4)

③ $\qquad 4r\dot{r}\dot{\theta} + 2r^2\ddot{\theta} = 2r^2\sin\theta\cos\theta\,\dot{\phi}^2 \qquad\qquad\qquad$ (17.5)

④ $\quad \sin^2\theta\left(4r\dot{r}\dot{\phi} + 2r^2\ddot{\phi}\right) + 4r^2\sin\theta\cos\theta\,\dot{\theta}\dot{\phi} = 0 \qquad$ (17.6)

$$-\ddot{x}^a = \Gamma^a_{\mu\nu}\dot{x}^\mu\dot{x}^\nu$$

① $\Rightarrow \qquad \Gamma^t_{rt} = \Gamma^t_{tr} = \dfrac{A'}{2A} \qquad\qquad\qquad\qquad\qquad$ (17.7)

② $\Rightarrow \qquad \Gamma^r_{tt} = \dfrac{A'}{2B}, \quad \Gamma^r_{rr} = \dfrac{B'}{2B}, \quad \Gamma^r_{\theta\theta} = -\dfrac{r}{B}, \quad \Gamma^r_{\phi\phi} = -\dfrac{r\sin^2\theta}{B} \quad$ (17.8)

③ $\Rightarrow \qquad \Gamma^\theta_{r\theta} = \Gamma^\theta_{\theta r} = \dfrac{1}{r}, \quad \Gamma^\theta_{\phi\phi} = -\sin\theta\cos\theta \qquad\qquad$ (17.9)

④ $\Rightarrow \qquad \Gamma^\phi_{r\phi} = \Gamma^\phi_{\phi r} = \dfrac{1}{r}, \quad \Gamma^\phi_{\theta\phi} = \Gamma^\phi_{\phi\theta} = \cot\theta \qquad\qquad$ (17.10)

$$R_{tt} = \frac{A''}{2B} - \frac{A'\beta_+}{4B} + \frac{A'}{rB} \qquad \text{where} \quad \beta_\pm \equiv \left(\frac{A'}{A} \pm \frac{B'}{B}\right)$$

$$R_{rr} = -\frac{A''}{2A} + \frac{A'\beta_+}{4A} + \frac{B'}{rB}$$

$$R_{\theta\theta} = 1 - \frac{r\beta_-}{2B} - \frac{1}{B}$$

$$R_{\phi\phi} = R_{\theta\theta}\sin^2\theta$$

$$R_{\mu\nu} = 0 \quad \text{for} \quad \mu \neq \nu \qquad\qquad\qquad\qquad\qquad (17.11)$$

We require $R_{ab} = 0$. By multiplying the first equation by B/A and adding to the second we find

$$\frac{A'}{A} = -\frac{B'}{B} \qquad \Rightarrow \qquad AB = \text{const}$$

If our metric describes spacetime around a finite object then as $r \to \infty$ we expect the metric to become that of flat spacetime, for which $A = c^2$, $B = 1$. Thus the constant must be c^2 and so we have

$$B(r) = \frac{c^2}{A(r)}.$$

The equation $R_{\theta\theta} = 0$ now becomes $rA' + A = c^2$. Therefore $(rA)' = c^2$, hence

$$A(r) = c^2 \left(1 - r_s/r\right),$$

where r_s is a constant, and we have chosen the sign of this constant so that in the weak field limit we shall get Newtonian gravity. We now have the famous Schwarzschild or Schwarzschild–Droste metric:

$$\mathrm{d}s^2 = -\left(1 - \frac{r_s}{r}\right) c^2 \mathrm{d}t^2 + \frac{\mathrm{d}r^2}{1 - r_s/r} + r^2 \left(\mathrm{d}\theta^2 + \sin^2\theta\,\mathrm{d}\phi^2\right). \tag{17.12}$$

We introduced r_s—the Schwarzschild radius—as a parameter. In order to relate this parameter to easily observed quantities, we compare the metric to the result in the Newtonian (weak field) limit, which is $g_{00} = -(1 + 2GM/rc^2)$ (eqn (5.46)) for a spherical body of mass M. Thus we find

$$r_s = \frac{2GM}{c^2}. \tag{17.13}$$

Let us define[2]

$$m \equiv GM/c^2 \tag{17.14}$$

then $r_s = 2m$ and the metric can be written

Schwarzschild–Droste metric

$$\mathrm{d}s^2 = -\left(1 - \frac{2m}{r}\right) c^2 \mathrm{d}t^2 + \frac{\mathrm{d}r^2}{1 - 2m/r} + r^2 \left(\mathrm{d}\theta^2 + \sin^2\theta\,\mathrm{d}\phi^2\right). \tag{17.15}$$

We have now derived an important result which was quoted in Volume 1 without derivation.

In the region $r > 2m$ we have static conditions, so we can imagine constructing a lattice made of rods whose intersections are permanently fixed at given Schwarzschild coordinate locations. We shall call such a lattice the **Schwarzschild lattice**.

17.1.1   Connection coefficients for Schwarzschild–Droste metric

As we have already remarked, in practice the 'geodesic equation' often is not used for finding geodesics: it is used for finding Christoffel symbols! That is to say, if we have the metric but do not yet know the Christoffel symbols, then it is usually quicker to write down the Euler–Lagrange equations than it is to carry out the derivatives and sums that are in the defining expression for Γ^a_{bc}. Having found the Euler–Lagrange equations, we examine the a'th one and expect to find an expression for $\ddot{x}^a$ in terms of a sum, each term of the sum involving a product of velocities $\dot{x}^b \dot{x}^c$. The coefficient of the term at given b, c is Γ^a_{bb} for $b = c$ and $\Gamma^a_{bc} + \Gamma^a_{cb} = 2\Gamma^a_{bc}$ for $b \neq c$.

---

[2] This parameter is often employed because it is equal to M if one adopts units in which $G = c = 1$.

This method is presented in Table 17.1. Using those results, and the formulas we have found for $A(r), B(r)$, we can write down:

$$\Gamma^r_{tt} = \frac{mc^2}{r^2}\left[1 - \frac{2m}{r}\right], \qquad \Gamma^r_{rr} = \frac{-m}{r^2(1 - 2m/r)}, \qquad \Gamma^r_{\theta\theta} = -(r - 2m), \quad \Gamma^r_{\phi\phi} = -(r - 2m)\sin^2\theta,$$

$$\Gamma^\theta_{\phi\phi} = -\sin\theta\cos\theta, \qquad \Gamma^\theta_{r\theta} = \Gamma^\theta_{\theta r} = \frac{1}{r}, \qquad \Gamma^\phi_{r\phi} = \Gamma^\phi_{\phi r} = \frac{1}{r}, \qquad \Gamma^\phi_{\theta\phi} = \Gamma^\phi_{\phi\theta} = \cot\theta.$$

$$\Gamma^t_{tr} = \Gamma^t_{rt} = \frac{m}{r^2(1 - 2m/r)}, \tag{17.16}$$

17.2 Orbits

To find test particle orbits in the Schwarzschild spacetime first we require the equations for the geodesics; they are readily obtained from (13.27), (17.5) and (13.25) (or by developing (17.3)–(17.6)):[3]

$$c^2(1 - 2m/r)\dot{t} = \text{const} \equiv E \tag{17.17}$$

$$\ddot{\theta} + \frac{2}{r}\dot{r}\dot{\theta} - \sin\theta\cos\theta\,\dot{\phi}^2 = 0 \tag{17.18}$$

$$r^2\sin^2\theta\,\dot{\phi} = \text{const} \equiv L \tag{17.19}$$

$$\dot{r}^2 + \left(1 - \frac{2m}{r}\right)\left(\frac{L^2}{r^2\sin^2\theta} + r^2\dot{\theta}^2\right) - \frac{2mc^2}{r} = \frac{E^2 - c^4}{c^2} \tag{17.20}$$

where the dot signifies differentiation with respect to proper time for timelike geodesics. Null geodesics are treated in Section 17.3. We can immediately simplify by orienting the axes of the coordinate system so that a given geodesic starts out in the plane $\theta = \pi/2$; by symmetry it will then remain always in that plane, hence (17.18) can be replaced by

$$\theta = \pi/2 \tag{17.21}$$

and $\dot{\theta} = 0$.

E and L are constants of the motion. Their role in equations (17.19) and (17.20) suggests that L is a form of angular momentum per unit mass, and E relates to energy per unit mass. In order to interpret them more precisely, introduce $v_r \equiv \dot{r}/\dot{t}$ and $v_\phi \equiv r\dot{\phi}/\dot{t}$ (these are components of the velocity) and examine the behaviour in the limit $r \to \infty$. In that limit we have $\dot{t} = E/c^2$ hence $(L/r) = v_\phi E/c^2$ and $\dot{r} = v_r E/c^2$. Then (17.20) gives $E = (1 - (v_r^2 + v_\phi^2)/c^2)^{-1/2}c^2$. This is precisely the special relativistic energy per unit mass. If $E < c^2$, on the other hand, then the particle is on a bound orbit and cannot reach infinity; in this case there is no direct connection between E and the properties the particle may have at infinity, but it is suitable to refer to E still as the energy per unit mass.

---

[3]c.f. Chapter 14 of Volume 1.

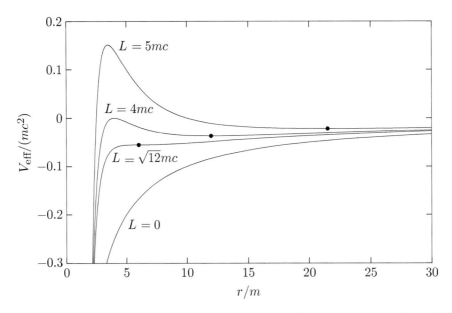

Fig. 17.1 Effective potential for Schwarzschild solution for some example values of L.

Eqn (17.20) has the same form as the equation for particle motion in one dimension in an effective potential well given by

$$V_{\text{eff}} = \left(1 - \frac{2m}{r}\right) \frac{L^2}{2r^2} - \frac{mc^2}{r} = \frac{L^2}{2r^2} - \frac{mc^2}{r} - \frac{mL^2}{r^3} \qquad (17.22)$$

This function is plotted in Fig. 17.1 for some example values of L. For large L there is a potential barrier, much like the centrifugal barrier in the Newtonian problem, but here it is overcome at small r by the inverse cubic term. As a result, for any given L a particle with enough energy can surmount the barrier and will continue to small r and hit the central body or enter the central black hole, as the case may be. At low L the barrier disappears completely and then an incoming particle of any energy will be absorbed.

The stationary values of V_{eff} are at the radii satisfying

$$mc^2r^2 - L^2r + 3mL^2 = 0. \qquad (17.23)$$

The inner stationary value corresponds to an unstable circular orbit, the outer to a stable circular orbit (see Fig. 17.1). When $L < \sqrt{12}mc$ this equation has complex roots and consequently this condition is the case where the barrier has disappeared and V_{eff} has no stationary values. At $L = \sqrt{12}mc$ one finds $r = 6m = 3r_s$; this is the radius of the smallest stable circular orbit (to be precise, it just becomes unstable at this radius).

17.2.1   Radial worldlines

Let us obtain the timelike geodesics in the radial direction (having first noted, by symmetry, that any geodesic that starts off in the t, r coordinate plane stays in that plane). Using (17.17)–(17.20) we obtain $L = 0$ and

$$c^2 \dot{t} = E(1 - 2m/r)^{-1} \tag{17.24}$$

$$c^2 \dot{r}^2 = E^2 - c^4(1 - 2m/r) \tag{17.25}$$

We begin with the simplest case, $E = c^2$, which corresponds to a particle dropped from rest at infinity. Then for an infalling trajectory one has

$$\dot{r} = -c\sqrt{2m/r} \tag{17.26}$$

and hence

$$\frac{dr}{dt} = \frac{\dot{r}}{\dot{t}} = -c(1 - 2m/r)\sqrt{\frac{2m}{r}}. \tag{17.27}$$

By combining this with (17.24) and integrating, we find

$$\tau = \frac{1}{3}\sqrt{\frac{2}{mc^2}}\left(r_0^{3/2} - r^{3/2}\right), \tag{17.28}$$

$$r = \left(\tfrac{3c}{2}(\text{const.} - \tau)\right)^{2/3}(2m)^{1/3}, \tag{17.29}$$

$$t = \frac{2m}{c}\left[-\frac{2}{3}\sqrt{\frac{r}{2m}}\left(3 + \frac{r}{2m}\right) + \ln\left|\frac{\sqrt{r/2m} + 1}{\sqrt{r/2m} - 1}\right|\right]_{r_0}^{r} \tag{17.30}$$

(exercise 11.6 of Volume 1), where $r = r_0$ and $\tau = 0$ at $t = 0$. Thus we can trace such a geodesic in the t, r plane, and also note the increase of proper time along it—see Fig. 17.2.

Next consider a particle dropped from rest at some radius $r = R$. Then (17.25) gives $E = c^2(1 - 2m/R)^{1/2}$ and hence $\dot{r} = c\sqrt{2m}(1/r - 1/R)^{1/2}$. To solve this it is helpful to introduce a change of variable to the *cycloid parameter* η defined by

$$\cos\eta = (2r/R) - 1. \tag{17.31}$$

After some straightforward algebra one finds that the equation for $\dot{r}$ takes the form $c\, d\tau/d\eta = (R^3/8m)^{1/2}(1 + \cos\eta)$ which is easily integrated. Thus we arrive at a worldline described by the parametric equations

$$r = \tfrac{1}{2}R(1 + \cos\eta), \qquad \tau = \frac{1}{c}\sqrt{\frac{R^3}{8m}}(\eta + \sin\eta). \tag{17.32}$$

$\eta = 0$ at $r = R$. We would also like to relate r to the coordinate time t. The analysis requires a non-trivial integral but is tractable; one finds

$$t = \frac{2m}{c}\left[\ln\left|\frac{X + \tan(\eta/2)}{X - \tan(\eta/2)}\right| + X\left(\eta + \frac{R}{4m}(\eta + \sin\eta)\right)\right] \tag{17.33}$$

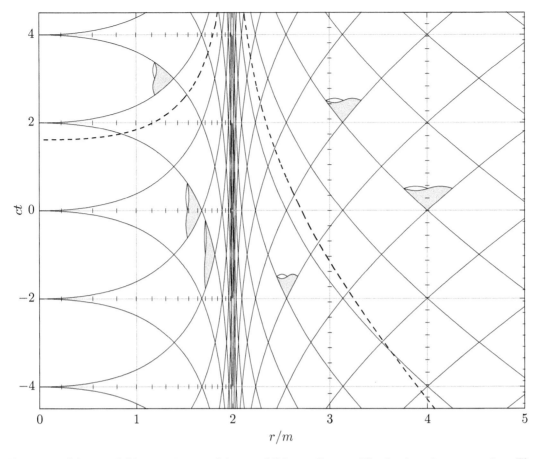

Fig. 17.2 Schwarzschild spacetime in Schwarzschild coordinates. The horizon is at $r = 2m$. The singularity is at $r = 0$. The full lines are null geodesics, and hence the worldlines of light rays. The bold dashed line is the worldline of an infalling massive particle that fell from rest at infinity. The tick marks show equal increments of proper time. In order to get a good impression of the situation in the inner region ($r < 2m$), turn the diagram through 90 degrees with the singularity at the top.

where $X = (R/2m - 1)^{1/2}$. Equations for the final case, $E > c^2$, can be obtained by introducing a quantity R which expresses the energy through $E = c^2(1 + 2m/R)^{1/2}$ and using a new parameter η; one finds now

$$r = \tfrac{1}{2}R(\cosh \eta - 1), \qquad \tau = -\frac{1}{c}\sqrt{\frac{R^3}{8m}}(\sinh \eta - \eta), \tag{17.34}$$

$$t = \frac{2m}{c}\left[\ln\left|\frac{Y + \coth(\eta/2)}{Y - \coth(\eta/2)}\right| - Y\left(\eta + \frac{R}{4m}(\sinh \eta - \eta)\right)\right] \tag{17.35}$$

where $Y = (R/2m + 1)^{1/2}$.

An equation for the null radial geodesics can be obtained by setting $ds^2 = 0$ in (17.15), from which $dr/dt = \pm c(1 - 2m/r)$. Integrating, one obtains

$$\pm c(t - t_0) = r + 2m \ln |(r/2m) - 1| \tag{17.36}$$

where the signs correspond to outgoing and incoming paths, and t_0 is a constant of integration. The result is shown in Fig. 17.2.

17.2.2 Orbit precession

For $L \neq 0$ the shape of the orbits is more conveniently found by introducing the variable $u \equiv 1/r$; the method was explained in Section 4.2.1 of Volume 1. One obtains

$$\frac{d^2 u}{d\phi^2} = -u + 3mu^2 + \frac{mc^2}{L^2} \tag{17.37}$$

Some examples obtained by numerical integration of this equation are shown in Fig. 17.3. An analytical solution can also be found in terms of elliptic functions, or else an approximate solution using perturbation theory; see exercise 11.16 of Volume 1. It is there shown that

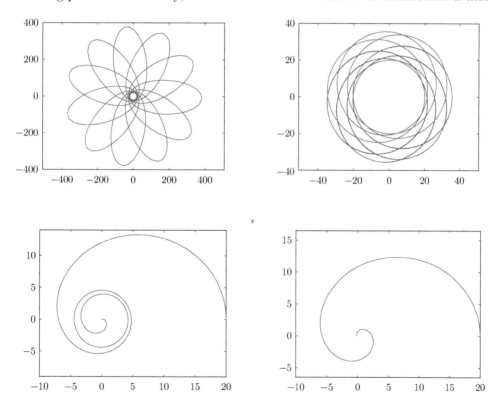

Fig. 17.3 Example orbits. The initial conditions have been chosen to illustrate four bound cases, of which two are permanently orbiting and two are captured. In each case a particle is launched from $r = 20$ in the azimuthal direction, with $L = 6.5$, 5.4, 3.6975, 3.5 respectively, for $G = M = c = 1$.

$$u = \frac{mc^2}{L^2}\left[1 + e\cos\left(\phi(1 - \epsilon)\right) + O(\epsilon^2)\right] \tag{17.38}$$

where e (the ellipticity) depends on initial conditions, and

$$\epsilon = 3m^2c^2/L^2. \tag{17.39}$$

Since u oscillates with respect to ϕ at the frequency $(1 - \epsilon)$, the radial motion returns to its innermost point after ϕ has advanced by $2\pi/(1 - \epsilon)$. This excedes 2π by

$$\delta = \frac{2\pi}{1 - \epsilon} - 2\pi \simeq 2\pi\epsilon = \frac{6\pi GM}{a(1 - e^2)c^2} \tag{17.40}$$

where we have expressed L^2 in terms of the ellipticity and the semi-major axis a of the quasi-ellipse by using (17.38). This formula gives the advance of the perihelion of planets in the solar system (to be added to the Newtonian effects), and more generally the advance of the periastron for any stellar system. The effect has been measured for the Hulse–Taylor binary to a fractional precision 1.2×10^{-6}. (Weisberg, Nice and Taylor, 2010)

17.2.3   Circular obits

A stable circular orbit is obtained when the radius satisfies (17.23), which gives $L^2 = mc^2r^2/(r - 3m)$, and $\dot{r} = 0$. Substituting this into (17.20) one finds

$$E = \frac{c^2(1 - 2m/r)}{\sqrt{1 - 3m/r}} \tag{17.41}$$

and therefore

$$\frac{dt}{d\tau} = \frac{1}{\sqrt{1 - 3m/r}}. \tag{17.42}$$

Introduce

$$\omega \equiv \frac{d\phi}{dt} = \frac{\dot{\phi}}{\dot{t}} \tag{17.43}$$

then we have $r^2\omega = r^2\dot{\phi}/\dot{t} = L\sqrt{1 - 3/r}$. After squaring this equation and substituting for L^2, one finds

$$r^3\omega^2 = mc^2 = GM. \tag{17.44}$$

This is Kepler's third law, and a useful result for connecting angular motion to r for circular orbits.

The 4-velocity is

$$[u^a] = \left(\frac{dt}{d\tau}, \ 0, \ 0, \ \frac{d\phi}{d\tau}\right) = (\gamma, \ 0, \ 0, \ \gamma\omega) \tag{17.45}$$

Table 17.2 Efficiency of mass-energy conversion (i.e. rest energy/kinetic energy conversion) by various processes. The entries labelled 'accretion' list the energy given up by an object initially at rest at infinity as it is transferred (e.g. via collisions) to the smallest stable circular orbit. 'Black hole plumbing' refers to the work extracted at the top of a strong rope when it is used to lower an object towards a horizon. The main lesson is that black holes are more efficient mass-energy converters than stars.

process	efficiency (%)
nuclear fusion	1
Schwarzschild accretion	6
Kerr accretion	32
Kerr jets	96
black hole plumbing	$\to 100$
Hawking radiation	$\simeq 100$
matter-antimatter annihilation	100

where $\gamma = \dot{t} = (1 - 3m/r)^{-1/2}$ using (17.42). An alternative way to derive (17.44) is to start with this statement and then assert $g_{\mu\nu}u^{\mu}u^{\nu} = -c^2$.

Equation (17.41) shows that energy losses of a non-negligible fraction of the rest energy are involved when particles reach circular orbits near the smallest stable one at $r = 6m$. Consequently large amounts of energy can be radiated by accretion discs around compact objects (Table 17.2). When $r \to 3m$ the energy diverges. This shows that in order to occupy the (unstable) circular orbits near $r = 3m = (3/2)r_{\mathrm{s}}$ the particle has to move fast. The spherical surface at radius $(3/2)r_{\mathrm{s}}$ is the **light sphere** where photons can orbit (albeit unstably).

Example 17.1 Astra is freely falling in a circular orbit around a neutron star at $r = 4m$. Her twin Adam meanwhile is launched from the surface of the star in the radial direction, such that he passes Astra on his way out, and meets her again as he falls back down, after she has completed 10 orbits. Who ages the most between the two meetings, and by what ratio?

Solution.

First consider Astra. Her orbit has $\omega = (mc^2/(4m)^3)^{1/2} = c/8m$ so her elapsed coordinate time is $\Delta t = 20\pi/\omega = 160\pi m/c$. Eqn (17.42) gives $\mathrm{d}t/\mathrm{d}\tau = 2$ so her elapsed proper time is $\Delta\tau = 80\pi m/c$. Next consider Adam. His elapsed coordinate time between the meetings must match Astra's, i.e. $160\pi m/c$, and since he accumulates most of his proper time while he is both high up and moving slowly, one can take this coordinate time as a rough estimate of his net proper time. Thus one concludes that he ages more, by about a factor 2. For a precise value one may employ eqns (17.32), (17.33). The elapsed coordinate time on either the way up or down is $80\pi m/c$. The maximum height R is to be derived; the final position is at $r = 4m$. Solving for R and η one finds $(R/2m) \simeq 17.917$, $\eta \simeq 2.460$ and hence overall $\Delta\tau \simeq 468.7\,m/c$ (which is 1.86 times the value for Astra).

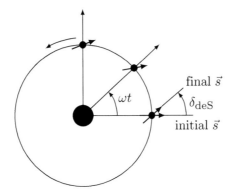

Fig. 17.4 The de Sitter precession for a circular orbit of a gyroscope whose spin vector is initially in the radial direction.

17.2.4   de Sitter (geodetic) precession

De Sitter precession was mentioned in Section 6.2. Here we shall calculate it by using the concept of parallel transport.

We consider a test particle whose spin is non-zero but sufficiently small not to affect its motion under gravity. Therefore the equation of motion under gravity is the geodesic equation. This tells us how the 4-velocity evolves with respect to proper time. Let **s** be the spin 4-vector. This must be orthogonal to the 4-velocity in any given LIF, and therefore it is orthogonal to the 4-velocity *tout court*:

$$\mathbf{s} \cdot \mathbf{u} = g_{\mu\nu} s^{\mu} u^{\nu} = 0. \tag{17.46}$$

In free fall the 4-velocity undergoes a parallel transport, and so does the spin (c.f. the Fermi–Walker transport, eqn (13.42)). Hence

$$\frac{Ds^a}{d\tau} = \frac{ds^a}{d\tau} + \Gamma^a_{\mu\nu} s^{\mu} u^{\nu} = 0. \tag{17.47}$$

Our task is to find the evolution of the spin by solving the above two equations.

We shall restrict to the case of circular orbits in the plane $\theta = \pi/2$. In this case u^a is given by (17.45), in which r, γ and ω are all constants. Substituting this into (17.46) we have $s_t \gamma + s_\phi \gamma \omega = 0$ and therefore, by raising indices using the metric,

$$s^t = -\omega (g_{\phi\phi}/g_{tt}) s^{\phi} = \frac{\omega r^2}{(1 - 2m/r)c^2} s^{\phi}. \tag{17.48}$$

Since this gives the temporal component of **s** in terms of the ϕ component, we now only need three rather than all four of the equations in (17.47). They are

$$\frac{\mathrm{d}s^r}{\mathrm{d}\tau} + \Gamma^r_{tt}s^t u^t + \Gamma^r_{\phi\phi}s^\phi u^\phi = 0 \tag{17.49}$$

$$\frac{\mathrm{d}s^\theta}{\mathrm{d}\tau} = 0 \tag{17.50}$$

$$\frac{\mathrm{d}s^\phi}{\mathrm{d}\tau} + \Gamma^\phi_{r\phi}s^r u^\phi = 0 \tag{17.51}$$

where we used that many of the connection coefficients are zero, and so are all but two of the components of **u**. After substituting in the connection coefficients listed in (17.16) and using (17.45) one finds

$$\frac{\mathrm{d}s^r}{\mathrm{d}\tau} - (r - 3m)\gamma\omega s^\phi = 0, \qquad\qquad \frac{\mathrm{d}s^\phi}{\mathrm{d}\tau} + \frac{\gamma\omega}{r}s^r = 0. \tag{17.52}$$

It is convenient to convert from τ to t by dividing by $\dot{t} = \gamma$, thus obtaining

$$\frac{\mathrm{d}s^r}{\mathrm{d}t} - (r - 3m)\omega s^\phi = 0, \qquad\qquad \frac{\mathrm{d}s^\phi}{\mathrm{d}t} + \frac{\omega}{r}s^r = 0. \tag{17.53}$$

These can be solved by differentiating the first and substituting $\dot{s}^\phi$ using the second. One finds $\mathrm{d}^2 s^r/\mathrm{d}t^2 = -(\omega/\gamma)^2 s^r$ whose solution is straightforward. We shall take as initial condition that the spin is along the radial direction at $t = 0$, then the full solution is

$$[s^i] = s^r(0) \begin{pmatrix} \cos(\omega t/\gamma) \\ 0 \\ -(\gamma/r)\sin(\omega t/\gamma) \end{pmatrix} \tag{17.54}$$

Thus the spin vector rotates relative to the radial direction in the negative sense with a coordinate angular velocity ω/γ. Thus is less than ω—c.f. Fig. 17.4. Relative to the Schwarzschild lattice the precession is therefore in a positive sense at the rate (in coordinate time):

$$\Omega_{\mathrm{deS}} = (1 - 1/\gamma)\omega = \left(1 - (1 - 3m/r)^{1/2}\right)\omega. \tag{17.55}$$

This is the exact formula for the de Sitter precession for the orbit we chose to examine.

17.3   Light in Schwarzschild spacetime

To find the gravitational **redshift** for a light wave moving between an emitter and a receiver both fixed in the Schwarzschild lattice, apply eqn (13.48):

$$\frac{\omega_B}{\omega_A} = \sqrt{\frac{1 - 2m/r_A}{1 - 2m/r_B}} \tag{17.56}$$

For moving emitters/receivers, apply also the special relativistic Doppler effect in a suitable LIF at emitter or receiver, noting that in order to do this one will need to know the direction of travel of the light wave. See Section 17.3.2 and exercise 17.12 for examples.

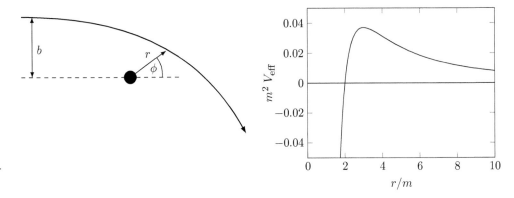

Fig. 17.5 Left: the impact parameter for non-captured photon trajectories. Right: effective potential for photons, eqn (17.61).

17.3.1 Photon motion: null geodesics

The **Shapiro effect** is a time delay in the arrival of photons which have travelled near to a gravitating body. Its observation was the first test of GR beyond the three 'classical tests of GR' proposed by Einstein (precession of Mercury's orbit; deflection of light; redshift of light). It is described in Section 11.3.5 of Volume 1.

Gravitational lensing is a generic term for the effect of gravity on the direction of propagation of light, especially its use in astronomy. It was discussed in Section 11.4 of Volume 1. Here we will describe the null geodesics in more detail in the strong field limit.

The equations for null geodesics in the equatorial ($\theta = \pi/2$) plane are

$$(1 - 2m/r)\dot{t} = \text{const} \equiv k \tag{17.57}$$

$$r^2\dot{\phi} = \text{const} \equiv h \tag{17.58}$$

$$\dot{r}^2 + \frac{h^2}{r^2}\left(1 - \frac{2m}{r}\right) = k^2 c^2 \tag{17.59}$$

where the dot is now differentiation with respect to an affine parameter λ, and k and h are constants of the motion. In order to interpret them it is useful to examine the $\dot{r}$ equation in the limit $r \to \infty$. Then one finds that, in company with (17.58), the equations have the same form as the equation of motion of a free particle in empty space in Newtonian physics, with h playing the role of angular momentum per unit mass. The value of this observation is that we can draw on known properties of the Newtonian problem, and thus obtain a useful way to express the constant k. In the Newtonian problem for a particle of mass m_0, a particle on a trajectory with impact parameter b would have angular momentum $m_0 h = m_0 bv$ about the origin and kinetic energy $m_0 v^2/2$, where v is the speed at infinity. When we interpret (17.59) as a Newtonian equation of motion, then $k^2 c^2$ is playing the role of twice the kinetic energy per unit mass, so we have $k^2 c^2 = v^2 = h^2/b^2$ and we can write (17.59) as

$$\dot{r}^2 + \frac{h^2}{r^2}\left(1 - \frac{2m}{r}\right) = \frac{h^2}{b^2}. \tag{17.60}$$

This version offers a clear interpretation of the parameter b for those trajectories that come from or go to infinity: it is the impact parameter (see Fig. 17.5). The parameter h can now be set equal to 1 by rescaling the affine parameter. Hence we have

$$\dot{r}^2 + V_{\text{eff}}(r) = b^{-2}, \qquad V_{\text{eff}} = r^{-2} - 2mr^{-3} \tag{17.61}$$

The 'effective potential' V_{eff} has a single stationary point—a maximum—at $r = 3m$ (the light sphere) and its maximum value is $V_{\text{max}} = 1/(27m^2)$. Consequently if $b^{-2} < 1/(27m^2)$, i.e. $b > \sqrt{27}\,m$, then the equation $V_{\text{eff}} = b^{-2}$ has a real solution and this means an incoming photon's trajectory will arrive at a point of closest approach where $\dot{r} = 0$ and then escape again to infinity. If $b < \sqrt{27}\,m$, on the other hand, then an incoming photon will cross the light sphere and be absorbed. This can be seen on Fig. 17.6.

In order to find a photon's trajectory in space (i.e. relative to the Schwarzschild lattice) it is best to write $\mathrm{d}r/\mathrm{d}\phi = \dot{r}/\dot{\phi}$ and use this to replace $\dot{r}$ in (17.60), using $\dot{\phi} = 1/r^2$ from (17.58). Then one introduces $u = 1/r$ and differentiates once so as to obtain

$$\frac{\mathrm{d}^2 u}{\mathrm{d}\phi^2} = u(3mu - 1). \tag{17.62}$$

Some example trajectories obtained by numerical integration of this equation are shown in Fig. 17.6. Notice the following features:

1. Any given ray only crosses the light sphere at most once.
2. Light rays starting out from a point and passing on the same side of the gravitating body are caused to diverge more.
3. An observer standing directly 'behind' the gravitating body and looking towards it can see some light arriving from a point source situated on the other side of the body. This light will have the appearance of a ring around the body.
4. If you shine a wide-angle torch or flash lamp at a black hole then some of the light will come back and hit you!
5. An observer looking towards the vicinity of the light sphere can in principle see light arriving at his location from sources situated almost anywhere, including behind him.
6. In so far as the body acts as a lens for incoming collimated light, its focal length increases with the impact parameter.

For further discussion of gravitational lensing see Section 11.4 of Volume 1.

17.3.2 Emission from an accretion disc

We mentioned in connection with Table 17.2 that accretion disks around compact objects can contain large amounts of kinetic energy, and consequently the particles in them may become

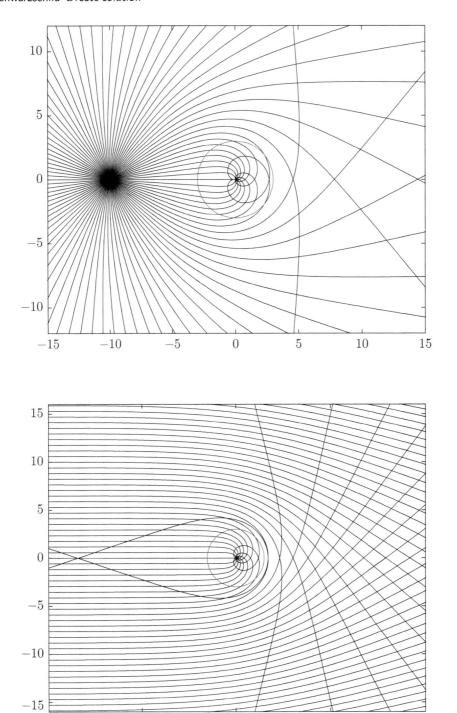

Fig. 17.6 An illustrative selection of photon trajectories in Schwarzschild spacetime.

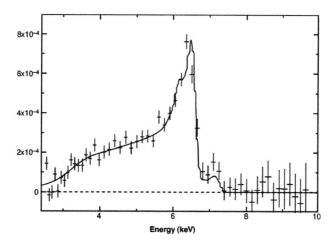

Fig. 17.7 X-ray emission spectrum from MCG-6-30-15 taken with the XMM-Newton satellite. The spectrum is associated with the Kα emission line in iron at 6.4 keV. The main features are consistent with emission from an accretion disc involving motion at a significant fraction of the speed of light. The central object may be rapidly rotating, in which case a Schwarzschild treatment is not precise, but it gives the main features. [Fig. 3 of (Fabian *et al.*, 2002)]

very hot and emit X-rays. This emission is a useful probe of the gravitational environment. Fig. 17.7 shows the received spectrum from an object called MCG-6-30-15 (a Seyfert 1 galaxy) (Fabian, Vaughan, Nandra, Iwasawa, Ballantyne, Lee, Rosa, Turner and Young, 2002). This spectrum is believed to arise from an X-ray transition in iron at 6.4 keV. Note the unusual shape and large width of the spectrum. The shape can be fitted by modelling the emission as coming from an accretion disc; the width implies that the Doppler and gravitational shifts are large.

Let us model an accretion disc as a collection of circular orbits. The material at radius r then has a 4-velocity given by (17.45) and the frequency shift for light emitted at A and received at B is given by (13.44). Let us take as receiver a detector at rest at infinity in Schwarzschild coordinates. Its 4-velocity is then $u^a_{(B)} = (1, 0, 0, 0)$. Hence (13.44) gives

$$\frac{\nu_B}{\nu_A} = \frac{p_0(B)}{\gamma(p_0(A) + \omega p_3(A))} = \sqrt{1 - 3m/r} \frac{p_0(B)}{p_0(A) + \omega p_3(A)} \qquad (17.63)$$

where p^a is the 4-momentum of the photon which moves from A to B. Now the metric is independent of time, therefore p_0 is conserved along the photon's geodesic, so we have

$$\frac{\nu_B}{\nu_A} = \frac{\sqrt{1 - 3m/r}}{1 + \omega p_3(A)/p_0(A)} \qquad (17.64)$$

Thus for a given emission frequency, the received frequency depends on the radius of the circular orbit of the emitter and the initial direction of travel of the emitted light.

For a photon emitted with $p^\phi = 0$ and therefore $p_3 = 0$ (e.g. emission in the radial direction or in the direction perpendicular to the accretion disc) the frequency ratio is $\sqrt{1 - 3m/r}$.

For a photon emitted in the ϕ direction, i.e. with $p_r = p_\theta = 0$, we can find the ratio p_3/p_0 by using $p_\mu p^\mu = 0$ from which $g^{\mu\nu} p_\mu p_\nu = 0$ so

$$-c^{-2}(1 - 2m/r)^{-1}(p_0)^2 + r^{-2}(p_3)^2 = 0 \tag{17.65}$$

which gives $p_3/p_0 = \pm(r/c)(1 - 2m/r)^{-1/2}$. Upon substituting this into (17.64) we obtain

$$\frac{\nu_B}{\nu_A} = \frac{\sqrt{1 - 3m/r}}{1 \pm (r/m - 2)^{-1/2}} \tag{17.66}$$

where we used 'Kepler's law' (17.44) for ω.

For the lowest stable circular orbit at $r = 6m$ one thus obtains the values

$$\frac{\nu_B}{\nu_A} = \begin{cases} 1/\sqrt{2} = 0.71 & \text{disc viewed face-on} \\ \sqrt{2} = 1.4 & \text{forward emission, disc viewed edge-on} \\ \sqrt{2}/3 = 0.47 & \text{backward emission, disc viewed edge-on} \end{cases}$$

Owing to the headlight effect, emission which is isotropic in the rest frame of the emitter will be preferentially in the forward direction, making the blue-shifted emission brighter. The data in Fig. 17.7 is nevertheless mostly redshifted, suggesting that the disc in that case is being viewed from a direction nearer to face-on than edge-on. By fitting the data one obtains information about the strong-field region of gravity.

17.4   Tidal stress tensor

We shall obtain the components of the tidal stress tensor defined in (15.57), as they are observed by an observer at rest relative to a given particle of 4-velocity $\mathbf{u}$. We indicate components in the LIF of such an observer by a prime. We have, for example, that $u'^a = (c, 0, 0, 0)$, and therefore

$$S'^a{}_b = c^2 R'^a{}_{00b} \tag{17.67}$$

To be more specific, we shall examine the case of a particle at rest (momentarily or otherwise) relative to the Schwarzschild lattice. In order both to orient the spatial axes of our observer, and also with a view to finding $R'^a{}_{00b}$, we invoke the tetrad method introduced in Section 11.2. A useful tetrad for our purposes is one made of vectors pointing in the t, r, θ, ϕ directions (since these directions are orthogonal, says the Schwarzschild metric), and chosen such that each has unit length. The contravariant components of these vectors are as follows:

$$\begin{aligned} \hat{e}^a_{(0)} &= (c^{-1}(1 - 2m/r)^{-1/2},\ 0,\ 0,\ 0) \\ \hat{e}^a_{(1)} &= (0,\ \ (1 - 2m/r)^{1/2},\ 0,\ 0) \\ \hat{e}^a_{(2)} &= (0,\ \ 0,\ \ \ 1/r,\ \ 0) \\ \hat{e}^a_{(3)} &= (0,\ \ 0,\ \ \ \ 0,\ \ 1/(r\sin\theta)) \end{aligned} \tag{17.68}$$

(c.f. (11.20)). Using these, we can obtain the Riemann tensor in the LIF from the Riemann tensor in Schwarzschild coordinates via

$$R'^a_{\ bcd} = \eta^{a\alpha} R'_{\alpha bcd} = \eta^{a\alpha} R_{\sigma\lambda\mu\nu} \hat{e}^{\sigma}_{(\alpha)} \hat{e}^{\lambda}_{(b)} \hat{e}^{\mu}_{(c)} \hat{e}^{\nu}_{(d)} . \tag{17.69}$$

Fortunately, we do not need all the components, and symmetries of the curvature tensor also reduce the calculation. Nevertheless, some work (by you) is required. One finds

$$\left[S'^i_{\ j} \right] = \frac{GM}{r^3} \begin{pmatrix} -2 & 0 & 0 \\ 0 & 1 & 0 \\ 0 & 0 & 1 \end{pmatrix} . \tag{17.70}$$

Thus there is tension in the radial direction and pressure in the transverse direction, as one should expect from the Newtonian result (treated in Section 9.1 of Volume 1). That is to say, the tidal forces are telling particles in free fall to become more separated in r and less separated in θ, ϕ as they fall. The stress in any object which refuses to allow its constituent particles to obey these instructions is consequently as we have just indicated.

Notice the $1/r^3$ dependence, which is characteristic of tidal effects. Also notice that nothing special happens as r approaches the horizon at $r_s = 2m$. For a one solar mass black hole, $r_s \simeq 3\,\mathrm{km}$ and then at the horizon the radial component of the stress tensor is $10^{10}\,\mathrm{ms}^{-2}$ per metre. Multiplying this by 10 kg, one obtains an estimate of 10^{11} newtons of tension for an object similarly proportioned to the human body. Clearly no hapless astronaut will survive a fall to such a horizon. In order to allow a falling astronaut to pass unharmed within the horizon, we need the force to be smaller by a factor $\sim 10^8$ and therefore the black hole mass to be greater than $\sim 10^4 M_\odot$.

Exercises

In the following, it should be taken for granted in every case that Schwarzschild–Droste spacetime is being discussed.

17.1 Starting from the metric, show that a particle falling freely in a radial direction has the equation of motion

$$\frac{1}{2}\left(\frac{\mathrm{d}r}{\mathrm{d}\tau}\right)^2 = \frac{GM}{r} + k \tag{17.71}$$

where k is a constant of the motion. Find the proper time for a particle released from rest at r_0 to reach the horizon.

17.2 In the 'Lemaître coordinates' τ, ρ defined such that

$$\mathrm{d}\tau = \mathrm{d}t + \sqrt{2m/r}\,(1 - 2m/r)^{-1}\mathrm{d}r,$$
$$\mathrm{d}\rho = \mathrm{d}t + \sqrt{r/2m}\,(1 - 2m/r)^{-1}\mathrm{d}r,$$

(taking $c = 1$) show that the line element is

$$\mathrm{d}s^2 = -\mathrm{d}\tau^2 + \frac{2m}{r}\mathrm{d}\rho^2 + r^2\mathrm{d}\Omega^2 \tag{17.72}$$

where $r = (\frac{3}{2}(\rho - \tau))^{2/3}(2m)^{1/3}$. Interpret the lines of constant ρ, θ, ϕ.

17.3 By how much is light from the rest of the universe blue-shifted when it is received by detectors at rest on the surface of a typical neutron star?

17.4 Show that $p_\theta^2 + p_\phi^2/\sin^2\theta$ is a constant of the motion for test particles.

17.5 Show that a massive particle moving on the innermost stable circular orbit has speed $c/2$ as measured by an observer fixed in the Schwarzschild lattice. Find the period of the orbit as observed by the latter, and the period of this same orbit as determined by an observer fixed at infinity (using an appropriate notion of simultaneity which you should specify).

17.6 A mountain on Earth's equator has worldline $(t, r, \theta, \phi) = (\gamma u, R, \pi/2, \omega u)$ where u is a parameter, γ, ω are constants and R is the coordinate radius of the Earth. Show that the orbital speed as determined by a distant fixed observer is $v = R\omega/\gamma$. By considering the 4-velocity magnitude, or otherwise, show that $\gamma = \left(1 - v^2/c^2 - 2m/R\right)^{-1/2}$.

17.7 Obtain (17.70), e.g. by the method suggested in the text.

17.8 Compare the gravitational tidal forces at the surfaces of the Earth, moon and Sun.

17.9 How closely could you approach a neutron star of one solar mass in comfort in free fall? What mass of black hole would be required for you to pass within its horizon without noticing any discomfort?

17.10 Show that the principal stresses in an object in circular orbit around a Schwarzschild black hole are $mc^2 r^{-3}(r - 3m)^{-1} \times (2r - 3m, -r, -r + 3m)$.

17.11 Show that under parallel illumination the shadow cast by a massive orbit is smaller than the object itself, and an optical image of a massive object makes the object appear larger than it is. Show that light grazing the surface of a sphere of coordinate radius $r > 3m$ arrives at infinity with impact parameter $b = r(1 - 2m/r)^{-1/2}$.

17.12 An atom is released from some coordinate radius r_1 and falls radially inward. When it reaches r_2 it emits a photon radially outwards. Find the redshift of this photon when it enters a detector at r_1, by the following steps:
(i) Noting (13.44), write the two 4-velocities $u^a_{(1)} = (\gamma, 0, 0, 0)$, $u^a_{(2)} = (\dot{t}, \dot{r}, 0, 0)$ where $\dot{t}$, $\dot{r}$ are given by (17.24), (17.25) and explain why $\gamma = (1 - 2m/r_1)^{-1/2}$.
(ii) By using a conservation law, or otherwise, show that the redshift is

$$\frac{\nu_1}{\nu_2} = \frac{\gamma}{\dot{t} + \left(p_1^{(2)}/p_0^{(2)}\right)\dot{r}}$$

where **p** is the 4-momentum of the photon.
(iii) Show that the emitted photon's null geodesic satisfies $p_1/p_0 = -1/c(1 - 2m/r)$.
(iv) Hence obtain

$$\frac{\nu_1}{\nu_2} = \frac{f(r_2)}{f(r_1) + [2m(1/r_2 - 1/r_1)f(r_1)]^{1/2}} \quad (17.73)$$

where $f(r) = 1 - 2m/r$.
(v) Find this value, and also (17.56), for the case $r_1 = \infty$, $r_2 = 50m/9$.
[*Ans.* 2/5, 4/5]

17.13 Show that a photon emitted at an angle α to the outward radial direction will escape to infinity provided $\sin \alpha < 3m\sqrt{3V_{\text{eff}}}$ where V_{eff} is given by (17.61). [Hint: invariants]. Find the limiting values of α for $r = 2m$ and $r = 3m$.

17.14 Show that the ratio between the proper time of one orbit on board a satellite on a circular orbit and the proper time elapsed for an observer fixed in the Schwarzschild coordinates is $((r-2m)/(r-3m))^{1/2}$. Which observer is aging less rapidly? How does this compare with the ordinary twin paradox?

18

Further spherically symmetric solutions

The Schwarzschild–Droste solution provides the foundation for understanding empty spacetime around spherically symmetric bodies such as non-rotating stars, planets and black holes. Non-empty spacetime, where $T^{ab} \neq 0$, raises the issue of the mutual interdependence of G^{ab} and T^{ab}. For a given stress-energy tensor we can (in principle) find the metric which solves the field equation. This metric then presents the gravitational forces which must be acting on the matter described by T^{ab}. If T^{ab} does not describe a possible energy and stress distribution for matter subject to such forces, then we have an inconsistency. The way to a solution is therefore to present T^{ab} not as a given, but as a function of parameters to be determined. For example, we could try the perfect fluid, eqn (14.18), which has the parameters density, pressure and velocity. The Einstein field equation will then impose conditions on these quantities. These conditions can be interpreted as energy and momentum conservation in a fluid subject to gravitational tidal forces. They do not uniquely specify the density and pressure, but they restrict the possible functional forms. We may then introduce a further equation called the *equation of state* to find a unique solution. The equation of state may be simply some assumption such as uniform density, or some more physically motivated relation between pressure and density, coming from the physics of materials or stellar interiors.

We shall limit the discussion here to the case of spherical symmetry, in which case the generic metric form (17.1) and the results in Table 17.1 can still be used. Hence the equations (17.11) for the Ricci tensor still apply. In the first instance we will explore a simple choice for the equation of state: $\rho = $ const. This results in the *interior* Schwarzschild solution, discovered by Schwarzschild in 1915. It teaches some useful general lessons about gravitation inside material bodies, and it can serve as a rough model for the interior of a star. We will then present the equations for a more realistic model of stellar interiors. We then turn to a case where there is a non-zero electrostatic field, deriving the Reissner–Nordström metric which describes spacetime outside a charged spherically symmetric body. We conclude by showing the way a non-zero cosmological constant modifies the Schwarzschild–Droste solution.

Relativity Made Relatively Easy: General Relativity and Cosmology. Volume 2. Andrew M. Steane,
Oxford University Press. © Andrew M. Steane 2021. DOI: 10.1093/oso/9780192895646.003.0018

18.1 Interior Schwarzschild solution

We seek a solution of the field equation for a static isotropic distribution of matter. For the sake of simplicity, suppose the proper matter density ρ is uniform. Then, from eqn (16.6), we must have that the Gaussian curvature in any spatial direction and at any point is

$$K = \frac{8\pi G}{3c^2}\rho. \tag{18.1}$$

A space with K the same everywhere and in all directions is said to be a space of *constant curvature*. In three dimensions and positive K this space is called the 3-sphere; some of its properties are presented in appendix B.

The metric for the 3-sphere is given by (B.17):

$$ds^2 = \frac{dr^2}{1 - Kr^2} + r^2 d\Omega^2 \tag{18.2}$$

This is the spatial part of the interior Schwarzschild solution. It remains to find the temporal part, which is done by essentially the same method as we used for the standard Schwarzschild metric, but now R_{ab} is not zero and the pressure has to be found. The result is (exercise 18.2), now setting $c = 1$,

$$ds^2 = -\frac{1}{4}\left(3a_0 - a(r)\right)^2 dt^2 + \frac{dr^2}{a(r)^2} + r^2 d\Omega^2 \tag{18.3}$$

where

$$a(r) = \sqrt{1 - Kr^2}, \qquad a_0 = a(r_0) = \sqrt{1 - Kr_0^2}. \tag{18.4}$$

Here r_0 is a constant and K is given by (18.1). The pressure is

$$p(r) = \rho\frac{a(r) - a_0}{3a_0 - a(r)} \tag{18.5}$$

and the equation of state is $\rho = \text{const}$.

The assumption of uniform density seems at first to be unphysical, because it appears to imply that the material of the star is incompressible. An incompressible fluid is strictly impossible because it breaks causality (the group velocity of sound would be infinite). However, this problem can be avoided if we use a two-fluid model in which both fluids are compressible and the denser material (e.g. dust) exerts less pressure than the less dense material (e.g. radiation).[1]

The pressure goes to zero at $a = a_0$, which is at $r = r_0$. We can interpret this by asserting that r_0 is the radius of the star. The pressure must fall to zero at the surface of a star since there is no further material to present a pressure from the outside (assuming we have a star surrounded by vacuum). If there were internal pressure at the surface then there would be a

[1]M. M. Som, M. A. P. Martins and A. A. Morégula, *Physics Letters A* Volume **287**, 50–52 (2001).

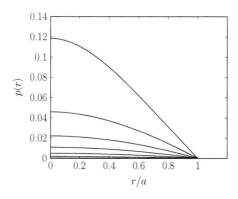 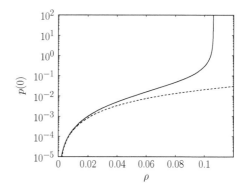

Fig. 18.1 Interior Schwarzschild solution: pressure as a function of radius for various densities (left), and central pressure as a function of density (right), with units such that $G = c = 1$. The dashed line on the right-hand graph shows the Newtonian prediction. The solution gives a rough first approximation for a neutron star. It has a runaway behaviour once the density is high enough.

finite unbalanced force on an arbitrarily thin layer of material at the surface, for which gravity could not compensate.

Treating the solution as a rough model of a star, we shall find that, owing to the constant curvature, all circles of a given size inside the star have the same radius excess. This does not mean the conditions are completely homogeneous, because g_{tt} and the pressure both depend on r. Therefore the gravitational field (Christoffel symbols) has a sense of direction towards the centre of the star, and the matter of the star is not homogeneous.

Since the field is given by the gradient of the metric, there is a further constraint. At the star's surface the metric must match smoothly onto the exterior Schwarzschild solution, such that g_{ab} and $\partial_r g_{tt}$ are continuous at $r = a$. This is achieved if $K r_0^3 = r_{\mathrm{s}} = 2MG/c^2$. Substituting this into (18.1) gives the active gravitational mass of the star:

$$M = \frac{4}{3} \pi r_0^3 \rho. \tag{18.6}$$

This looks just like 'volume' times density, though of course it is the coordinate volume not the true volume, therefore this mass is somewhat less than the total proper mass of the material in the star. This point merits a little elaboration.

ρc^2 is the energy per unit proper volume. Therefore the total proper mass (defined as energy$/c^2$) is

$$\tilde{M} = \int \rho \, \mathrm{d}V = \int_0^{r_0} \rho \frac{4\pi r^2 \mathrm{d}r}{a(r)} = \frac{4}{3}\pi r_0^3 \rho + \frac{2}{5}\pi K r_0^5 \rho + \cdots \tag{18.7}$$

The difference $\tilde{M} - M$ can be interpreted as gravitational binding energy, and indeed the $O(r_0^5)$ term in the expression for $\tilde{M}$ is equal to the binding energy of a uniform sphere in Newtonian gravity. This is an example of the subtle way in which GR performs its energy accounting (c.f. remarks in Sections 9.4, 10.5.2 of Volume 1, and also Section 5.5 of this volume). It should not surprise us that the material gravitates by less than $\tilde{M}$, because the material is situated at the bottom of a gravitational potential well, and it had to give up some energy in order to arrive there in equilibrium.

If we wished to recover all the material of the star, separating it by raising it slowly using light ropes until it is far dispersed, then we will have to supply the energy $\tilde{M} - M$ at the top of our ropes. Thus we (the observer at infinity) have to invest $\tilde{M} - M$ in order to receive back $\tilde{M}$. At the end of such a process we will overall have gained Mc^2 amount of energy at infinity.

A long way away from the star, the orbits and other phenomena must match those of Newtonian gravity for a spherical object of some given mass—the active gravitational mass. This mass is given by

$$M_\infty = \int \sqrt{|g_{00}|}(\rho + 3p)\mathrm{d}V = \int_0^{r_0} a\rho \frac{4\pi r^2 \mathrm{d}r}{a} = \frac{4}{3}\pi r_0^3 \rho = M. \tag{18.8}$$

(c.f. (16.18)).[2] The result is equal to the mass given by (18.6), i.e. that which makes the solution join smoothly with the exterior Schwarzschild solution, and of course this is precisely what we must expect. This serves to confirm that the first integral in (18.8) is indeed the correct expression. The $\sqrt{g_{00}}$ factor can be interpreted as the effect of the gravitational potential, or equally as the way the theory accounts for the binding energy. It is noteworthy that the pressure disappears from the result. This shows that it is not correct to say that pressure gravitates exactly in the same way as energy density does.

The denominator of the pressure equation (18.5) goes to zero at some $r > 0$ unless $Kr_0^2 < 8/9$. This is both a restriction on the range of validity of the solution, and also a physical prediction. The prediction is that if a star of constant density has a radius less than or equal (9/8) of its Schwarzschild radius, then the star will undergo gravitational collapse. *Buchdahl's theorem* states that this same limit will apply for more general conditions, as long as the energy density is non-increasing outwards and the pressure is isotropic:

Buchdahl's theorem (static, spherically symmetric body)

$$r_0 > \frac{9}{8}r_s = \frac{9}{4}\frac{GM}{c^2}. \tag{18.9}$$

The proof of this from the Einstein equation is beyond the scope of our discussion, but one may note that it is to be expected from the interior Schwarzschild result, since one would not expect a non-uniform density to allow a smaller r_0 at given M, because it would require a higher density somewhere. Buchdahl's theorem has the interesting implication that the gravitational redshift of light emitted from the surface of such a star cannot exceed 2 (where redshift z is defined as $z = (\omega_A/\omega_B) - 1$). However, stars are not usually isotropic and the radius of an anisotropic body can approach more closely to the Schwarzschild radius associated with its mass; in this case a higher redshift is possible.

---

[2]I never cease to be amazed by the way the pressure cancels out between g_{00}, $(\rho + 3p)$ and $\mathrm{d}V$ here!

18.1.1   Stellar structure equations

In order to construct a more general and more realistic model, we treat the interior of a star as an ideal fluid, where now the pressure and density may be functions of r. We therefore have

$$T_{ab} = (\rho + p/c^2)u_a u_b + p g_{ab} \tag{18.10}$$

and $T^\lambda_\lambda = 3p - \rho c^2$. Hence the field equation (2.7) reads

$$R_{ab} = \kappa(T_{ab} - \tfrac{1}{2}g_{ab}T^\lambda_\lambda) = \kappa\left((\rho + p/c^2)u_a u_b + \tfrac{1}{2}(\rho c^2 - p)g_{ab}\right) \tag{18.11}$$

where $\kappa \equiv 8\pi G/c^4$.

We adopt once again the form

$$g_{ab} = \text{diag}(-A(r),\ B(r),\ r^2,\ r^2\sin^2\theta). \tag{18.12}$$

and we can use (17.11). Now, notice from (17.11) that R_{ab} is diagonal, and we have also that g_{ab} is diagonal. It follows from (18.11) that so is $u_a u_b$. Hence $u_0 u_i = 0$, but $u_0 \neq 0$, so we have $u_i = 0$ and consequently $u_a = (u_0, 0, 0, 0)$. Now use that $u_\lambda u^\lambda = -c^2$. Therefore $u_0(-1/A)u_0 = -c^2$ hence $u_0 = c\sqrt{A}$. Hence (18.11) can be written

$$R_{tt} = \kappa(\alpha + 2p)A, \quad R_{rr} = \kappa\alpha B, \quad R_{\theta\theta} = \kappa\alpha r^2, \quad R_{\phi\phi} = \kappa\alpha r^2\sin^2\theta, \tag{18.13}$$

where $\alpha \equiv (\rho c^2 - p)/2$. Now let us form a combination of eqns (17.11) so as to elliminate A and its derivatives. We find

$$\frac{R_{tt}}{A} + \frac{R_{rr}}{B} + \frac{2R_{\theta\theta}}{r^2} = \frac{2}{r^2}\left(1 - \frac{1}{B}\right) + \frac{2B'}{rB^2} \tag{18.14}$$

and this same combination of the right-hand sides of eqns (18.13) gives $2\kappa\rho c^2$. Hence we obtain

$$\left(1 - \frac{1}{B}\right) + r\frac{B'}{B^2} = \kappa\rho c^2 r^2 \tag{18.15}$$

which can also be written

$$\frac{\mathrm{d}}{\mathrm{d}r}\left(r(1 - 1/B)\right) = \kappa\rho c^2 r^2. \tag{18.16}$$

This has the solution

$$B(r) = \frac{1}{1 - 2m(r)/r} \tag{18.17}$$

where

$$m(r) = \frac{G}{c^2}\int_0^r \rho(r)4\pi r^2 dr. \tag{18.18}$$

This function has an obvious interpretation in terms of mass (but keep in mind the discussion after (18.6)).

In order to find $A(r)$, one can now substitute this solution into the $R_{\theta\theta}$ equation. This yields a differential equation for $A(r)$ in terms of $\rho(r)$ and $m(r)$, which can then be solved. This is a perfectly acceptable mathematical route but it does not yield much physical insight. It is more instructive to appeal to the hydrostatic equilibrium, eqn (14.29). From this equation one finds immediately that

$$\frac{A'}{A} = \frac{-2p'}{\rho c^2 + p}. \tag{18.19}$$

This equation cannot be solved until one has specified the relationship between p and ρ—the *equation of state*. In practice it is convenient to begin by specifying some given equation of state and then recast the above equations as follows. First one provides an equation of state:

$$p = p(\rho) \tag{18.20}$$

(for example, $p = a\rho^\gamma$ is often used; this is called a *polytropic* equation of state, and if $\gamma = 1+1/n$ then n is called the *polytropic index*).

Next one adopts the differential form of (18.18):

$$\frac{\mathrm{d}m(r)}{\mathrm{d}r} = (G/c^2)4\pi r^2 \rho \tag{18.21}$$

Finally one uses (18.19) and (18.17) to express A and B in the equation for $R_{\theta\theta}$, and hence, after some algebra:

Tolman–Oppenheimer–Volkoff equation

$$\frac{\mathrm{d}p}{\mathrm{d}r} = \frac{-(\rho c^2 + p)\left(m(r) + 4\pi G r^3 p/c^4\right)}{r(r - 2m(r))} \tag{18.22}$$

This is the equation of hydrostatic equilibrium. If we write it in the form

$$\frac{\mathrm{d}p}{\mathrm{d}r} = -\frac{GM\rho}{r^2}\left(1 + \frac{p}{\rho c^2}\right)\left(1 + \frac{4\pi r^3 p}{Mc^2}\right)\left(1 - \frac{2GM}{c^2 r}\right)^{-1} \tag{18.23}$$

then we see that the right-hand side has the form of the Newtonian result multiplied by three factors, all of which are greater than one. Thus the general relativistic effects all act to steepen the pressure gradient relative to the Newtonian case.

Equations (18.21) and (18.22) are also called the first and second equations of stellar structure. These are the starting point for any stellar model in terms of an ideal fluid and in which Newtonian gravity is insufficiently precise.

18.1.2 Mass limits for stars

As the mass grows, so does the gravity, so the question arises whether there are limits to the mass that a star can attain. The answer will depend on the type of star. For a hot star in which the gravitation is opposed by the pressure owing to classical thermal motion, it is currently thought that a star will not form if the total mass is more than a few hundred solar masses.

There are at least two types of star for which further limits apply: the *white dwarf* and the *neutron star*. In both cases the limiting mass is obtained from a competition between gravitational attraction and *Fermi degeneracy pressure*. The latter is owing to the fact that since identical fermions (such as electrons or neutrons) cannot occupy the same quantum state, in a cold gas of fermions many of the particles have to occupy states of motion in which they have significant amounts of kinetic energy. The lowest available states have higher energy when the system has a smaller size; this is manifested as pressure.

An order-of-magnitude estimate proceeds as follows. The pressure at the core of a star is of order $p \sim \rho g r \sim GM^2/R^4$ where M is the mass. A white dwarf has roughly equal numbers of neutrons, protons and electrons, so $M \simeq 4000 n m_e$ where m_e is the mass of an electron and n is the number density of electrons. The degeneracy pressure for a relativistic Fermi gas at low temperature is $p \simeq \hbar c n^{4/3}$. Equating this to the pressure at the core gives $M \simeq (\hbar c/G)^{3/2}(4000 m_e)^{-2} \simeq 0.4\,M_\odot$. A more thorough analysis gives $1.4\,M_\odot$. This is called the *Chandrasekhar limit*, after Subrahmanyan Chandrasekhar who first obtained it to reasonable accuracy by solving the stellar structure equations with an equation of state furnished by a model of a relativistic Fermi gas.

The analogous result for a neutron star is called the *Tolman–Oppenheimer–Volkoff limit*. Now the main source of pressure is neutron degeneracy pressure. The rough calculation suggests the mass limit is a few times higher than the one for white dwarfs. A precise estimate is difficult because it is not completely known what nuclear processes may be possible in the interior of a neutron star. The mass limit is currently believed to be in the range $1.5\,M_\odot$ to $3\,M_\odot$.

These mass limits bear on the question of evidence for black holes; see Section 20.6.

18.2 Reissner–Nordström metric

Schwarzschild's solution is the unique vacuum solution for a spherically symmetric scenario, where by vacuum we mean $T^{ab} = 0$. If we keep the spherical symmetry but allow a static electric field, then the Reissner–Nordström metric is obtained:

$$\mathrm{d}s^2 = -\left(1 - \frac{2m}{r} + \frac{q^2}{r^2}\right)c^2\mathrm{d}t^2 + \left(1 - \frac{2m}{r} + \frac{q^2}{r^2}\right)^{-1}\mathrm{d}r^2 + r^2\mathrm{d}\Omega^2 \qquad (18.24)$$

with $m = GM/c^2$ as before, and $q^2 = GQ^2/(4\pi\epsilon_0 c^4)$ where Q is the charge on the spherical body. The body could, for example, be a charged non-rotating black hole, or a planet or a star, or any charged spherical lump of matter, or a charged particle such as a proton, and this is the metric outside it.

Hans Reissner and Gunnar Nordström discovered this solution in 1918, soon after Schwarzschild's work. Our main aim in this section is to derive their result.

Towards deriving the metric, first note that the energy-tensor of an electromagnetic field has zero trace, and consequently the Einstein field equation takes the form $R_{ab} = \kappa T_{ab}$. We require also that the electromagnetic (e-m.) field tensor $\mathbb{F}^{ab}$ satisfies the GR form of Maxwell's equations, (2.25), (2.26) (with zero 4-current). It is not hard to convince oneself that in static spherically symmetric conditions the e-m. field must take the form

$$[\mathbb{F}_{ab}] = E(r) \begin{pmatrix} 0 & -1 & 0 & 0 \\ 1 & 0 & 0 & 0 \\ 0 & 0 & 0 & 0 \\ 0 & 0 & 0 & 0 \end{pmatrix} \tag{18.25}$$

where $E(R)$ is a function to be discovered. We adopt again the static spherically symmetric form of the metric in terms of functions $A(r) = -g_{tt}$, $B(r) = g_{rr}$ to be discovered. These enter the Maxwell equations via the covariant derivative, and $E(r)$ enters the Einstein field equation via the energy tensor. Hence we have coupled equations for the three functions A, B, E.

First let us find the energy tensor in terms of E. For an electromagnetic field one has

$$T_{ab} = \epsilon_0 c^2 \left(-\mathbb{F}_{a\mu}\mathbb{F}^{\mu}{}_{b} - \tfrac{1}{2}g_{ab}D\right) \tag{18.26}$$

where $D = \tfrac{1}{2}\mathbb{F}_{\mu\nu}\mathbb{F}^{\mu\nu}$. By using the inverse metric to raise indices, one finds

$$[\mathbb{F}^a{}_b] = E(r) \begin{pmatrix} 0 & 1/A & 0 & 0 \\ 1/B & 0 & 0 & 0 \\ 0 & 0 & 0 & 0 \\ 0 & 0 & 0 & 0 \end{pmatrix} \tag{18.27}$$

and $F^{ab} = (-1/AB)F_{ab}$. Hence $D = -E^2/AB$ and

$$[T_{ab}] = \frac{\epsilon_0 E^2 c^2}{2} \begin{pmatrix} 1/B & 0 & 0 & 0 \\ 0 & -1/A & 0 & 0 \\ 0 & 0 & r^2/AB & 0 \\ 0 & 0 & 0 & (r^2\sin^2\theta)/AB \end{pmatrix} \tag{18.28}$$

Therefore the Einstein field equation gives

$$R_{tt} = \frac{\kappa\epsilon_0 E^2 c^2}{2B}, \qquad R_{rr} = -\frac{\kappa\epsilon_0 E^2 c^2}{2A}, \qquad R_{\theta\theta} = \frac{\kappa\epsilon_0 E^2 c^2 r^2}{2AB}. \tag{18.29}$$

The first two of these give $BR_{tt} + AR_{rr} = 0$. Using eqns (17.11) this yields $BA' + B'A = 0$ and therefore $AB = \text{const}$, just as we found for the Schwarzschild case. By requiring that the solution gives Newtonian physics in the limit of large r, we find that this constant is c^2. We can now substitute $B = c^2/A$ into the $R_{\theta\theta}$ equation in (17.11) which leads to

$$R_{\theta\theta} = 1 - (A + rA')/c^2. \tag{18.30}$$

We would like to set this equal to (18.29). Before doing so, we will use the Maxwell equations to relate E to A and B. We have

$$\nabla_\mu \mathbb{F}^{\mu b} = \frac{1}{\sqrt{-g}}\partial_\mu \left(\sqrt{-g}\,\mathbb{F}^{\mu b}\right) = 0 \tag{18.31}$$

where we used (11.41) to write the divergence of this antisymmetric tensor. Using that $g = -ABr^4 \sin^2\theta$ and that only the derivative with respect to r is non-zero, we find

$$\frac{\mathrm{d}}{\mathrm{d}r}\left(\frac{r^2 E}{\sqrt{AB}}\right) = 0 \tag{18.32}$$

and therfore $r^2 E = k\sqrt{AB}$ where k is a constant. By requiring that the electric field be that of a charge Q at large r, one finds $k = Q/4\pi\epsilon_0 c$. We can now substitute $E^2 = k^2 AB/r^4$ into (18.29) and set the result equal to (18.30), yielding

$$A + rA' = c^2(1 - q^2/r^2). \tag{18.33}$$

The solution is straightforward, yielding the Reissner–Nordström metric, eqn (18.24).

Owing to the energy-density of the electric field, spacetime curvature is affected (i.e. the electric field gravitates) and therefore even neutral particles and light rays notice the difference between a charged and a non-charged body. Charged particles experience in addition the Lorentz force.

At first sight one might argue, from the difference in the signs between the m and q^2 terms in (18.24), that the charge has reduced the net gravitation, compared to a body of the same mass but no charge. But how can this be? Is there a sort of gravitational repulsion happening? Not at all. It is simply that for any finite r one is still inside the gravitating entity—the electric field in this case. One should expect, then, that as one goes to larger r the net gravitation will fall away more slowly, as a function of r, than it would in a vacuum empty of all matter and e-m. fields. The situation is loosely comparable to moving from one radius to another in a diffuse gas or a stellar interior. The parameter we have labelled m and named 'mass parameter' can be interpreted quantitatively by measuring orbits at large r; in that limit the gravitation matches the Newtonian result and we choose to call the quantity appearing in Newton's formula the 'mass' of the source. One should not assume that this mass bears any simple relation to the integral over space of the mass density of the central body.

The Reissner–Nordström metric has only modest significance to astronomy since astrophysical bodies are electrically neutral to good approximation. However, it presents some interesting possibilities for the geometric structure of spacetime, especially in the context of black holes. The metric (18.24) yields horizons[3] at $r_{\pm} = m \pm \sqrt{m^2 - q^2}$ when $q^2 < m^2$, i.e. for low charge. The situation then has many similarities with the Schwarzschild black hole. The main difference is that the singularity lies on a timelike line and therefore timelike lines inside the black hole do not necessarily have to terminate on the singularity: with a sufficiently powerful rocket motor (and a sufficient strong constitution in view of the tidal forces) one could survive indefinitely, exploring the inner region but never emerging from it. The case $|q| > m$ is discussed briefly in Section 20.2.2; it is thought not to arise in practice, at least via processes adequately described by classical physics. The case $|q| = m$ yields the *extreme Reissner–Nordström black hole*. Though physically unlikely, this is a solution of theoretical interest, with the remarkable property that a pair of such black holes, having charges of the same sign, neither attract nor repel one another: the gravitational attraction just balances the electrostatic repulsion. Indeed one can construct

---

[3]These are defined and discussed more fully in Chapter 20.

an exact solution of the Einstein and Maxwell equations together, corresponding to any number of extreme charged black holes distributed through space and remaining static: exercise 18.7.

18.3   de Sitter–Schwarzschild metric

If we allow a non-zero cosmological constant, then the vacuum field equation becomes

$$R_{ab} = \Lambda g_{ab}. \tag{18.34}$$

In this case it may be shown (exercise) that the spherically symmetric solution is

$$\mathrm{d}s^2 = -\left(1 - \frac{2m}{r} - \frac{\Lambda}{3}r^2\right)c^2\mathrm{d}t^2 + \left(1 - \frac{2m}{r} - \frac{\Lambda}{3}r^2\right)^{-1}\mathrm{d}r^2 + r^2\mathrm{d}\Omega^2 \tag{18.35}$$

This result has applications chiefly to cosmology and it is revisited in Chapter 23 where, for $m = 0$, it appears under the guise of an FLRW metric. The most notable property is that when $\Lambda > 0$ the Λ term causes g_{tt} and g_{rr} to change sign at some large enough r, and consequently at large r the spacetime is dynamic, since then r is the timelike coordinate and t spacelike, but the metric depends on r. More generally, for positive Λ the influence of the cosmological constant represents a form of gravitational repulsion at any given r. For example, the light rays setting out from a point at some given r (in the region where r is spacelike) can be divided into those that spiral into the central body and those that escape to infinity. The proportion that escape to infinity grows with Λ. (Stuchlík and Hledík, 1999)

Exercises

18.1 Show that if we explicitly include the cosmological constant term described in (2.31), then (5.50) becomes

$$\nabla^2\Phi = 4\pi G(\rho_0 + 3p/c^2) - \Lambda c^2. \tag{18.36}$$

Hence show that the Newtonian potential outside a spherical mass is

$$\Phi = -GM/r - \Lambda c^2 r^2/6. \tag{18.37}$$

Outline the qualitative features of the motion of test particles in this potential.

18.2 Obtain the interior Schwarzschild solution by either or both of the following two methods: (i) follow the procedure used to obtain the stellar structure equations, making use of the constant value of ρ_0 to simplify where possible. (ii) Start from (18.21) and the Oppenheimer-Volkoff equation, and integrate.

18.3 Show that (18.8) agrees with (16.18). [Hint: (11.52)]

18.4 Find the Kretschmann scalar $R_{abcd}R^{abcd}$ for the Reissner–Nordström metric and hence show that the singularity at $r = 0$ is intrinsic.

18.5 Show that the radial null geodesics of the Reissner–Nordström spacetime are given by

$$ct = \pm r \mp (r_-^2/\Delta)\ln|r/r_- - 1|$$
$$\pm (r_+^2/\Delta)\ln|r/r_+ - 1|$$

where $\Delta = r_+ - r_-$ and $r_\pm$ are the two horizons.

18.6 Show that a particle with charge/mass ratio e/M moving in Reissner–Nordström spacetime has a constant of the motion

$$k = (1 - 2m/r + q^2/r^2)\dot{t} + q(e/M)/r.$$

18.7 **Multiple extreme charged black holes** (Majumdar–Papapetrou).
(i) Show that with the change of coordinate to $\bar{r} \equiv r - m$ the extreme ($q = m$) Reissner–Nordström metric becomes (taking $c = 1$)

$$ds^2 = -U^{-2}dt^2 + U^2(d\bar{r}^2 + \bar{r}^2 d\Omega^2)$$

where $U = 1 + m/\bar{r}$, and the electric field can be written $E = F_{rt} = \partial_r A_0$ with $A_0 = (4\pi\epsilon_0 G)^{-1/2}((1/U) - 1)$.
(ii) Since the metric is isotropic we can conveniently switch to rectangular coordinates defined such that $d\bar{r}^2 + \bar{r}^2 d\Omega^2 = dx^2 + dy^2 + dz^2$. Show that if we now take U to be some unknown function to be discovered, then the above metric, along with the 4-vector potential, satisfies the Einstein and Maxwell equations together, as long as U is a solution of Laplace's equation $(\partial_x^2 + \partial_y^2 + \partial_z^2)U = 0$.
(iii) We can now write down solutions such as

$$U = 1 + \sum_{i=1}^{N} \frac{m}{|\vec{x} - \vec{x}_i|}. \tag{18.38}$$

Physically interpret such a solution. Do the black holes attract one another? What happens to a test body (charged or uncharged) moving in their vicinity?

18.8 **Cosmic strings**. A cosmic string is an infinitely long cylindrical configuration of matter that is invariant under translations and Lorentz boosts along the symmetry axis; such objects are suggested by some calculations in quantum field theory. They have cylindrical (not spherical) symmetry but are included here since they are static and we we will apply to them methods similar to those of this and the previous chapter.
(i) Show that for the energy tensor of a perfect fluid to be invariant under Lorentz boosts, one requires $p = -\rho c^2$.
(ii) Writing the metric $ds^2 = -A(r)dt^2 + dr^2 + B(r)d\phi^2 + C(r)dz^2$, show that the nonzero connection coefficients are

$$\Gamma^t_{tr} = A'/2A, \quad \Gamma^r_{\phi\phi} = -B'/2,$$
$$\Gamma^r_{zz} = -C'/2, \quad \Gamma^r_{tt} = A'/2, \tag{18.39}$$
$$\Gamma^\phi_{\phi r} = B'/2B, \quad \Gamma^z_{zr} = C'/2C$$

(ii) Show that for a cosmic string the line element outside the matter can be written in the above form with $A = c^2$, $B = (\alpha + \beta r)^2$, $C = 1$ where α, β are constants. Hence show that for $\alpha = 0$ the spatial hypersurface at each $t = $ const, $z = $ const is a cone.
(iii) Taking now $A = c^2$, $C = 1$ with $B(r)$ to be discovered, show that inside the matter $B(r)$ satisfies $B(r) = b^2(r)$ where $b'' = -(8\pi G/c^2)\rho b$.
(iv) Hence obtain the line element for the case of uniform density inside the string.

19

Rotating bodies; the Kerr metric

In this chapter we begin with general properties of spacetime around rotating bodies, and then we consider an exact solution of the vacuum field equation called the Kerr metric. This describes spacetime around a rotating black hole, and it approximately describes spacetime around some other rotating objects. Since most astronomical bodies rotate, and some rotate very rapidly, this is a very important metric. However, it is also significantly more complicated than the ones we have treated in previous chapters, and we will discuss it less thoroughly than we did for the non-rotating (Schwarzschild–Droste) case.

19.1 The general stationary axisymmetric metric

A stationary metric has the property that there exists a coordinate t which is timelike throughout the region deemed stationary, and the metric does not depend on this coordinate. We wish to consider the case where the metric is both stationary and also has an axis of rotational symmetry. This means there is a coordinate ϕ (the azimuthal angle) which increases from 0 to 2π around the axis, and the metric does not depend on ϕ. In four-dimensional spacetime there will be two further coordinates which we shall call r and θ, though we make no assumption at the outset about the nature of these coordinates (they do not necessarily bear any relation to spherical or cylindrical polar coordinates for example). Hence we have

$$g_{ab} = g_{ab}(r, \theta). \tag{19.1}$$

Next let us consider whether we can guarantee to find coordinates in which some of the components of g_{ab} are zero. To this end, impose a further restriction on the physical scenario: we suppose the line element is invariant under simultaneous inversion of both t and ϕ, i.e. the transformation

$$t \to -t, \qquad \phi \to -\phi.$$

This implies that

$$g_{tr} = g_{t\theta} = g_{r\phi} = g_{\theta\phi} = 0$$

(since otherwise there would be a change in the line element under the considered transformation, through a sign change of quantities such as $dt\,dr$). What this amounts to is that the

Relativity Made Relatively Easy: General Relativity and Cosmology. Volume 2. Andrew M. Steane,
Oxford University Press. © Andrew M. Steane 2021. DOI: 10.1093/oso/9780192895646.003.0019

motion of the gravitating body is not such as to produce effects relating r and t (so it is not moving radially) nor θ and t (so it is not wobbling or otherwise evolving in the θ direction), etc. In short, we are assuming the motion of the gravitating body is purely rotational about the fixed axis of symmetry. We have thus determined that the non-zero elements of g_{ab} are at the locations indicated by dots in the following:

$$[g_{ab}] = \begin{pmatrix} \cdot & & & \cdot \\ & \cdot & \cdot & \\ & \cdot & \cdot & \\ \cdot & & & \cdot \end{pmatrix} \tag{19.2}$$

where we ordered the coordinates (t, r, θ, ϕ). Now the (r, θ) part is two-dimensional and depends only on r, θ. We can always find orthogonal coordinates for a two-dimensional manifold, so we can always find coordinates such that the metric has the form

$$[g_{ab}] = \begin{pmatrix} \cdot & & & \cdot \\ & \cdot & & \\ & & \cdot & \\ \cdot & & & \cdot \end{pmatrix} \tag{19.3}$$

Furthermore the (t, ϕ) part of the line element can always be written $-A\mathrm{d}t^2 + B(\mathrm{d}\phi - \omega\mathrm{d}t)^2$ by completing the square. Hence the line element has the form

$$\mathrm{d}s^2 = -A\mathrm{d}t^2 + B(\mathrm{d}\phi - \omega\mathrm{d}t)^2 + C\mathrm{d}r^2 + D\mathrm{d}\theta^2 \tag{19.4}$$
$$= -(A - B\omega^2)\mathrm{d}t^2 - 2B\omega\mathrm{d}t\mathrm{d}\phi + C\mathrm{d}r^2 + D\mathrm{d}\theta^2 + B\mathrm{d}\phi^2 \tag{19.5}$$

where A, B, C, D, ω are functions of r and θ. Note in particular that

$$g_{tt} = -(A - B\omega^2), \quad g_{t\phi} = -B\omega, \quad g_{\phi\phi} = B. \tag{19.6}$$

It will be useful to have also the contravariant components g^{ab}. The matrix $[g_{ab}]$ can be divided into two 2×2 matrices which makes its inversion easy. Observe that

$$\frac{-1}{A}\begin{pmatrix} 1 & \omega \\ \omega & \omega^2 - A/B \end{pmatrix}\begin{pmatrix} -A + B\omega^2 & -B\omega \\ -B\omega & B \end{pmatrix} = \begin{pmatrix} 1 & 0 \\ 0 & 1 \end{pmatrix} \tag{19.7}$$

therefore

$$g^{tt} = -1/A, \quad g^{t\phi} = -\omega/A, \quad g^{\phi\phi} = (A - B\omega^2)/(AB). \tag{19.8}$$

19.1.1 The physical meaning of ω

The form (19.4) reminds us of the metric of the rotating lattice in flat spacetime, eqn (4.12). In (4.12) the parameter ω refers to the rotation of a rigid disc; in the present context the parameter $\omega(r, \theta)$ refers to a property of spacetime. We can ascribe a direct physical significance to this ω in two ways, as follows.

By examining the line element in the form (19.4) we observe that if $d\phi = \omega dt$ then the B term vanishes. Let us interpret this. Consider a pair of timelike-separated events at the same r, θ, separated by some given coordinate time dt and some value $d\phi$ to be determined. We seek the value of $d\phi$ such that the proper time $d\tau$ between the events shall be maximized. The coordinate separations between these events satisfy $dr = d\theta = 0$ by construction and dt is fixed. It is obvious that if $B > 0$ then the maximum occurs when $d\phi = \omega dt$, since the B term can only contribute negatively to $d\tau$ (recall $d\tau^2 = -ds^2$). Therefore the elapsed proper time is maximized, on a worldline at fixed r, θ, if the particle in question (whose worldline it is) is moving around the central body with coordinate angular velocity $d\phi/dt = \omega$. (Note, we have not assumed that the particle is in free fall, only that its worldline is as described; this will be brought about by whatever forces are acting on it—think, for example, of an astronaut equipped with a rocket-pack).

A further observation concerns particles in free fall. We have that the metric does not depend on ϕ, and therefore the covariant momentum component p_ϕ is conserved along geodesics. It can be interpreted as the covariant component of angular momentum along the rotation axis:

$$p_\phi = \text{constant} = L_z. \tag{19.9}$$

The corresponding contravariant component is

$$p^\phi = g^{\phi t}p_t + g^{\phi\phi}p_\phi = -\frac{\omega}{A}p_t + \frac{A - B\omega^2}{AB}p_\phi. \tag{19.10}$$

We consider the worldline of a particle in free fall. By definition p^a is parallel to the 4-velocity (which is the tangent to the worldline), so we have $p^t \propto dt/du$ and $p^\phi \propto d\phi/du$ where u is an affine parameter along the worldline. Therefore for events along the worldline, ϕ increases with coordinate time according to

$$\frac{d\phi}{dt} = \frac{p^\phi}{p^t} = \frac{g^{\phi t}p_t + g^{\phi\phi}p_\phi}{g^{tt}p_t + g^{t\phi}p_\phi}. \tag{19.11}$$

Now specialize to the case $L_z = 0$. This is the case of a particle moving inward or outward on a trajectory that has no angular velocity in the limit where it is far from the central body—see Fig. 6.3. For this case we find

$$\frac{d\phi}{dt} = \frac{g^{\phi t}}{g^{tt}} = \omega. \tag{19.12}$$

Thus we find, as noted in connection with Fig. 6.3, that a particle falling from infinity is swept in the direction of rotation of the central body, and a particle moving outwards has its velocity steered in the opposite sense, such that in both cases $d\phi/dt$ is positive for the case of a trajectory having $L_z = 0$. We have furthermore a quantitative statement about ω: it is equal to the rate of change of ϕ with coordinate time t for such a trajectory.

19.2   Stationary limit surface and ergoregion

A further important property of metrics of the form (19.4) is that g_{tt} can change sign. By inspecting the form (19.5) one can immediately determine that if g_{tt} becomes positive then a

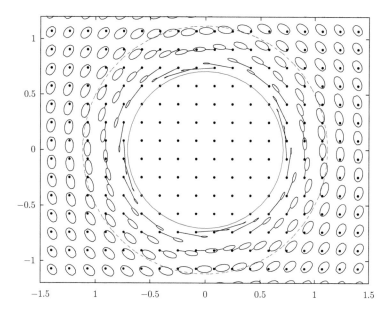

Fig. 19.1 Light cone structure near an ergoregion. The diagram shows a spatial plane at right angles to the axis of symmetry, with a set of dots marking points in the plane. Around each dot a cross-section through the light cone of that dot (after some small elapsed time) has been plotted. Where the dot falls outside this cross-section, a line is drawn from the dot to the cone as an aid to interpreting the information. The dashed circle is a stationary limit surface; the full circle is a horizon. The ergoregion is between these circles. No information has been plotted for points inside the horizon.

worldline can only maintain its timelike status if the $-2B\omega \mathrm{d}t\mathrm{d}\phi$ term contributes negatively to $\mathrm{d}s^2$ (assuming $C, D > 0$), and this requires

$$\mathrm{d}\phi > 0 \tag{19.13}$$

assuming that also $B > 0$. This important observation means that *all worldlines have positive* $\mathrm{d}\phi/\mathrm{d}t$ *in regions of spacetime where* $g_{tt} > 0$ (for $B, C, D > 0$). In other words, we have a region where a particle cannot 'sit still' nor can it wander in either direction around the central body; it has no choice but to make its way in the correct sense. It has been caught in a kind of whirlpool in space; see Fig. 19.1.

Since the limiting velocity is the speed of light, it is useful to consider null worldlines in order to explore the limits of this behaviour. We consider photons emitted from some location such that their velocity is initially in the purely ϕ direction. Then the worldline initially satisfies

$$g_{tt}\mathrm{d}t^2 + 2g_{t\phi}\mathrm{d}t\mathrm{d}\phi + g_{\phi\phi}\mathrm{d}\phi^2 = 0 \tag{19.14}$$

(the condition for an interval to be null). This can be converted into a quadratic equation for $\mathrm{d}\phi/\mathrm{d}t$ whose solution is

$$\frac{\mathrm{d}\phi}{\mathrm{d}t} = -\frac{g_{t\phi}}{g_{\phi\phi}} \pm \left[\left(\frac{g_{t\phi}}{g_{\phi\phi}} \right)^2 - \frac{g_{tt}}{g_{\phi\phi}} \right]^{1/2}. \tag{19.15}$$

In the special case $g_{tt} = 0$ the two solutions are

$$\dot{\phi} = 2\omega \quad \text{or} \quad 0. \tag{19.16}$$

Hence the photon under consideration must either proceed at an angular velocity 2ω in the first instance, or else it must stand still. But in the latter case it will then continue to stand still! It has been emitted against the 'flow' of the 'whirlpool' and, as it moves at the speed of light relative to any massive particle in its vicinity, it makes no progress at all relative to the ϕ coordinate.

The set of spatial locations for which $g_{tt}(r, \theta) = 0$ is called a *stationary limit surface*. A region where $g_{tt}(r, \theta) > 0$ is called an *ergoregion*. A stationary limit surface is also an infinite redshift surface; c.f. eqn (13.45). The name 'ergoregion' is a reference to work and energy; we will show in Section 21.1 that energy has interesting properties in such regions.

19.3 The Kerr metric

An important step was taken by Roy Kerr in 1963 when he discovered a new exact vacuum solution of the Einstein field equation for the axially symmetric case. In the case of a rotating black hole, the Kerr metric is the unique and exact solution. It is presumably also the exact result for the spacetime outside some sort of rotating non-collapsed object, but it is not known exactly what rotating mass distribution would give rise to it. It can be applied approximately to the spacetime around a rotating star.

Let us introduce a system of coordinates (r, θ, ϕ) related to ordinary rectangular coordinates by

$$x = \sqrt{r^2 + a^2} \sin\theta \cos\phi$$
$$y = \sqrt{r^2 + a^2} \sin\theta \sin\phi$$
$$z = r \cos\theta \tag{19.17}$$

where a is a constant. Such (r, θ, ϕ) coordinates are called *Boyer–Lindquist coordinates*; they are plotted in Fig. 19.2 for the case of flat space where x, y, z can be taken to be orthonormal. r is, roughly speaking, the distance from a disc of radius a. The whole disc is at $r = 0$; its edge is at $(r = 0,\ \theta = \pi/2)$. The coordinates tend to spherical polar coordinates in the limit $r \gg a$.

We will now write down the Kerr metric, and then show that the above provides a useful interpretation of the system of coordinates in the limit where the gravitating mass vanishes and spacetime reverts to Minkowski form.

In Boyer–Lindquist coordinates the Kerr metric is[1]

---

[1] For guidance on the derivation of this metric, see, for example, (d'Inverno, 1992; Misne, Thorne and Wheeler, 1975; Wald, 1984).

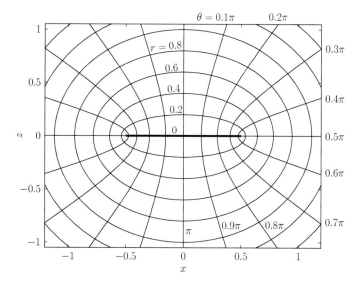

Fig. 19.2 Lines of constant r and constant θ in the xz plane, as decribed by eqns (19.17), for the case $a = 1/2$.

$$ds^2 = -A dt^2 + g_{\phi\phi}(d\phi - \omega dt)^2 + g_{rr} dr^2 + g_{\theta\theta} d\theta^2 \tag{19.18}$$

where

$$A = c^2 \rho^2 \Delta / \Sigma^2, \qquad\qquad \omega = 2mcra/\Sigma^2$$
$$g_{rr} = \rho^2 / \Delta, \qquad\qquad g_{tt} = -A + \omega^2 g_{\phi\phi} \;=\; -(1 - 2mr/\rho^2)c^2$$
$$g_{\theta\theta} = \rho^2, \qquad\qquad g_{\phi\phi} = (\Sigma^2 \sin^2\theta)/\rho^2 \tag{19.19}$$

in which m and a are constants and we have used

$$\Delta \equiv r^2 + a^2 - 2mr$$
$$\rho^2 \equiv r^2 + a^2 \cos^2\theta$$
$$\Sigma^2 \equiv (r^2 + a^2)^2 - a^2 \Delta \sin^2\theta \quad \left[= \rho^2\Delta + 2mr(r^2 + a^2) \right]. \tag{19.20}$$

Notice that

$$\Sigma^2 \geq 0, \qquad \rho^2 \geq 0 \tag{19.21}$$

whereas Δ can have either sign.

The result can also be written

$$ds^2 = -\left(1 - \frac{2mr}{\rho^2}\right) c^2 dt^2 + \frac{\rho^2}{\Delta} dr^2 + \rho^2 d\theta^2 + X^2 d\phi^2 - 2X^2 \omega dt d\phi \tag{19.22}$$

where $X = (\Sigma/\rho) \sin\theta$.

The metric has two parameters: m and a, and it depends on these and on two of the coordinates: r and θ.

19.3.1   The interpretation of the parameters

In order to interpret the parameters m and a, we use the asymptotic limit. As $r \to \infty$ the metric tends to the metric of flat Minkowski spacetime, and before it becomes totally flat the metric takes a weak field, linearized form which we can compare with metrics having known properties. To be precise, we note that in the limit $r \gg m, a$,

$$\Delta \to r^2 - 2mr, \qquad \rho^2 \to r^2, \qquad \Sigma^2 \to r^4, \qquad \omega \to 2mca/r^3,$$

and then the metric is

$$ds^2 = -(1 - 2m/r)c^2 dt^2 + \frac{dr^2}{1 - 2m/r} + d\Omega^2 - \frac{4mca\sin^2\theta}{r}\,dt d\phi. \tag{19.23}$$

This is the Schwarzschild metric plus a $dt d\phi$ term. By comparing with the weak field result (6.12) we can make the deductions

$$m = GM/c^2 \tag{19.24}$$
$$a = J/(Mc) \tag{19.25}$$

where M is the mass and J is the angular momentum of the source. Notice the reasoning. We *define* the mass of a source of gravity not by integrating over its matter distribution (a procedure which in any case is not possible in the case of a black hole) but by noting its effect on spacetime (c.f. eqn (16.21)). Before GR was developed, it was already the standard practice to 'weigh' astronomical bodies such as the Sun by noting their effect on orbits around them. In GR this is the only way to get a satisfactory definition—see Section 16.6. The new feature in the Kerr metric is that we adopt the same practice for angular momentum. We *define* the angular momentum J of the source via its impact on the metric at locations far away. In this reasoning we are not assuming that the Kerr metric is the metric for a rotating spherical mass; we are noting only that it is a solution of the vacuum field equation and we can determine what amount of angular momentum would give the same gravitational effects in the asymptotic limit.

If we retain a but set $m = 0$, then the metric becomes

$$ds^2 = -c^2 dt^2 + \frac{\rho^2 dr^2}{r^2 + a^2} + \rho^2 d\theta^2 + (r^2 + a^2)\sin^2\theta\, d\phi^2. \tag{19.26}$$

In the absence of any mass the metric should revert to that of flat spacetime, and indeed this is the Minkowski metric but written in the coordinates given by (19.17). Therefore in the limit of vanishing mass, we can interpret the coordinates as indicated in Fig. 19.2. Notice that r is not the radius of any sphere, and surfaces of constant r are not spherical (except in the limit $r \gg a$). When $m \neq 0$ a plot such as Fig. 19.2 can no longer be assumed to be drawn in Euclidean space. In this case the coordinates have no simple exact interpretation, but the diagram still provides a good general intuition.

19.3.2   The structure of the Kerr black hole

The Kerr metric has singularities of one kind or another at $\rho = 0$, at $\Delta = 0$ and at $g_{tt} = 0$. By studying the curvature one may determine that the first is a curvature singularity and the

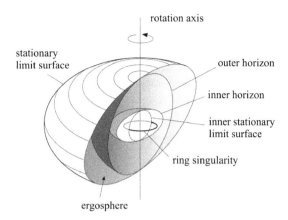

Fig. 19.3 The main features of the Kerr black hole.

second is not; it is a coordinate singularity. We will show in Section 20.2 that $\Delta = 0$ describes a pair of null surfaces, giving two nested event horizons. The third case ($g_{tt} = 0$) is a stationary limit surface (Section 19.2). Thus the main items of the structure of the Kerr spacetime are (Fig. 19.3):

1. Curvature singularity at $\rho = 0$, which therefore occurs when both

$$r = 0, \quad \text{and} \quad \theta = \pi/2. \tag{19.27}$$

   Thus we find that there is a ring-shaped singularity located at the edge of a disc of coordinate radius a in the (x, y, z) coordinates defined in (19.17).

2. Horizons at $\Delta = 0$ (and therefore $g_{rr} = \infty$). These are located at

$$r_{\pm} = m \pm \sqrt{m^2 - a^2} \tag{19.28}$$

   Hence they are surfaces of constant r, one inside the other. Neither is spherical—that is to say, the metric of the surface defined by $r = $ const., $t = $ const. is not the metric of a 2-sphere. Notice that in the limit $a \to 0$, one horizon collapses to the curvature singularity at $r = 0$, and the other moves outwards until it is situated at the location of the Schwarzschild horizon. (We are here assuming $a < m$; we will comment on the case $a \geq m$ in Chapter 20.)

3. Stationary limit surface at $g_{tt} = 0$ and therefore at $r^2 - 2mr + a^2 \cos^2 \theta = 0$. This can be solved to give

$$r_{0\pm} = m \pm \sqrt{m^2 - a^2 \cos^2 \theta} \tag{19.29}$$

   One of these surfaces lies inside the inner horizon (touching it at the poles $\theta = 0, \pi$). The other lies outside the outer horizon (touching it at the poles). Between r_{0+} and r_+ lies the ergoregion, often called ergosphere. This name has the connotation 'sphere of influence'; the ergosphere is not spherical but is shaped approximately as the region between two oblate spheroids in the coordinates defined in (19.17).

The outer horizon is the more important of the two horizons; once something has crossed the outer horizon it is lost to the outside world and the details of its behaviour on its further journey are not important to any other physical system (except possibly physicists). The area of this horizon is easy to calculate (exercise 19.6); one finds

$$A_H = 4\pi(r_+^2 + a^2) = 8\pi m r_+ = 8\pi m(m + \sqrt{m^2 - a^2}). \qquad (19.30)$$

We noted in Section 19.2 that in the ergoregion, in order to be timelike a worldline must have $\dot{\phi} > 0$. By writing down a 4-velocity of the form $u^a = u^0(1, 0, 0, \Omega)$ and then requiring $u_\mu u^\mu = -c^2$ one can constrain the value of $\dot{\phi} = \Omega$ for a worldline which is (momentarily or permanently) in the ϕ direction. In Section 19.2 we saw that at the stationary limit surface, Ω is constrained to lie between the values 0 and 2ω. Inside the ergoregion, the range is more and more constrained as r decreases, until for a worldline touching the horizon r_+ one finds Ω is constrained to take the single value

$$\Omega_H \equiv \Omega(r_+) = \omega(r_+) = \frac{ac}{r_+^2 + a^2}. \qquad (19.31)$$

It is sometimes said that this is 'the angular velocity of the horizon'.

The case $a = m$ is called an *extreme* Kerr black hole. It is believed that near-extreme black holes can develop naturally, as they acquire angular momentum from matter falling in. Calculations suggest that the limiting value $a \simeq 0.998m$ can be achieved this way, and furthermore it should not surprise us to find such values of a being realized. For example, the X-ray source Cygnus X-1 is part of a binary star system in which multiple evidences suggest one partner is a black hole, which is now thought to rotate with $a > 0.9m$. (Gou *et al.*, 2011)

The case $a > m$, on the other hand, probably does not occur in nature. That is, it is conjectured that the angular momentum J will in practice always fall below the value GM^2/c. Matter having more angular momentum than this will tear itself apart and not collapse to form a black hole. This is an example of a conjecture called the *cosmic censorship hypothesis*; see Section 20.2.2.

19.4 Freefall motion in the plane $\theta = \pi/2$

Orbital motion in the Kerr geometry is considerably more complicated than in the spherically symmetric case, because non-equatorial orbits do not stay in one plane: they wander over some range of θ. Remarkably, there is a conserved quantity associated with the θ motion which considerably simplifies the treatment of a general orbit. However, we will restrict our discussion to motion lying wholly in the equatorial plane, $\theta = \pi/2$. The reflection symmetry in this plane suffices to prove that a geodesic lying in it will not depart from it. For this motion the line element simplifies to

$$ds^2 = -\left(1 - \frac{2m}{r}\right)c^2 dt^2 - \frac{4mac}{r} dt d\phi + \frac{r^2}{\Delta} dr^2 + \left(r^2 + a^2 + \frac{2ma^2}{r}\right)d\phi^2. \qquad (19.32)$$

We will seek equations for the worldlines in terms of x^a and $\dot{x}^a$ where the latter is a derivative with respect to proper time for timelike geodesics and with respect to an affine parameter for null geodesics. Let us define the momentum $p^a = \dot{x}^a$; this means it is the momentum per unit mass for massive particles, and for a null geodesic it is the tangent vector. We have three variables: (t, r, ϕ) and the metric depends on only one of them (r). Consequently p_t and p_ϕ are both constants of the motion:

$$p_t = \text{const} \equiv -E, \qquad p_\phi = \text{const} \equiv L. \tag{19.33}$$

Using $p_a = g_{a\mu}\dot{x}^\mu$ we thus obtain

$$-c^2(1 - 2m/r)\dot{t} - (2mac/r)\dot{\phi} = -E \tag{19.34}$$
$$-(2mac/r)\dot{t} + (r^2 + a^2 + 2ma^2/r)\dot{\phi} = L. \tag{19.35}$$

These simultaneous equations are easily solved for $\dot{t}$ and $\dot{\phi}$, giving:

$$c^2\dot{t} = \left[(r^2 + a^2 + 2ma^2/r)E - (2mac/r)L\right]/\Delta \tag{19.36}$$
$$\dot{\phi} = \left[(2ma/cr)E + (1 - 2m/r)L\right]/\Delta \tag{19.37}$$

It remains to obtain an equation for $\dot{r}$. This is most conveniently done by recalling that the 4-velocity is a constant, so we can write $p_\mu p^\mu = \text{const} \equiv -\epsilon^2$ where $\epsilon^2 = c^2$ for a timelike geodesic and $\epsilon^2 = 0$ for a null geodesic. The use of ϵ^2 here is a convenient way to treat both types of geodesic by the same equation. We have $g^{\mu\nu}p_\mu p_\nu = -\epsilon^2$ so we require the contravariant metric components. Using (19.8) we find

$$g^{tt} = \frac{-1}{c^2\Delta}\left(r^2 + a^2 + 2ma^2/r\right), \qquad g^{t\phi} = -\frac{2ma}{cr\Delta} \tag{19.38}$$

$$g^{\phi\phi} = \frac{1 - 2m/r}{\Delta}, \qquad g^{rr} = \frac{\Delta}{r^2} \tag{19.39}$$

and the equation for $\dot{r}$ is to be extracted from

$$g^{tt}p_t^2 + 2g^{t\phi}p_t p_\phi + g^{\phi\phi}p_\phi^2 + g^{rr}p_r^2 = \epsilon^2. \tag{19.40}$$

After using (19.33) and $p_r = g_{rr}\dot{r}$ this gives

$$\begin{aligned}\dot{r}^2 &= g^{rr}\left(\epsilon^2 - g^{tt}E^2 + 2g^{t\phi}EL - g^{\phi\phi}L^2\right) \\ &= E^2/c^2 - \epsilon^2 - 2V_{\text{eff}}\end{aligned} \tag{19.41}$$

where

$$V_{\text{eff}} = -\frac{\epsilon^2 m}{r} + \frac{L^2 - a^2((E/c)^2 - \epsilon^2)}{2r^2} - \frac{m(L - aE/c)}{r^3}. \tag{19.42}$$

Eqns (19.36), (19.37) and (19.41) can be solved numerically to find the geodesics. In practice the numerical method goes better if one first differentiates (19.41) so as to obtain a second-order equation for r.

We can also use the function V_{eff} to get some general insights. Observe that when $a \to 0$, V_{eff} goes to the Schwarzschild effective potential, and when $a \neq 0$ it has the same functional dependence on r. It follows that much of what we said about the radial motion in the Schwarzschild case applies here too; the radial acceleration is given by $\ddot{r} = -\mathrm{d}V_{\text{eff}}/\mathrm{d}r$ for example, and circular orbits occur when the energy satisfies $E^2 = 2V_{\text{eff}}c^2 + \epsilon^2 c^2$ and r is at a stationary point of V_{eff}. However, V_{eff} is not in all respects like an effective potential because it depends on E.

One finds that the innermost stable circular orbit occurs at r satisfying

$$r^2 - 6mr - 3a^2 \pm 8a\sqrt{mr} = 0 \qquad (19.43)$$

where the upper sign corresponds to the co-rotating orbit, and the lower to the counter-rotating orbit. In the extreme Kerr limit $a = m$ this gives $r = m$ for the co-rotating and $r = 9m$ for the counter-rotating case. This means that accretion discs around rapidly rotating black holes can approach closer to the central body than if it were not rotating, and consequently a greater fraction of the energy of matter gradually spiralling in can be emitted as X rays (Table 17.2).

A particle dropped from infinity with $L = 0$ will be steered towards the positive ϕ direction, as we would expect from our earlier study of gravimagnetism and the ergoregion. One should note that *both* t and ϕ coordinates become 'bad' or unusable at the horizon, and the motion inside the horizon should be interpreted with caution in these coordinates. A worldline approaching a horizon will spiral around the black hole an infinite number of numbers in t, ϕ coordinates, but this does not imply that a traveller on that worldline sees the universe rotating around her infinitely many times; it is a pathology of the coordinate system. Other geodesics, such as null ones, undergo the same spiralling and the result is that the traveller sees the outer universe undergoing some more modest amount of rotation, as judged by received signals. She also experiences a finite proper time on her inward journey, just as for the Schwarzschild case. Such issues are clarified by adopting *Doran coordinates*.[2]

19.4.1 Near-radial null geodesics

There are no purely radial null geodesics in the equatorial plane, but there are incoming and outgoing geodesics having the condition $L = aE/c$. These are called the *principal null geodesics* and they provide both example photon trajectories and also the main radial features of the light cone structure. The condition $L = aE/c$ leads to $V_{\text{eff}} = 0$ in (19.42), and the equations of motion become

$$\dot{r} = \pm\frac{E}{c}; \qquad \dot{\phi} = \frac{L}{\Delta} = \frac{aE}{c\Delta}; \qquad \dot{t} = \frac{(r^2 + a^2)E}{c^2\Delta}. \qquad (19.44)$$

Hence

$$\frac{\mathrm{d}t}{\mathrm{d}r} = \pm\frac{r^2 + a^2}{c\Delta}; \qquad \frac{\mathrm{d}\phi}{\mathrm{d}r} = \pm\frac{a}{\Delta} \qquad (19.45)$$

where the plus(minus) sign corresponds to an outgoing(incoming) ray. These equations are readily integrated; see exercise 19.9 and Fig. 19.4.

[2]C. Doran, Phys. Rev. D61 (2000) 067503; Hamilton and Lisle, Am. J. Phys 76, 519 (2008).

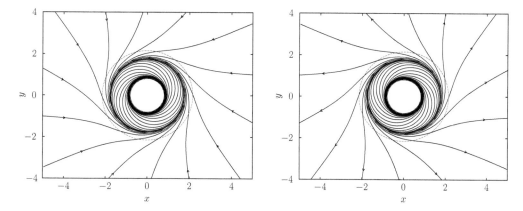

Fig. 19.4 Example light rays in the equatorial plane for a Kerr black hole with $m = 1$, $a = 0.8$. Dashed circles show the stationary limit surface, horizon and singularity. In the x, y coordinates adopted for the diagram these have radii 2.154, 1.789, 0.8 respectively in this example. The chosen coordinates do not handle the horizon very well; at the horizon $d\phi/dr$ changes sign but p_0 is constant.

19.4.2 Equatorial photon orbits

A very nice insight into the null geodesics in the equatorial plane can be obtained by factorizing the equation for $\dot{r}$. Observe that when $\epsilon^2 = 0$ the first equation in (19.41) has terms in E^2, EL and L^2 on the right-hand side. After some algebra, one finds that the equation can be written

$$\dot{r}^2 = \frac{(r^2 + a^2)^2 - a^2\Delta}{c^2 r^4}(E - V_+)(E - V_-) \tag{19.46}$$

where

$$V_\pm = \left(\omega \pm \sqrt{w^2 - g^{\phi\phi}/t^{tt}} \right) L = \frac{2mra \pm r^2\sqrt{\Delta}}{(r^2 + a^2)^2 - a^2\Delta}cL \tag{19.47}$$

Notice that $V_\pm$ do not depend on E, and that $\dot{r} = 0$ when either one of them is equal to E. Clearly if $E > \max(V_+, V_-)$ then the orbit has no turning point. Also, if $\Delta < 0$ then $V_\pm$ becomes complex, which implies there are no real roots to $\dot{r} = 0$ and hence no turn-around; such a photon has entered the outer horizon of the black hole.

If E is between V_+ and V_- then $\dot{r}^2$ is negative, which again is ruled out so we can infer that the geodesic will not reach values of r where $V_- < E < V_+$. The region $E < V_-$ is also ruled out when L is positive, because this implies E is less than the minimum that is required to guarantee that a LIF observer finds a positive energy for the photon, see exercise 19.10. Similarly the region $E < V_+$ is ruled out when L is negative. Overall, then, we find that possible worldlines are those for which

$$E \geq \max(V_+, V_-). \tag{19.48}$$

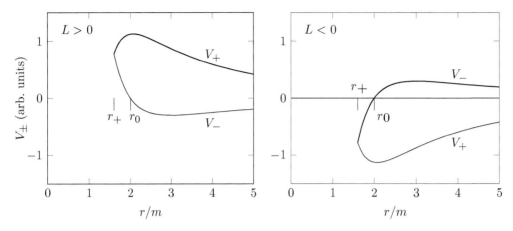

Fig. 19.5 $V_{\pm}$ functions defined in (19.47). The plots have been adjusted to bring out the main features. Allowed null geodesics have E at or above the higher of the two curves (bold line). The Penrose process is associated with the region where $V_- < 0$ on the right-hand ($L < 0$) graph.

This evaluates to V_+ when L is positive, V_- when L is negative (if we assume that the coordinates were chosen in the first place so as to make a positive); see Fig. 19.5.

The case $L > 0$ yields behaviour qualitatively similar to the Schwarzschild black hole (compare Fig. 19.5(a) with Fig. 17.5(b)). The case $L < 0$—a photon moving against the rotation of the hole—yields an interesting new possibility. For, now there are allowed geodesics with $E < 0$. Photons on such geodesics are sometimes said to have 'negative energy', though in fact they have $E_{\text{obs}} > 0$. The physical meaning of $E < 0$ here is that the photon cannot escape to infinity, and furthermore when such a photon falls into the black hole, the mass of the black hole *decreases*. The notion of negative energy is a somewhat loose way of speaking. Strictly speaking the creation of such a photon involves the extraction of momentum and energy from the ergosphere; consequently a falls a little, as well as m, and the horizon grows. We will prove these statements in Section 21.1. The interesting result is that one can imagine an object falling into the ergosphere, emitting a photon with $E < 0$, and thus (by conservation of energy) gaining an increased p_t and subsequently emerging from the ergosphere with *more* energy than it had when it fell in! This is an example of a more general process called the *Penrose process* which will be discussed in Section 21.1.

Exercises

19.1 Consider the general axisymmetric metric (19.4). A system of mirrors is arranged to guide photons around a circular path at $\theta = \pi/2$ and fixed r. Show that the angular velocity around such a trajectory (not usually geodesic) is

$$\frac{d\phi}{dt} = \omega \pm \sqrt{A/B} \qquad (19.49)$$

Hence show that a clock at fixed coordinates emitting two such photons in opposite directions receives them back separated by the proper time interval

$$\delta\tau = \frac{4\pi\omega B}{c(A - B\omega^2)^{1/2}}. \qquad (19.50)$$

This is the Sagnac effect for an arbitrary axisymmetric spacetime.

19.2 Are the photons in the previous exercise received with the same wavelength they had when they set out?

19.3 For a spaceship on a circular (not necessarily geodesic) trajectory with $u_\phi = 0$, show that photons sent out in opposite directions around a circular mirror track as in exercise 19.1 are received back simultaneously. What can be said about their wavelengths?

19.4 Obtain (19.22) from (19.18).

19.5 Confirm, by a coordinate transformation, that (19.26) is a metric of flat spacetime.

19.6 Show that the area of the event horizon of a Kerr black hole is

$$A = 4\pi(r_{\text{horizon}}^2 + a^2).$$

and that this can also be written

$$A = 8\pi m \left(m + \sqrt{m^2 - a^2} \right) \qquad (19.51)$$

where $m = GM/c^2$.

19.7 Show that the circumference around the equator at $r = r_+$ is larger than the circumference passing through the 'poles'.

19.8 Show that the angular velocity for a circular orbit in the equatorial plane satisfies

$$\frac{d\phi}{dt} = \frac{cm}{am \pm (mr^3)^{1/2}} \qquad (19.52)$$

[Hint: standard geodesic methods are applicable, or the rotating frame method of exercise 6.8 of Chapter 4]

19.9 **Principal null geodesics.** Show that if we take the plus signs then (19.45) gives

$$c\frac{dt}{dr} = 1 + \frac{2mr}{(r - r_-)(r - r_+)}$$

$$= 1 + \frac{2m}{r_+ - r_-} \left(\frac{r_+}{r - r_+} - \frac{r_-}{r - r_-} \right),$$

$$\frac{d\phi}{dr} = \frac{a}{2\sqrt{m^2 - a^2}} \left(\frac{1}{r - r_+} - \frac{1}{r - r_-} \right)$$

and therefore

$$ct = r + \frac{m}{\sqrt{m^2 - a^2}} \left[r_+ \ln|(r/r_+) - 1| \right.$$
$$\left. - r_- \ln|(r/r_-) - 1| \right], \qquad (19.53)$$

$$\phi = \frac{a}{2\sqrt{m^2 - a^2}} \ln \left| \frac{r - r_+}{r - r_-} \right| + \text{const.}$$
$$(19.54)$$

19.10 (i) An observer furnished with a rocket engine is hovering at fixed r, θ in the equatorial plane of a Kerr black hole, making their way along a circular trajectory with 4-velocity $u^a = u^t(1, 0, 0, \Omega)$. Show that in order that this vector be timelike, Ω must lie between the limits given by (19.15).
(ii) The observer receives a photon on a null geodesic with parameters E, L. Show that the observed energy is $E_{\text{obs}} = u^t(E - \Omega L)$.
(iii) Using the condition on Ω from part (i), show that the condition for E_{obs} to be non-negative is (19.48). [Hint: if the limits on Ω are $\Omega_\pm$ then notice that eqn (19.47) reads $V_\pm = \Omega_\pm L$.]

19.11 Show that for a circular prograde equatorial orbit the time T elapsed at a distant observer between the arrival of signals sent from successive orbits, divided by proper time τ along the orbit, is

$$\frac{T}{\tau} = \frac{r^{3/2} + am^{1/2}}{r^{3/4}\sqrt{(r - 3m)r^{1/2} + 2am^{1/2}}}.$$

20

Black holes

In this chapter we discuss black holes and geometric ideas associated with them, and some physical processes such as gravitational collapse. We will use the Schwarzschild black hole for detailed examples, and also invoke more generic concepts. The level of discussion is similar to that of the rest of the book, which means we aim as far as possible at a full derivation of the aspects we discuss, but we will not bring in sophisticated topological methods that are used at the research frontier.

20.1 Birkhoff's theorem

Let us address the question: what is the most general spherically symmetric solution of Einstein's field equation in vacuum? In Chapter 17 we obtained a particular solution—the Schwarzschild–Droste solution—and in order to get it we could assume from the outset whatever we liked; for example we assumed that the metric was diagonal and independent of the coordinate t (as well as θ and ϕ). The field equation then assured us that a solution with those properties exists. Now we wish to drop those assumptions, and thus seek more general possibilities, such as a radially pulsating solution, or something like that. We still seek a vacuum solution, however.

It turns out that if we seek a *vacuum* solution to the Einstein field equation having *spherical symmetry*, then we must obtain the Schwarzschild–Droste solution even if we do not restrict to static conditions at the outset. In other words, the link between space and time forged by the field equation constrains a spherically symmetric vacuum to be of Schwarzschild form and no other. This important result is called *Birkhoff's theorem* and now we shall prove it.

To begin the proof we need a way to formalize the notion of spherical symmetry. What we mean by this idea is that it will be possible to apply rotations about either of two intersecting axes with no observable effect on the manifold. This in turn implies that it must be possible to find coordinates (u, v, θ, ϕ) in which the metric takes the form

$$\mathrm{d}s^2 = g_{uu}(u,v)\mathrm{d}u^2 + 2g_{uv}(u,v)\mathrm{d}u\mathrm{d}v + g_{vv}(u,v)\mathrm{d}v^2 + f^2(u,v)\left(\mathrm{d}\theta^2 + \sin^2\theta\,\mathrm{d}\phi^2\right) \qquad (20.1)$$

where we have shown explicitly that all the functions g_{uu}, g_{uv}, g_{vv}, f are functions of u and v alone: they are independent of θ, ϕ. Notice that we here make two types of claim: first that various parts do not depend on θ, ϕ, and also that various cross terms are zero.

Relativity Made Relatively Easy: General Relativity and Cosmology. Volume 2. Andrew M. Steane,
Oxford University Press. © Andrew M. Steane 2021. DOI: 10.1093/oso/9780192895646.003.0020

We are here asserting (20.1) without a derivation. We can motivate it by observing that if one of the functions did depend on θ, ϕ then on the face of it we appear to have lost the spherical symmetry. Also, if there were a non-zero cross term such as $g_{u\phi}\mathrm{d}u\mathrm{d}\phi$ then it would change sign under $\phi \to -\phi$, which implies a sense of direction for rotations in the ϕ direction, and thus again a lack of symmetry. Equation (20.1) can be obtained more rigorously, but here we shall content ourselves with proving that *if* the metric has the form (20.1) *then* there exist coordinates in which it has the form we shall derive.

The first step is to note that we can define a coordinate r whose value is equal to $f(u, v)$. Thus we have a relationship $r = f(u, v)$ and we can in principle invert this equation so as to find an expression for v in terms of r and u. In general we can only guarantee thus to get an expression for r over some range of values that may be limited, but we shall accept this restriction. Equally, if $f(u, v)$ were in fact a function of u alone then we cannot use it to find v, but in that case we can always switch to coordinates r, v and then rename them. Therefore, after starting out from (20.1), it is always possible to find coordinates (u, r, θ, ϕ) such that

$$\mathrm{d}s^2 = g_{uu}(u, r)\mathrm{d}u^2 + 2g_{ur}(u, r)\mathrm{d}u\mathrm{d}r + g_{rr}(u, r)\mathrm{d}r^2 + r^2\mathrm{d}\Omega^2 \tag{20.2}$$

where $\mathrm{d}\Omega^2 = \mathrm{d}\theta^2 + \sin^2\theta\,\mathrm{d}\phi^2$ as usual.

The metric under discussion has the form of a pair of 2×2 blocks, one of which is diagonal. We want to make the first block diagonal too. Introduce a function $t(u, r)$. Then we have

$$\mathrm{d}t = \frac{\partial t}{\partial u}\mathrm{d}u + \frac{\partial t}{\partial r}\mathrm{d}r \equiv U(u, r)\mathrm{d}u + R(u, r)\mathrm{d}r \tag{20.3}$$

where U and R are the partial derivatives of t. Hence

$$\mathrm{d}t^2 = U^2\mathrm{d}u^2 + 2UR\,\mathrm{d}u\mathrm{d}r + R^2\mathrm{d}r^2. \tag{20.4}$$

Now seek two functions $A(u, r)$ and $B(u, r)$ such that

$$g_{uu}(u, r)\mathrm{d}u^2 + 2g_{ur}(u, r)\mathrm{d}u\mathrm{d}r + g_{rr}(u, r)\mathrm{d}r^2 = -A\mathrm{d}t^2 + B\mathrm{d}r^2. \tag{20.5}$$

The issue is, do there necessarily exist any such functions A, B? By equating coefficients of $\mathrm{d}u, \mathrm{d}r$ we have the requirements

$$g_{uu} = -AU^2, \qquad g_{ur} = -AUR, \qquad g_{rr} = -AR^2 + B. \tag{20.6}$$

This is a set of three equations for three unknowns (A, B, t) which is just sufficient to find the unknowns, if we allow that t has to be determined from its partial derivatives so there will be a constant of integration. Once this is done, we can express A and B as functions of t, r so we have the form

$$\mathrm{d}s^2 = -A(t, r)\mathrm{d}t^2 + B(t, r)\mathrm{d}r^2 + r^2\mathrm{d}\Omega^2. \tag{20.7}$$

Note that we introduced the minus sign in the above for convenience in what follows. We have not yet specified the sign of either of the functions A or B, so although we are hinting that t will turn out to be a timelike coordinate, at least in some parts of the manifold, we have not assumed this. On the other hand, if the metric overall has Minkowski signature then it must be that g_{tt} and g_{rr} have opposite sign, so our functions A and B are either both positive or both negative at any given point.

That is as far as we can go based on spherical symmetry; the rest of the argument requires an appeal to the Einstein field equation. So now we have to compute the Christoffel symbols and thence the Ricci tensor, and set it equal to zero. It will suffice to examine the components R_{tr} and $R_{\theta\theta}$ and it is found that they are equal to (exercise):

$$R_{tr} = \frac{1}{rB} \frac{\partial B}{\partial t}, \tag{20.8}$$

$$R_{\theta\theta} = 1 - \frac{1}{B} - \frac{r}{2B} \left(\frac{A'}{A} - \frac{B'}{B} \right) \tag{20.9}$$

where the prime signifies $\partial/\partial r$; the second equation is the same as the corresponding result in (17.11). From the field equation we have $R_{tr} = 0$, which implies that B is independent of t. Using this fact, we can now differentiate the equation $R_{\theta\theta} = 0$ with respect to time, obtaining

$$\frac{\partial^2 A}{\partial t \partial r} = \frac{1}{A} \frac{\partial A}{\partial t} \frac{\partial A}{\partial r}. \tag{20.10}$$

In order to understand the significance of this result, write $A = e^Z$ for some function $Z(t,r)$, and then the equation becomes

$$\frac{\partial^2 Z}{\partial t \partial r} = 0 \tag{20.11}$$

which implies that Z must have the form $Z = \alpha(t) + \beta(r)$, and therefore A is a product. Therefore the first term in the line element has the form $-e^{\beta(r)} e^{\alpha(t)} dt^2$. We can now redefine the t coordinate such that $dt \to e^{-\alpha/2} dt$ and then we have

$$ds^2 = -A(r)dt^2 + B(r)dr^2 + r^2 d\Omega^2. \tag{20.12}$$

where we have first introduced $\bar{A} \equiv e^\beta$ and then dropped the bar since we no longer need the original function.

We still have not made an explicit statement of the signs of the functions A and B (the mathematical excursion via e^Z for A would work just as well if Z were complex-valued). However, we can now appeal to our earlier work on the Schwarzschild metric where we found an explicit formula for $A(r)$ and $B(r)$. That same derivation will apply exactly as it did before, and then we discover that t is indeed timelike for values of r larger than the Schwarzschild radius. Overall, we thus find that the assumption of spherical symmetry together with the field equation are sufficient to guarantee that the metric in vacuum takes the Schwarzschild form. This is Birkhoff's theorem. QED

In the above derivation, by 'vacuum' we mean $T_{ab} = 0$ in the field equation, where we include the cosmological constant in T_{ab}. If instead one allows the cosmological constant term by asserting $R_{ab} = \Lambda g_{ab}$, then the argument proceeds mostly unchanged, and one arrives at the de Sitter–Schwarzschild metric (18.35).

Here are some applications of Birkhoff's theorem:

1. A radially pulsating star does not emit gravitational waves. Indeed, the spacetime outside a spherically symmetric radially pulsating star is strictly Schwarzschild and shows no change whatsoever owing to such motion of the star (and note, the field equation guarantees that the star's mass parameter will not change in such motion). For example, to the extent that a supernova explosion is spherically symmetric, it will not radiate gravitationally. A similar result applies to the electromagnetic field outside a spherically symmetric sphere in classical electromagnetism.
2. If a spherically symmetric star or other finite object collapses in a spherically symmetric manner, then we can calculate the worldlines of particles in the outermost layer by using geodesics of Schwarzschild spacetime.
3. In an empty spherical region in a spherically symmetric body, spacetime is strictly flat. This is because it must take the Schwarzschild form with mass parameter $m = 0$. This is like the Newtonian result.
4. The spacetime around a spherically symmetric star that is situated at the centre of such a spherically symmetric region is Schwarzschild (or de Sitter Schwarzschild for $\Lambda \neq 0$), and this remains true even if the material outside the region is moving radially, such as for example in the cosmological expansion.

20.1.1   A note on spherically symmetric spaces

It will be well to think about what is meant by spherical symmetry in the context of Riemannian and pseudo-Riemannian geometry. From everyday experience one has a natural notion of spherical symmetry in terms of spherical objects and images in three-dimensional Euclidean space. What we need is a way to improve our intuition about the impact of Gaussian curvature on such pictures. In Volume 1 we described two useful tools: the use of the 'expansion field' idea, and the use of embedding. In the 'expansion field' picture one seeks coordinates in which the metric takes the form

$$ds^2 = g_{00}(\bar{r})dt^2 + f(\bar{r})(d\bar{r}^2 + \bar{r}^2 d\Omega^2) \tag{20.13}$$

where the bar indicates that $\bar{r} = \bar{r}(r)$ is some new coordinate adopted for this purpose. In this case the space is just like Euclidean space in terms of the coordinates $(\bar{r}, \theta, \phi)$ except that all distances are affected by the factor f which is itself a function of $\bar{r}$.

In the embedding picture, one embeds the given three-dimensional manifold in a four-dimensional Euclidean space. Now one has the task of imagining a three-dimensional object sitting in four-dimensional space—still a difficult thing to do. But the symmetry can help a lot. We take a *slice* through this space—specifically we consider the coordinate plane $\theta = \pi/2$, and then we argue that everything we discover about this plane will apply equally well to other planes through the point $r = 0$. Note that a flat slice through a four-dimensional *Euclidean* space is a perfectly well-defined idea. Note also that such a slice results in a three-dimensional object in three-dimensional Euclidean space. It is the surface of this object that concerns us. The

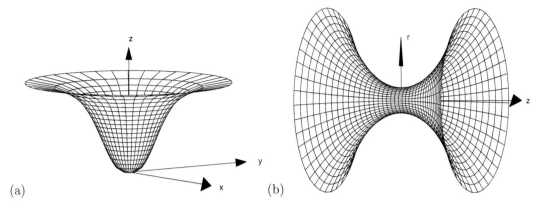

(a)   (b)

Fig. 20.1 Embedding diagrams for two spherically symmetric manifolds. Both manifolds are three-dimensional; one dimension has been suppressed, resulting in a two-dimensional curved surface which is then embedded in Euclidean space to yield the diagram. Each circle on the embedded surface corresponds to a sphere in the manifold.

surface will have axial symmetry about an axis in the direction which corresponds to the extra or fictitious dimension which was introduced purely for the mathematical purpose of embedding. One must insist, to ones own imagination and intuition, that this direction has nothing to do with the manifold itself. It is merely a way to translate properties of the manifold into a language or image that we may or may not find helpful. Fig. 20.1 shows two examples. The diagram is called *an embedding diagram with one dimension suppressed*. In such a diagram each circle on the surface about the overall axis of symmetry corresponds to a spherical surface in the full manifold. Movement from one circle to the next in the radial direction corresponds to movement from one sphere to the next. Movement around a circle corresponds to movement on the surface of a sphere.

Fig. 20.1(a) is reasonably straightforward. It can be used to find the shortest path between two points on the manifold, for example, and hence a spacelike geodesic if the situation is stationary. For the surface shown one can also see at a glance that the circumference of any given circle centred on the origin is less than 2π times the distance along the manifold from the origin. This means the surface area of the corresponding sphere is smaller than 4π times its proper radius squared. The circumference of a small circle not centred at the origin, on the other hand, is more than 2π times its radius, indicating that the local spatial curvature of the manifold is negative (c.f. (2.18)).

Fig. 20.1(b) presents a greater challenge to the imagination. What is going on at the throat of the tube? Let us see: if we pass from the right to the left part of the manifold as shown, we observe a sequence of circles whose radii first reduce, down to some minimum, and then increase again. This means that in the full manifold one can find a spacelike path which crosses a sequence of nested spherical surfaces in the inwards direction, and which continues perfectly smoothly on to a region of the manifold where now this same path is crossing spherical surface after spherical surface in the outwards direction, finally arriving at a flat region far from where it started out. Note, the manifold we are investigating is three-dimensional. Also, the path we just followed was spacelike, therefore it is not a path which any worldline can trace. More generally, *no influence whatsoever* can extend from the first to the second region of such a space unless spacetime has the given geometry for a sufficiently long period of time. It is currently unknown whether this can happen in our universe, as we shall explain in Section 20.5.

20.2   Null surfaces and event horizons

A *null surface* is a hypersurface (i.e. a three-dimensional submanifold of spacetime) whose normal 4-vector is null. We will show that such a surface is also a one-way membrane, because timelike lines can only cross it in one direction. Any light cone is an example of a null surface. After we have studied these surfaces a little, we will define the term *event horizon*. All event horizons are null surfaces, but not all null surfaces are event horizons.

Let a surface be defined by

$$f(x^0,\ x^1,\ x^2,\ x^3) = \text{const} \tag{20.14}$$

for some scalar function f. The normal to the surface is given by

$$n_a = \nabla_a f. \tag{20.15}$$

Proof: Let Δx^a be a tangent vector to the surface. This means that in the limit of small quantities, it lies in the surface and therefore is a movement between two points having the same value of f, i.e. for which $df = 0$. But

$$df = (\nabla_\mu f)\Delta x^\mu \tag{20.16}$$

so we have

$$(\nabla_\mu f)\Delta x^\mu = 0 \qquad \Rightarrow \quad n_\mu \Delta x^\mu = 0, \tag{20.17}$$

therefore n_a is orthognal to any tangent to the surface; QED.

One can equally see that any vector orthogonal to n_a is tangent to the surface, since it results in $df = 0$ in (20.16).

At every point on a null surface there is at least one null tangent vector, because the normal to the surface is itself such a vector. *Proof.* A null vector is orthogonal to itself and we already showed that all vectors orthogonal to **n** are tangent to the surface.

Now any vector orthogonal to a null vector is itself either spacelike or null. *Proof.* Let the null vector be **n** and adopt a LIF whose x-axis is aligned with the spatial part of **n**. Therefore $[n^a] = (n^0,\ n^0,\ 0,\ 0)$. Let some other general vector **v** have components $[v^a] = (u, w, a, b)$. If this is orthogonal to **n** then

$$\mathbf{n} \cdot \mathbf{v} = -n^0 u + n^0 w = 0 \qquad \Rightarrow \quad u = w. \tag{20.18}$$

Hence $[v^a] = (u,\ u,\ a,\ b)$. It follows that $\mathbf{v} \cdot \mathbf{v} = a^2 + b^2$. This is zero if $a = b = 0$ and positive otherwise. QED. Combining this with the argument of the previous paragraph, we have that, for a Lorentzian spacetime, *all vectors tangent to a null surface are either spacelike or null, and*

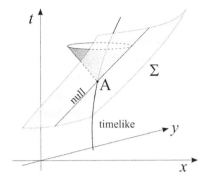

Fig. 20.2 A worldline crossing a null surface, shown in LIF coordinates near the crossing point.

there is one and only one null tangent direction at each point. For an example of this, consider a light cone.

We can now prove the one-way property of null surfaces. Suppose a worldline crosses a null surface Σ at some event A. Adopt a LIF at A, with the x axis aligned with the spatial part of the local normal to Σ (which is also a null vector tangent to Σ). Now perform a local Lorentz transformation (a boost), so as to adopt the LIF in which the direction of the worldline in question has no x component at A. We thus arrive at the situation shown in Fig. 20.2. It is obvious from the diagram that such a worldline can only cross in one direction from past to future. It also follows that the whole of the interior of the future light cone at A is on one side of the surface Σ.

Some spacetimes contain one or more **event horizons**. The concept of event horizon is defined by reference to the global structure of the spacetime:

Definition 20.1 *For an asymptotically flat spacetime, the future horizon H^+ is defined as the boundary of the causal past of future null infinity, $\mathcal{I}^+$.*

The notion indicated formally by $\mathcal{I}^+$ is defined below in the context of Penrose diagrams (see Fig. 20.4(b)), but we can explain what the definition means using more everyday language. An event horizon is a boundary (a hypersurface) between those events that are connected to infinity by timelike paths and those that are not, where by 'infinity' we mean the region of spacetime where geodesics and other lines extend indefinitely, without coming to a stop. An event horizon of an asymptotically flat spacetime is a boundary between events from which a null geodesic can reach future null infinity, and events from which no such geodesic exists. The null (and the timelike) geodesics setting out from such events are confined to stay within a region of finite spatial extent.

There are plenty of null surfaces distributed through any spacetime, not all of them holding any special interest. However, when a null surface is spatially closed with an area that remains finite into the infinite future, so that it has an 'inside' and an 'outside' that persists over time, then since timelike lines can only cross it in one direction, it must be an event horizon. Therefore *for a stationary metric a spatially closed null surface is an event horizon.*

The Schwarzschild–Droste spacetime has such a horizon; in Schwarzschild coordinates it is the surface $r = 2m$. In order to prove this, define a scalar field f whose value is everywhere equal to r, then the normal vector to a surface of constant f is

$$n_a = \nabla_a f = \partial_a f = \partial_a r = (0,\ 1,\ 0,\ 0). \tag{20.19}$$

Hence $n_\mu n^\mu = g^{\mu\nu} n_\mu n_\nu = g^{rr} = 1 - 2m/r$. This is well-defined at all $r \neq 0$, and equal to zero when $r = 2m$. This confirms that $r = 2m$ is a null surface. However, in view of the fact that the Schwarzschild coordinates break down at $r = 2m$, one might wish to repeat the calculation in another system of coordinates—see exercise 20.6.

By a similar calculation you can readily show that for the Kerr metric in Boyer-Lindquist coordinates the surfaces $r^2 + a^2 - 2mr = 0$ are null and hence they are horizons as was claimed in Chapter 19.

For dynamic spacetimes the above definition of horizon is logical and consistent, but it is hard to work with in practice since one does not generally know all about the future. Therefore for numerical work physicists often fall back on the *locally trapped surface* mentioned below in Section 20.4.2. The outwardly directed null rays of such a surface are not expanding at some particular moment. If they are not contracting either then the surface often offers an excellent approximation to a horizon.

20.2.1 Killing horizon and surface gravity

A null surface having the property that some Killing vector field is everywhere normal to it is called a **Killing horizon**. In stationary conditions there is a close relationship between event horizons and Killing horizons but in general they are distinct concepts. Hawking was able to show that the event horizon for any stationary black hole must be a Killing horizon for some Killing vector field. This is useful because Killing horizons are a small subset of null surfaces, and they are defined in terms of local information. An important property of such horizons is called the surface gravity, defined as follows. First we note that the Killing horizon has some Killing vector K^a everywhere null on it, so we can use the function $f(x^a) = K^\mu K_\mu = 0$ to define the surface, and hence the normal to the surface can be written, using (20.15),

$$n_a = \nabla_a (K^\mu K_\mu). \tag{20.20}$$

Given that the surface is a null surface, this vector n_a is null. And at events on the surface the vector K_a is also null. Could they be in fact none other than the same vector? The answer is yes, up to some multiplying factor, because $\mathbf{K} \cdot \mathbf{n} = 0$ as you can easily prove using the antisymmetry of $\nabla_a K_b$, therefore $\mathbf{K}$ is a tangent vector, but it is also null, so it is the unique null tangent vector (proved above), which is also the normal vector. Hence we have

$$\nabla_a (K^\mu K_\mu) = 2\kappa K_a \tag{20.21}$$

where κ is some multiplying factor. Using Killing's equation, this can also be written

$$K^\mu \nabla_\mu K_a = -\kappa K_a. \tag{20.22}$$

This equation defines the quantity κ which is called the **surface gravity**. (20.22) is saying that the change of K_a as one moves along its own direction is parallel to itself; in other words it is an example of the geodesic equation with a non-affine parameter, see (13.17), (13.18). In order to complete the definition of κ one must agree a normalization for K^a. This is done in an asymptotically flat spacetime by taking $K_\mu K^\mu = -c^2$ at infinity.

The surface gravity gets its name because in the case of a static spacetime, one can relate κ to the 'fishing rod equation' (3.15). κ is then equal to the force per unit mass which would be required at infinity to dangle a massive object at the horizon using a massless rope. For stationary but non-static spacetimes this interpretation no longer holds, but κ is important because it can be shown that it is constant over any continuous horizon; this is discussed further in Chapter 21.

20.2.2   Extreme black holes and the cosmic censorship hypothesis

The Reissner–Nordström metric (18.24) yields horizons at $r_\pm = m \pm \sqrt{m^2 - q^2}$, when $|q| \leq m$, as was noted in Chapter 18. The case $|q| > m$, on the other hand, has no horizons. In this case the coordinate t is always timelike and the coordinate r is always spacelike, and there exist both null and timelike lines extending from the singularity to positions infinitely far away. This situation is called a *naked singularity* and it is troubling because GR offers no guide at all as to what is going on at the singularity, but what is going on can now influence the rest of the universe, so at every event in spacetime in the future lightcone of an event at the singularity (i.e. a large part of spacetime) we have the possibility of an unknown influence coming in.

The Kerr metric (19.22) yields a naked singularity when $a > m$.

Naked singularities, if they were realized in nature, would be physical situations which required, for their correct description, a well-established theoretical framework involving quantum physics in extremes of spacetime curvature. We currently lack such a theoretical framework. Attempts to apply existing knowledge to this scenario are sometimes said to imply many sorts of wild possibilities that are hard to make sense of. However, studies of the evolution over time of ordinary matter suggest that although it can collapse and form singularities, this never results in naked singularities. For example, studies of processes by which black holes may acquire charge and angular momentum suggest that the dynamics do not allow the system to cross over from the 'hidden' to the 'naked' parameter regime. Penrose's study of this problem led him to propose:

Proposition 20.1 The cosmic censorship conjecture (Penrose)*: Naked singularities cannot form in gravitational collapse from generic, initially nonsingular states in an asymptotically flat spacetime satisfying the dominant energy condition.*

The proof of this (or its disproof) is one of the outstanding research problems in GR. Effort to construct counterexamples has not had success and the conjecture is very plausible. The word 'generic' is serving to indicate that the statement is not concerned with some special scenario with finely tuned initial conditions. If the conjecture is correct then one may reasonably extend it to the assertion that naked singularities do not occur in the natural world—all singularities which come about in the course of time are 'safely hidden' behind horizons.

The application of GR to cosmology implies that the universe itself has evolved from an extreme condition (at the Big Bang) which classical GR cannot adequately describe. This is also commonly referred to as 'a singularity' in the absence of further understanding. This will be discussed further in Chapters 22 and 26.

20.3   The Schwarzschild horizon

Let us restate, for convenience, the Schwarzschild–Droste metric as given in (17.15):

$$ds^2 = -\left(1 - \frac{2m}{r}\right)c^2 dt^2 + \frac{dr^2}{1 - 2m/r} + r^2\left(d\theta^2 + \sin^2\theta\, d\phi^2\right). \tag{20.23}$$

For this metric clearly the coordinate location $r = 2m$ is significant, and we established above that it is the location of a horizon. If the gravitating body is smaller that this radius then it must collapse to a black hole because for $r < 2m$ the future light cones are all directed 'inwards', i.e. towards reducing r.

The Schwarzschild coordinates describe spacetime for $r > 2m$ in a straightforward manner. They do not on their own clearly indicate what significance one should assign to the region with $r \leq 2m$. However, once we develop other coordinates that do not have irregular behaviour at $r = 2m$, we can acknowledge that the region $r \leq 2m$ is smoothly connected to the rest of spacetime, and the Schwarzschild metric still satisfies the field equation there.

In order to train our intuition about the situation at $r < 2m$ it may be useful to introduce the coordinate relabelling $\bar{t} \equiv -r/c$, $\bar{r} \equiv ct$, and hence write the metric

$$ds^2 = -\frac{1}{-2m/c\bar{t} - 1}c^2 d\bar{t}^2 + \left(\frac{2m}{-c\bar{t}} - 1\right)d\bar{r}^2 + c^2\bar{t}^2\left(d\theta^2 + \sin^2\theta\, d\phi^2\right). \tag{20.24}$$

The horizon is at $\bar{t} = -2m/c$; the singularity is at $\bar{t} = 0$ and allowed values of $\bar{t}$ are negative. The idea of this relabelling is merely that after the horizon, $\bar{t}$ is a timelike coordinate and $\bar{r}$ is a spacelike coordinate, so we can see at a glance the thoroughly non-static form of the metric in this region. Also, it is clear that the inward-going worldlines shown in Fig. 17.2 are indeed going forwards in time as they approach the singularity.

Returning now to the region outside the horizon, one can notice the following effects associated with it. Let r be the location of some event outside the horizon. Then, as $r \to 2m$:

1. The gravitational redshift of light emitted at r and observed by a distant observer tends to infinity: eqns (17.56), (17.73).
2. The force that would be required to maintain one's location fixed at r tends to infinity: Section 11.2 of Volume 1 and eqn (20.26) below.
3. The speed of an infalling massive particle, as observed by an observer fixed at r, tends to the speed of light: eqn (20.29).
4. The tidal stress increases smoothly and does not diverge: (17.70).
5. The proper time of an infalling massive particle does not diverge: (17.28).
6. The Schwarzschild coordinate time on the worldline of an infalling massive particle does diverge: (17.30).
7. Spacetime itself is smooth and in this sense 'ordinary' at the horizon (c.f. (17.72), (20.34), (20.49)), but is not well charted by Schwarzschild coordinates there.

To calculate item 2, let us note that the 4-velocity of an observer at fixed r is $(u^0, 0, 0, 0)$ with $u_\mu u^\mu = -c^2$ so $(u^0)^2 g_{00} = -c^2$ hence $u^0 = (1 - 2m/r)^{-1/2}$. Therefore the 4-acceleration of such an observer is

$$a^b = \frac{Du^b}{d\tau} = \frac{du^b}{d\tau} + \Gamma^b_{\mu\nu} u^\mu u^\nu = 0 + \Gamma^b_{00}(1 - 2m/r)^{-1} = (0,\ mc^2/r^2,\ 0,\ 0) \qquad (20.25)$$

where we used (17.16) for the connection coefficients. This expression gives the contravariant components of the 4-acceleration $\mathbf{a}$ in Schwarzschild coordinates and these quantities do not diverge at $r = 2m$. However, at fixed t, any given change dr corresponds to a ruler distance $dl = \sqrt{g_{rr}} dr = (1 - 2m/r)^{-1/2} dr$ and this does diverge as $r \to 2m$. Indeed if we examine the magnitude of the 4-acceleration we obtain

$$(\mathbf{a} \cdot \mathbf{a})^{1/2} = (g_{\mu\nu} a^\mu a^\nu)^{1/2} = \frac{mc^2}{r^2} \frac{1}{\sqrt{1 - 2m/r}}. \qquad (20.26)$$

To calculate item 3, let us adopt a tetrad suited to an observer at a fixed r, as given by (17.68), in order to discover how the velocity $\mathbf{v}$ of an infalling particle would appear to such an observer. The covariant components of the observed 4-vector are given by (11.18):

$$v'_a = \hat{e}^\lambda_{(a)} v_\lambda = \left(\frac{1}{c\sqrt{1 - 2m/r}} v_0,\ \sqrt{1 - 2m/r}\, v_1,\ 0,\ 0 \right) \qquad (20.27)$$

$$= \left(-c\sqrt{1 - 2m/r}\, \dot{t},\ \frac{1}{\sqrt{1 - 2m/r}} \dot{r},\ 0,\ 0 \right) \qquad (20.28)$$

where we used $v^a = (\dot{t}, \dot{r}, 0, 0)$ and $v_a = g_{a\lambda} v^\lambda = (g_{00}\dot{t}, g_{11}\dot{r}, 0, 0)$. The components v'_a are those observed in a LIF, so they have the form $(-\gamma c, \gamma \vec{v})$ as in Special Relativity, where γ is the Lorentz factor and $\vec{v}$ is the observed 3-velocity. Therefore by taking the ratio of the spatial and temporal parts we can factor out γ and obtain $\vec{v}$. It is in the radial direction, with radial component equal to

$$(\vec{v})_r = \frac{1}{1 - 2m/r}\frac{\dot{r}}{\dot{t}} = -(c/E)\sqrt{E^2 - c^2(1 - 2m/r)} \tag{20.29}$$

using (17.24),(17.25) (and see exercise 20.10 for an alternative derivation). The sign shows it is directed inward, the magnitude tends to the speed of light as $r \to 2m$ (and an infalling observer would agree that the horizon moves past him at the speed of light, relative to his LIF). In the case of a particle falling from rest at infinity one has $E = c$ and then $(\vec{v})_r = -c\sqrt{2m/r}$, $\gamma = (1 - 2m/r)^{-1/2}$.

Watching a particle fall in. Now let us take the point of view of a distant observer watching what happens as a particle falls towards the horizon. We suppose the particle emits a sequence of light pulses at events (t, r) and these pulses are received at events (t_R, r_R) where r_R is constant. Using the equation of an outgoing null geodesic, (17.36), we find

$$c(t_R - t) = r_R - r - 2m\ln(r/2m - 1) + \text{const} \tag{20.30}$$

where the constant depends on m and r_R. We would like to solve this equation so as to obtain r as a function of t_R. To eliminate t, we use the equation for the particle's worldline, (17.30), which for r very close to $2m$ has the form

$$ct \simeq -2m\ln(r/2m - 1) + \text{const} \tag{20.31}$$

After substituting this into (20.30), and neglecting $(r/2m - 1)$ in comparison to $\ln(r/2m - 1)$, one obtains

$$r \simeq 2m + (\text{const})e^{-ct_R/4m}. \tag{20.32}$$

This result tells us that the observer at r_R sees the location of the falling particle approach the horizon rapidly: the received light comes from a location whose coordinate distance from the horizon falls exponentially with a time constant $4m/c = 4GM/c^3$. This is a very fast approach (approximately 20 microseconds for one solar mass). There are many processes involving exponential decay in physics (spontaneous emission from atoms, thermal relaxation, etc.) and we do not normally say that such processes never finish; we say that the atom has decayed, or the system has relaxed, once the elapsed time is large compared to the relevant time constant. In a similar way, the above calculation shows that we should not imagine that particles falling into a black hole linger on their journey to the horizon. As far as observable information is concerned, the falling matter disappears into blackness in a fraction of a blink of an eye. The frequency and intensity of the received light also fall exponentially owing to the gravitational redshift and related effects—exercise 20.12.

20.4 Black hole formation

20.4.1 Advanced Eddington–Finkelstein coordinates

In view of the fact that Schwarzschild coordinates are problematic at $r = 2m$, it is useful to seek a different coordinate chart in which there is no singular behaviour of the coordinates at the horizon. An example is given by the Lemaître coordinates which lead to the line element (17.72). Eddington and Finkelstein proposed that another good way to proceed is to note the equation of a null radial worldline, (17.36), and pick a coordinate p which is arranged to be constant all the way along such a worldline. In other words it is the constant of integration involved in obtaining (17.36) and it can be written

$$p = ct + r + 2m \ln |r/2m - 1|. \tag{20.33}$$

Therefore $dp = cdt + (1 - 2m/r)^{-1}dr$ and by substituting this into (20.23) we find the line element is

$$ds^2 = -(1 - 2m/r)dp^2 + 2dpdr + r^2d\Omega^2. \tag{20.34}$$

This is regular at $r = 2m$ so we have successfully removed the coordinate singularity. Note that the mathematical route to (20.34) does not have to go via (20.23) and (20.33). One can propose (20.34) as an ansatz and then confirm that it satisfies the Einstein field equation in vacuum,[1] without reference to the Schwarzschild coordinates, and therefore without encountering the singularity in the relationship between p and r at $r = 2m$ which is expressed in (20.33). This shows that there is no question about whether some irregular or singular behaviour at the horizon has been ignored or overlooked (c.f. exercise 20.1). Eddington–Finkelstein coordinates, and also other possible choices such as Lemaître coordinates, show that spacetime is perfectly well-behaved at the horizon, as are all the geodesics which cross it. One can also define perfectly well-behaved LIFs at the horizon.

The coordinate p is called a *null coordinate* because $ds = 0$ for events separated only in p. Since we are more familiar with a timelike coordinate, one now introduces

$$t' \equiv (p - r)/c \tag{20.35}$$

and then the line element is

$$ds^2 = -c^2(1 - 2m/r)dt'^2 + (4m/r)cdt'dr + (1 + 2m/r)dr^2 + r^2d\Omega^2. \tag{20.36}$$

The coordinates either (p, r, θ, ϕ) or (t', r, θ, ϕ) are called *advanced Eddington–Finkelstein coordinates*.

In these coordinates the equation for a radial infalling null geodesic is straightforward: it is simply $p =$ constant, and therefore $ct' = -r+$ constant. For an outgoing null geodesic, one finds from (17.36) (or by writing down and solving a differential equation in the new coordinates):

$$ct' = r + 4m \ln |(r/2m) - 1| + \text{constant}. \tag{20.37}$$

This leads to the lightcone structure shown in Fig. 20.3(a). This diagram is more useful than Fig. 17.2 for understanding the behaviour around the horizon, and it also does a good job for

---

[1] Indeed, one can prove this without doing any more work, because if a real analytic function satisfies the Einstein field equation throughout some open set of events then, by standard methods in differential analysis, it must satisfy the equation everywhere. But we already know that (20.34) satisfies the field equation for $r > 2m$. Therefore it must also satisfy it for $0 < r \leq 2m$.

the rest of spacetime. Notice that the lightcones tip over in such a way that no future-going timelike worldline crosses the horizon in the outward direction.

20.4.2 Gravitational collapse

When the density of a star is high, the forces required to maintain the star in hydrostatic equilibrium against its own gravitation become very large. They become infinitely large as the radius of the star approaches the Buchdahl limit. Notwithstanding this, one may argue that the spherically symmetric case is only ever an approximation—perhaps the angular momentum of real stars somehow allows them to avoid the regime where they are overwhelmed by their own gravity? In a series of important theorems, Penrose showed that even in more realistic situations a *trapped surface* (also called locally trapped surface) must form and that there must be a singularity inside it. That is,

Theorem 20.1 *In situations where the strong energy condition holds, if there is a trapped surface then, for generic metrics, there must be either a closed timelike curve or a singularity (manifested by incomplete timelike or null geodesics).*

A generic metric is one for which every timelike geodesic has at least one event at which $R_{\mu b c \nu} t^\mu t^\nu \neq 0$ where t^μ is the tangent vector, and a further condition[2] holds for null geodesics. These conditions merely exclude some very unusual metrics which happen to have vanishing curvature everywhere along some geodesic. A trapped surface is a closed spacelike surface whose area decreases as the parts of the surface are transported in any future timelike direction. Such a surface must be inside an event horizon. Penrose's result establishes that black hole formation is possible in a wide range of scenarios, so the spherically symmetric case is not special in this regard. Therefore the Schwarzschild metric is not just a mathematical curiosity: we can use it to get many correct insights into black holes in general.

One useful observation is Birkhoff's theorem. Outside a spherically symmetric object (which we may as well call a star) spacetime is Schwarzschild, so the worldline of any particle in freefall at or just outside the outermost layer of the star is given by the equations we developed in Section 17.2.1. By using those results we obtain the picture shown in Fig. 20.3(a) in Eddington–Finkelstein coordinates.

Inside a homogeneous and isotropic collapsing star, the metric takes the FLRW form described in Chapter 22.

A distant observer receiving light emitted during the collapse receives the light red-shifted and dimmed. The last light received with non-negligible dimness is not that emitted near the horizon, but that emitted somewhat earlier into a just-escaping trajectory near the light sphere; such light first orbits one or more times then escapes to infinity with a redshift of order 2.

[2] $t_{[a} R_{b]\mu\nu[d} t_{e]} t^\mu t^\nu \neq 0.$

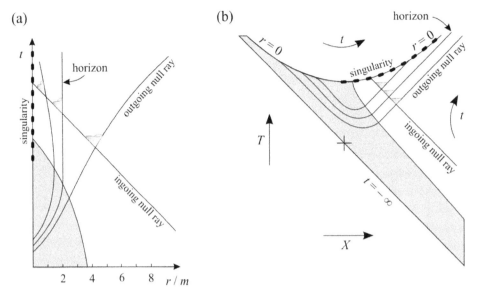

Fig. 20.3 Gravitational collapse. The figure shows a spacetime diagram for matter (shaded) collapsing and bringing into being a Schwarzschild black hole, first (a) in Eddington–Finkelstein coordinates, then (b) in Kruskal–Szekeres coordinates. The geometry *outside* the matter is Schwarzschild. The geometry in the shaded region is of some other form and has no singularity at the start. The outer layer of the collapsing material follows a Schwarzschild geodesic. Some null geodesics are also shown. The one which just fails to escape to infinity marks the growth of the horizon (Section 20.4.3). The reader should peruse (b) while reading Section 20.5; it is included here for comparison but the coordinates are unusual; they are well-matched to the Schwarzschild geometry but their relation to spacetime in the shaded region can be chosen in more than one way; the diagram is merely indicative in this region.

20.4.3   The growth of a horizon

When a black hole forms by gravitational collapse, a horizon comes about where previously there was none. How is the horizon born? Does it spring into being at the Schwarzschild radius, or grow from a point, or come about some other way?

This question is answered by considering the null geodesics during and after the gravitational collapse. For the sake of simplicity, let us consider the case where the collapse is spherically symmetric. Then, setting out from the point (in Schwarzschild coordinates) where eventually a singularity will form, there are, during the collapse, various null rays, some of which are drawn in Fig. 20.3. The earlier of these rays will escape to infinity; the later ones will not. Somewhere there is a ray which just manages to get to the eventual Schwarzschild horizon and no further. This ray, and the light cone of which it is a part, bounds a finite region within which there is no escape, therefore it marks the location of the horizon both during and after the formation of the black hole. We thus infer that the horizon grows from a point.

This argument illustrates that the property of being a horizon is not a local property. It is a statement about the whole future history of the given spacetime. Indeed, if the black hole later acquires some more mass and thus grows its horizon, then it emerges that the light ray we just considered was not the true horizon after all, but lay just inside it. For a Schwarzschild black hole whose mass is m_1 for a while and then m_2 later on and into the infinite future, the true horizon is set by the ray which asymptotically approaches $r = 2m_2$ after the mass has become m_2.

20.4.4 Causality structure and the Penrose diagram

We noted in Section 20.3 that there is an infinite elapsed Schwarzschild coordinate time for geodesics approaching the Schwarzschild horizon. The notion of simultaneity that is expressed by the Schwarzschild coordinates struggles to deal with the situation at the horizon. The coordinates are in effect being asked to provide an answer to the question, 'which tick of some clock located outside the horizon is simultaneous with the tick of an infalling clock as it crosses the horizon?' Simultaneity is relative, so there is no absolute answer to such a question; the answer provided by Schwarzschild coordinates is 'it is the tick at $t = +\infty$', which is one legitimate answer and is consistent with the fact that the outside observer never receives any information from within the horizon. However, we are at liberty to adopt other coordinate frames and then we may assert that objects reach and cross the horizon in a finite time. The *Penrose diagram*, also known as a Penrose–Carter diagram or conformal diagram, is a useful way to capture the global causality structure of spacetime and answer questions of this type.

The idea of the Penrose diagram is to introduce a combination of coordinate and conformal transformations using functions such as $\tanh^{-1}$ which can map an infinite to a finite range, and thus allow a diagram in which an infinitely extended spacetime can be mapped to a finite diagram such as Fig. 20.4(a). The use of a conformal transformation implies that the diagram shows a manifold different to the one we started with, but having the same causal structure.

Typically we suppress two spatial dimensions and thus obtain a two-dimensional diagram, and we always choose the transformation such that light has coordinate speed 1. In this way lines at 45 degrees on the diagram are null, steep lines are timelike and shallow ones are spacelike. Timelike geodesics running back into the infinite past are squashed together by the conformal transformation, such that they all converge at one coordinate location called i^-; in the infinite future they arrive at i^+. Spacelike curves run off to spatial infinity and thus all arrive at coordinate location i^0. Null lines are bounded by $\mathcal{I}^-$ and $\mathcal{I}^+$ called *null infinity*. There can be non-geodesic timelike curves which begin or end at null infinity.

When the gravitational wave signal of a black hole merger was first reported, I asked myself the question, 'but if the black holes have not even formed yet, then how can they merge?' This thought is based on the treatment of gravitational collapse using Schwarzschild coordinates—the collapse takes an infinite Schwarzschild time to reach and pass the horizon. The Penrose diagram shown in Fig. 20.4(b) can resolve questions of this kind. The diagram shows a spacetime with a single Schwarzschild black hole in it. The shaded region shows matter collapsing to form the

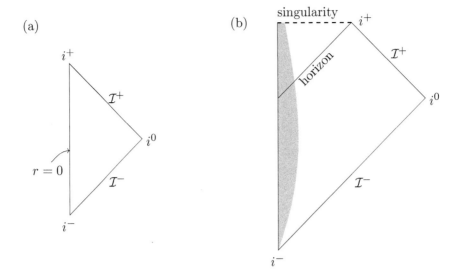

Fig. 20.4 (a) Penrose diagram of Minkowski spacetime. Polar coordinates have been adopted, and θ, ϕ suppressed. Hence each point on the diagram is a representative point on a spherical surface. The origin of coordinates (where $r = 0$) is not a special place in spacetime; worldlines arriving there continue to $r \geq 0$. Worldlines arriving at any of the other boundaries have reached infinity and end there. (b) Penrose diagram of a spacetime with a Schwarzschild black hole formed by gravitational collapse. Timelike geodesics finish either at i^+ or at the singularity.

hole; the horizon (null) and the singularity are also shown. Since the horizon is a null surface (Section 20.2) we have no choice but to show it as a line at 45 degrees. The Schwarzschild singularity is spacelike, so we can show it as a line at any shallower slope; we may as well use a horizontal line.

Now we can imagine ourselves situated at some event outside the horizon, and asking the question, 'has the horizon formed yet?' To answer this, first note whether or not you are in the past light cone of the event where the horizon first appears. If you are in the past light cone, then that event has not happened yet, so the answer to the question is 'no'. If you are not in the past light cone, but you are outside the horizon, then you are not in the future light cone either, so you are spacelike-separated from the event of the horizon starting to form. Therefore you can answer either 'yes, it has formed' or 'no it has not' depending on your choice of which spatial surface to call simultaneous. Similar statements apply to the formation of the singularity.

The answer to the puzzle about the black hole merger is that we should think of a black hole as a *spacetime* phenomenon—a region of spacetime. Two such regions can have worldlines which approach and circle around one another, till their horizons coalesce. To calculate what happens is fiendishly difficult, but the theory makes sense and in principle the calculation could be done. When we then ask about directly observable phenomena, such as signals arriving at a given observer, then we get answers with sensible timescales, as (20.32) illustrates.

The Penrose diagram and the Kruskal–Szekeres diagram (Fig. 20.3(b)) can also serve as an aid to our understanding of the Schwarzschild singularity. These diagrams help us to see clearly that the Schwarzschild singularity is not a point in space but a moment in time; or, more precisely, a spacelike edge to spacetime. Events near this edge lie along a line of infinite length, and yet it all fits inside the horizon! This is possible because the region after the horizon is itself like a future light cone. Timelike geodesic bundles are stretched and thinned like spaghetti as they approach the singularity. One should not necessarily expect that spacetime really tends to infinite curvature inside black holes; more likely is that the truth of the physical world is not captured in full by GR, and something else is going on, which would require quantum physics for its correct description.

20.5 Kruskal–Szekeres spacetime

We shall now introduce a series of coordinate transformations in order to obtain further insight into the Schwarzschild geometry. We have already described the advanced Eddington–Finkelstein coordinate p introduced in (20.33). Next we introduce a null coordinate q (called retarded Eddington–Finkelstein coordinate) which is defined so that it is constant along each *outgoing* null geodesic:

$$q = ct - r - 2m \ln |r/2m - 1|. \tag{20.38}$$

Then in terms of p, q, θ, ϕ the line element is

$$ds^2 = -(1 - 2m/r)\mathrm{d}p\mathrm{d}q + r^2 \mathrm{d}\Omega^2 \tag{20.39}$$

where r should be understood to be a function of p and q. Now, we prefer to have coordinates in which no part of the metric changes sign as r varies. To this end, introduce (suggests Kruskal):

$$\tilde{p} = e^{p/4m}, \qquad \tilde{q} = e^{-q/4m} \tag{20.40}$$

which does the job admirably:

$$ds^2 = (32m^3)(r)e^{-r/2m}\mathrm{d}\tilde{p}\mathrm{d}\tilde{q} + r^2\mathrm{d}\Omega^2. \tag{20.41}$$

Finally, take a timelike and a spacelike combination:

$$
\begin{array}{ll}
\text{for } r > 2m : & \text{for } r < 2m : \\
T \equiv \frac{1}{2}(\tilde{p} - \tilde{q}) & T \equiv \frac{1}{2}(\tilde{p} + \tilde{q}) \\
X \equiv \frac{1}{2}(\tilde{p} + \tilde{q}) & X \equiv \frac{1}{2}(\tilde{p} - \tilde{q})
\end{array}
\tag{20.42}
$$

and we arrive at the Schwarzschild geometry expressed in *Kruskal–Szekeres coordinates*:

$$ds^2 = \frac{32m^3}{r}e^{-r/2m}(-\mathrm{d}T^2 + \mathrm{d}X^2) + r^2\mathrm{d}\Omega^2, \tag{20.43}$$

where r is defined implicitly by

$$X^2 - T^2 = (r/2m - 1)e^{r/2m}. \tag{20.44}$$

You can show (exercise) that

$$ct = \begin{cases} 4m \tanh^{-1} T/X & r > 2m \\ 4m \tanh^{-1} X/T & r < 2m \end{cases} \tag{20.45}$$

and the relationship between X, T and Schwarzschild coordinates x, t is

for $r > 2m$:
$$T = \sqrt{r/2m - 1}\, e^{r/4m} \sinh(ct/4m)$$
$$X = \sqrt{r/2m - 1}\, e^{r/4m} \cosh(ct/4m)$$
for $r < 2m$:
$$T = \sqrt{1 - r/2m}\, e^{r/4m} \cosh(ct/4m)$$
$$X = \sqrt{1 - r/2m}\, e^{r/4m} \sinh(ct/4m). \tag{20.46}$$

In terms of these new coordinates, the gravitational collapse shown in Fig. 20.3(a) takes the form shown in 20.3(b). The metric (20.43) applies to the spacetime *outside* the collapsing body, but not to region of spacetime occupied by the collapsing matter which is shown shaded on diagram. In the shaded region some other metric applies and then lines of constant r are timelike, which is not true in the vacuum region for $r < 2m$. (One could also choose to adopt a different relation between (T, X) and (t, r) in this region, and then the diagram would look different). The collapsing matter falls down to $r = 0$. The diagram shows some outward-going null rays. One starts out before the horizon and escapes; the next is the horizon; the last starts out after the horizon and hits the singularity. The location $r = 0$ is a hyperbola in Kruskal coordinates.

Although Kruskal coordinates can always be adopted, no matter what form the metric may have, they are not a convenient a choice for the purpose of describing the collapse itself. However, outside the collapsing matter, where spacetime has the Schwarzschild geometry, Kruskal coordinates are convenient because null radial rays are straight in these coordinates, and there is no coordinate singularity at the horizon. Hence light cones in the vacuum region of the diagram are all at 45 degrees, and in this region all horizontal lines are spacelike, all vertical lines are timelike (in this respect the diagram is like a Penrose diagram, but in other respects it is not).

Note that the line $T = -X$ (which corresponds to $t = -\infty$) is one of the boundaries of spacetime in Fig. 20.3(b). This results in a spacetime diagram with a triangular shape, as shown.

We shall now discuss an aspect of the spacetime described by the Kruskal line element that has much mathematical interest, but whose application to physics is not yet clear. The idea is to eliminate the collapsing matter and suppose that the geometry described by the Kruskal line element (20.43) might apply to the whole coordinate plane $-\infty < X < \infty$, $-\infty < T < \infty$.

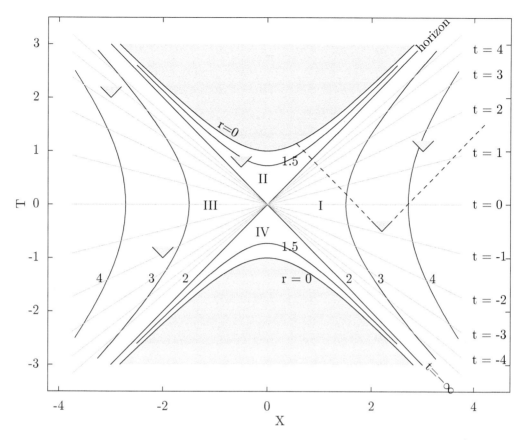

Fig. 20.5 Kruskal diagram, showing the X, T plane of the manifold with metric (20.43). The manifold does not extend into the shaded regions; these are bounded by a pair of singularities, both at $r = 0$. The hyperbolas are lines of constant r, plotted for $r = 0, 1.5, 2, 3, 4$ in units where $m = c = 1$. The horizon is at $r = 2$ and appears as a pair of straight lines. The ordinary Schwarzschild horizon runs from $(X, T) = (0, 0)$ to $(+\infty, +\infty)$. The line from $(0, 0)$ to $(+\infty, -\infty)$ is a different type of horizon, which can only be crossed in the outward rather than inward direction. Region I is outside both these horizons; region II is inside the Schwarzschild horizon. Regions III and IV are further parts of the manifold. The Schwarzschild t coordinate runs from $-\infty$ at the lower part of the horizon to $+\infty$ at the upper part. The shaded triangles are future-directed light cones. The dashed lines show an ingoing and outgoing light ray.

Under this supposition we arrive at the diagram shown in Fig. 20.5. In order to understand what the coordinates, line element and diagram are telling us, care is needed. First note that lines of constant r are described by (20.44) and appear as hyperbolas on the diagram. For this reason one must *not* imagine that the diagram is a cross-section through a three-dimensional diagram that could be obtained by rotating the whole diagram about the T axis. The Kruskal diagram is strictly two-dimensional. This is in contrast to the Eddington–Finkelstein diagram where lines of constant r are vertical so in those coordinates it is possible to proceed to a three-dimensional diagram in which ϕ is shown explicitly, simply by rotating the diagram about the line $r = 0$. In the Kruskal diagram this avenue is not available. One must simply take the plane diagram and then note that each point in the diagram corresponds to the whole of a spherical surface in the four-dimensional manifold.

Next, note that the origin of X, T coordinates is *not* at $r = 0$ but at $r = 2m$, that is, at the horizon (see (20.44)). In X, T coordinates the singularity at $r = 0$ lies on two spacelike hyperbolas at the top and bottom of the diagram.

Lines of constant t appear as straight lines through the origin. The line with a negative 45 degree slope corresponds to $t = -\infty$; this overlaps part of the hyperbola at $r = 2$ (this hyperbola is in an extreme condition where it takes the form of two straight lines). The range $-\infty < t < \infty$ fills up region I and also extends to region III which we will describe in a moment. The line at positive 45 degrees corresponds to $t = \infty$ and overlaps the horizon.

The horizon $r = 2m$ lies at $X^2 - T^2 = 0$, for which there are two solutions: $T = \pm X$. This divides the diagram into four regions, numbered I, II, III, IV on Fig. 20.5. Regions I and II are the familiar outer and inner regions we already encountered in the previous discussion of Schwarzschild spacetime. All the light cones in region II have their tops sealed by the singularity, indicating that everything that passes into region II must arrive at the singularity. The horizon is shown to be the null surface that it is: it is 'zooming outwards' at the speed of light. This was hinted at but somewhat obscured in Schwarzschild coordinates. In consequence, nothing can pass from region II to region I.

How are regions III and IV to be interpreted? Region IV is a region into which it is impossible to enter, and out of which every timelike line must pass as time goes on. This is the opposite behaviour to that of a black hole. It is called a white hole. Region III is a region much like region I, but where is it? How can one go there? The answers are *it is located at a different part of the manifold* and *one cannot travel to there starting from anywhere in regions I and II.*

The movement from Schwarzschild coordinates to Kruskal coordinates is called an *extension* because the Kruskal coordinates, with the associated line element, chart a manifold that matches the ordinary Schwarzschild regions I and II exactly, and they also extend smoothly to further regions, III and IV, showing that the original manifold can be considered to be part of a larger whole. This is like the movement from a coordinate system that only charts one hemisphere of a spherical surface, to one that charts the whole surface. The Kruskal extension is furthermore a *maximal* extension because there are no further possibilities: every geodesic either extends to infinity or finishes at a singularity, and there are no coordinate singularities that are not also at a singularity of the manifold itself.

Formally, a spacetime is said to be *extendible* if it is isometric to a proper subset of another spacetime.

We have, then, an interesting manifold that satisfies Einstein's field equation in vacuum and possesses spherical symmetry. So far we have learned about its causality structure by examining the light cones. However, a diagram of the type of Fig. 20.5 does not show what the curvature of the manifold is doing—one could be forgiven for thinking one was looking at a flat space! In order to understand the geometry, it is helpful to consider some further diagrams.

20.5.1 Einstein–Rosen bridge

We begin by considering the spacelike hypersurface at $T = 0$. Let us take the coordinate plane $T = 0$, $\theta = \pi/2$, which is two-dimensional, and then embed this two-dimensional region in a three-dimensional Euclidean space. The result is shown in Fig. 20.1(b). This is Flamm's paraboloid (Volume 1 Section 10.3) which has the equation

$$z^2 = 8m(r - 2m) \tag{20.47}$$

where z is the extra dimension introduced for embedding purposes. Here we allow both solutions for z at any given r. Region I of the manifold lies on one side of the 'tube', region III on the other. This tube is so visually appealing that one immediately begins to imagine marvellous possibilities, such as travelling through the tube, but as we already warned in Section 20.1.1, this will not be possible for a spacelike tube unless it persists for a sufficiently long period of time. But we already know from the lightcone structure (Fig. 20.5) that it does not: there are no timelike lines passing between regions I and III.

In order to understand this, one may examine a few more spacelike surfaces before and after $T = 0$ (it does not matter precisely which), and one finds that the generic behaviour is as indicated in Fig. 20.6. The tube, which is called an *Einstein–Rosen bridge* or more commonly a *wormhole*, is not present on spacelike surfaces which cross the T axis before $T = -1$ or after $T = 1$. The spacetime is dynamic; the tube opens briefly and then closes again before there is time for anything, including light, to pass through. This suggests that regions I and III have strictly no influence on one another. However, in calculations of particle interactions in quantum field theory, spacelike paths are included in the integration over all possibilities. This suggests that quantum physics near a horizon might be sensitive to the presence of region III if it is there. This is an open question.

That is all we shall say here about such wormholes, except to finish with a comment on the possible relevance or otherwise to physics. The first thing to note is that whereas gravitational

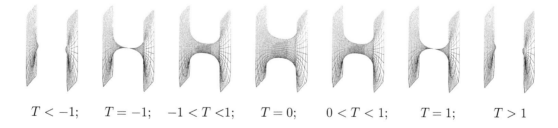

$T < -1$; $T = -1$; $-1 < T < 1$; $T = 0$; $0 < T < 1$; $T = 1$; $T > 1$

Fig. 20.6 A sequence of spacelike sections through the extended Schwarzschild manifold (embedded), showing the time evolution of the Einstein–Rosen bridge.

collapse can give rise to a black hole, there is no known process whereby a white hole, and the associated wormhole, can come to exist in a region of spacetime which did not already contain one. The mathematical tool of *analytic extension* which gave us the Kruskal geometry is a form of extrapolation into the past. From a physical point of view, it might be compared, for example, with observing the falling of an apple and on this basis extrapolating back into the past and 'discovering' that the apple was originally moving upwards. Such an extrapolation ignores the fact that at a finite moment in the past, another body (the apple tree) was present and influencing the apple. In the case of a black hole which has been formed by gravitational collapse, the star or other material lies in the past, in a region of spacetime not described by the Schwarzschild metric. For this reason it is not clear whether the wormhole is a mathematical curiosity only, or a physically possible thing. This is somewhat like the consideration of tachyons or of impossible things such as negative mass in Newtonian gravity: one can study the mathematics of such things but that does not in itself guarantee that one is exploring that part of mathematics which applies directly to the physical world.

Analytic extension can also be applied to Kerr and Reissner–Nordström geometries, where new possibilities arise. Again, the method is mathematically consistent and yields vacuum solutions to the field equation, but if it involves extrapolation into a past situation different from the one that actually happened then its physical relevance is questionable. In the Kerr case, extension of a timelike worldline through the coordinate location $r = 0$ (the interior of the ring singularity) emerges into a new asymptotically flat region of the manifold, different from the starting one. Its metric is Kerr with $r < 0$ and it has no horizons. There are then *closed timelike curves*, a situation which it is hard to make sense of physically, as we discussed in Section 13.7.

The Reissner–Nordström extension leads to a white hole somewhat like the Schwarzschild case, but now the singularity lies on a timelike line, and a wormhole-like structure persists, such that it can be traversed by timelike lines. Once again, it does not follow that such structures in spacetime have been created in our universe, and classical (in the sense of not quantum) dynamics suggest they have not. Furthermore, the metric giving rise to the wormhole region is unstable and liable to change dramatically when perturbed by matter moving in this region; ordinary matter is liable to close the very wormhole it was 'trying' to enter. Hence stability considerations make it questionable whether travel through such wormholes is possible. However, quantum mechanics opens up sufficient richness that these exotic possibilities ought not to be ignored. Research in the combination of quantum theory with gravitational theory is now suggesting beautiful connections between the structure of spacetime and the notion of entanglement in quantum fields. These studies sometimes suggest that traversable wormholes are not as implausible as they are in purely classical models.

20.6   Astronomical evidence for black holes

Ever since black holes were theoretically suggested, the hunt was on to find one by astronomical observation. (Straumann, 1989; Begelman, 2003) The presence of a black hole is inferred from its gravitational effects. There are four main lines of evidence: deducing the mass and size of a compact object from the motion of other nearby bodies; gravitational lensing; gravitational waves; X ray and other sources. The latter are important because if there is matter in orbit around a black hole then one expects to find an accretion disc with both a high temperature and a high orbital velocity; the resulting X ray (and other) emission then has a spectrum which encodes information about the velocity distribution in the disc; this can be compared with modelling of the orbits (Section 17.3.2). The emission may also exhibit temporal fluctuations which convey a lot of further information.

The evidence that various individual candidates are indeed black holes is now so strong as to be almost indubitable. It follows that there is a reasonable probability that whole classes of astronomical object are in fact black holes.

An example of an almost certain black hole is an X-ray source called Cygnus X-1 (the first strong black hole candidate to be discovered). This is a member of a binary star system lying in the Cygnus constellation (Fig. 20.7) in our galaxy, approximately 6070 light-years away. It is one of the brightest X-ray sources in the sky. The companion star is a blue supergiant; the orbit has a period 5.6 days and semi-major axis 0.2 AU. This data suffices to establish that the mass of Cygnus X-1 is above about 5 $M_\odot$, and multiple further studies place the mass at approximately 15 $M_\odot$. There are fluctuations in the X-ray emission on a timescale of milliseconds, implying that the emitting region is of the order of 300 km radius. An ordinary star of this mass would be much larger and would emit strongly in the visible region. The only way we know for an object to be this compact is for it to be a neutron star or a black hole, but the mass is far above the Tolman–Oppenheimer–Volkoff limit. The evidence is consistent with a black hole and inconsistent with every other known type of astronomical object. There is also indirect evidence that this object does not possess a surface onto which other material can fall and stick, because if it did then the falling material would glow strongly at longer wavelengths as it struck the surface, which is not observed.

A few tens of broadly similar binary systems are known. The masses of these black holes are a few to a few tens of $M_\odot$; they are known collectively as 'stellar mass' black holes. Further evidence for black holes in this mass range comes from microlensing events, and from gravity wave signals. The shape of the latter can be used to constrain both the mass and time/distance scales of the process giving rise to them (Section 7.5.2). At the time of writing, (2019) about 10 such signals are almost certainly from binary black hole mergers. The rate of these mergers implies a large overall population of such black holes.

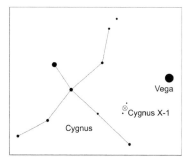

Fig. 20.7 Location of the black hole Cygnus X-1.

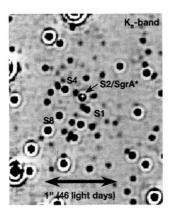

Fig. 20.8 Image of the galactic centre taken on UT4 (Yepun) of the VLT at the European Southern Observatory, after a deconvolution algorithm has been applied, from (Schödel *et al.*, 2002). The rings are artefacts of this algorithm. The orbit of S2 presents a semimajor axis of 0.12 seconds of arc, which is similar to the size of the overexposed black blobs (stars) in this image.

Intermediate mass black holes have mass in the range 100–$10^5 \, M_\odot$. There is growing evidence that black holes in this mass range exist, but it is unclear how they could form and the evidence is less strong than that for stellar mass black holes.

Black holes with mass above $10^5 \, M_\odot$ are called *supermassive black holes*. No one would be bold enough to suggest a black hole with such a vast mass were it not for compelling evidence from stellar orbits, especially near the centre of our galaxy and the galaxy NGC 4258, and the phenomenon of *quasars*. The latter are extremely brightly emitting compact sources, with a luminosity 100–1000 times that of a large galaxy; the most likely mechanism requires a large black hole at the centre of each quasar. Visible light in the universe comes mostly from stars and thus owes its origin ultimately to the liberation of energy by nuclear fusion. In the X ray region of the spectrum, on the other hand, it now appears that gravitation is the main power-source.

An 8-metre telescope of the European Southern Observatory operating at $2.18 \, \mu\mathrm{m}$ has a diffraction limit of 2.7×10^{-7} radians and can locate sources to 5×10^{-8} radians after deconvolution and averaging. This corresponds to a distance of 80 AU at galactic centre[3] (8 kpc from Earth); see Fig. 20.8. Observations of a star called S2 determined that it follows an orbit with period 15.2 years, pericentre distance 120 AU and eccentricity 0.87. Using a Newtonian treatment (appendix A) one deduces that the central mass is $M \simeq 4 \times 10^6 \, M_\odot$. How can this much mass be contained within a region of radius less than 120 AU? Most suggestions can be ruled out by simple arguments from Newtonian gravity. For example a dust cloud or a collection of millions of solar-mass-sized objects would not survive along enough against its own gravitation and collisional processes to be a viable candidate. By far the most likely candidate, to account for at least a large fraction of M, is a black hole. There are now sufficient examples of supermassive black hole candidates near galaxy centres for a correlation between their mass and that of the

---

[3]Sagittarius A*, lying just outside the 'bow' of Sagittarius, on the Scorpius side; right ascension 17h 45m 40.0409s, declination $-29°0'28.118''$.

central part of the galaxy to be discerned. This suggests that the formation of these black holes may be an ordinary rather than exceptional part of the process of formation of galaxies. The formation process itself remains mysterious, but is likely via mergers and acquisitions rather than the collapse of a single large cloud.

Exercises

20.1 The metric $ds^2 = \rho^{-1}d\rho^2 + dy^2$ blows up at $\rho \to 0$. Discover whether or not this manifold is irregular by introducing a coordinate change to $x = 2\sqrt{\rho}$.

20.2 Is a stationary limit surface of necessity also a null surface?

20.3 Show that in Minkowski spacetime a light cone is not a Killing horizon, but the surface $x^2 - c^2t^2 = 0$ is a Killing horizon.

20.4 Introduce a coordinate T related to Schwarzschild coordinates by

$$c\,dT = c\,dt + (1 - 2m/r)^{-1}\sqrt{2m/r}\,dr. \tag{20.48}$$

Show that the line element of Schwarzschild spacetime is then

$$ds^2 = -c^2dT^2 + (dr + c\sqrt{2m/r}\,dT)^2 + r^2d\Omega^2 \tag{20.49}$$

Now examine the hypersurfaces at $T=$const. We have found a sequence of Euclidean spaces smoothly filling the Schwarzschild spacetime! Is spacetime not curved after all then? Sketch these surfaces on a copy of Fig. 17.2. Show that this T is proper time along an infalling radial worldline starting from rest at infinity. The coordinates (T, r, θ, ϕ) are called *Gullstrand–Painlevé coordinates*.

20.5 Are (20.36) and (20.49) counterexamples to Birkhoff's theorem, and therefore disprove it? Explain.

20.6 (i) Obtain the Eddington–Finkelstein metric (20.34) from the Schwarzschild metric, first by a coordinate transformation applied to g_{ab}, and then by the method of simply replacing cdt by $dp - (1 - 2m/r)^{-1}dr$ in the line element. Prove that the methods are equivalent.
(ii) Find the inverse of the Eddington–Finkelstein metric and hence show that $r = 2m$ is a null surface.

20.7 Fig. 20.3(a) suggests that a horizon can form before a singularity does, and therefore one might find oneself in a region of space such that one is inside a horizon but there is no singularity (yet). Is that right?

20.8 An observer stands on the surface of a star as it collapses. Once the observer passes the horizon, will they still be able to see their feet beneath them?

20.9 A healthy man goes to visit a doctor's surgery located just inside the horizon of a black hole of mass $10^{11}M_\odot$, in order to enquire how long he has left to live. What answer does he receive? [*Ans.* approx 18 days]

20.10 A particle moves radially inwards towards a Schwarzschild black hole with coordinate speed u_0 at infinity. Use (11.50) to find the speed relative to a local observer fixed in the Schwarzschild lattice as the particle reaches any given r. Hence show that the relative speed tends to c as r tends to $2m$, irrespective of the value of u_0.

20.11 A particle falls as in the previous exercise and then is brought to rest at some r. Show that the energy released, as observed by an observer at r, is

$$E = mc^2\left(\gamma_0(1 - 2m/r)^{-1/2} - 1\right)$$

where $\gamma_0 = (1 - u_0^2/c^2)^{-1/2}$. If this energy is converted into photons which propagate upwards, what is their observed energy at infinity? What is the answer as $r \to 2m$? What happens to the mass of the black hole if the object subsequently falls in after giving up its kinetic energy in this way?

20.12 A body is dropped from infinity towards a Schwarzschild black hole; when it reaches r it emits a photon in the outward radial direction. Show that the redshift of this photon when it is received at infinity is $\nu_{(R)}/\nu_{(E)} = 1 - \sqrt{2m/r}$ [Hint: this is straightforward if you have completed the exercises of Chapter 17.] Hence show that if the body continuously emits light of fixed frequency in its rest frame, then as the emission point approaches the horizon, the frequency of the received radiation falls exponentially with time, as $\nu_{(R)} \propto \exp(-ct_R/4m)$. Explain why the received flux decays faster still, as $\exp(-ct_R/2m)$.

20.13 Obtain (20.46) from (20.42).

20.14 How could a voyager starting in region I of a maximally extended Schwarzschild spacetime learn about events in region III?

20.15 If a black hole of mass 10^9 kg fell to Earth, would we notice? Would it produce an impact crater?

20.16 How large would a black hole having the same mass as Earth be? Is it possible to manipulate black holes? Could one store and transport them, in principle? How? Could you cut one in half with a knife?

20.17 A puzzled student makes the following argument. 'Information does escape out through the horizon, even in classical GR, because when matter falls in the mass grows and as a result of this the surrounding spacetime is changed; thus spacetime itself transmits some information out of the black hole.' Explain what is wrong with this argument. [Hint: consider spacetime inside and outside a spherically symmetric shell of matter collapsing towards a pre-existing horizon.]

20.18 For a sufficiently large central mass, the tidal effects described by (17.70) are mild for $r \simeq r_s$. Consider, then, an astronaut equipped with a backpack which enables him to hover 50 centimetres outside the horizon of a black hole of mass $M = 10^9 \, M_\odot$. One might imagine that such an astronaut could reach out his hand to touch the horizon, or even dip it into the horizon, and then pull his hand back again with no ill effect. Is that right? If not, then how or why has the intuition failed? To provide a thorough answer, describe the experiment using a Kruskal diagram, or using a Minkowski spacetime diagram for a LIF of diameter a few metres. Do not forget to calculate the acceleration which the astronaut's backpack has to provide relative to such a LIF.

21

Black hole thermodynamics

In this chapter we explore a number of processes at the interface between thermodynamics, quantum physics and classical General Relativity. The first process (Penrose process) is purely classical. It motivates some thermodynamic ideas, and these in turn lead to important discoveries in quantum physics in a curved spacetime, especially the theoretical discovery that Black holes can emit electromagnetic (and other) radiation.

21.1   The Penrose process

A beautiful piece of reasoning and an important physical process associated with the Kerr black hole is called the Penrose process.

Suppose a pair of rocks is prepared with a compressed spring between them, with a latch that prevents the spring from expanding and which can be released at a chosen moment. The rocks are dropped towards a Kerr black hole, and when they are in the ergoregion the latch is released. The spring expands and the rocks are thrown apart in such a way that one of them is thrown outwards and eventually reaches infinity, and we will calculate what happens to the other. The Penrose process concerns the remarkable possibility that the rock which is thrown outwards can, in some circumstances, emerge with more energy than the total energy of *both* (including the spring) when they were first dropped!

In order to understand this, we need to think carefully about conservation of energy and momentum, and the nature of timelike vectors. The following notation will be adopted (Fig. 21.1): A superscript capital letter A, B, C refers to the three entities involved. A is the initially prepared system. We described it above as a pair of rocks with a spring; more generally it could be any particle which can decay. B and C are the decay products; B is the one which (we will show) gets buried in the black hole and C is the one which can be collected at infinity. These entities have 4-momentum $\mathbf{p}^A$, $\mathbf{p}^B$, $\mathbf{p}^C$ respectively. All the 4-momenta are functions of time. Let event $\mathcal{Q}$ be the release of the spring (or more generally the decay of A). Then conservation of 4-momentum at the decay event gives

$$\mathbf{p}^A(\mathcal{Q}) = \mathbf{p}^B(\mathcal{Q}) + \mathbf{p}^C(\mathcal{Q}). \tag{21.1}$$

Relativity Made Relatively Easy: General Relativity and Cosmology. Volume 2. Andrew M. Steane, Oxford University Press. © Andrew M. Steane 2021. DOI: 10.1093/oso/9780192895646.003.0021

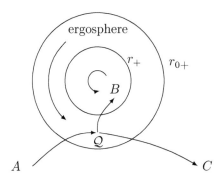

Fig. 21.1 The Penrose process.

Since the Kerr metric does not depend on the t coordinate, we also have conservation of the *covariant* component p_t for any particle in freefall. Therefore if we use (∞) to refer to the situation far from the black hole, we have

$$p_t^A(\infty) = p_t^A(\mathcal{Q}); \qquad p_t^C(\infty) = p_t^C(\mathcal{Q}) \tag{21.2}$$

and hence, by using this in the previous equation,

$$p_t^C(\infty) = p_t^A(\infty) - p_t^B(\mathcal{Q}). \tag{21.3}$$

Now the covariant component at infinity is equal to minus the energy observed by a detector at rest there, since the metric tends to that of flat space, and the energy of a particle of 4-momentum **p** observed by a detector with 4-velocity **u** is $E = -p_\mu u^\mu$. A detector at rest in the coordinates has $u^a = (1, 0, 0, 0)$ and therefore, applying this to (21.3) we find

$$E^C(\infty) = E^A(\infty) + p_t^B(\mathcal{Q}). \tag{21.4}$$

For a metric of signature $+2$, ordinarily the covariant temporal component of a momentum vector is negative, so this equation would give $E^C(\infty) < E^A(\infty)$. This means the escaping particle has less energy than the originally dropped particle, because some of the energy was carried away by particle B. However, inside the ergoregion we have $g_{tt} > 0$ and therefore it is not self-evident that $p_t^B(\mathcal{Q})$ will be negative. In fact, if $g_{tt} > 0$ then the coordinate basis vector $\mathbf{e}_{(t)}$ is spacelike (its length-squared is $\mathbf{e}_{(t)} \cdot \mathbf{e}_{(t)} = g_{tt}$). Therefore the component p_t^B is in a spacelike direction and may be either positive or negative, depending on the direction of travel relative to the coordinate frame. If in fact $p_t^B > 0$ then it will remain positive all the way along B's geodesic (owing to the conservation law for a metric with no t-dependence) and this implies that this geodesic cannot be one which emerges to the outside of the ergoregion, because if it did then its tangent would then be spacelike, which is not possible because the timelike/null/spacelike character of any geodesic is preserved all the way along the geodesic.

We have thus deduced that a necessary condition for $E^C(\infty) > E^A(\infty)$ in the process under consideration is that B either stays permanently in the ergoregion or else falls into the black hole. It only remains to add that such a case is in fact possible: there is nothing to prevent the worldline of B after the decay to be one with $p_t^B > 0$.

Let us suppose that B falls through the horizon. If it had fallen from infinity then we would say that the energy it delivers to the black hole is equal to $-p_t^B$. We must apply this same formula in the present case. That is, the mass of the black hole changes by

$$M \to M - p_t^B/c^2. \tag{21.5}$$

This expresses the conservation of energy overall:[1] the extra energy gained by C has come from the black hole itself!

It seems counter-intuitive that the mass of a black hole could decrease, because it appears to suggest that some energy has escaped out through the event horizon. It is not so. The energy provided to C has been extracted not from inside the event horizon but from the ergoregion. The mass parameter of any black hole is not a statement only about energy located inside the horizon because spacetime itself contains energy-density when it is curved. Some of the mass-energy associated with M is stored throughout the space around the gravitating body.

Similar reasoning applied to angular momentum conservation tells us that the angular momentum of the black hole changes by

$$J \to J + p_\phi^B. \tag{21.6}$$

We will now show that in the case under consideration, $p_\phi^B < 0$, so J gets smaller. To prove this, all we need to do is consider what is observed by an observer maintaining an orbit at fixed r, θ in the ergoregion (by means of a rocket motor for example). Such an observer has 4-velocity $u^a = u^t(1, 0, 0, \Omega)$ in which u^t must be positive by the definition of 4-velocity, and Ω must be positive in the ergoregion. This observer would detect B's energy to be

$$E_{\text{obs}} = -p_\mu^B u^\mu = -u^t(p_t^B + \Omega p_\phi^B). \tag{21.7}$$

This energy must be positive, so we find $(p_t^B + \Omega p_\phi^B) < 0$ and therefore

$$p_\phi^B < -p_t^B/\Omega, \tag{21.8}$$

which suffices to determine the sign of p_ϕ^B. The overall result is that when the black hole loses energy by the Penrose process, it also loses angular momentum.

Let us now go one step further and find the lower limit of $|\delta J|$. The above argument applies to any observer who can maintain an orbit at fixed r, θ. We found in Section 19.3.2 that the maximum Ω for such observers is Ω_H given in (19.31), so we have

$$\delta J \le \frac{\delta M c^2}{\Omega_H}. \tag{21.9}$$

An important observation about the Penrose process is that whereas the mass of the black hole decreases, *the event horizon grows*. To show this, first write the area (19.30) in terms of M and J:

---

[1] There is some subtlety involved in this assertion; the essential point is that once stationary conditions are achieved the field equation asserts overall conservation of energy, and we are assuming that the gravitational radiation is negligible.

$$A(M, J) = 8\pi \left(\frac{G}{c^2}\right)^2 \left(M^2 + \sqrt{M^4 - J^2 c^2/G^2}\right). \tag{21.10}$$

Using $dA = (\partial A/\partial M)dM + (\partial A/\partial J)dJ$ one obtains

$$\delta A = \frac{8\pi Ga}{c^3 \Omega_{\mathrm{H}} \sqrt{m^2 - a^2}} \left(c^2 \delta M - \Omega_{\mathrm{H}} \delta J\right) \tag{21.11}$$

and therefore the limit (21.9) implies that the area either grows or remains unchanged during the Penrose process.

We have treated the Penrose process applied to massive particles following timelike geodesics; the same mechanism can also apply to photons on null geodesics. It is a mechanism which can in principle extract large amounts of energy. A related but more complicated process called the *Blandford-Znajek process* involves electromagnetic fields and electron-positron currents in the vicinity of a rotating black hole; it is believed that this process could be the explanation for the immensely powerful relativistic jets observed to be emerging from some quasars.

21.2 Area theorem and entropy

Equation (21.11) can be written in the instructive form

$$dMc^2 = \frac{\kappa}{8\pi G/c^2} dA + \Omega_{\mathrm{H}} dJ \tag{21.12}$$

where

$$\kappa = \frac{c^2 \sqrt{m^2 - a^2}}{2m(m + \sqrt{m^2 - a^2})} \tag{21.13}$$

is the surface gravity defined in (20.22). Equation (21.12) is reminiscent of the thermodynamic statement

$$dU = TdS + \Omega dJ \tag{21.14}$$

where the second term is work done on a rotator, and we note that the non-decreasing behaviour of A for the Kerr black hole is like the non-decreasing behaviour of entropy S in the thermodynamic statement. This is not a proof but a suggestion that a form of thermodynamic reasoning may apply to black holes. This turns out to be a very illuminating suggestion.

The metrics of Schwarzschild, Reissner–Nordström and Kerr all exhibit a generic property of black holes: they exhibit their properties (such as mass, electric charge and angular momentum) by impressing them upon the surrounding spacetime (and also providing an electromagnetic field in the charged case). Other properties, such as the details of the matter which went to form them, are not expressed outside the horizon, but hidden inside it, says classical GR. The spacetime just outside the horizon must soon throw off by gravitational radiation any irregularities, and rapidly settle down to a simple form. Furthermore, the dynamics inside the horizon pushes all the material down to the singularity in a finite time, so that the simple solutions can be expected to hold exactly. This seems to suggest that black holes are sinks of information: many different microstates of infalling matter all lead to the same final microstate of the black hole. This would be a violation of the second law of thermodynamics.

The idea that black holes in steady state are completely described by only a small number of parameters is called the **no hair theorem**. Classical GR implies that the only parameters needed are mass, electric and magnetic charge,[2] and angular momentum, plus the position and linear momentum of the hole. However, once quantum effects are considered, the situation changes dramatically. It is now thought that the horizon of a black hole is extremely 'hairy', in the sense of having a huge number of microstates and a large entropy.

Something of a revolution in theoretical General Relativity took place in the period 1963–1974. Kerr's discovery of 1963 was followed by the introduction of topological techniques by Penrose in 1965. As already mentioned, he was able to show that collapsing matter inside an event horizon cannot somehow swirl around in such a way as to avoid the formation of a singularity. These topological methods describe the geodesics; they do not treat the curvature directly. They suffice to show that the geodesics have to come to a finite end; one may then reasonably suppose that they end at a curvature singularity. In 1970 Hawking and Penrose developed these techniques further, to prove the **area theorem**, which states that the area of a black hole horizon can never decrease:

Theorem 21.1 Area theorem (Hawking): *In classical GR, the area of a horizon cannot decrease if the weak energy condition (16.5) holds and cosmic censorship holds (no singularities on or outside the horizon).*

This is in contrast to the mass, for example, which can decrease, for the Kerr black hole, via the Penrose process. The significant observation is that the Penrose process lowers the mass but not the horizon area. Furthermore, when two black holes merge they may throw off large amounts of rest energy by gravitational radiation, but not so much as to break the area theorem: the area of the resulting black hole horizon always exceeds the sum of the previous areas.

This 'one-way' property of black hole horizon areas is reminiscent of the second law of thermodynamics. Bekenstein proposed that an entropy ought to be associated with a black hole; in particular, the entropy should be proportional to the area of the horizon. He gave arguments to suggest that the entropy of a black hole is of the order of the horizon area, in units of the Planck length:

$$S \sim k_{\mathrm{B}} \frac{A}{l_{\mathrm{P}}^2} = k_{\mathrm{B}} \frac{Ac^3}{G\hbar}. \tag{21.15}$$

If the horizon of a black hole has entropy, then it ought to have temperature as well, and that means it ought to radiate. But is not that just what black holes are supposed not to do?

---

[2]If there were further fields having long-range interactions, then further properties would accrue.

By exploiting techniques to handle quantum field theory calculations in a warped spacetime, Hawking made the amazing discovery that a black hole horizon *does* emit radiation, by a quantum mechanical process. And furthermore, the radiation has a thermal spectrum! So a black hole is also a (thermodynamic) black body! The temperature of the radiation, according to Hawking's calculation, is

$$T = \frac{\hbar c^3}{8\pi k_B GM} = \frac{\hbar\kappa/c}{2\pi k_B} \qquad (21.16)$$

where κ is the surface gravity ($c^4/4GM$ for Schwarzschild case). This gives 61 nK for a solar-mass black hole.

Motivated by the strong equivalence principle, Unruh (Unruh, 1976) then showed by a different calculation that even in flat spacetime, an observer undergoing uniform acceleration must expect to see thermal radiation incident on him from all directions[3] with a temperature given by

$$T = \frac{\hbar a_0/c}{2\pi k_B} \qquad (21.17)$$

where a_0 is the proper acceleration (Fulling, 2005; Crispino, Higuchi and Matsas, 2008).

A related process is called the Sokolov–Ternov effect. This concerns relaxation by spin-flips for electrons orbiting in a constant magnetic field. The spin-flips can occur by the emission of radiation, and one might expect that eventually the spin would be perfectly polarized in the lower energy state. In fact, the equilibrium state is incompletely polarized and this is owing to a non-zero probability to absorb energy; this aspect of the process is similar to what occurs in the Unruh effect, and has been experimentally observed.

21.3 Unruh and Hawking effects

The above summarizes the historical sequence of discoveries. With the benefit of hindsight, one may now make the Unruh effect the starting point, and proceed from there.

It would require a much longer discussion to develop the relevant quantum field theory, but in the absence of a complete derivation one can still grasp some basic truths and avoid some basic misconceptions. The first basic truth is that there is a considerable amount of subtlety involved in quantum field theory, and physicists struggle to find language to convey faithfully what the mathematics implies. An example of this is the use of the word 'fluctuation' (see box). The state called vacuum state in quantum field theory does have structure, a structure which responds to the presence of charge by vacuum polarization for example, and which can give rise to Casimir–Polder forces on reflecting surfaces, but this structure does not spontaneously change from one moment to the next, therefore the term 'vacuum fluctuation' is misleading. In flat spacetime the vacuum does not and cannot spontaneously produce particle-antiparticle pairs, nor does it in the least shake, interrupt, disperse or otherwise disturb the motion of sensitive objects such as electrons and photons. Their energy and momentum is precisely conserved as they move through the vacuum; if their surroundings were fluctuating in the ordinary sense of the word then this would not be so.

---

[3]The radiation comes up from the Rindler horizon and falls down again in such a way as to be isotropic at each point.

Does the quantum vacuum fluctuate? The answer to this question is *no*, but we need to understand why the term 'vacuum fluctuation' has become widespread in physics. It is owing to a similarity between some aspects of quantum superposition and classical effects involving fluctuation.

Consider first the quantum harmonic oscillator. In its ground state the mean excursion is zero, $\langle 0| \hat{x} |0\rangle = 0$, but the mean squared excursion is not zero, $\langle 0| \hat{x}^2 |0\rangle \neq 0$. The dynamics are simple: the state is an energy eigenstate so does not evolve at all, apart from a global phase factor. Hence it does not fluctuate. However, the statistical property known as *variance* is not zero for the position distribution, and if one were to make a position measurement on each of a set of such systems, all prepared in the ground state, then one will get a range of different answers, distributed as a normal distribution. It is this set of measured outcomes that leads to the use of the word 'fluctuate'; the position measurement outcomes fluctuate from one realization of the harmonic oscillator to the next. These points carry over to the discussion of quantum fields, where the amplitude of the field behaves like a harmonic oscillator.

A further issue also arises. When there are two or more fields that can interact, such as the electromagnetic field and the Dirac field (which describes electrons and positrons), then the energy eigenstates of the joint system are not energy eigenstates of the individual fields. Therefore the ground state of the complete system, called $|\Omega\rangle$, is not equal to the ground state of the 'free' (i.e. not interacting) fields, called $|0\rangle$. This means that if one could design a detector which could distinguish states of the free fields, and one used such a detector to measure the vacuum (i.e. the state $|\Omega\rangle$), then the detector will report that many different states of the free fields are contributing, and consequently its measurement outcomes will fluctuate from one realization of the experiment to the next. However, this observation is misleading, because such a detector is itself an abstraction. An ordinary detector (a photographic film, a Geiger-Muller tube, a photoelectric multiplier, etc.) discriminates states of the *interacting* fields, not the free fields. It will faithfully report that the state $|\Omega\rangle$ is indeed the ground state; no excitations are detected.

The mathematical analysis is often carried out in terms of integrals over energy eigenstates of the free fields, and this is what leads to talk of 'photons' and 'electrons' and 'positrons' 'existing in' the vacuum state $|\Omega\rangle$ or being 'created' via a 'fluctuation of the vacuum'. But this is merely a loose use of language adopted to discuss calculations involving $|\Omega\rangle$, and in fact the state $|\Omega\rangle$ is an energy eigenstate, which means it is a state with no dynamics other than a global phase. Therefore it does not fluctuate at all; it is utterly smooth in all spacetime directions, but its quantum state does have a non-zero variance in various interesting properties. In short, there is no fluctuation in the vacuum, but there is correlation in the vacuum.

The above remains mostly true in a more general spacetime, but new effects come in when spacetime is dynamic, and when horizons are involved. And of course if some physical system, such as an accelerating detector, interacts with the quantum fields then excitations can happen.

One lesson of the Unruh effect is that the vacuum state of the electromagnetic field, and other quantum fields, is not a state in which 'nothing at all is there'. Rather, the electromagnetic field is there, and in its ground state although the mean field amplitude is zero, there are many other properties, such as correlations involving products of field amplitudes at different events, which are not zero. Any physical system having a non-zero coupling to the field can therefore, in principle, be influenced by these correlations. A good way to understand the Unruh effect is to study the evolution of a quantum system S that is following an accelerating trajectory in Minkowski spacetime. This S could for example be an atom or an absorbent speck of dust. The interaction of S with the surrounding electromagnetic field (assumed to be in its ground state initially) is such that correlated influences which would vanish for inertial motion do not vanish for accelerated motion, with the result that S picks up internal energy, and photons are also emitted by S into the vacuum. The energy required for both these aspects of the behaviour is obtained from whatever further system is providing the force which causes the acceleration.

When one studies this same scenario from the point of view of an accelerated reference frame which keeps pace with S, one must again find that S becomes internally excited, but now the physical interpretation is different. Now the vacuum state of the field, viewed in the accelerated frame, is itself a thermal state.[4] This point of view is valuable because it exhibits the physics in great generality, independent of the details of whatever physical system (or 'detector') is undergoing excitation.

Using either method of calculation, one finds that for a detector with constant proper acceleration a_0, the internal excitation of the detector acquires a thermal distribution, with the temperature given by Unruh's formula (21.17).

The availability of two different ways to interpret what causes this effect is reminiscent of an issue which arises in classical field theory (classical electromagnetism and Special Relativity). For, if we consider a uniformly accelerating small charged body, then ordinarily we would say that it emits electromagnetic radiation and we could, for example, employ Larmor's formula $P = (2/3)(q^2/4\pi\epsilon_0)a_0^2/c^3$ to calculate the rate of emission of energy. However, if an observer boards a rocket which keeps pace with the accelerating body, then the charge is not accelerating (nor moving at all) relative to that observer. Does such an observer detect electromagnetic radiation from the charged body? Classical electromagnetism is perfectly adequate to calculate (up to some finite precision) both the electromagnetic field and its interaction with any detector such an observer may carry. A full discussion requires one to investigate the events in the past when the acceleration began and those in the future when it ceases (a strictly infinite period of acceleration being deemed unphysical). The usual interpretation is that the distinction between radiative and non-radiative contributions to a given electromagnetic field only becomes unambiguous in the sufficiently far future of any given source event. One thus argues that the interpretation of a given field depends on the state of acceleration of the observer. (Fulton and Rohrlich, 1960)

---

[4]The method of calculation is to study the quantum field through a coordinate change from Minkowski to Rindler coordinates. One determines the raising and lowering operators in the new coordinates, and then makes the argument that one should take an average (a trace) over the part of the field that is beyond the horizon of the Rindler region in which the detector is situated.

This purely classical (not quantum) observation has itself an important application to gravitational physics, because we have to give an account of what happens when a charged body sits at rest relative to a heavy static planet. Such a charge is accelerating relative to LIFs in its vicinity, but the situation is static for an observer standing on the planet. The charge does not radiate energy to such an observer. The role of this classical example for present purposes is to show that *the presence or absence of radiative electromagnetic energy is not an absolute property of the field, but is a statement about the interaction of the field with detectors, a statement whose form can depend on the state of motion of those detectors.* The energy tensor of the field is, meanwhile, just that—a tensor—so it is what it is, irrespective of coordinate frames (though we will often wish to know its components relative to one frame or another).

Returning now to the Unruh effect, we have a quantum (not classical) result whose physical interpretation is frame-dependent.

The methods of calculation ordinarily employed in the Hawking and Unruh effects are different, and one should not assume that the two effects are related in all respects. They differ on distance- and time-scales where the influence of spacetime curvature is non-negligible. However, one can argue that given the Unruh result, the Hawking effect must follow, as we now show.

Consider a location near but outside the horizon of a Schwarzschild black hole. Introduce a coordinate $\rho = 2\sqrt{r_{\rm s}(r - r_{\rm s})}$. Then

$$\frac{r_{\rm s}}{r} = \left(1 + \frac{\rho^2}{4r_{\rm s}^2}\right)^{-1} \simeq 1 - \frac{\rho^2}{4r_{\rm s}^2}$$

for $\rho \ll r_{\rm s}$. Taking $c = 1$, the Schwarzschild metric gives

$$\mathrm{d}s^2 \simeq -\frac{\rho^2}{4r_{\rm s}^2}\mathrm{d}t^2 + \mathrm{d}\rho^2 + \mathrm{d}\sigma^2 = -\rho^2\mathrm{d}\tau^2 + \mathrm{d}\rho^2 + \mathrm{d}\sigma^2 \tag{21.18}$$

where $\tau = t/2r_{\rm s}$ and $\mathrm{d}\sigma^2 = (r_{\rm s} + \rho^2/4r_{\rm s})^2\mathrm{d}\Omega^2$. An observer held suspended at ρ is undergoing constant proper acceleration relative to a LIF, and the local metric, eqn (21.18), matches that of the uniformly accelerated reference frame in flat spacetime—the Rindler metric. We now invoke Unruh's result. Since quantum field theory is local, then to the extent that the approximate metric is sufficient to describe the relevant range of r, such an observer will find the vacuum to be in a thermal state at the Unruh temperature associated with his proper acceleration, i.e.

$$T(r) = \frac{1}{2\pi\rho} = \frac{1}{4\pi\sqrt{r_{\rm s}(r - r_{\rm s})}} \tag{21.19}$$

where we set $\hbar = k_{\rm B} = 1$.

Given that the observer at r detects such thermal radiation, an observer at another radius $r_{\rm D}$ detects the radiation after a gravitational redshift. The black body spectrum has the property that if all the frequency components change frequency by the same factor, then the spectrum remains thermal, with a temperature reduced by that factor. Hence

$$T(r_\mathrm{D}) = T(r)\sqrt{\frac{1 - r_\mathrm{s}/r}{1 - r_\mathrm{s}/r_\mathrm{D}}} \simeq \frac{1}{4\pi\sqrt{r_\mathrm{s}r(1 - r_\mathrm{s}/r_\mathrm{D})}}.$$

This result is approximate, but as we allow r to approach the horizon at r_s, the accuracy improves since the range of r significant to the calculation gets smaller. In the limit $r \to r_\mathrm{s}$

$$T(r_\mathrm{D}) = \frac{1}{4\pi r_\mathrm{s}\sqrt{1 - r_\mathrm{s}/r_\mathrm{D}}}.$$

The temperature observed at infinity is

$$T(\infty) = \frac{1}{4\pi r_\mathrm{s}}.$$

This agrees with Hawking's result. This argument also implies that the significant property of the horizon is the surface gravity. This allows the result to be readily extended to all types of black hole.

Note that if we had adopted the Rindler metric throughout, then we would be describing flat spacetime, and then the redshift at infinity would be different, and in fact it would reduce the observed temperature at infinity all the way to zero.[5] The Schwarzschild metric imposes a more modest redshift, with the result that the radiation reaches infinity with a non-zero temperature.

21.3.1   A heuristic argument for Hawking radiation

So far we simply quoted the Unruh result. We can go a little further by presenting a heuristic argument for Hawking's formula (21.16). The idea is to present properties of the vacuum state $|\Omega\rangle$ that are relevant near a horizon. Here the state $|\Omega\rangle$ is the ground state of the complete set of interacting quantum fields, and all of them can in principle be involved, but we will only treat the electromagnetic field. This is a good approximation because photons are by far the dominant contribution in Hawking radiation except in extremes of high temperature.

Hawking radiation can be seen as a consequence of a coupling between the gravitational field and the electromagnetic field. If the electromagnetic (e.m.) field is not itself in a thermal state at the Hawking temperature, then the consequence of this coupling is a steady movement of energy between the two fields. In particular, if the quantum vacuum surrounding a black hole is initially in its ground state then energy moves from the black hole to the surrounding e.m. field (and other fields); this energy movement takes the form of radiation with a thermal spectrum and is called Hawking radiation.

---

[5]The Rindler frame has the property that redshift of the Unruh radiation propagating from any given height reproduces the Unruh radiation at any other given height (an example of the Ehrenfest–Tolman effect, exercise 13.11 of Chapter 13), and since the proper acceleration tends to zero at high heights, so does the Unruh temperature.

Fig. 21.2 Interaction vertex for two photons near a horizon.

In order to understand how energy can move in the 'wrong' direction across a horizon, we note that in quantum physics it is not unusual for a particle to have a non-zero probability to access a region of space which is classically inaccessible. For example, a particle having a well-defined total energy E can have a non-zero wavefunction in regions of space where $E < V$, if V is the local potential energy. In such a region we may say the particle has a negative kinetic energy $(E - V)$ and sometimes this is called tunnelling. In such a case the wavefunction is not zero but it does decay exponentially with position and/or time.

Now let us investigate the observations of an observer fixed in the Schwarzschild coordinates near but outside the horizon of a Schwarzschild black hole. In order to calculate the properties of the electromagnetic field, an integral is required over all contributing processes. We can conveniently write the integral in terms of interaction vertices and free propagation, as in the Feynman path integral method. Gravitation comes in by influencing the metric. Consider, for example, an interaction vertex in which there is no incoming line, but simply two outgoing photon lines—Fig. 21.2. (One may say that we have a particle and an antiparticle, but for photons that distinction is not needed.) In ordinary circumstances, processes of this kind do not make a non-zero contribution to the total integral, so they can safely be ignored. Near a horizon, however, things can go differently.

In quantum field theory, each interaction vertex strictly conserves energy and momentum, so we have

$$\mathbf{p}^A + \mathbf{p}^B = 0 \tag{21.20}$$

where $\mathbf{p}^A$ and $\mathbf{p}^B$ are the 4-momenta of the two lines (particles). An observer with 4-velocity $\mathbf{u}$ would find that the energies (kinetic energy plus rest energy) of these two particles are $-\mathbf{p}^A \cdot \mathbf{u}$ and $-\mathbf{p}^B \cdot \mathbf{u}$. Hence if one particle (say A) has positive energy E then the other (B) has negative energy $-E$. Since we are talking about the sum of kinetic energy and rest energy, a negative result would suffice to say the process does not happen in classical physics. In quantum physics what happens is that the associated evolution has built into it an exponential decay with a time-constant of order $\tau \simeq \hbar/2E$. If, within this time, the particle B crosses the horizon, an observer employing Schwarzschild coordinates will determine that it still has the same p_t^B (a conserved quantity for free-fall motion in Schwarzschild spacetime), but inside the horizon p_t^B corresponds to a *spatial* component so it can have either sign; the overall quantum amplitude for the process now need not decay exponentially any further. Consequently it can make a non-zero contribution to the integral which describes the complete process over long times. Note that the changing role of p_t here is similar to what happens in the Penrose process, but the present case is different in that it is thoroughly quantum mechanical whereas the Penrose process is classical (also, one concerns the ergosphere, the other concerns the horizon).

A precise value for both position and momentum cannot be given for the overall wavefunction, but here we calculate an example path which contributes to an integral over all paths. It is convenient to consider a non-null path. Suppose the interaction vertex is situated at Schwarzschild coordinate $R = 2m + \delta$ and particle B is initially at rest relative to the coordinates (note that even a null geodesic can satisfy this condition in the limit as the starting point approaches the horizon). The proper time elapsed during radial freefall from R to the horizon is

$$\Delta\tau = \int_{2m+\delta}^{2m} \frac{\mathrm{d}\tau}{\mathrm{d}r}\mathrm{d}r = \frac{1}{\sqrt{2mc^2}} \int_{2m}^{2m+\delta} \left(\frac{1}{r} - \frac{1}{2m+\delta}\right)^{-1/2} \mathrm{d}r \qquad (21.21)$$

where we used (17.25) with the initial condition $\dot{r} = 0$. We only need to estimate the integral for values of δ not large compared to m. One finds to lowest order $\Delta\tau \simeq \sqrt{8m\delta}/c$. Therefore the process has non-negligible quantum amplitude for particle energies up to

$$E \simeq \frac{\hbar}{2\Delta\tau} \simeq \frac{\hbar c}{4\sqrt{m\delta}}. \qquad (21.22)$$

This energy is the one given by $E = -\mathbf{p}\cdot\mathbf{u}$ for an observer fixed in the Schwarzschild coordinates *at or near the event in question.*

Once we have satisfied ourselves that there is a non-negligible probability for the process to happen, the quantum calculation has done its job, and we have an outgoing photon A with positive energy. Quantum field theory on a curved spacetime background can also describe the subsequent outwards motion of the photon, but we can already calculate that perfectly well using GR alone. The photon propagates outwards with conserved p_t. Let $E^A(r)$ be its energy as observed by an observer fixed at Schwarzschild radial coordinate r. The relationship between $E^A(r)$ and E is precisely that of gravitational redshift:

$$E^A(r) = \left(\frac{1 - 2m/(2m+\delta)}{1 - 2m/r}\right)^{1/2} E. \qquad (21.23)$$

Therefore

$$E^A(\infty) \simeq \sqrt{\frac{\delta}{2m}} E = \frac{\hbar c}{8m} = \frac{\hbar c^3}{8GM}. \qquad (21.24)$$

It is remarkable that the distance δ has dropped out of the calculation. The only distance scale remaining is the Schwarzschild radius, and this gives the typical wavelength in the radiation observed at infinity.

The average energy of a photon in thermal radiation is approximately $2.7k_\mathrm{B}T$ so our estimate suggests the observed temperature is $T \simeq \hbar c^3/22k_\mathrm{B}GM$. This is close to Hawking's result.

Note that we do not require $\delta \ll m$; the particles originate outside the horizon and not necessarily very close to it. If one extends the above argument by performing the integral (21.21) numerically for all δ, and also allowing for a range of directions and energies of the particles, then one concludes that the Hawking radiation originates throughout a region whose radial thickness is of the order of m, from the horizon to approximately the light sphere. This argument is merely indicative. At the time of writing (2019) the distribution of the source of Hawking radiation remains a subject of active investigation in quantum field theory. (Dey, Liberati and Pranzetti, 2017; Greenwood and Stojkovic, 2009; Brynjolfsson and Thorlacius, 2008; Giddings, 2016)

In view of the relationship between Hawking and Unruh radiation, it is natural to ask whether or not an observer in free fall outside a black hole will register Hawking radiation. An observer in free fall at $r \gg m$ will detect Hawking radiation, since that is what the whole calculation asserts and means, whereas the Equivalence Principle suggests that an inertial observer located near the horizon will not, on distance and timescales small compared to the local radius of Gaussian curvature of spacetime, i.e. $r_\mathrm{s} = 2m$. However, the non-local aspect of the quantum effects extends over this distance scale so the argument from the Equivalence Principle applies only approximately. A recent calculation suggests that the temperature observed by a detector in free fall and momentarily at rest at r tends to $2T_\mathrm{H}$ as $r \to 2m$. However, there would not be time for the detector to register this value precisely. Notice that there is no question of experiencing diverging temperature when freely falling near the horizon: the freely falling observer has no such experience, and a distant observer would not expect them to, because the radiation received by the latter has not come from the horizon itself, but from distances of order r_s outside the horizon.

If instead we consider an observer hovering at a fixed distance outside the horizon, then we have a different scenario and now there is a diverging proper acceleration and a diverging Unruh temperature as one approaches r_s.

For Hawking radiation the energy tensor of the electromagnetic field outside the horizon is non-zero, and all observers will agree this. All will also agree that there is a flow of energy from the black hole to the electromagnetic field around it, with the result that the mass of the black hole decreases over time.

21.4   Laws of black hole mechanics

Thermodynamic entropy is related to temperature and internal energy by

$$\frac{\partial S}{\partial E} = \frac{1}{T}.$$

If we apply this to a Schwarzschild black hole, then since the temperature is a function of $M \ (= E/c^2)$ alone, the expression can be integrated to give

$$S = \frac{8\pi k_\mathrm{B} G}{\hbar c^3} \int M \mathrm{d}E = \frac{4\pi k_\mathrm{B} G}{\hbar c} M^2 = k_\mathrm{B} \frac{A}{4} \frac{c^3}{G\hbar}. \tag{21.25}$$

In order to supply the constant of integration, we assumed that the entropy goes to zero when the area does (a non-trivial assumption because the temperature diverges when the mass tends to zero). The combination $G\hbar/c^3$ is the square of the Planck length. The entropy is huge: one unit of k_B for each $10^{-69}\,\mathrm{m}^2$ of horizon. This is much greater than the entropy associated with other types of astronomical object of similar mass. The entropy of a single black hole of mass $10^6\,M_\odot$ would exceed the entropy of all the stars in all the galaxies within one Hubble radius.

In presenting (21.25) we relied on the Unruh and/or Hawking result (21.16) and the assumption that we can apply classical thermodynamics to the horizon. This is legitimate because classical thermodynamics supplies correct reasoning about relations between entropy, temperature and energy in the thermodynamic limit—roughly speaking, the limit of large systems with many degrees of freedom. Nevertheless, further calculations in quantum field theory have been devoted to understanding this entropy more fully, and the result of all this work is that the ideas and the quantitative result seem secure.

We have now justified the interpretation of (21.12) in thermodynamic terms. The temperature is given by the surface gravity; the entropy is given by the horizon area. One can also formulate laws called the 'laws of black hole thermodynamics' or 'black hole mechanics'. The analogy is essentially perfect. The zeroth law of ordinary thermodynamics states that if two bodies are in equilibrium with a third, then they must be in equilibrium with each other. This implies that the temperature is uniform throughout a system in thermal equilibrium, and the law is often stated this way. The corresponding (true) statement for black holes is that stationary black holes have uniform surface gravity everywhere on the horizon.

The first law of thermodynamics is energy conservation; for black holes the corresponding statement is the generalization of (21.12) to electrically charged holes:

$$\mathrm{d}Mc^2 = \frac{\kappa}{8\pi G/c^2}\mathrm{d}A + \Omega_{\mathrm{H}}\mathrm{d}J + \Phi\mathrm{d}q \qquad (21.26)$$

where Φ is the electrostatic potential. We have shown that this relates mass, area and angular momentum for a Kerr black hole in stationary conditions; further studies have shown that it also applies to situations where the metric is perturbed from the Kerr form in very general ways.

The second law is that the area of the horizon does not decrease, which holds in classical GR. Finally, the third law is that entropy tends to zero as T does, and this implies (in ordinary thermodynamics) that no finite adiabatic change will achieve $T = 0$. The corresponding law for black holes is that it will not be possible to achieve $\kappa = 0$. It turns out that $\kappa = 0$ corresponds to the extremal black holes. It seems therefore that cosmic censorship is somehow related to black hole thermodynamics, but exactly why or how remains unknown.

Black hole entropy was at first regarded as a curiosity whose relevance to the rest of physics was unclear, but more recently it has been seen as having a deeper significance. It offers a hint that black holes, and horizons generally, may have a microstructure, at the Planck scale, that would be described by a theory that went beyond General Relativity. Also it is striking that the entropy is proportional to the area not the volume as one might have guessed. This was the first pointer towards a wider idea called the *holographic principle* in quantum gravity. By taking horizon temperature and entropy as a starting point, one can also derive or strongly constrain the gravitational field equations, among the class of equations that might possibly describe pseudo-Riemannian manifolds. (Padmanabhan, 2010)

21.4.1 Black hole evaporation

In principle a black hole that does not receive any incoming material will eventually emit all its rest energy, a process called evaporation. The end of such an evaporation is a run-away process that finishes violently (Fig. 21.3). In practice the black hole temperature would have to be greater than that of the cosmic microwave background radiation for this to happen, and even in a cold universe the total lifetime of a black hole of ordinary size (i.e. one produced by stellar collapse) would be of order 10^{69} years (scaling as M^3). However, there remains the possibility that small black holes may have been produced in the Big Bang, with a mass just sufficient for them to last until the present and be now exploding. That mass is of order 10^{12} kg, giving a Schwarzschild radius $\sim 10^{-15}$ m: a pinch in space having the mass of a mountain and the radius of an atomic nucleus. The rate of energy release can be estimated from the formula for the energy flux in thermal radiation from a surface, the Stefan–Boltzmann law:

$$P = \sigma A T^4 \tag{21.27}$$

where P is power, A is area and σ is the Stefan–Boltzmann constant. In the present case the relevant area is of order $4\pi r_{\mathrm{s}}^2$, so we find

$$P \simeq 4\pi\sigma \left(\frac{2GM}{c^2}\right)^2 \left(\frac{\hbar c^3}{8\pi k_{\mathrm{B}}GM}\right)^4 = \frac{\hbar}{240\pi}\left(\frac{c^3}{8GM}\right)^2. \tag{21.28}$$

Let us apply this formula when the mass is sufficiently small that approximately one second remains until it evaporates completely. We solve $Pt \simeq Mc^2$ for M when $t = 1$ second and thus obtain $M \sim 10^5$ kg and $P \simeq 10^{22}$ watts. This would manifest as a brief burst of gamma rays, but with a luminosity far below that of the gamma ray bursts that have been detected from other sources. An evaporating black hole would probably have to be in our galaxy in order to be detected. No such bursts have been observed to date.

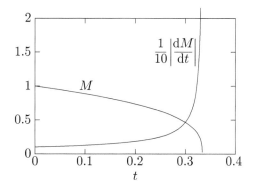

Fig. 21.3 The function $M = (1 - 3t)^{1/3}$ and its derivative. This is a solution to $\mathrm{d}M/\mathrm{d}t = -1/M^2$ and thus illustrates the process of evaporation: an initially slow decay finishing in an abrupt explosion.

Black hole information paradox. Since the mass of a Schwarzschild black hole is reduced by Hawking radiation, so is the horizon area. Thus the theorem that the area can only grow is true of classical GR but not of the quantum case. The area decrease owing to Hawking radiation does not represent a net drop in the entropy of the universe since (21.25) indicates that the entropy of the black hole is carried away in the thermal radiation. However, it is not clear whether the whole process of birth, life and death of a black hole conserves entropy overall, as would be required if it were described by quantum field theory, where all processes are strictly *unitary* (i.e. entropy conserving). This question is called the *black hole information paradox*. The word 'paradox' is serving to indicate that it is difficult to understand how black hole processes can be unitary overall,[6] and it is also difficult to understand what is implied for the rest of physics if in fact such processes are non-unitary. The situation is important because it implies that there can be whole regions of spacetime (the future lightcones of evaporated black holes) for which GR and quantum field theory are not able to give a complete account of how the future relates to the past. But for all we know there may not be any fully deterministic behaviour in such cases: maybe this is the lesson.

Exercises

21.1 Compare and contrast the Penrose process with the 'gravity-assist' which is often exploited in planning the trajectories of space probes in the solar system.

21.2 It is proposed to turn a Kerr black hole with $a < m$ into an extremal Kerr black hole ($a = m$) by throwing matter into it in such a way as to increase the angular momentum parameter a as much as possible. Prove that the attempt cannot succeed.

21.3 Show that the total horizon area of two Kerr black holes of masses M_1, M_2 and angular momentum J_1, J_2 is smaller than the area of a single black hole of mass $M_1 + M_2$ and angular momentum $J_1 + J_2$.

21.4 The Kerr metric in Boyer–Lindquist coordinates is independent of t and ϕ, therefore both $(1, 0, 0, 0)$ and $(0, 0, 0, 1)$ are Killing vectors, and consequently so is $(\alpha, 0, 0, \beta)$ for any constants α, β. Show that the horizon at r_+ is a Killing horizon for the Killing vector $K^a = (1, 0, 0, \Omega_{\mathrm{H}})$ where Ω_{H} is given by (19.31).

21.5 If two Schwarzschild black holes, each of mass M, come together and merge, show that the mass of the resulting black hole is $\geq \sqrt{2}M$ (and therefore the energy radiated by gravitational radiation is at most $(2 - \sqrt{2})Mc^2$).

21.6 Describe the visual appearance (from a distance) of a black hole of mass 10^{20} kg.

21.7 How large a volume of water at ordinary pressure and temperature would have the same entropy as a Schwarzschild black hole of radius one centimetre?

---

[6]Attempts to relate information falling past the horizon directly to entropy in the Hawking radiation do not succeed because the rates do not match.

Part IV

22

Cosmology

We now turn to the application of GR to the large-scale structure of the universe: the subject of cosmology. According to Einstein's equation, spacetime responds dynamically to the motion of matter, and every mass (and associated pressure etc.) curves the space around it, like dimples in a sheet. The net effect of all these motions and dimples will be large-scale dynamic behaviour, and a net curvature of the whole sheet—the curvature of spacetime on a cosmic scale. What is the size of this overall curvature? What is its sign? And how has the universe evolved over time? How old is it? These questions can in principle be answered if we know enough about the distribution of matter in the universe, and the state of motion of that matter. They form the starting point of much of modern cosmology.

Associated questions concern the topology of the universe. For example, if it turned out that the metric on a large enough scale were homogeneous with a spatial part of uniform positive curvature, then we would guess the topology to be that of a sphere. If the curvature came out negative, we would still have the right to guess that the topology is closed spatially, but that would now be a less natural guess (Section 22.2.3).

22.1 Observed properties of the universe

The first thing to say about our universe is that it is vast. When we gaze up into the sky, there is no sign of any boundary, and there is no evidence (yet) of a finite volume, up to distance scales of the order of ten billion light years. The total volume of space might be finite, but in any case it is vast. (A finite volume need not imply that there is any boundary, just as the finite surface area of the Earth does not imply that it has an 'edge'.)

Within our own galaxy there are about 10^{11} stars, and this is typical. One hundred billion is a number hard to imagine. It is similar to the number of grains of rice that you could fit into a large hall. Furthermore, and this is one of the striking discoveries of the twentieth century,[1]

[1] It was conjectured as long ago as 1576 (Thomas Digges; Giordano Bruno; Nicholas of Cusa) that the stars were spread uniformly through space. In the mid-eighteenth century Thomas Wright, Immanuel Kant and others argued that there may be galaxies besides our own, but it was not confirmed until 1925, and the sheer number of further galaxies was wholly unexpected.

Relativity Made Relatively Easy: General Relativity and Cosmology. Volume 2. Andrew M. Steane, Oxford University Press. © Andrew M. Steane 2021. DOI: 10.1093/oso/9780192895646.003.0022

Fig. 22.1 A visual impression of the distribution of galaxies as indicated by data from the Sloan Digital Sky Survey. The image shows a large fraction of the northern part of the sky. Roughly speaking, each pixel corresponds to one galaxy. (Courtesy Sloan Digital Sky Survey)

our galaxy is just one among vast numbers of galaxies. A 200-inch telescope can detect about 10^{11} galaxies. None of us can hope even to glimpse them all in our limited human lifespan.

There is structure among the galaxies: they mostly cluster in gravitationally bound groups of 1000 or more, and on larger scales they lie on a kind of spongy or filamentary structure, with edges and cell walls and huge voids in between. These cells have dimensions of order 10^8 light years. Above this scale there appears to be no further structure and an approximate homogeneity sets in. There is room for about a million such cells in the observable universe (to be defined later), so at a very large scale it is reasonable to treat the matter in the universe as approximately forming a homogeneous fluid; see Fig. 22.1.

The next thing to say is that the universe is ancient. Modern estimates put this age at 13.8 billion years (this is the elapsed proper time at any place since a very early extremely-high-energy-density phase in which the notion of *time* probably breaks down; more on this later). The universe is certainly (i.e. it is beyond reasonable doubt that it is) older than 10 Gyr. It is possible to assert this confidently because several independent strands of evidence agree on it. That planet Earth is itself very old is signalled by all sorts of evidences, from geology, plate tectonics, biological evolution, etc. Radioactive dating methods put the age of the Earth and other solar system objects at around 4.5 Gyr. The universe is older than this, of course, but this already signals that timescales of many Gyr should be expected. Radiometric dating can be extended to the material from which the solar system was formed. For example, the two isotopes of uranium, ^{238}U and ^{235}U, are created in roughly equal amounts in stellar interiors and then released into space through explosions, but they are found now in a ratio of about 140:1. This can be ascribed to the known shorter lifetime of ^{235}U, and after accounting for the fact that the production is spread out over time, one finds that some stars in our (cosmic) vicinity were formed 13 ± 1.5 Gyr ago. Studies of old stars in the halo of our galaxy also yield age estimates based on their observed luminosity and reasonably well-tested stellar models; an example recent result is 12.5 ± 1.5 Gyr for the age of one such group of stars. Finally, estimates of the age of the universe as a whole come from applying General Relativity and particle physics calculations to spacetime and the matter in it, taking some quite accurate guidance from large-scale empirical observations, especially nuclear abundances, galaxy structures, and the other primary source of cosmological evidence: the cosmic microwave background radiation. We shall examine some of these calculations in the next few chapters.

The main further remarks to be made about the universe as a whole are that it is *expanding, consistent with General Relativity, cooling, isotropic and homogeneous on a large scale, mostly dark, and profoundly mysterious.* We shall comment on each of these in turn, but of course like all scientific descriptions, the overall picture is not built up piecemeal but has multiple internal connections and consistency checks.

Expanding. As soon as large numbers of distant galaxies became detectable using large telescopes, it also became apparent that the galaxy clusters are all moving away from us with recessional velocities approximately proportional to their distance from us. This is called the *expansion of the universe.* Edwin Hubble[2] originally described his observations in the form $v = H_0 d$ where v is a recessional velocity (deduced from redshift), d is distance from Earth (deduced from distance calibration techniques; Section 22.7) and H_0, called the Hubble constant, is the coefficient of proportionality in the observations. In a more thorough approach based on GR, we prefer to speak of a dimensionless scale factor $a(t)$ which calibrates distances in the metric of spacetime, and then define the

---
Hubble parameter

$$H(t) = \frac{\dot{a}}{a} \tag{22.1}$$

Then H_0 is equal to the value of the Hubble parameter at present:

$$H_0 \equiv H(t_0) \tag{22.2}$$

where t_0 is the present time on Earth. Since $H(t)$ only changes appreciably over billions of years, t_0 does not need to be specified very accurately; any time in human history will do. To derive the Hubble law from this, use that $a(t)$ is defined in such a way that the distance d varies as $d = a(t)d_0$ where d_0 is constant, so $\dot{d} = \dot{a}d_0 = aHd_0 = Hd$; this is Hubble's law.

This is a good moment to remove a possible confusion with the terminology. $H(t)$ is called the Hubble *parameter* and it can change with time. H_0 is called the Hubble *constant* and it cannot change with time because by definition it is the value of the Hubble parameter at one particular time. However, this terminology is not always adhered to, and you may encounter statements asserting that the Hubble constant had a different value in the past. They mean the Hubble parameter.

[2]The law now named after Hubble was first proposed by Lemaître, who also gave the first estimate of the constant now called the Hubble constant; Hubble's observational work both pioneered important astronomical techniques and also established the expansion beyond reasonable doubt.

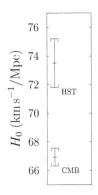

Fig. 22.2 Measured values of H_0 using two largely independent methods. The discrepancy may be owing to chance, or may hint at new physics to be discovered. (Riess *et al.*, 2018*b*; Riess *et al.*, 2018*a*; Planck Collaboration: Ade, P. A. R. *et al.*, 2014; Planck Collaboration: N. Aghanim *et al.*, 2018)

Two recent measurements of H_0 are shown in Fig. 22.2. Measurements based on observed redshift of stars and galaxies give $H_0 = 73.5 \pm 1.7\,\mathrm{km\,s^{-1}/Mpc}$; measurements based on the cosmic microwave background radiation and other information give $H_0 = 66.93 \pm 0.62\,\mathrm{km\,s^{-1}/Mpc}$. Thus there is a good precision, of order 2%, but the two measurements differ by 9%. This both confirms the broad lines of our understanding of cosmology, and also suggests there is more to be discovered if this discrepancy persists as measurements grow more precise. The value $70 \pm 3\,\mathrm{km\,s^{-1}/Mpc}$ is $H_0 = (2.3 \pm 0.1) \times 10^{-18}\,\mathrm{s^{-1}}$ in SI units.

By extrapolating the observed motion back into the past, it is obvious that the galaxies used to be closer to each other than they are now. By assuming that basic physical laws and fundamental constants are unchanging (an assumption that has been experimentally tested to high precision) one is driven to the conclusion that, of the order of 14 billion years ago, conditions were very different and, in particular, the average energy density in both matter and radiation was much much higher. Running time forward again from that epoch, we have the picture of a very dense, extremely hot plasma exploding in all directions at all places. This overall picture is called the *Hot Big Bang*.

To get a feel for the average density of ordinary matter in the universe, one can use the rough figure 'one proton per cubic metre' ($10^{-27}\,\mathrm{kg/m^3}$) for conditions now. The average value within any given galaxy is 10^7 times higher than this, but the voids between galaxy clusters bring the average down. To compare with a more everyday density: hydrogen gas at standard temperature and pressure has approximately 10^{25} molecules per cubic metre. The average density of matter in the universe would reach that value if the linear dimensions of the universe were reduced by a factor 2×10^8. The cosmological reasoning to be discussed in this chapter suggests that this occurred when the time elapsed since the Big Bang was of the order of ten minutes, and the ambient temperature was 10^8 K (i.e. 30 times hotter than the centre of the Sun). Earlier times had higher densities and yet higher temperatures.

Consistent with General Relativity. By treating the matter, radiation and vacuum in the universe as a homogeneous fluid, it is possible to apply Einstein's field equation to the large-scale content and behaviour of the universe. According to General Relativity, we should not regard spacetime as a passive backdrop in which the expansion of the universe plays out. Rather, spacetime itself is dynamic, and has been expanding—we shall elucidate the precise meaning of this phrase in this chapter. The main point is that this approach can make perfect sense of the expansion and other main properties, and because of this one has a sound framework in which to apply other calculations, such as electromagnetic theory to describe matter-radiation interactions, whether now or in the early universe. In such calculations, classical theory is applicable at temperatures below about 10^9 K (0.5 MeV), nuclear physics at temperatures up to 10^{13} K (1 GeV), and particle physics and quantum field theory at higher energies. Eventually, at the earliest times, our scientific knowledge runs out, and this is a fascinating frontier area of fundamental science, about which we shall say more in a moment.

It turns out that the average curvature of the space of our universe is small—the universe is flat or almost flat—so the impact of gravity on the processes in any 'small' region (say, just a lightyear in diameter) is dominated by the effect of local mass concentrations, not global curvature. This makes the calculations tractable (indeed, one can safely ignore gravitation altogether when calculating reaction rates etc., as the Equivalence Principle says, in all but the most curved circumstances.)

Cooling. After combining the previous two points (expansion and GR) one may infer that at sufficiently early times the universe was not cold (on average) as it is today: the matter and radiation in the universe had large amounts of random kinetic energy. At a time before stars and galaxies formed, the matter was spread out more uniformly and everywhere glowed brightly: the whole universe was filled with hot plasma. At least this is what one may conjecture, and then one may seek evidence now that this was the case. That evidence has been found in two main ways.

It is found that there is about one helium atom for every ten hydrogen atoms distributed throughout the universe. If we suggest that this helium came from fusion of hydrogen, then there have not been enough stellar furnices in which the fusion could have happened. However, according to the Hot Big Bang model the whole universe was once a thermonuclear furnace, and by using a self-consistent model of the temperature, density and duration of the nucleosynthesis 'epoch' (from about 3 minutes to 20 minutes after the Big Bang) one may predict relative abundances not only of ^{1}H and ^{4}He but also of ^{2}H, ^{3}He and ^{7}Li. It is found that these predictions match the observations (Fig. 22.3). This success is both powerful evidence for the correctness of the Big Bang model, and also a remarkable example of the long reach of scientific reasoning.

After the nucleosynthesis epoch the Big Bang model proposes that the universe continued to cool, but for a long time, about 250,000 years, it remained too hot for atoms to form. Any electrons that became briefly bound to protons would have been quickly liberated again due to excitation by ultra-violet- or higher-energy photons. In this state the universe everywhere glowed brightly but was opaque: a given photon only travelled a short distance before undergoing Thomson scattering[3] off an electron. However, as the plasma expanded it also cooled (says the GR hydrodynamics equation (14.23)—you can interpret this as a transfer from random kinetic energy to gravitational potential energy if you like), until a temperature of approximately 3500 K was reached. At this temperature the equilibrium position of the reversible reaction

---

[3]Thomson scattering is the low energy limit of Compton scattering.

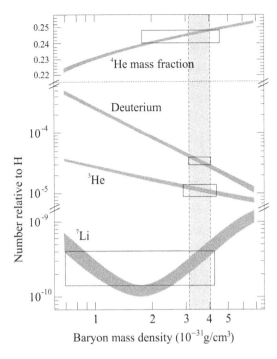

Fig. 22.3 Big Bang nucleosynthesis. The curves show the relative abundances of the lighter nuclei, as a function of the average mass density in the present universe, as calculated from nuclear physics on the assumption that these nuclei were formed by fusion in the early universe. The boxes indicate the empirical values, estimated from studies of stars and other bodies, including Earth. The fact that there is a value of density where the calculations are consistent with all the data (dashed lines) is a remarkable success. The fact that this value also agrees with other estimates further enhances our level of confidence in the Big Bang model.

$$e^- + H^+ \leftrightarrow H + \gamma$$

shifted abruptly because the rate of the process from right to left fell rapidly. The resulting process of combination of electrons and protons into neutral atoms is called **recombination**.[4] A rough calculation of this process can be made by using the **Saha equation**, which relates the density of free electrons to hydrogen atoms in a plasma in static thermal equilibrium:

$$\frac{n_e n_p}{n_H} = \left(\frac{m_e k_B T}{2\pi \hbar^2} \right)^{3/2} e^{-E_0/k_B T} \tag{22.3}$$

where E_0 is the binding energy (approximately 13.6 eV, but we will comment further on this in a moment) and n_e, n_p, n_H are the number densities of electrons, protons and hydrogen atoms. Since whenever an electron combines with a proton exactly one hydrogen atom appears, the quantity $n_B = n_p + n_H$ is constant under changing degrees of ionization. So a useful measure of the degree of ionization is

$$x \equiv \frac{n_e}{n_p + n_H}. \tag{22.4}$$

For a neutral plasma we have $n_p = n_e$ so we can also write $n_p = n_B x$ and one finds

$$\frac{n_e n_p}{n_H} = \frac{x^2}{1-x} n_B. \tag{22.5}$$

Using the Saha equation this can be solved for x at any given T, if we know n_B. This can be obtained by measuring the value now and extrapolating back using the known dependence on

---

[4]This is a slight misnomer because the process was a union of things never previously stably bound together, and therefore a combination not a recombination, but the name has stuck.

the universal expansion, eqn (23.24). One finds $n_B \simeq 10^8 \, \mathrm{m}^{-3}$ at recombination. Substituting this into (22.5) and using (22.3) one finds that x becomes small (i.e. recombination happens) when $T \simeq 3600 \, \mathrm{K}$. This is a reasonable estimate, and owing to the exponential term in the Saha equation the temperature value is not sensitive to imprecision in n_B. However, the full process is more complicated: one should take account of the internal structure of the hydrogen atom, because the typical photon energy at recombination is much lower than E_0. A more sophisticated treatment would also allow for the dynamic background spacetime.

It took approximately 70,000 years for the ion fraction to fall from near 1 to near 0. The plasma being electrically neutral overall, the consequence was that when the electrons and protons combined into neutral hydrogen (and some helium etc.) atoms, almost no free charged particles were left. At all but resonant wavelengths, the cross-section for photon scattering off a neutral atom is much less than that for Thomson scattering off a charged particle. Therefore the opaque plasma turned into a transparent gas; this is called **decoupling**. A suitable definition of this decoupling is when the scattering rate falls to the value H (the Hubble parameter) or less. This should not be confused with recombination although the two effects are related. Neither process is completely abrupt but it is safe to say that the temperature at decoupling is somewhat lower than at recombination: $T_{\mathrm{decouple}} \simeq 3000 \, \mathrm{K}$ (a calculation is presented in Section 23.5). After decoupling, any photons already in existence simply continued to propagate; some of them for billions of years! *We now detect some of those same photons, arriving in our microwave detectors from every direction in the sky.* This is the second major line of evidence for the hot Big Bang.

The first direct detection of the radiation now called cosmic microwave background (CMB) radiation was a triumph of experimental physics achieved by Arno Penzias and Robert Wilson in 1964.[5] To explain the observation now, on Earth, of microwaves that have been travelling for billions of years, we claim that the radiation was produced everywhere in the universe, and subsequently journeyed from all points towards all other points. We now receive the part that happened to set off in our direction from points on a huge sphere around our location. Observers elsewhere receive radiation from a sphere around them. This radiation is called the **cosmic microwave background** (CMB) radiation; its detailed properties present a host of data that can be used to test and constrain cosmological theories.

After a dipolar anisotropy has been taken into account, the CMB radiation is observed to be accurately thermal (Planckian) at a temperature $T = 2.725 \, \mathrm{K}$ (hence a peak wavelength of 2 mm). This is consistent with the above process if the radiation has undergone a redshift by a factor $3000/2.725 = 1100$ since being produced. We will show in Section 22.4.1 that this redshift may be attributed to the 'stretching' of space during cosmological expansion. To be precise, there exists a sensible choice of reference frame (the comoving frame) in which the redshift of the CMB radiation is wholly owing to a change in the metric between that at the events long ago when the photons last scattered and that at the detection events in our measuring devices. If you like, you can call this a difference in gravitational time dilation for observers at different positions and times in an evolving spacetime. (This is a more useful physical picture than one

---

[5]Andrew McKellar had previously used observations of absorption lines to determine a rotational temperature of interstellar molecules in the vicinity of 2.3 K but the significance of this was unclear at the time.

that attributes the shift to a Doppler effect, but gravitational and Doppler effects are to some extent interchangeable by choice of coordinate system.)

Most photons propagating in the universe today are CMB photons. It might seem odd to account for continuous electromagnetic radiation by counting photons but it is a well-defined idea; the number at any given frequency in any given volume of space can be obtained from the energy density in the radiation at that frequency ω, divided by $\hbar\omega$. To find this number density (averaged over frequency) one can use standard properties of thermal radiation:

$$n_\gamma = 0.243 \left(\frac{k_B T}{\hbar c} \right)^3 \qquad \simeq 4.095 \times 10^8 \text{ m}^{-3} \text{ at } 2.725 \text{ K}. \qquad (22.6)$$

Isotropic and homogeneous

An observer who moved with respect to the centre of our galaxy at 368 ± 2 km s^{-1} in the appropriate direction[6] would see the CMB spectrum to be the same in all directions, to within a few parts in 10^5. This extreme isotropy is an important clue to the large-scale structure of the universe, and one that greatly simplifies the task of cosmology.

It seems on the face of it unlikely that our galaxy would happen to occupy a special position in the universe from a cosmological point of view. Obviously in order to support life, planet Earth must have many special features, including for example its 'just right' distance from a stable, long-lived star in a quiet part of the galaxy. Other planets elsewhere may have similar life-encouraging properties, but there is no reason we know of to think that the large scale structure of the universe has anything to do with it. Therefore we make the conjecture that our galaxy has no special location, or to be precise, *the universe would look much the same from any other location*. This idea is called the *cosmological principle* or *Copernican principle*. The word 'principle' is perhaps a little too strong, since this supposition does not have the same status as ideas like the Principle of Relativity or the Second Law of Thermodynamics. It is just the reasonable assumption that if there is some special place in the universe, then we probably do not live there, since it is more likely that our local cluster of galaxies will be at an 'ordinary' rather than 'extraordinary' location.

Of course our location is extraordinary from many points of view—there are living things here, for example—but we suspect this is not connected with any local special feature of the large scale structure of the universe.

The assumption that our location is not special, combined with the observed near isotropy around our location, together imply that the universe will be isotropic at other locations too, and this is sufficient to imply homogeneity.[7] This does not imply that homogeneity is itself a principle, or something one can confidently expect without looking. The argument requires the observed isotropy in order to function as an argument.

---

[6]The dipole anisotropy in the CMB at Earth implies that the solar system is moving in the direction of galactic longitude 263.85°, latitude 48.25° relative to a local frame in which the CMB would be isotropic.

[7]Isotropic means uniform with respect to direction; homogeneous means uniform with respect to position.

The reason why large-scale homogeneity is not a principle one should claim to know *a priori*, or to confidently expect without looking, is that there is no good reason to either expect it or not expect it. There are, after all, many aspects of the natural world that display structure at multiple scales—a tree, a coastline and a phase transition are examples. It is reasonable to expect the universe also to display structure at multiple scales, and so it does: galaxies, clusters of galaxies, filaments of galaxy clusters, etc. There are two possibilities, then. Either, as one goes to larger and larger distances scales, one continues to encounter non-uniformity at larger and larger distance scales, like the movement from the twigs to the branches to the trunk of a tree, or else one does not. The way to determine which case holds is the empirical way.

The empirical observations indicate homogeneity. There is now evidence from two main sources: galaxy surveys (Fig. 22.1) and the CMB. The very high degree of isotropy of the CMB (we already noted it is isotropic in temperature to a few parts in 10^{-5}) is direct evidence of isotropy of the universe about our location, and it strongly suggests homogeneity as well, because the CMB photons we see have travelled over large distances. It beggars belief that these photons have passed through inhomogeneities that happen to be spherically symmetric and centred on our location. In any case, General Relativity allows a homogeneous model in which the isotropy is to be expected for an observer located anywhere in the universe. Since such a model is simple and natural it is to be preferred over others (by Occam's razor) unless some empirical evidence should accumulate against it. This is not to say it is the whole truth, since it is an approximation. The fact that the universe is not quite perfectly homogeneous will have an effect on the expansion we observe, but one which we can safely ignore to first approximation.

Mostly dark

Astronomers like to use the word 'dark' to refer to any type of observed entity which is not glowing like a star. Most everyday objects are 'dark' in this sense, but in cosmology almost all the 'dark' matter and energy, whose presence is inferred from gravitational effects, is 'dark' in a further sense—*we do not know what it is, and suspect that it is unlike any of the known types of particle species.*

The term *dark matter* refers to matter which is not glowing and is distributed in large quantities around and between galaxies, and which does not exhibit much tension. Its presence is inferred from five stands of evidence: rotation curves of galaxies; gravitational lensing; study of structure formation; modelling of nucleosynthesis; and cosmological models bringing together galactic clustering, microwave background and related measurements. We shall discuss these briefly in turn.

The stars and interstellar dust in any given galaxy orbit around the galactic centre, and by determining the velocity from the Doppler effect, one builds up the velocity profile (called rotation curve), which in term implies the mass profile of the gravitating matter producing the orbits. Both the amount and distribution of this matter differs greatly from that of the average over the visible (chiefly stellar) matter. The distribution of the inferred dark matter is roughly spherical, even in otherwise roughly flat galaxies.

The gravitational lensing effect of entire galaxies can be measured; this gives information on the mass content and to some extent the mass distribution. Again it is found that the inferred mass exceeds by an order of magnitude that which can be attributed to the stars. One can also use lensing to detect the presence of gravitating material in filaments between galaxies, where there is almost no visible matter at all.

Structure formation refers to the gravitational collapse of initially dispersed matter to form such things as galaxies and stars over billions of years. This is now modelled numerically with great sophistication; such models can attain reasonable agreement with the statistics of observed structures and timescales in the universe, but only if the gravitation at galactic scales is largely owing to dark matter rather than ordinary (baryonic) matter.

The measures of mere mass coming from lensing and rotation curves do not indicate the nature of the mass; it might be ordinary matter (usually called baryonic matter in cosmology discussions). Structure formation calculations can distinguish ordinary from other matter, and the nucleosynthesis calculations can do this more directly. The various abundances indicated in Fig. 22.3 line up with the observed values at one particular value of the baryon mass density, enabling it to be extracted with confidence. Its value is considerably lower than that of the total mass density inferred from the other studies. Therefore dark matter is mostly non baryonic.

The combination of several observational techniques, to be described in Chapter 25, gives further and precise information. It is thus determined that the ratio of dark matter to baryonic matter (mass density averaged over the largest scales) is $(5.3 \pm 0.2) : 1$ (Table 25.1). Overall the case that most of the matter in the universe is dark matter is now so strong as to be almost incontestable. The interesting contemporary research problem is to observe its distribution in more detail, and to discover more about it. It is known that it cannot be in motion at high velocity or it would by now have spread out more uniformly; for this reason it is called cold dark matter. Proposed extensions to the Standard Model of particle physics offer several candidate types of particle which might form dark matter; experiments are underway in hopes of detecting some such particles here on Earth.

The word 'dark' is also employed as a name for the further contribution to the average energy tensor called 'dark energy', mentioned in Chapters 2 and 16. The dynamics and curvature of the universe on the largest scales of time and distance cannot be accounted for by the observed amounts of baryonic and dark matter alone. The evidence is discussed in Chapter 25. The simplest way to account for the observed behaviour is to include a non-zero cosmological constant term in the Einstein field equation. This is not the only possibility. There are three main hypotheses about the nature of dark energy: it might be a property of the vacuum state of all the quantum fields which together fill or constitute spacetime, or it might be a property of some unknown type of matter, or it might be a feature of large-scale geometry imposed on General Relativity for some reason we do not currently understand.

Profoundly mysterious

Any scientific investigation touches on unknown territory when one pursues it far enough. In the case of cosmology this is more immediately apparent than in most subjects. Although it is right to celebrate the great progress of the last century, and indeed of the last few decades, it is also scientifically appropriate to keep two things clearly in view. The first is that we do not know what most of the universe is made of. The second is that we do not know how or why the initial conditions came about, nor what happened at the earliest moments and highest densities, which lie outside the domain of applicability of all our physical theories.

About 5% of the mass-energy content of the universe is reckoned to be 'familiar' matter such as stars, planets and ordinary interstellar dust, the other 95% is thought to be in the two forms discussed above: a quarter of it called 'dark matter' which has energy but little or no pressure or tension, and three quarters of it called 'dark energy' which has energy and also a large amount of tension. It is hard to overstate how little we know about these things, in comparison to our knowledge of the simplest everyday matters such as ordinary liquids, solids and gases. Glance up at the sky at night and you are peering through an invisible cloud of we know not what.

Regarding the very early conditions and the initial conditions, high-energy physics, based on quantum field theory and collider experiments, can make reliable calculations up to the 'electroweak' scale (100–1000 GeV), and arguments may be made to suggest that it remains broadly reliable up to the so-called 'grand unification' energy scales of order 10^{15} GeV. One should not forget, however, that particle physics experiments have often led the way and caused the rejection of promising theoretical models and stimulated the discovery of better ones, so to apply current models right up the grand unification scale should not be regarded as a precise science. Experimental tests have only been possible so far to $\sim 10^4$ GeV.

Beyond the 10^{15} GeV energy scale, all claims to be able to describe conditions in the early universe should be regarded as tentative. Quantum field theory, or something like it, may be able to handle physical conditions up to the Planck energy (10^{19} GeV), where gravitational curvature effects dominate. In conditions more extreme still, possibly some sort of quantum geometrical model is appropriate.

The Big Bang theory is not a theory of creation in the metaphysical sense. It does not describe the creation of anything from nothing. Rather, it asserts that the universe has evolved in a consistent way since its earliest moments, and at some very early stage the universe was extremely hot and dense, and also configured in such a way as to expand very smoothly. Everything which the study of physics can uncover about this evolution starts out from the assertion that the universe was already constituted, in the form of a collection of interacting quantum fields in a very special state. No physical model can provide a basis for discovering why there is any physical realization of mathematical concepts in the first place, nor why a long-lived combination of quantum fields should exist rather than not exist, nor why symmetry—but not perfect symmetry—is preferred over the lack of it, nor why entropy was initially so low, etc. etc.

Against 'infinity'.   As a final preparatory comment, I would like to mention here a personal prejudice which you can ignore if you like. This is that I feel it is too great an extrapolation from experience to call the universe 'infinite'. If the universe were infinite, we could never know it, of course. One thing we do know is a sensible way in which it could be finite and unbounded—it just needs to have a closed topology (which is possible whatever the spatial curvature, more on this later). It seems to my prejudices that this is a more likely scenario than that there is an infinite number of galaxies and an infinite volume of space. It is simply that the mind boggles at this kind of infinity. Mathematical abstract notions of infinity are one thing, but physical entities are another.

I do not have any logical proof to offer, no *reductio ad absurdum*, but I do want to insist that we really do not have any evidence that the physical cosmos is infinite, and the idea that it is infinite seems to me to involve a wild degree of extrapolation. The difference between finite and infinite is, after all, not a small adjustment. It is a huge claim. It is the claim that a collection of 10^{22} stars, or even $100^{100^{100}}$ stars, would be but as an infinitessimal mote swimming in infinite surroundings. I doubt this—but I do not know it one way or the other.

I have included this section merely to allow the reader to know that there is no strong scientific reason to compel us to think that the physical universe is infinite, and there is a natural way in which it could be finite without the need to have a boundary.

22.2 Cosmic time and space

Consider a block of cheese in flat space. By using a curved cheese-cutter, you can cut out a non-flat slice of cheese—for example, a section of a spherical surface. This illustrates the fact that a given manifold in N dimensions does not determine the curvature of all sub-manifolds of $N-1$ dimensions that can be sliced out of it (but it does determine the curvature of geodesic planes). In particular, in a four-dimensional spacetime manifold, the phrase 'the curvature of space' has no unique meaning: it depends on how the spacetime is sliced up (foliated) into 'time' and 'space'. This and other facts makes the task of thinking about the spacetime of our cosmos rather difficult. However, the task is grealy simplified by the observed isotropy and homogeneity at large scales. At any point in the Universe a natural standard reference frame can be singled out as that of an observer whose cosmic microwave background is isotropic. Such observers are called **fundamental observers**. A spatial coordinate system defined such that all fundamental observers are not moving (relative to the coordinate grid) is called a system of **comoving coordinates**. Such a coordinate system provides a natural definition of what to call 'space' and what to call 'time': the time axis at any event is parallel to the worldline of a local fundamental observer; the spatial hyperplane is perpendicular to this. You can imagine a fundamental observer as corresponding roughly to one galaxy or cluster of galaxies.

This definition amounts to saying that at any event in spacetime there is a preferred reference frame. Does this break the Principle of Relativity? Not at all. The situation can be compared to that of living in a house. For someone who lives in a house, there is a natural reference frame in which to describe their daily life, namely, the reference frame in which the house is at rest. Does the existence of this preferred reference frame violate the Principle of Relativity? No. Similarly, there is a natural (accelerating) reference frame which may be used to describe our cosmic home, and no violation of relativity principles is made by this assertion. It is an assertion about the distribution of matter in the universe, and the large-scale regularity of its motion. In short, *the manifold does not dictate what you should call space, but the matter distribution gives a very strong hint.*

Once one has accepted this, the language of cosmology simplifies a great deal. One can talk of 'cosmic time'—it is the proper time of a fundamental observer since the Big Bang—and 'cosmic distance' (often called 'proper distance')—it is the integral of the ruler distance from one fundamental observer to another along the spatial hypersurface at the same cosmic time.

22.2.1 Synchronous comoving coordinates

We shall now formalize the above. If we define a fundamental observer as one whose CMB is isotropic, then *Weyl's postulate* is the assumption that the worldlines of these observers form a bundle or *congruence* in spacetime. That is to say, these worldlines are non-intersecting, except possibly at a singular point in the past or the future or both. We say that such a group of worldlines provides a *threading* of spacetime. We may now construct spatial hypersurfaces which are everywhere orthogonal to these worldlines. Cosmic time t is then defined such that each such surface gets the same t everywhere, and furthermore we choose this t equal to proper time along the worldline of one fundamental observer. Then, since all the worldlines of fundamental observers are orthogonal to the hypersurfaces, all of them are accumulating the same amount of proper time from one hypersurface to the next (in case you doubt this, we shall demonstrate it in a moment.)

Next, introduce spatial coordinates x^i such that they are constant along the worldline of each fundamental observer. Under these definitions, and Weyl's postulate, the metric of spacetime, at the largest cosmic scales, must have the form

$$ds^2 = -c^2dt^2 + g_{ij}dx^i dx^j, \tag{22.7}$$

where t is cosmic time, x^i are spatial coordinates fixed at the fundamental observers, and g_{ij} may be a function of all the coordinates (t, x^i). Note that in this metric we have not yet inserted any notion of homogeneity or isotropy. We have merely asserted that g_{00} is constant and $g_{0i} = 0$, so there are no mixed time/space terms.

We now claim that the lines $dx^i = 0$ are geodesics. To prove it, note that any contribution from a non-zero spatial journey can only increase ds^2 when the spatial surfaces are strictly orthogonal to the line in question, so the length of the line is stationary when $dx^i = 0$; QED ('proof by twin paradox'). If you want to see it spelled out in algebra, then use that the 4-velocity along any one of these lines, in our chosen coordinate system, is $[u^a] = (c, 0, 0, 0)$, so we only need to find the Γ^a_{00} term in the geodesic equation. We find $\Gamma^a_{00} = \frac{1}{2}g^{a\lambda}(2\partial_0 g_{0\lambda} - \partial_\lambda g_{00}) = 0$ by using that g_{00} is constant and $g_{0i} = 0$. It follows that the lines at constant x^i are indeed geodesics, so they are the worldlines of observers in free fall. t is the proper time for all these observers.

22.2.2 How to think about the Big Bang

It is common in cosmology to speak of what has happened 'since the Big Bang' or to refer to conditions at such-and-such a time 'after the Big Bang singularity'. Such language introduces

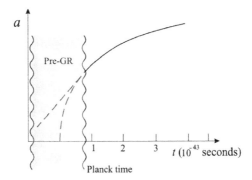

Fig. 22.4 The nature of universe in its earliest moments is largely unknown to us, but this does not affect the precision of calculations based on extrapolating back from later periods, as long as the extrapolation is not pushed too far. Statements such as 'x seconds after the Big Bang' should be interpreted to mean 'x seconds after the early little-known period' not 'x seconds after a singularity'.

a significant problem of interpretation, because at sufficiently early moments, when the conditions were sufficiently extreme, we do not know whether or not any particular physical and mathematical model we may have developed will apply. To deal with this lack of knowledge, it is useful to separate in our thinking a very early epoch in which we do not know whether terminology such as 'spacetime' or 'quantum field' is appropriate, and subsequent evolution where those terms are appropriate (Fig. 22.4). This earliest epoch precedes the era commonly called the inflationary era in modern-day cosmology, because the process of inflation, if it happened, is reckoned to fall within the competence of GR and particle physics as we know them. We shall refer to this earliest epoch as the 'pre-GR' epoch. It is often called the 'Planck era'. We do not even know whether words such as 'time' apply to it.

The transition between the pre-GR epoch and the spacetime evolution era (i.e. all the evolution since) need not be imagined to be abrupt, but everything we shall say about spacetime will be based on the idea of starting out from an object of non-zero dimensions, not a point. This object is spacetime and all the matter and radiation in it. Whatever the pre-GR epoch was, it finished by producing a universe of non-infinitesimal volume and finite curvature. It also had finite temperature and finite density. It is this universe which we shall be discussing using the framework of GR.

With this in mind, we can now assert that fundamental observers located on different galaxies in the present were, at the earliest cosmic times accessible to GR, close to one another, but not infinitesimally close. Therefore they could not communicate with one another at first: it took a finite time for light or other signals to propagate from any one of them to any another. However, this does not rule out that the physical conditions such as density and temperature at different fundamental observers may have been uniform before such communication could happen. Those conditions will have been brought about in the pre-GR epoch which we do not claim to understand, so we have no reason to say whether uniformity is more or less to be expected than non-uniformity.

We can build a rough time-line of the early universe (Fig. 22.5) by using the model we shall be discussing in this and the next chapters. The model is based on Einstein's field equation applied to a homogeneous distribution of matter. We will show that the predicted behaviour is such that when the temperature is high enough the universe evolves much like adiabatic expansion of thermal radiation. That is to say, the temperature T is proportional to the typical frequency and inversely proportional to the linear expansion, and the energy density goes as T^4 in the first few minutes. It follows that whereas it is true to say that the density was high, the temperature was also high enough that the number of de Broglie wavelengths between any given pair of places stays fixed during such an expansion; in this sense there was still 'plenty of room' in the early high-density period. If we measure the linear expansion by a linear scale factor a, then the field equation predicts that $a \simeq (2H_0 t)^{1/2}$ in such a 'radiation-dominated' epoch (Section 23.2.3).

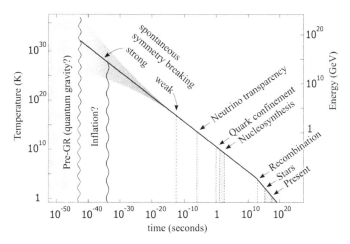

Fig. 22.5 A rough timeline of the very early universe. The shaded regions above 10^{17} K indicate increasingly unknown regimes of physics.

Table 22.1 Some significant events and epochs.

time	redshift (z)	
1–10 s	4×10^9	protons and neutrons form
10 s – 3 minutes	1.3×10^9	nucleosynthesis, formation of helium, some lithium
years after Big Bang:		
47,000	3391	matter-dominated era begins
250,000	1370	recombination (ion fraction = 1/2)
378,000	1090	decoupling (scattering rate = H); last scattering
0.4 to 0.7 billion	~ 10	reionization; first stars and galaxies form
...		supernovae and mergers create heavier elements
9.2 billion	0.42	solar system forms
13.8 billion	0	now

The shaded regions on the left of Fig. 22.5 act to show where our present knowledge of particle physics runs out; it does so increasingly as the energy scale exceeds that available to accelerator experiments. Most of the discussion in this and the next few chapters does not require this extrapolation beyond 10^{17} K; we shall be concerned almost entirely with the more ordinary scales of temperature and density. The main events are listed in Table 22.1.

22.2.3 Topology

The metric determines the curvature but not the global topology of a space. For example, the flat plane, the surface of a cylinder and the surface of a cone (excluding the apex) all have the same metric. It might be imagined that only curved spaces such as the 3-sphere can be *closed*, i.e. finite but unbounded in all directions, but this is not true. To make a finite unbounded flat space, just draw a rectangle on a flat piece of paper and declare that opposite edges are *identified*, that is, the points at coordinate $x = x_0$ are none other than the same points as those at $x = 0$, and the points at $y = y_0$ are none other than the same points as those at $y = 0$. If you like, to remind yourself of this identification define new coordinates $\theta \equiv 2\pi x/x_0$ and $\phi \equiv 2\pi y/y_0$ and say to yourself '2π is equivalent to zero'. There is now nothing special about the points at the 'corners': just shift the coordinate system by a displacement in x and y (or θ and ϕ) and what was a 'corner' point is revealed to be no different from any other point. It is surrounded in all directions by neighbouring points in the space. Such a space is called a flat torus, the word 'flat' being a reference to its zero Gaussian curvature, and the word 'torus' being applied in a topological sense.

Similarly, the hyperbolic sphere metric naturally suggests an open topology, but it does not have to. A universe that has everywhere a negative Gaussian curvature for the spatial surfaces may be topologically either open or closed in the spatial direction. However, whereas such closure is natural for the 3-sphere, it might be regarded as a mere mathematical curiosity for the other cases, since they lose their global isotropy when closed (they remain isotropic in any local region).

The terms 'open' and 'closed' have in the past often been misapplied in discussions of cosmology, though this is less so now. For example, it was thought that a universe which is temporally open, i.e. set to expand for ever, was necessarily also spatially open, i.e. infinite in spatial extent, but this is not necessarily true.

22.3 Friedmann–Lemaître–Robertson–Walker metric

The metric (22.7) captured the idea of a common cosmic time, and thus a certain sort of uniformity in the universe, but it did not require or assume spatial homogeneity nor isotropy. But given that, as we have said, the universe does appear to be homogeneous and isotropic at large scales, we shall now constrain the metric so as to assert those properties. It is easy to do: just put the metric for a space of constant curvature for the spatial part, and allow the scale to be a function of time, and you are done. However, we shall motivate this a little further before doing it.

First suppose we construct a triangle between three fundamental observers at some particular cosmic time. If conditions are isotropic, the triangle formed by those same observers at other times must be similar to the first one, since otherwise how is one to tell which corner gets a larger or smaller angle? So the relation between these triangles must be just a magnification factor.

Now let us bring in homogeneity: this says the magnification factor must be the same everywhere at any given time. So now the metric is constrained to take the form

$$ds^2 = -c^2dt^2 + a^2(t)\sigma_{ij}dx^i dx^j \tag{22.8}$$

where $a(t)$ is the **scale factor** and σ_{ij} is constant. This σ_{ij} has to incorporate homogeneity and isotropy. The terminology is that such a σ_{ij} is said to be the metric of a *maximally symmetric space* (c.f. (15.87)).

Maximally symmetric spaces can form a subject of study in their own right (exercise 15.12 of Chapter 15). All we need here is the three-dimensional version, our old friend the 3-sphere from appendix B, and its hyperbolic 'cousin' which has the same form of metric but with a constant negative curvature. We thus obtain[8]

Friedmann–Lemaître–Robertson–Walker metric)

$$ds^2 = -c^2dt^2 + a^2(t)\left(\frac{dr^2}{1 - K_0 r^2} + r^2\left(d\theta^2 + \sin^2\theta\, d\phi^2\right)\right) \tag{22.9}$$

---

where K_0 is a constant. We are free to chose the scale represented by the coordinates. We choose that $a(t_0) = 1$, where t_0 indicates the present.[9] This means that $rd\theta$ indicates distance around the circumference of a circle centred at the origin in the present, and the constant K_0 is equal to the present Gaussian curvature of space.

Now notice that if $K_0 \neq 0$ then we can introduce

$$k = K_0/|K_0|$$
$$\bar{r} = \sqrt{|K_0|}\, r$$
$$R = a/\sqrt{|K_0|}$$

and obtain

$$ds^2 = -c^2dt^2 + R^2(t)\left(\frac{d\bar{r}^2}{1 - k\bar{r}^2} + \bar{r}^2\left(d\theta^2 + \sin^2\theta\, d\phi^2\right)\right) \tag{22.10}$$

---

[8]All four names are included because these authors developed the ideas independently. Friedmann and Lemaître independently developed the theory of an expanding universe governed by GR in the 1920s. Lemaître also found further solutions and connected the theory to the observed redshifts. He also first clearly framed the concept now known as the Big Bang. Robertson and Walker later showed that the metric (22.9) is unique for homogeneous isotropic conditions, irrespective of the dynamics.

[9]i.e. the year 2020 according to our calendar, which is 13.8 billion years after recombination (or any other early-universe event).

where the possible values of k are 0, ± 1. This has precisely the same form as (22.9), so if one regards the coordinates as mere labels, then nothing has changed. However, coordinates do not have to be mere labels; they can also serve a useful role as distance and time measures, and this makes each version of the metric more or less useful than the other, depending on what one wants to investigate. The second version asserts that we do not need to find the size of K_0; that information can be absorbed into the as-yet-unspecified function $R(t)$. This is useful when thinking about the global geometry. All that matters, as far as the geometry is concerned, is the sign of k, and there are now just the three possibilities $k = -1$, 0 or 1, corresponding to negative, zero or positive curvature respectively (we use $k = 0$ when $K_0 = 0$). This rescaling is thus useful for some purposes, but one does not have to do it.

Since k is dimensionless in (22.10), then so is $\bar{r}$, so the length scale is in the factor R. If K has the dimensions of curvature, on the other hand (inverse length squared) then r is a length and then a is dimensionless. This can be more convenient for discussing physical observations. Finally, one can adopt (22.10) and then define a scale factor

$$a(t) = \frac{R(t)}{R_0},$$
(22.11)

where $R_0 = R(t_0)$ is the value of R in the present. Then one has $r = R_0\bar{r}$. The Gaussian curvature of space at time t is

$$K(t) = \frac{k}{R^2(t)} = \frac{k}{R_0^2 a^2(t)},$$
(22.12)

therefore if $k \neq 0$ then R has a natural interpretation as the radius of curvature of the universe.

The Friedmann–Lemaître–Robertson–Walker (FLRW) metric appears to single out one spatial point, $r = 0$, for special consideration, but this is an illusion. For $k = 0$ we have the metric for Euclidean space in polar coordinates; for $k > 0$ we have the 3-sphere (i.e. a three-dimensional space which is everywhere curved positively by the same amount) and for $k < 0$ the hyperbolic 3-space. In all cases one can move the spatial origin and the metric will not change. The best way to read the metric is to say 'let the origin of spatial coordinates be wherever you like; this is then the metric you will get'.

Let us examine the three cases a little further. We shall use

$$d\Omega^2 \equiv d\theta^2 + \sin^2\theta\, d\phi^2.$$

For $k = 0$ the spatial metric is

$$d\sigma^2 = a^2(dr^2 + r^2 d\Omega^2) = a^2(dx^2 + dy^2 + dz^2)$$
(22.13)

which is simply flat Euclidean space. The global topology could be infinite, or else closed such as the 3-torus.

For $k = 1$ we can replace the $\bar{r}$ coordinate using $\bar{r} = \sin\chi$ which gives the form

$$d\sigma^2 = R^2(d\chi^2 + \sin^2\chi\, d\Omega^2). \tag{22.14}$$

This is the 3-sphere, and the only possible topology is also the 3-sphere. As discussed in appendix B, a surface of given χ is a 2-sphere with radius $R\chi$ and surface area $4\pi R^2 \sin^2\chi$, so the radius excess is positive. The entire space has a finite volume given by (B.21), which gives $V = 2\pi^2 R^3$.

For $k = -1$ we can use $\bar{r} = \sinh\chi$ to obtain

$$d\sigma^2 = R^2(d\chi^2 + \sinh^2\chi\, d\Omega^2). \tag{22.15}$$

This is the space of constant negative curvature, called hyperbolic space because it relates to spheres somewhat as hyperbolas relate to circles. Globally the space could extend infinitely, but it could also be non-simply connected and finite (*compact*, in the jargon), so it should not necessarily be called 'open'. A surface of given χ is a 2-sphere with radius $R\chi$ and surface area $4\pi R^2 \sinh^2\chi$, so the radius excess is negative.

The above versions in terms of χ can be summarized:

FLRW metric again

$$ds^2 = -c^2 dt^2 + R^2(t)\left(d\chi^2 + S^2(\chi)\, d\Omega^2\right). \tag{22.16}$$

---

where $S = \sin\chi,\ \chi,\ \sinh\chi$ for $k = 1, 0, -1$ respectively. This form is useful for calculating various observable effects, to be described.

22.4 Redshift in an expanding space

22.4.1 Cosmological redshift

Let us sit somewhere in the universe and set up local inertial coordinates τ, l giving proper time and ruler distance between local events. We notice two brick walls floating in space, each fixed in the cosmic comoving coordinates, and we attach a rigid frame to one of them, extending to, but not attached to, the other. We notice that the other brick wall is located at

$$l(\tau) = a(\tau)\rho \tag{22.17}$$

where ρ is constant (and has dimensions of length) and a is the dimensionless cosmic scale factor. You can think of ρ as $(1 - K_0 r^2)^{-1/2}\delta r$ if you like, where r is the coordinate in the FLRW metric (22.9), but we will not need that metric for the argument. However, you should note that the argument does not require an assumption of globally flat space.

The speed at which the second wall is receding from the first, as measured by our LIF coordinates, is

$$v = \frac{\mathrm{d}l}{\mathrm{d}\tau} = \frac{\mathrm{d}a}{\mathrm{d}\tau}\rho = \frac{1}{a}\frac{\mathrm{d}a}{\mathrm{d}\tau}l. \tag{22.18}$$

Now suppose a photon is emitted by the first wall and travels towards the second. It will take a time $\delta\tau = l/c$ to travel the proper distance l. When it arrives at the second wall, an observer fixed to that wall will consider that it has a Doppler shift given by the exact (in the limit) relativistic formula

$$\frac{\delta\lambda}{\lambda} = \frac{v}{c} = \frac{1}{a}\frac{\mathrm{d}a}{\mathrm{d}\tau}\delta\tau. \tag{22.19}$$

Therefore, taking the limit of small quantities, we have

$$\frac{\dot{\lambda}}{\lambda} = \frac{\dot{a}}{a} \tag{22.20}$$

which could equally well be written $\mathrm{d}\lambda/\mathrm{d}a = \lambda/a$. In any case the solution is

$$\lambda \propto a. \tag{22.21}$$

This simple argument has given us a remarkable and very useful result. It asserts that if a photon sets out with wavelength λ_E as observed by a comoving observer E, then when it arrives at some other comoving observer R the latter will find the wavelength to be given by

---

cosmic redshift

$$1 + z = \frac{\lambda_R}{\lambda_E} = \frac{a(t_R)}{a(t_E)} \tag{22.22}$$

---

where the **redshift** z is defined as

$$z \equiv \frac{\lambda_R - \lambda_E}{\lambda_E} = \frac{\nu_E}{\nu_R} - 1. \tag{22.23}$$

The ratio of scale factors is all that is involved in the redshift formula. This implies that the cosmic redshift should not be thought of as a statement about motion of one source or another, but rather a statement about expanding distance scales. Consider for example a universe where $a(t)$ was first unchanging, and then increased for a while, and then was constant again: there would still be a redshift for light emitted in the first constant period and received in the second. Notice also that the result did not require any treatment of the dynamics of $a(t)$; it only assumed homogeneity, so that we could assume the same $a(t)$ wherever the photon happened to be on its journey at each cosmic time t (which is also proper time for comoving observers). z can be directly observed by astronomers, and its connection to the scale factor is remarkably direct and simple.

Do not allow the simplicity of eqn (22.22) to obscure from you its significance. This simple linear relationship is a piece of *precision cosmology*. It is precise to the degree that homogeneity is precise, and it applies at all z, not just small z. If the reception event is in the present and we adopt the standard scaling $a(t_0) = 1$, then we have

$$1 + z = \frac{1}{a} \tag{22.24}$$

where a is the scale factor at the emission event. If you measure a redshift of 2, then, in any homogeneous cosmological model, it has come from a time when the universe had one-third of its current size. But the photon does not tell you how long ago that time was.

Calculation from the metric. For further comfort, we shall now derive (22.22) by two more arguments, using the FLRW metric. Both are based on the fact that a photon propagating from one place to another follows a null geodesic.

We can place the spatial origin at the location of an observer who receives a photon. By symmetry, the radial null lines are null geodesics. Putting $ds^2 = d\Omega^2 = 0$ and using the χ version of the metric (22.16) for convenience, gives that the photon's worldline satisfies

$$c\,dt = -R(t)\,d\chi \tag{22.25}$$

(where we choose the sign for an incoming photon). We consider a wavefront emitted at some location χ_E at cosmic time t_E and arriving at the receiver at cosmic time t_R. The next wavefront is emitted at time $t_E + \delta t_E$ and received at $t_R + \delta t_R$ where δt_E, δt_R are the periods at emitter and receiver respectively. Both wavefronts satisfy (22.25) throughout their journey, and they travel between the *same* locations χ_E, 0, so we have

$$-\int_{\chi_E}^{0} d\chi = \int_{t_E}^{t_R} \frac{c}{R(t)} dt = \int_{t_E + \delta t_E}^{t_R + \delta t_R} \frac{c}{R(t)} dt \tag{22.26}$$

The middle part of these two integrals over time is the same in the two cases, therefore if the whole integrals agree then the small parts at the ends must equal one another (see Fig. 22.6). Therefore we have

$$\frac{\delta t_E}{R(t_E)} = \frac{\delta t_R}{R(t_R)} \tag{22.27}$$

which is the redshift formula (22.22). QED.

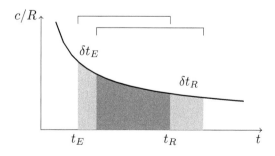

Fig. 22.6 Obtaining (22.27) from (22.26). The lines at the top show the two ranges of integration. The two lightly shaded areas must be the same.

Next we will show how the above relates to the general formula for gravitational redshift that was derived in Chapter 13. From (22.25) we have

$$c\dot{t} = -R(t)\dot{\chi} \tag{22.28}$$

where the dot signifies differentiation with respect to an affine parameter along the geodesic. The metric does not depend on χ, therefore the corresponding covariant component of the tangent to the geodesic is constant along the geodesic, eqn (13.27):

$$\text{const} = \dot{x}_1 = g_{11}\dot{x}^1 = R^2(t)\,\dot{x}^1 \tag{22.29}$$

But $\dot{\chi}$ is itself the contravariant component $\dot{x}^1$, so we have $\dot{\chi} = \text{const}/R^2(t)$, and substituting this into (22.28) gives

$$c\dot{t} = \frac{\text{const}}{R(t)} \tag{22.30}$$

Now we recall that 4-momentum is proportional to 4-velocity, and therefore, in the case of a photon, the 4-momentum can be set equal to the tangent to the worldline, when the latter has been treated by an affine parameter. In the present case, therefore, the zeroth component of the 4-momentum of the photon, at any point on the worldline, is

$$p^0 = c\dot{t} = \frac{\text{const}}{R(t)} \tag{22.31}$$

It remains merely to substitute this into the general redshift formula for observers fixed in the coordinates, (13.45), and we obtain (22.22).

22.5 Visualizing the evolution of a curved space

In the previous section we have begun to make reference to the dynamics of the universe, associated with the changing scale factor. Before we treat the dynamics more fully, it will be useful to get some intuition about what we will be talking about.

Observe the three pictures in Fig. 22.7. The first picture shows a three-dimensional space. There are dots distributed through the space. If I now tell you that the dots are particles of equal rest mass, and the distribution is on average uniform in terms of ruler distance, then you can deduce that the space is homogeneous and isotropic on a large scale, but you cannot determine its curvature. However, I have drawn a geodesic triangle in the space, and from the angles of this triangle you can deduce that the space has positive curvature.

The next picture has taken a slice through the space shown in the first picture, in the plane of the triangle. In the coordinate system of the first picture, this is a coordinate plane. It is also a geodesic surface, which one can see by symmetry: we can arrange the coordinates in the FLRW

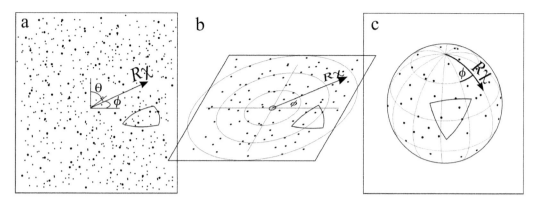

Fig. 22.7 (a) A picture of a three-dimensional homogeneous universe with positive curvature; (b) a slice through the universe; (c) the same slice shown in an embedding diagram.

metric such that this surface is the equatorial plane $\theta = \pi/2$. Geodesics in this surface will not deviate out of it to either higher or lower values of θ. But the surface is not flat in the Gaussian sense. Its metric is the spatial part of

$$ds^2 = -c^2 dt^2 + R^2(t)(d\chi^2 + \sin^2(\chi)d\phi^2). \tag{22.32}$$

In order to get an intuition about the warped geometry of this surface at any given t, we now proceed to embed it into a three-dimensional Euclidean space. That means, we trace out a surface in Euclidean 3-space such that Euclidean distances on the curved surface match the metric distances we are interested in. We thus get the third part of Fig. 22.7, and we get also some insight and some possible confusion. I repeat: *Fig. 22.7c shows a* plane *from Fig. 22.7a.* It is a 'slice through space'. 'But the slice looks spherical' you say: yes, I reply, that is because I have introduced a mathematical abstraction, a further dimension on Fig. 22.7c which has no significance to the space we are talking about, except to visualize warping. It also has the added bonus of letting us visualize topology too.

Now return to Fig. 22.7a and suppose the space is uniformly expanding. After a while all the dots will be further apart.

In Fig. 22.7c this corresponds to the sphere growing like a balloon.

Now we take an interest in the question, how does light propagate in this spacetime? In metric terms, we are seeking the null geodesics. Using the symmetry of the FLRW metric under spatial rotations, we can infer that each t, χ plane at $d\theta = d\phi = 0$ is a geodesic plane (if a geodesic wanted to turn out of this plane, how would it decide in which sense to deviate?). Each such plane has the metric

$$ds^2 = -c^2 dt^2 + R^2(t)d\chi^2 \tag{22.33}$$

so geodesics of this metric are geodesics of the full four-dimensional metric. In particular, the null lines satisfying

$$c\,dt = \pm R(t)d\chi \tag{22.34}$$

are geodesic (as we already said in the previous section). In other words the radial lines starting out from the origin in Fig. 22.7a and proceeding at the speed of light are null geodesics, and furthermore, owing to the isotropy and homogeneity, this is true in all directions and at all places as we move the origin of coordinates around. It does not follow that these lines will all look straight on a diagram drawn in Euclidean space. Translating this into the balloon picture, we find that light travels like a beetle crawling over the surface of the balloon on a great circle.

Now let us consider the speed. The crawling beetle representing a pulse of light travels at a fixed speed as measured by 'amount of local balloon surface traversed per unit time'. This remains true even if the balloon is expanding. Note that such beetles may be carried, via the balloon expansion, such that the proper distances between any pair of them may change at any rate at all, including rates greater than the speed of beetle.

The purpose of introducing the 'expanding balloon' picture is not just to indicate some qualitative features. When it is used correctly, the balloon picture is *a precise and correct* indicator of what the FLRW metric is telling us. We only need to keep in mind that the spatial curvature of our universe might be negative rather than positive, so the sphere might have to be replaced by a surface of negative curvature. In fact measurements to date suggest the curvature is slightly positive or zero.

Long rods and measuring tape.   Draw two dots on the balloon. They represent two galaxies at the present cosmic moment. Get hold of a steel measuring tape and extend it over the surface of the balloon, following a great circle. You are measuring the *ruler distance* or *proper distance* between the galaxies. This is the total spacetime interval along a geodesic of the spacelike surface extending between events at those two galaxies at the same t. It can be calculated using the FLRW metric by performing an integral along the path.

Now inflate the balloon a little (or a lot). The ruler distance between your pair of galaxies changes. This means the following. We can in principle measure proper distances in the cosmos by using a set of steel rods. Suppose each rod has a length of 10^{13} km (about 1 lightyear). Occupants of various planets dotted around the cosmos furnish themselves with similar steel rods. Suppose we all lay our light-year-long rods end to end at the same cosmic moment, along a huge line, as straight as we can make it, between the Milky Way galaxy and some other galaxy not in any local gravitationally bound assembly. To synchronize the measurement, all observers do it when the CMB radiation they receive has the same agreed temperature. Count the number of rods required. Let us say it is 100 million. Now wait for a time $\delta t = 0.01/H_0 \simeq 144$ million years, and repeat the measurement. The number of rods required to reach between the same two galaxies will now be 101 million.

Now let us do another experiment. Draw on the balloon a sequence of short marks, all equally spaced, along a great circle between two given dots. The number of such marks represents the comoving coordinate separation of the dots. Now inflate the balloon. After doing that, do not get any measuring tape, just count again the marks you already made, between the same two dots. The number has not changed. *The comoving coordinate distance between any pair of fundamental observers is constant.* In this sense, the galaxies are not moving. They are sitting still—relative to a certain well-defined and physically motivated coordinate frame.

Redshift. Now mark on your balloon a blue dot, and next to the blue dot draw a small raster of 10 equally spaced short parallel lines. These represent the wavefronts of a pulse of light. Measure the length of your raster using a steel measuring tape. Let us say it is 1 cm long. We will arrange for the speed of light in our balloon universe to be 1 cm per second. Next place an exact copy of the balloon over the first one, and draw on the second balloon a similar raster of lines, with the same spacing, such that it overlaps half of the first one, and half extends beyond it. The light pulse has travelled 0.5 cm, so this second balloon is going to represent the situation half a second after the first one. Also draw a blue dot on the second balloon, at the same location as the first blue dot: it will represent a fixed location in comoving coordinates.

Now inflate the second balloon, by as much or as little as you like, leaving the first one unchanged inside it. This represents the change in the scale factor during the half second that the light pulse has travelled so far. The raster of lines will spread out and move away from the blue dot. (In the real process both light propagation and cosmic expansion take place continuously and simultaneously; the present description will match the result in the limit of many small time intervals.)

Now repeat the exercise: put a third balloon over the second one, and copy the raster from one balloon onto the other, but moved along a little. You can move it by x cm each time, in which case the next balloon represents the universe x seconds later (pick small values of x for the method to work).

After repeating this exercise many times, we have a set of nested spheres, and we can trace the progress from one sphere to the next of our pulse of light. In principle we could also allow the spheres to get smaller, or to oscillate in size. But in any case the pulse will make its journey, and when it arrives at any given destination we can measure it with a steel measuring tape. We shall find that the spacing of the raster—the wavelength of the pulse—will be in exact proportion to the amount by which the surface of the balloon has stretched. This is eqn (22.22). See Fig. 22.8 for an example.

Hubble's law. For our final experiment, draw three dots as shown in the first picture of Fig. 22.9). Let the dot at the lower left represent planet Earth. Then draw 5 concentric circles around the upper dot, with uniformly spaced radii, say 1 to 5 mm. This is a set of 5 spherical wavefronts.

Now repeat the exercise of nested spheres in order to track the propagation of these waves over the surface of the balloon. They will spread out in expanding circles until, at some later time, they arrive at positions as shown in the middle picture of Fig. 22.9). Notice that the wavelength (distance from one wavefront to the next) increases as the balloon expands.

Next draw a similar group of waves round the right-hand dot, as shown in the middle picture of Fig. 22.9), and then expand the balloon some more and allow all these waves to propagate.

Eventually the waves reach Earth (third image of Fig. 22.9). As each group of waves arrives, its wavelength indicates the net change of scale factor over its journey. If the balloon has indeed expanded at every stage, then the waves arriving from the more distant source will have the

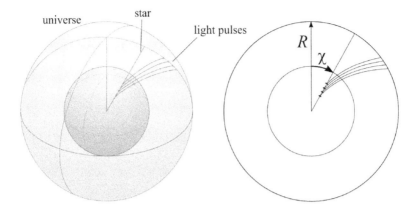

Fig. 22.8 Cosmic redshift. A star pulses four times and emits light in the $\phi = 0$ direction. The light pulses propagate along the surface of the sphere at c relative to any local comoving observer, and meanwhile the sphere grows, with the overall result as shown. The two spheres on the left show the situation at two moments of cosmic time; the diagram on the right shows a cross-section through them. The light pulses are spread out more widely at the larger sphere, in proportion to the diameter of the sphere. This can be proved by noting the circular symmetry: it results in the fact that all the photon worldlines have the same shape; each one can be brought on top of another by rigid rotation about the origin. A different expansion history would change this shape, but it would still be in common for the four worldlines.

longer wavelength. This is Hubble's law. Note that the linear relationship between velocity and distance in Hubble's law is approximate, whereas the relationship between redshift and scale factor is exact.

Cosmic free-fall.   The one reservation I have about the balloon model is that it presents somewhat too concrete a view of space—the rubber of the balloon. One should not regard the expansion of the universe as if the galaxies were being carried along by some inexorable force, like mountains riding on tectonic plates. The (clusters of) galaxies are just falling. Their motion is every bit as free and easy as anything else in free fall, and they can easily be disturbed out of their free fall by other forces such as electromagnetic ones. When looking at Fig. 22.9, the best way to view the dotted squares is to imagine them as sets of test particles floating in space, which happen to be lined up in squares. They are freely falling along with the galaxies and everything else.

The wavefronts of electromagnetic waves in empty space are in free fall too. An interesting property, implied by our study of redshift, is that a given light wave does not 'hold itself together' in the longitudinal direction. Each wavefront is not pulled along by the one ahead; the wavefronts travel along independently of one another, like a set of beads lined up along a line but *not* exerting forces on one another.

Finally, the motion we have studied with our balloon model only applies to the average motion at the largest scales in our universe. On smaller scales, the universe is neither homogeneous nor

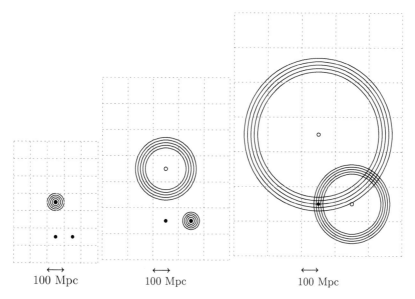

Fig. 22.9 Hubble's law: the relationship between received wavelength and expansion since the emission. The diagrams show spatial situations on the surface of the balloon (for a large radius of curvature so the diagrams appear flat) at three different times. The scale of proper distance is the same throughout all three diagrams. The increase of the scale factor is shown by the grid squares: their proper diameter increases, but their comoving coordinate size does not. The bottom left dot is Earth. There are two stars at different distances from Earth. In the first picture, the first has exploded in a supernova and emitted some spherical wavefronts. As time goes on, the wavefronts propagate outwards, and the wavelength remains a constant fraction of the grid square diameter. After a while the second star explodes in the *same type* of supernova (second picture), emitting waves of the same wavelength and duration in its LIF as the first one did in its LIF. When all these waves arrive at Earth (third picture) the wavelength for each group is proportional to the change in scale factor since the emission of that group. Notice also that the proper size of Earth does not change: the binding forces of rock are easily sufficient to prevent the Earth's surface drifting outwards in cosmic free-fall.

isotropic. We can model this by modifying the balloon: allow its surface to be dimpled, and furthermore allow that small parts of it need not expand at all from one cosmic moment to the next. Within any given galaxy or solar system, space is not uniformly filled with matter but rather consists of dense entities such as stars surrounded by comparatively empty space. The metric in such regions can easily be stationary to first approximation, thus resulting in no expansion at all for such quantities as the radii of galaxies and solar systems. That this is likely to be a good first approximation is implied by Birkhoff's theorem. To imagine this in terms of the balloon model, imagine placing little dots of rigid glue on the balloon.

22.5.1 The Milne universe

We will briefly present now a FLRW cosmological model which does not match the observed universe but has some useful lessons to teach in both Special and General Relativity. It is called the *Milne model* or Milne universe. It concerns a spacetime which is entirely flat everywhere, and

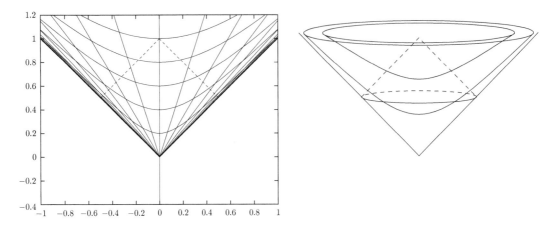

Fig. 22.10 Spacetime diagram in the t, r plane for the Milne universe. The straight lines are world-lines of fundamental observers. The spacelike hypersurfaces are loci of constant proper time along the worldlines; they are hyperbolas. The dashed lines show an example past light cone.

which can therefore be treated by Special Relativity. We suppose that a bunch of non-interacting particles is shot out from the origin of coordinates for some unknown reason, and they all move inertially outwards, such that their worldlines are straight lines on a standard spacetime diagram as shown in Fig. 22.10. The mass and density of these particles are sufficiently low that gravitational effects can be ignored (we will question this assumption shortly). We assume this explosion is isotropic and orchestrated in such a way that the particles having higher velocity v (relative to some given particle) also have higher number density, in proportion to $\gamma^2 n_0$ where n_0 is a constant and $\gamma = (1 - v^2/c^2)^{-1/2}$ is the Lorentz factor. It is an interesting feature of the model that this condition can be fulfilled for all the particles, because it is equivalent to the condition that the proper density of the cosmological fluid is uniform across a hypersurface at fixed proper time after the explosion.

The 'cosmic time' coordinate in the Milne universe is the proper time after the explosion, as registered along the worldline of any given particle. The surfaces of constant cosmic time have the form of hyperbolas on the diagram of Fig. 22.10. If we refer to each such surface as 'space' at some given time, then we discover that space in the Milne universe has constant negative curvature at any given time, and the amount of curvature falls as time goes on. (Rindler, 2000) Note that it is perfectly possible to find curved hypersurfaces in a flat spacetime. In comoving coordinates the metric is of FLRW form with $k = -1$ and $a \propto t$.

The Milne model has some interesting features. At any given cosmic time the spatial hypersurface has an infinite spatial extent, even though it all fits inside a single future light cone. Also, if we jump aboard one of the particles, regarding that particle as a fundamental observer in the Milne universe, then we find that the universe looks the same no matter which particle we pick. For to change the perspective from one such observer to another, one must apply a Lorentz transformation, and if one then plots a spacetime diagram in the new coordinates, it will look just like Fig. 22.10 once again. In order to make this possible for all observers, including pairs whose relative velocity approaches the speed of light, one must allow the density $\gamma^2 n_0$ to approach infinity as $v \to c$; this is a weakness of the model which shows that we cannot ascribe to the fundamental observers any mass at all if we are to obtain negligible gravitational effects.

An example past light cone is also shown on Fig. 22.10. This light cone intersects all the particle worldlines, and at its outer limits it reaches events at zero proper time after the explosion.

The main contrast between the Milne model, which describes an example of negative spatial curvature, and the balloon model, which describes an example of positive spatial curvature, is that in the former the worldlines of the matter content of the universe only occupy part of spacetime whereas in the latter they occupy all of spacetime. It is very tempting, when looking at Fig. 22.10, to suppose that this 'Big Bang' happened at one point in space, whereas we know that the balloon model does not bear that interpretation. In fact, if space is defined as we have specified for the Milne universe, then the Big Bang happened equally at all spatial locations. To make sense of this it is better to choose some early spatial hypersurface and announce the conditions on this hypersurface as having been brought about by a process which classical physics cannot account for—we do not trace the worldlines back any earlier than this. Then all places everywhere are equally rushing away from one another (in accordance with Hubble's law); there is no unique spatial centre to the explosion. One might also propose that the empty regions of spacetime on Fig. 22.10 have no physical meaning if this is a model of the whole universe. But all this is merely a lesson in relativity. Since there is no mass in the Milne universe, there is no thing in it.

22.6   Luminosity distance, angular diameter distance

In any FLRW model, at any given cosmic time, the proper distance d_R between a galaxy at the origin of coordinates and a galaxy at coordinate χ is

$$d_R = R(t)\chi, \tag{22.35}$$

but this cannot be directly measured and is rarely of immediate interest. Observations are all concerned with light arriving at Earth after travelling great distances for large amounts of time, and during the travel the scale factor of the universe, and the curvature of space, may have changed. What do we even mean by the distance to an event in the past? We mean whatever we chose to mean: we define distance measures which can be most directly extracted from observations. If we chose our definitions carefully, then there is no great mystery.

Luminosity.   In astronomy, the brightness of a shining object is normally indicated by *luminosity* L, which is a measure of power. The *absolute luminosity* or *intrinsic luminosity* is the total power emitted by the source (units $W = J\,s^{-1}$). If spacetime were static and Euclidean, and if the emission is isotropic, then the *flux* (power per unit area, units $W\,m^{-2}$) at distance d from the source would be $F = L/(4\pi d^2)$. So, with this in mind, we can *define* a notion of distance in terms of flux and luminosity:

$$d_L \equiv \left(\frac{L}{4\pi F}\right)^{1/2}.$$

(22.36)

This is a definition, so it is what it is, and applies in any geometry and any time-dependence of the universe. It is called the *luminosity distance* of the source. When someone says, of a given galaxy, 'it is at a luminosity distance of 100 Mpc' and you ask 'how far away is it then?' then they may reply, 'I do not know, but I measure its light arriving in my detector to be 10000 times dimmer than it would be if the source were at a luminosity distance of 1 Mpc'.

To answer our friend's question a little more fully, we need to ask by how much has the light spread out on its journey. In other words, what is the *surface area* of the spherical surface the light crosses as a small part of it enters our detector. Also there are two further effects: the redshift, which reduces the light's energy in proportion to its frequency, and the rate at which photons arrive (or, if you prefer, the inverse of the time it takes to receive a given wave train), which is reduced by the same factor. If the source was at cosmic coordinate χ when it emitted the light, then it is still at that coordinate now, and the sphere we are interested in is centred at χ. Let us move the origin to that place, in which case Earth is at coordinate χ and the spherical surface has coordinate radius χ. Transverse areas, such as surface areas of spheres centred on the origin, are easy to calculate in the FLRW metric (see eqn (22.16)): the answer is simply

$$A = 4\pi R(t)^2 S^2(\chi)$$

(22.37)

where the time we are interested in is $t = t_0$, giving $R(t_0) = R_0$. After taking into the account the two factors associated with redshift, we find that the received flux, in terms of the absolute luminosity L of the source, is:

Received flux

$$F = \frac{L}{4\pi R_0^2 S^2(\chi)} \frac{1}{(1+z)^2}.$$

(22.38)

Therefore the relationship between luminosity distance, coordinate distance and redshift is

$$d_L = R_0 S(\chi)(1+z).$$

(22.39)

This gives a practical way to figure out the coordinate location of galaxies in the cosmos. In the context of observational cosmology, the use of d_L is typically that one gathers all the following information

- Temporal behaviour of source $\rightarrow L$ (see next section)
- Received power into telescope aperture $\rightarrow F$
- Observed spectrum of source $\rightarrow z$

By inserting these items of information into (22.38), one can deduce $R_0 S(\chi)$ for that particular source. With sufficient such data for a range of values of z, one reconstructs the shape of the past light cone of Earth, which can be expressed first as a function $R_0 \chi(z)$ and then in any other convenient coordinates. This function is a distance–redshift relation and from it one can deduce the Hubble constant $H_0 = H(t_0)$, and also, in principle, the Hubble parameter $H(t)$ at all times t which the observations can probe; see eqn (25.12).

Angular diameter distance. You can guess how this is defined. If we have reason to know that a given astrophysical object has proper diameter l (for example, because it is the result of light travel for a known time during a known period of expansion), and the light from the edges of the object subtends an angle α as it arrives in our detector, then we say the *angular diameter distance* of the object is

$$d_A \equiv \frac{l}{\alpha}. \tag{22.40}$$

To visualize the situation, sit yourself at the north pole of the outer sphere shown in Fig. 22.8 and place an object of proper length l somewhere at some other latitude on the inner sphere. Let the centre of coordinates be at yourself, and take it that the surface of the balloon is the plane at $\theta = \pi/2$ in the FLRW metric, with the object lying in this plane and orthogonal to the radial direction χ. You can deduce from the metric that the coordinate angular separation $\Delta\phi$ between the ends of the object is $\Delta\phi = l/(R(t)S(\chi))$. Now let light propagate from there to yourself at the pole, while the balloon expands or shrinks in any way. You should be able to see that the light arrives with this same coordinate angular separation. Therefore $\alpha = \Delta\phi$ and

$$d_A = R(t)S(\chi) = R_0\frac{R(t)}{R_0}S(\chi) = \frac{R_0 S(\chi)}{1+z} = \frac{d_L}{(1+z)^2}. \tag{22.41}$$

Thus by discovering whether or not $d_L = (1+z)^2 d_A$ in observations, one can test whether the FLRW metric holds good.

The definitions of d_L and d_A are made merely in order to provide a useful language in which to treat physical location and the exchange of light signals in a curved dynamic spacetime. They are not intended to say what is the 'best' measure of distance between any two events.

The angular diameter distance is especially useful for extracting information from *baryon acoustic oscillations* which show up in the distribution of galaxies across space and time; this is discussed in Chapter 25.

22.7 The cosmic distance ladder

The calibration of luminosity is central to the practical value of any observations based on luminosity distance. How do we really know the power being emitted by a star or galaxy or other object visible through our telescopes far off in the depths of space (and thus also at an earlier cosmic time)? The basic idea is to form the reasonable hypothesis that objects that look or behave similarly, apart from an overall redshift and a change in brightness, are themselves similar and so have the same absolute luminosity. Then we determine the absolute brightness of some of these that happen to be near to us, using some other method to determine their distance, and thus calibrate the whole system. By 'look or behave similarly' we mean chiefly the *spectral type*, or overall shape of the spectrum of emitted light, and the *temporal behaviour*, especially periodic fluctuation in brightness or else a sudden glow and fade.

In more detail, the method begins at the scale of the solar system. Distances among the planets are measured by radar reflection and by combining optical observations with calculations of planetary motion. (Historically, careful timing of the transit of Venus across the face of the Sun was important to determining the distance of Earth from the Sun.) This provides a baseline: the diameter of Earth's orbit, which is a nearly circular ellipse of average diameter 1.496×10^{11} m.

Next, the method of triangulation, called parallax in the astronomical context, is used to determine distances to stars that lie within a thousand parsecs from us. The *parsec* is defined as that distance at which a line 1 AU long subtends 1 second of arc. Here 1 AU is an exact defined quantity (149 597 870 700 metres), and therefore so is the parsec (1 pc $\simeq 3.085678 \times 10^{16}$ m).

We can determine the intrinsic luminosity of any star whose parallax is measurable. Among these stars are the Cepheid variables. These are important to astronomical distance measurements partly because they are bright and chiefly because their brightness fluctuates regularly with a well-defined period–luminosity relationship. This was discovered by observations of a large number of Cepheid variables that lie in a group in the Magellanic Clouds.

It seems amazing, to me at least, but individual stars can be resolved by telescopes in the outer regions of galaxies other than our own. These can be used to determine the distance to such galaxies, especially if we can find Cepheid variables which make the luminosity information more precise. By studying sufficient galaxies in this way, one can make further discoveries. Tully and Fisher found that the rotation velocities of spiral galaxies are tightly correlated with the intrinsic luminosity of the entire galaxy. So now we have a method which can be used to much larger distances (the rotation velocity can be determined by Doppler shifts from different parts of the arms). Correlations can also be found for other types of galaxy.

Finally, with these distance measures in hand, it has been possible for astronomers to discover the intrinsic luminosity of large numbers of supernova events. As a result, the supernovae themselves have been classified into different types, based on their temporal behaviour, and in particular the Type Ia supernovae are especially regular. It is thought that they all originate from the collapse of white dwarf stars of similar mass (the Chandrasekhar limit). In any case, there is only a modest (40%) variation in the intrinsic luminosity of such explosions, and furthermore this variation is largely correlated with the timescale of the rise and fall of the emission, called the 'light curve'. By observing the light curve one can largely correct for the intrinsic variation and hence reduce the scatter to less than 15%. Thus these supernovae are said to be 'standard candles'. They have enabled the luminosity distance scale to be extended out to redshifts of order 1.

This whole ladder of measurements rests, to a significant degree, on the humble baseline set up by distance measurements within the solar system! But this is a very precise baseline, and it is not the only support. The sequence of deductions is also supported by the physical models and understanding of the stars and supernovae themselves.

Historical note on the cosmological principle.   We finish the chapter with an historical note.

The name of Copernicus is associated with the cosmological principle because the heliocentric model of the solar system that Copernicus proposed had the effect of displacing Earth from the centre of the model of the cosmos adopted by astronomers and others, and this is similar to the general idea that Earth should not be expected to have a special place. This is indeed a similarity, but it cannot correctly be construed as an indication that human culture has acquired greater humility since the seventeenth century. The pre-modern mindset was more cautious and humble before the accumulated wisdom of previous times. The modern mindset is more assertive and ready to look dismissively on previous times.

In the centuries before Copernicus, careful naked-eye surveys had been made of the night sky, with instruments such as the astrolabe and the quadrant, and calculation techniques developed. The physical model with Earth at the centre was based to a large extent on this observational evidence, not on psychological or religious resistance to another model. It appeared *to observation* that astronomical bodies orbited the Earth. It also appeared *to observation* that astronomical bodies were 'perfect', that is, of uniform and unchanging behaviour, whereas this is plainly not true of Earth. Therefore the model which proposed Earth to be a place both unlike any other, and also surrounded by other stuff which rotates around it, was consistent with the available data, and it would be unreasonable to expect anyone to develop another view in the absence of data to suggest it. The heliocentric model was developed once the data was sufficiently well established to warrant it. The resistance, at the time, to this change of perspective was not primarily based on prejudice. In public and private exchanges there was a certain amount of argumentative ill-feeling, but when it came to official assessments, opinions were sought and expressed in scientific language. There was another good model available, that of Tycho Brahe, and the two were explicitly compared. Brahe's model better fitted the observations, especially the observation that if Earth moved then the motion should be detectable through stellar parallax, but no such effect was found. It was therefore unclear whether the heliocentric model was better until such discrepancies with observational data could be resolved. To insist on this, as the court that tried Galileo did, is comparable to the standards of proof that are generally reckoned to be important in science. In recent times for example it has been important to assess whether claims that certain forms of vaccination can cause autism were in fact warranted. The issue in these types of disputes is not whether or not a new idea is feasible, but whether it has been presented to the general public correctly, with a fair statement of relevant material and the overall strength of the case. Galileo was not required by any court to abandon an idea, but to cease teaching it as an established fact to lay-people unacquainted with the whole evidence.

Two distinct issues are at stake in such cases. The first is the scientific evidence for the disputed point; the second is the degree of control that it is appropriate to exert in the area of public education. The latter remains and will probably always remain a difficult and delicate area. The main point for our present purpose is simply that the pre-Copernican view was not primarily a statement of psychological or religious disposition, nor based on hubris or anything like that. It was a reasonable response to the available evidence. The same can be said of the modern cosmological principle.

Exercises

22.1 Taking the average density of ordinary matter in the universe to be $10^{-27} \,\mathrm{kg/m^3}$, express this density (very roughly) in units of 'galaxies per galaxy volume', thus obtaining an impression of the relative sizes of typical galaxies and typical voids in the universe now.

22.2 Adapt the method of appendix B so as to obtain the metric of a 3-space with constant negative curvature, and hence obtain (22.15). Show that such a space has infinite volume if it has the simplest possible topology.

22.3 Obtain (22.22) (cosmic redshift) by your preferred method or methods.

22.4 Show that at any given cosmic time t the peculiar velocity (i.e. velocity relative to local fundamental observers) of a massive particle can be written $v = R_0 a(t) \,\mathrm{d}\chi/\mathrm{d}t$ if the motion is in a certain direction which you should specify. Hence show that if the particle is in free fall then the velocities at times t_1, t_2 are related by

$$\frac{\gamma(v_2)v_2}{\gamma(v_1)v_1} = \frac{a(t_1)}{a(t_2)} \qquad (22.42)$$

where $\gamma(v) = (1 - v^2/c^2)^{-1/2}$. By considering momentum, relate this to the cosmic redshift formula (22.22).

22.5 The solar system formed about 4.6 billion years ago. Use the measured value of the Hubble constant to make a rough estimate of the expansion of the universe since then. Was the Earth significantly closer to the Sun when it first formed? [This is a trick question]

22.6 The proper-motion distance is defined by $d_M = v/\dot{\theta}$ where v is a proper transverse velocity of some object or part of an object (assumed known), and $\dot{\theta}$ is the observed angular velocity. Show that $d_M = (1+z)d_A$.

22.7 *K*-**correction**. In this problem we find out how observations of radiation within some frequency range, as opposed to total flux, are affected by cosmic expansion.
(i) The total flux F in some frequency range $[\nu_1, \nu_2]$ is

$$F(\nu_1, \nu_2) = \int_{\nu_1}^{\nu_2} f(\nu) \,\mathrm{d}\nu$$

where $f(\nu)$ is the spectral density of the flux (units $\mathrm{W\,m^{-2}\,Hz^{-1}}$). If f_{em}, f_{obs} are the emitted and observed values of this quantity for some source at redshift z, show that $f_{\mathrm{obs}}(\nu) = f_{\mathrm{em}}((1+z)\nu)/(1+z)$. Expressed in terms of magnitudes (i.e. after taking a logarithm) this ratio leads to an additive correction known as the *K*-correction.
(ii) Hence, for a source with $f_{\mathrm{em}} \propto \nu^n$ for some n, show that

$$F_{\mathrm{obs}}(\nu_1, \nu_2) = (1+z)^{n-1} F_{\mathrm{em}}(\nu_1, \nu_2) \qquad (22.43)$$

(and note that this formula is independent of the expansion history $R(t)$).
(iii) The surface brightness Σ of an extended object is defined as the observed flux per unit solid angle. For a small circular object subtending angular diameter $\Delta\theta$ this is

$$\Sigma = \frac{F(\nu_1, \nu_2)}{\pi(\Delta\theta)^2/4}$$

Show that

$$\Sigma_{\mathrm{obs}} = \frac{4}{\pi} \frac{L_{\mathrm{em}}(\nu_1, \nu_2)}{4\pi l^2} \frac{K_z}{(1+z)^4} \qquad (22.44)$$

where l is the projected diameter of the object, L_{em} is its intrinsic luminosity, and $K_z = F_{\mathrm{obs}}/F_{\mathrm{em}}$ is the factor discussed in part (ii). Note that the 4th power here makes objects at high z hard to see, and this is independent of cosmological parameters.

22.8 Discuss whether or not physical evidence, combined with physical modelling, can give us warrant to give greater credence to a claim that the physical universe is infinite than to a claim that the physical universe is finite.

23

Cosmological dynamics

In this chapter we show how the evolution over time of a universe with an FLRW metric is controlled by the matter and radiation content, via the Einstein field equation, for given initial conditions. The initial conditions are those of hot, dense matter everywhere moving in synchrony such that the metric retains the FLRW form with an increasing scale factor. The matter and radiation then thins (becomes less dense) and cools, and also influences the rate of change of the scale factor. We first develop the mathematical tools needed to treat this in terms of a small number of parameters (densities and Hubble parameter), and then use observations to check for overall consistency and to determine the parameter values.

23.1 The Friedmann equations

Let us substitute the FLRW metric into the Einstein field equation, in order to discover how the scale factor depends on the matter content of the universe. First we need the Ricci tensor, obtained from

$$R_{bd} = \partial_\lambda \Gamma^\lambda_{bd} - \partial_d \Gamma^\lambda_{\lambda b} + \Gamma^\lambda_{\lambda \mu} \Gamma^\mu_{db} - \Gamma^\lambda_{b\mu} \Gamma^\mu_{d\lambda}. \tag{23.1}$$

The connection coefficients are obtained from the metric in the usual manner; they are displayed in Table 23.1. The calculation can adopt any coordinates; it is convenient to adopt (t, r, θ, ϕ) so that the metric has the form (22.9).

It turns out that it is sufficient to obtain just R_{00} and R_{11}, but for good measure one may as well obtain all the other components in order to check for consistency. One finds

$$R_{00} = -3\ddot{a}/a \tag{23.3}$$

$$[R_{ij}] = \frac{1}{c^2} \left(a\ddot{a} + 2\dot{a}^2 + 2c^2 K_0 \right) \begin{pmatrix} (1 - K_0 r^2)^{-1} & 0 & 0 \\ 0 & r^2 & 0 \\ 0 & 0 & r^2 \sin^2\theta \end{pmatrix} \tag{23.4}$$

Notice that R_{ij} has the form $R_{ij} = f(t)g_{ij}$; this is a hint that the space is homogeneous.

Next we must specify the right-hand side of the field equation in some way, and then seek a self-consistent solution. We proceed by modelling the *content* of the universe, i.e. that which contributes to the energy tensor, as an ideal fluid, so that T^{ab} has the form $T^{ab} = (\rho + p/c^2)u^a u^b +$

Relativity Made Relatively Easy: General Relativity and Cosmology. Volume 2. Andrew M. Steane,
Oxford University Press. © Andrew M. Steane 2021. DOI: 10.1093/oso/9780192895646.003.0023

Table 23.1 Connection coefficients for the FLRW metric (c.f. Table 17.1). In the 'Lagrangian' $\mathcal{L}$ the dot signifies a derivative with respect to an affine parameter; in the Γ equations $\dot{a} = \mathrm{d}a/\mathrm{d}t$.

$$\mathcal{L} = -c^2\dot{t}^2 + a^2(t)\left(\frac{\dot{r}^2}{1 - K_0 r^2} + r^2\dot{\theta}^2 + r^2\sin^2\theta\,\dot{\phi}^2\right)$$

$$\Gamma^t_{rr} = \frac{a\dot{a}}{c^2(1 - K_0 r^2)}, \quad \Gamma^t_{\theta\theta} = \frac{a\dot{a}r^2}{c^2}, \quad\quad \Gamma^t_{\phi\phi} = \frac{a\dot{a}r^2\sin^2\theta}{c^2},$$

$$\Gamma^r_{tr} = \dot{a}/a,$$

$$\Gamma^r_{rr} = \frac{K_0 r}{1 - K_0 r^2}, \quad \Gamma^r_{\theta\theta} = -(1 - K_0 r^2)r, \quad \Gamma^r_{\phi\phi} = -(1 - K_0 r^2)r\sin^2\theta,$$

$$\Gamma^\theta_{t\theta} = \dot{a}/a, \quad\quad \Gamma^\theta_{r\theta} = 1/r, \quad\quad \Gamma^\theta_{\phi\phi} = -\sin\theta\cos\theta,$$

$$\Gamma^\phi_{t\phi} = \dot{a}/a, \quad\quad \Gamma^\phi_{r\phi} = 1/r, \quad\quad \Gamma^\phi_{\theta\phi} = \cot\theta. \tag{23.2}$$

pg^{ab} and we shall add the cosmological constant term separately. Since in our coordinates the cosmological fluid is not moving, one finds $u^a = (1,0,0,0)$ and $u_a = (-c^2,0,0,0)$, and therefore

$$T_{00} - \tfrac{1}{2}g_{00}T^\lambda_\lambda = \tfrac{1}{2}(\rho c^2 + 3p)c^2, \quad\quad T_{ij} - \tfrac{1}{2}g_{ij}T^\lambda_\lambda = \tfrac{1}{2}(\rho c^2 - p)g_{ij} \tag{23.5}$$

Hence the Einstein field equation reads

$$\left.\begin{array}{r} -3\ddot{a}/a = \tfrac{1}{2}\kappa(\rho c^2 + 3p)c^2 - \Lambda c^2 \\ a\ddot{a} + 2\dot{a}^2 + 2c^2 K_0 = c^2 a^2\left(\tfrac{1}{2}\kappa(\rho c^2 - p) + \Lambda\right) \end{array}\right\} \tag{23.6}$$

where $\kappa = 8\pi G/c^4$. By now eliminating $\ddot{a}$ from the second equation, and using $K_0 = k/R_0^2$, we obtain

---

Friedmann equations

$$\frac{\dot{a}^2 + kc^2/R_0^2}{a^2} = \frac{8\pi G\rho + \Lambda c^2}{3} \tag{23.7}$$

$$\frac{\ddot{a}}{a} = -\frac{4\pi G}{3}\left(\rho + 3p/c^2\right) + \frac{\Lambda c^2}{3} \tag{23.8}$$

Sometimes the first equation here is called the Friedmann equation, and then the second is called the *acceleration equation* or the *Raychaudhuri equation*. These equations are essentially a statement of the Einstein field equation: they state what form the field equation takes under the assumptions of homogeneity and the ideal fluid. We have still to solve them! The solution will depend on the values of the constants k, Λ and on the way ρ and p depend on conditions—the equation of state.

Even without a full solution, we can immediately make a simple general observation based on the acceleration equation:

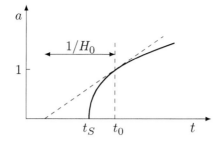

Fig. 23.1 The scale factor as a function of time when the strong energy condition holds. H_0 is equal to the slope at t_0, where $a = 1$. Note, however, that observational evidence suggests this condition does not always hold.

Proposition 23.1 *If the strong energy condition always holds then there was a singularity at a finite time t_S in the past, with $(t_0 - t_S)H_0 \leq 1$.*

Proof: The strong energy condition implies $\ddot{a}/a \leq 0$ in the acceleration equation. If $\ddot{a} < 0$ then the function $a(t)$ is concave so it must hit the $a = 0$ axis at a time less than $1/H_0$ before present—see Fig. 23.1. For $\Lambda \neq 0$ the strong energy condition does not necessarily always hold, but the observation remains useful in that it shows that a universe of finite age is the 'ordinary' behaviour.

A Newtonian diversion. It is useful to pause now to get some physical insight into the Friedmann equations. The first equation has to do with T^{00} and therefore it has to do with energy. The second equation has to do with T^{ij} and therefore it has to do with force and momentum. It is a useful exercise to compare these results with a Newtonian model of an expanding homogeneous gravitating fluid, as follows.

We suppose we have a Euclidean space filled with matter of uniform density $\rho(t)$ which is everywhere expanding. That is to say, if the matter is in the form of many small particles, then each particle moves outwards from any given origin with a velocity equal to

$$\mathbf{v} = \frac{\dot{a}}{a}\mathbf{r} \tag{23.9}$$

where r is the distance from the origin and $a(t)$ is the scale factor—which is to say, a factor which plays the role it plays in this equation. The equation describes motion having the form of a universal expansion, preserving homogeneity.

Now consider a spherical region of radius $R = R_0 a$ where R_0 is a constant. Any given particle of mass m at the outer edge of this region experiences a net inwards force GMm/R^2 owing to the gravitation of the mass $M = (4/3)\pi R^3\rho$ inside the region, and we shall assume that the forces from matter outside the region cancel (as they would in a spherically symmetric case). This is a non-trivial assumption to which we shall return in a moment. The equation of motion (Newton's second law) for such a particle is

$$-\frac{GMm}{R^2} = m\ddot{R} \qquad \Rightarrow \quad \ddot{a} = -\frac{4}{3}\pi G\rho a. \qquad (23.10)$$

After replacing ρ by $\rho + 3p/c^2$ (which makes perfect sense when we recall eqn (5.50) for the Newtonian limit), and then noting that the Λ term can be interpreted as a contribution to energy density and pressure, we thus obtain the second Friedmann equation exactly! The fact that this Newtonian result precisely matched the GR result is fortuitous, but it does provide a sound physical insight into what the second Friedmann equation is telling us. The equation is saying that the pull of gravity is causing the expansion to slow (in the presence of ordinary matter). Also, an argument in the other direction would not be fortuitous: the Friedmann equation applied in a Newtonian limit must reproduce a suitable Newtonian result.

The main difficulty with the Newtonian picture just presented is that we had to ignore the force owing to matter outside any given sphere, which requires a symmetry argument, and to maintain this argument throughout space it appears we shall require an infinitely extended volume of matter-filled space, and I do not like infinity.[1] One can avoid this infinity by suggesting a compact topology, such as for example the flat 3-torus. One can picture this as a cubic region, with anything leaving one face of the cube entering again on the opposite face.[2] It is noteworthy that in this case one must allow space itself (i.e. the cube) to expand along with the matter in it. Thus we have arrived at a model with thoroughly non-Newtonian features, but one which makes perfect Einsteinian sense.

Next let us consider energy in the Newtonian picture. Consider a particle of mass m at coordinate radius r, and suppose that at any moment its velocity is equal to the gravitational escape velocity associated with the total mass M inside that radius. This is a statement of energy conservation, as long as the particle just keeps ahead of the matter inside the ball, neither overtaking further matter (by going too fast) nor falling behind, so that M is a constant in the calculation. Then we find

$$\frac{1}{2}mv^2 = \frac{GMm}{r}$$

$$\Rightarrow \qquad \frac{1}{2}r^2\left(\frac{\dot{a}}{a}\right)^2 = \frac{G}{r}\frac{4}{3}\pi r^3 \rho$$

$$\Rightarrow \qquad \frac{\dot{a}^2}{a^2} = \frac{8\pi G\rho}{3} \qquad (23.11)$$

After allowing for Λ, this exactly reproduces the first Friedmann equation for the case $k = 0$, which makes sense for a calculation which assumed flat space. In order to check that the assumed motion is consistent with constant M, one may take the derivative of (23.11) with respect to time; one finds that one reproduces (23.10) as long as $\rho \propto 1/a^3$, which is what was assumed. This second Newtonian calculation has confirmed the intuition that the first Friedmann equation may be loosely regarded as a statement of energy conservation.

---

[1] To make this objection more formally, observe that an infinite homogeneous matter-filled space is at best questionable in a Newtonian picture, because it requires the perfect balance of infinitely large forces if the net gravitational force at any location is to be finite.

[2] This does not imply any lack of isotropy for the intrinsic geometry which remains flat, but it does involve a non-isotropic global topology.

Table 23.2 A selection of metrics studied in this book.

isotropy	homogeneity	$T^{ab} = 0$	static	metric
✓	✗	✓	✓ / ✗	Schwarzschild
✓	✗	✗	✓	Schwarzschild interior
✓	✓	✗	✗	FLRW

23.2 Solving the Friedmann equations

In order to find the FLRW dynamics, our previous study of spherically symmetric metrics (Table 23.2) cannot directly help because the Schwarzschild metric applies to vacuum whereas we want to treat $T^{ab} \neq 0$, and the stellar structure solutions are static. On the other hand, it is useful to recall that we have already studied the motion of an ideal fluid in a general spacetime, in our discussion of hydrodynamics in Chapter 14. That discussion was based on the equation $\nabla_\mu T^{a\mu} = 0$, from which we obtained (14.23) and (14.22) which we repeat here:

$$\left(\rho + \frac{p}{c^2}\right) u^\mu \nabla_\mu u^a = -\left(g^{a\mu} + \frac{u^a u^\mu}{c^2}\right) \nabla_\mu p, \tag{23.12}$$

$$c^2 \frac{D\rho}{d\tau} \equiv c^2 u^\mu \nabla_\mu \rho = -\left(c^2 \rho + p\right) \nabla_\mu u^\mu. \tag{23.13}$$

The first of these is the equation of motion of the fluid. In a homogeneous situation, the gradient of the pressure is zero, so the right-hand side is zero, and then the left-hand side tells us that the parts of the fluid (i.e. the galaxies or clusters of galaxies) are following geodesics. This is what we should expect: in the absence of a pressure that is larger on one side than another, the pressure forces on any given part cancel out, and so the parts of the fluid move inertially. As they rush away from one another (or, if you prefer, as they stay still in the comoving coordinate frame), the galaxies are in free fall, as we already noted in Section 22.2.1.

Now proceed to eqn (23.13). The left-hand side can be written $c^2 \dot\rho$ in comoving coordinates. If you need to convince yourself of this, use that $u^\mu = \delta_0^\mu$ in these coordinates, and $\nabla_a \rho = \partial_a \rho$ since ρ is a scalar. The right-hand side can be evaluated either by using the Christoffel symbols, or by using the formula (11.38) for the covariant divergence of a vector field. One finds $\nabla_\mu u^\mu = 3\dot a/a$ (exercise 23.2). Hence, when applied to a uniform fluid and the FLRW metric, (23.13) gives the

---

homogeneous continuity equation

$$\dot\rho + 3\left(\rho + p/c^2\right) \frac{\dot a}{a} = 0. \tag{23.14}$$

This useful equation can also be derived from the Friedmann equations by differentiating the first and then eliminating $\ddot a$ between them. The name arises from the fact that it is the continuity equation for a flow which keeps the fluid homogeneous, c.f. (24.5).

From now on we will employ only (23.14) and (23.7) and drop (23.8). One can assert, in other words, that we do not need the second Friedmann equation at all, if we already have the fluid equation. Alternatively one can say that we have completed the first step towards the solution of the Friedmann equations as a pair, by forming from them a simpler but equivalent pair.

23.2.1 Relation of the density to the scale parameter

It does not take long to discover that (23.14) can be written

$$\frac{d(\rho a^3)}{da} = -\frac{3pa^2}{c^2}. \tag{23.15}$$

This equation can also be read as '$dU = -pdV$' where $U = \rho c^2 a^3$ and $V = a^3$. This is an example, in the context of GR, of the thermodynamic statement

$$dU = TdS - pdV. \tag{23.16}$$

In the model we are pursuing, the cosmological expansion is adiabatic, in any finite region, and therefore in the whole. It is adiabatic because it is reversible and there is no temperature gradient and therefore no heat flow. Hence $dS = 0$ and we have (23.15). In this sense, (23.14) can be interpreted as a statement of energy conservation in a cosmological setting. But be warned: in dynamic situations energy conservation can be a slippery concept, and we already remarked the difficulties of associating energy with gravity itself.

Next we bring in the fact that in physics generally, pressure is often related in a simple way to energy density u. For example, in an ideal gas one has $p \propto u$ and in thermal radiation one has $p = u/3$. In many solids and liquids the relationship is equally simple to good approximation. Therefore we propose, for the cosmological fluid, the

---
equation of state

$$p = w\rho c^2 \tag{23.17}$$

---

where w is a constant. In fact, we shall shortly treat a more general model where the cosmological fluid is composed of several species with different values of w, but let us look at a single species to begin with. After substituting (23.17) into (23.15) one has a differential equation which can be solved[3] for $\rho(a)$:

$$\rho \propto a^{-3(w+1)}. \tag{23.18}$$

This is good progress! We now know how the density of stuff (matter, radiation, etc.) in the universe depends on the scale factor, if all the stuff is of one type. We have not yet found $a(t)$.

We can immediately generalize to a multi-component fluid.

---

[3]To solve the equation, introduce the variable $y = \rho a^3$ and solve for $y(a)$, then write $\rho = y/a^3$.

For any system composed of several different parts, the energy tensor of the whole is given by

$$T^{av} = \sum_i T^{ab}_{(i)} + \sum_{i>j} I^{ab}_{(i,j)} \tag{23.19}$$

where $I^{ab}_{(i,j)}$ gives the influence of interactions between the parts. If we now make the further assumption of non-interacting parts, then $I^{ab}_{(i,j)} = 0$ and we can formulate the separate $T^{ab}_{(i)}$ separately. If each of the parts is a fluid then each $T^{ab}_{(i)}$ has the form (14.18), and if all these fluids have the same flow velocity at any given place, then we have

$$T^{av} = \sum_i \left[(\rho_i + p_i/c^2)u^a u^b + p_i g^{ab} \right]$$

$$= \left(\sum_i \rho_i + p_i/c^2 \right) u^a u^b + \left(\sum_i p_i \right) g^{ab} \tag{23.20}$$

Therefore the whole can be modelled as a single fluid with:

---

multi-component non-interacting fluid

$$\rho = \sum_i \rho_i, \qquad p = \sum_i p_i. \tag{23.21}$$

When we say 'non-interacting' here, we mean not interacting other than gravitationally. The parts do interact gravitationally, since they each contribute to the total energy tensor, which in turn shapes the metric, and then each part of the fluid moves as the equation of motion tells it to move. In the present example the parts all follow geodesics and therefore move together in synchrony, in agreement with what has been assumed in obtaining (23.21).

Now note that the differential equation (23.14) is linear, and therefore any given solutions can be superposed in order to construct further solutions. Therefore if each part of the multi-component fluid has an equation of state

$$p_i = w_i \rho_i c^2 \tag{23.22}$$

where the constants w_i are different for the different components, then we find

$$\rho = \sum_i \rho_{i,0} \, a^{-3(w_i+1)} \tag{23.23}$$

where the notation $\rho_{i,0}$ signifies the value of ρ_i at the moment when $a(t) = 1$.

It is the standard practice in cosmology to treat the universe as made of three basic types of stuff. These are called matter, radiation, and vacuum. The term 'matter' refers to anything whose energy tensor is like that of a collection of dust particles freely floating in space but not rushing too and fro colliding with each other, or exerting pressure in other ways. Thus 'matter' has zero pressure and therefore $w_{\mathrm{m}} = 0$ in the equation of state. This describes the net cosmological effect of planets, stars, galaxies and, it is thought, dark matter, quite well. The

type of fluid	w
radiation	1/3
matter (dust)	0
Λ	-1

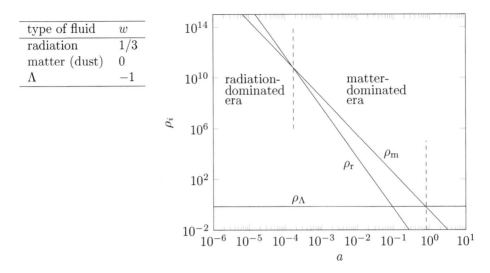

Fig. 23.2 The dependence of the densities on a in the three-component cosmological fluid. The table on the left shows the value of the state parameter w for the three types of fluid. The graph shows the evolution of the three densities. The relative values of ρ_{r0}, ρ_{m0}, $\rho_{\Lambda 0}$ have been chosen roughly appropriate to our universe ((5×10^{-5}, 0.3, 0.7)ρ_{crit} respectively). The vertical dashed lines show the time when the dominant contribution swaps over; this is not an abrupt transition but, owing to the third and fourth powers in (23.24), it does not take much change in a for the largest ρ_i to take over in each case. The present corresponds to $a = 1$; the second dashed line is just before this, suggesting the universe recently entered a vacuum-dominated era.

term 'radiation' refers to anything whose energy tensor is like that of isotropic electromagnetic radiation. This has $p = u/3$ and therefore $w_r = 1/3$. It describes photons, fast neutrinos, gravitons, and anything else travelling at or near the speed of light—so in the very early very hot universe it describes practically everything other than vacuum. The term 'vacuum' refers to anything whose energy tensor is given by a constant multiplied by the metric tensor.[4] It therefore has $w_\Lambda = -1$. These values are summarized in the table on the left of Fig. 23.2. We thus find:

Density of cosmic fluid under assumption of constant $\rho_{r,0}$, $\rho_{m,0}$, Λ

$$\rho = \rho_{r,0}\, a^{-4} + \rho_{m,0}\, a^{-3} + \rho_\Lambda \tag{23.24}$$

where

$$\rho_\Lambda = \rho_{\Lambda,0} = \Lambda c^2/8\pi G. \tag{23.25}$$

Notice that the behaviour of matter (dust) is described by $\rho_m a^3 = $ constant; this has a natural physical interpretation as the conservation of the amount of matter. The behaviour of radiation is $\rho_r a^4 = $ constant, which also has a natural interpretation since the volume term a^3 accounts for dilution of the number of photons per unit volume, and the other factor of a accounts for

---

[4]Other models of the vacuum can also be suggested, with w varying or not exactly -1.

the redshift which lowers the energy along with the frequency. Thus in homogeneous conditions the gravitational dynamics has the same effect on radiation as would the adiabatic expansion of a cavity with reflective walls containing the radiation in flat spacetime.

A consequence of these two behaviours is that the ratio of the number of baryons to the number of photons in the universe stays constant (since the photons are mostly those in the background radiation). This number is therefore a useful way to indicate something about the global character of the cosmos; its value is found by observation to be approximately:

Baryon–photon ratio

$$\eta = \frac{n_{\mathrm{B}}}{n_\gamma} \simeq 6 \times 10^{-10}. \tag{23.26}$$

This number plays a crucial role in both nucleosynthesis and recombination and thus is fundamental to cosmology. We do not have a theoretical way to say why or how it came to have the value it does, but if we had to describe our universe to some interested visitor, this is one of the numbers it would be appropriate to mention.

Eqn (23.24) gives a lot of useful insight into cosmology, even before we find the values of the various constants, or the behaviour of $a(t)$. For if we assume none of the constants are zero, then from (23.24) it is clear that radiation will dominate at sufficiently small a, whereas at sufficiently large a, vacuum will dominate. This is indicated in Fig. 23.2. Since there is good evidence that a was small in the early universe, we may deduce that the dynamics of the cosmos at early times were dominated by radiation. Since there is also evidence that a is permanently growing and Λ is non-zero, we may deduce that the dynamics of the cosmos at late times will be dominated by vacuum.

The period when radiation had a higher energy-density than matter is called the *radiation-dominated era*. The subsequent period when matter dominates the energy-density is called the *matter-dominated era*. If Λ is non-zero then eventually we reach a vacuum-dominated era; it is currently thought that this is the case in our universe now.

All these conclusions only follow when the parts of the fluid are not evolving in themselves, or interacting with one another, but merely moving under gravity. It is currently suggested that there was a very early period called *inflation* during which Λ was very much larger. We will return to this suggestion in Chapter 26, and ignore it for now.

23.2.2 Relation of spatial curvature to density

We now turn to the solution of the remaining Friedmann equation, (23.7). Recalling that $H = \dot{a}/a$, the equation can be written

$$H^2 = \frac{8\pi G}{3}\rho + \frac{1}{3}\Lambda c^2 - \frac{c^2 k}{R_0^2 a^2}. \tag{23.27}$$

This hides the fact that it is a differential equation, but helps in getting insight into the role of the various terms. A useful way to organize the solutions is to note that there are three cases to consider for k, namely $k = +1, 0, -1$. These are commonly called 'closed, flat, open' but since there is in fact no direct link between curvature and compactness, nor between the sign of k and the long-term evolution of the universe, this terminology can be misleading. We will simply refer to positive, zero and negative spatial curvature.

By dividing through by H^2 and making the term involving k the subject of the equation, one obtains

$$\frac{c^2 k}{R_0^2} \frac{1}{a^2 H^2} = \frac{8\pi G}{3H^2} (\rho_m + \rho_r + \rho_\Lambda) - 1 \tag{23.28}$$

$$= \Omega_m + \Omega_r + \Omega_\Lambda - 1 \tag{23.29}$$

where

$$\Omega_i(t) \equiv \frac{\rho_i(t)}{\rho_{\text{crit}}(t)}, \qquad \rho_{\text{crit}}(t) = \frac{3H^2}{8\pi G}. \tag{23.30}$$

The Ω_i are called the *density parameters* and ρ_{crit} is the *critical density*. These definitions are motivated by observing that in (23.29) the sign of k is determined by whether or not the total density is larger or smaller than the critical density, and this can be expressed in terms of the density parameters as follows:

$$\Omega \equiv \Omega_m + \Omega_r + \Omega_\Lambda \tag{23.31}$$

$$\Omega < 1 \qquad k < 0: \text{ negative spatial curvature}$$

$$\Omega = 1 \qquad k = 0: \text{ flat}$$

$$\Omega > 1 \qquad k > 0: \text{ positive spatial curvature}$$

We here introduce the symbol Ω without subscript to indicate the sum of the density parameters associated with three parts of the three-fluid model. The density parameters can all depend on time, as can the critical density, but k does not. So if Ω is above, at, or below 1 at any time (such as the present), then it will remain so and will always have been so, since k is a constant and $(a^2 H^2) > 0$. In short, *the sign of $(\Omega - 1)$ does not change with time.*

Sometimes the curvature term is itself written to look like a 'density parameter', by defining

$$\Omega_k(t) \equiv \frac{-c^2 k}{R_0^2 a^2(t) H^2(t)} \tag{23.32}$$

and then we can write the easily remembered formula

$$\Omega_m + \Omega_r + \Omega_\Lambda + \Omega_k = 1. \tag{23.33}$$

Hence another way to write the curvature 'density' is $\Omega_k = 1 - \Omega$.

Let Ω_{i0} be the value of Ω_i at the time t_0 (the present) when $a = 1$. Then using (23.22), (23.30) we have

$$\Omega_i = \frac{8\pi G}{3H^2}\rho_i = \frac{8\pi G}{3H^2}\rho_{i,0}a^{-3(w_i+1)} = \Omega_{i0}\left(\frac{H_0}{H}\right)^2 a^{-3(w_i+1)}. \tag{23.34}$$

Substituting this into (23.33) and multiplying through by H^2, one obtains the equation for H in terms of the other parameters:

$$H^2 = H_0^2\left(\Omega_{m0}a^{-3} + \Omega_{r0}a^{-4} + \Omega_{\Lambda0} + (1 - \Omega_0)a^{-2}\right) \tag{23.35}$$

where $\Omega_0 \equiv \Omega_{m0} + \Omega_{r0} + \Omega_{\Lambda0}$. In terms of redshift this is

$$H^2 = H_0^2\left(\Omega_{m0}(1 + z)^3 + \Omega_{r0}(1 + z)^4 + \Omega_{\Lambda0} + (1 - \Omega_0)(1 + z)^2\right). \tag{23.36}$$

Hence, once the constant parameter values are known, one can find the Hubble parameter at any given redshift.

23.2.3 Evolution of the scale factor

So far we have used the Friedmann equations to relate density to scale parameter and spatial curvature to density. Now we shall turn at last to solving the differential equation for $a(t)$. From (23.35) and $H = \dot{a}/a$ we have that the equation we seek to solve is

$$\frac{da}{dt} = H_0\left(\frac{\Omega_{m0}}{a} + \frac{\Omega_{r0}}{a^2} + \Omega_{\Lambda0}a^2 + 1 - \Omega_{m0} - \Omega_{r0} - \Omega_{\Lambda0}\right)^{1/2}. \tag{23.37}$$

One can solve this quite readily by numerical methods for any values of the parameters. In order to build some insight, it is useful also to consider some simple example cases. We will begin by treating the three cases where only one of the density parameters is non-zero. For each of these cases there are three possible values of the curvature index k, giving 9 cases in all. The results are summarized in Table 23.3 (and see exercise 23.5).

Radiation only.

$$k = 0: \quad a(t) = (2H_0t)^{1/2} \tag{23.38}$$

$$k = \pm1: \quad a(t) = \left(2H_0\sqrt{\Omega_{r0}}\,t\right)^{1/2}\left(1 + \frac{1 - \Omega_{r0}}{2\sqrt{\Omega_{r0}}}H_0t\right)^{1/2} \tag{23.39}$$

One expects a period in the early universe when radiation dominates, so these solutions are relevant to that early period, approximately the first ten to fifty thousand years, excluding the pre-GR epoch and any inflationary epoch.

Examples of the three types of solution are plotted in Fig. 23.3a. The cases $k \pm 1$ differ from one another since in one case $\Omega_{r0} > 1$, and in the other $\Omega_{r0} < 1$.

Table 23.3 A summary of the functional form of solutions $a(t)$ to the Friedmann equations in the cases where only one density is non-zero or all are zero. α is either a constant or a parameter; the details of each equation are given in the main text. No cases have been omitted (assuming Ω_m, $\Omega_r \geq 0$).

k	radiation $\Omega_m = \Omega_\Lambda = 0$ $\Omega_r > 0$	dust $\Omega_r = \Omega_\Lambda = 0$ $\Omega_m > 0$	vacuum $\Omega_m = \Omega_r = 0$ $\Lambda > 0, \quad \Lambda = 0, \quad \Lambda < 0$
-1	$t^{1/2}(1+\alpha t)^{1/2}$	$a \sim \cosh\alpha - 1$ $t \sim \sinh\alpha - \alpha$	$\sinh(\alpha t), \quad t, \quad \sin(\alpha t)$
0	$t^{1/2}$	$t^{2/3}$	$e^{\alpha t}, \qquad \text{const}, \qquad -$
1	$t^{1/2}(1-\alpha t)^{1/2}$	cycloid	$\cosh(\alpha t), \quad -, \quad -$

Matter only.

$$k = 0: \quad a = \left(\frac{3}{2}H_0 t\right)^{2/3} \tag{23.40}$$

$$k = 1: \quad a = \frac{\Omega_{m0}/2}{(\Omega_{m0} - 1)}(1 - \cos\alpha), \qquad t = \frac{\Omega_{m0}/2}{H_0(\Omega_{m0} - 1)^{3/2}}(\alpha - \sin\alpha) \tag{23.41}$$

$$k = -1: \quad a = \frac{\Omega_{m0}/2}{(1 - \Omega_{m0})}(\cosh\alpha - 1), \qquad t = \frac{\Omega_{m0}/2}{H_0(1 - \Omega_{m0})^{3/2}}(\sinh\alpha - \alpha) \tag{23.42}$$

The evidence suggests the energy density of the universe has been dominated by matter for most of its history (c.f. Fig. 23.2), so this case enables us to determine the main facts about that evolution. The three solutions are plotted in Fig. 23.3b. After contemplating Birkhoff's theorem one realizes that there is a close connection between this scenario and freefall motion in Schwarzschild–Droste spacetime. Both can be used to describe the expansion or collapse of a pressure-less spherically symmetric cloud.

The equations for $k = 1$ ($\Omega_{m0} > 1$) describe a cycloid. One should not extend the solution for this case beyond $\alpha = 2\pi$ (the 'Big Crunch') since one does not know what happens in the post-GR epoch of such a 'Crunch'.

The case $k = 0$ is called the *Einstein-de Sitter* (EDS) model. This is useful as a rough model for much of the evolution of our universe, since observations suggest the spatial curvature of our universe is small.

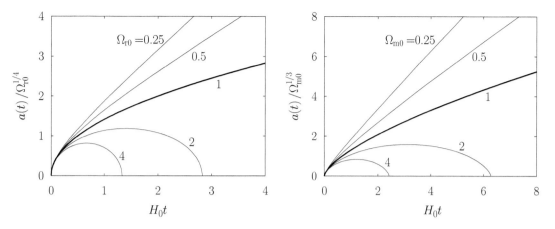

Fig. 23.3 The scale factor as a function of time for radiation-only (left) and matter-only (right) cases.

In any of the three cases, the present age of the universe is indicated by the value of t when $a = 1$. For $k = 0$ one finds

$$t_0 = \frac{2}{3H_0} \simeq 9.2 \times 10^9 \text{ years} \tag{23.43}$$

where we used the value $H_0 \simeq 70\,\text{km s}^{-1}/\text{Mpc}$ obtained from Fig. 22.2. Since there is observational evidence of stars of order 13 billions years old, clearly the estimate of 9.2 billion years for the age of the universe is inaccurate, which tells us that the matter-only case does not model the evolution very precisely. This is one of a number of pieces of evidence that in fact $\Lambda \neq 0$.

Vacuum only. A universe with no matter or radiation might be thought to be no universe at all, but nevertheless it is a useful case to consider, since it represents what happens when the matter and radiation is dispersed at very low density, or when Λ is so large that the other densities are negligible by comparison.

There are six cases, as noted in Table 23.3, and here are the details:

$$\Lambda = 0 \qquad\qquad k = 0: \quad a(t) = 1 \tag{23.44}$$

$$k = -1: \quad a(t) = t/t_0 \tag{23.45}$$

$$\Lambda > 0 \qquad k = 0,\ \Omega_{\Lambda 0} = 1: \quad a(t) = e^{H_0(t-t_0)} = e^{\alpha(t-t_0)} \tag{23.46}$$

$$k = -1,\ \Omega_{\Lambda 0} < 1: \quad a(t) = (1/\Omega_{\Lambda 0} - 1)^{1/2} \sinh \alpha t \tag{23.47}$$

$$k = 1,\ \Omega_{\Lambda 0} > 1: \quad a(t) = (1 - 1/\Omega_{\Lambda 0})^{1/2} \cosh \alpha t \tag{23.48}$$

$$\Lambda < 0 \qquad\qquad k = -1: \quad a(t) = (1 - 1/\Omega_{\Lambda 0})^{1/2} \sin \alpha t \tag{23.49}$$

where

$$\alpha = H_0 \sqrt{|\Omega_{\Lambda 0}|} = \sqrt{|\Lambda| c^2/3}. \tag{23.50}$$

The case $\Lambda = 0$, $k = 0$ is the familiar Minkowski spacetime. The case $\Lambda = 0$, $k = -1$ is surprising at first, since when all the densities, including Ω_Λ, are zero, then we must have flat

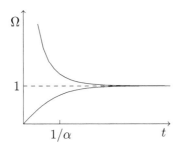

Fig. 23.4 Evolution of the density parameter in vacuum-only (or vacuum dominated) models with $\Lambda > 0$.

spacetime, i.e. Minkowski spacetime again, so how can space be curved? What has happened in this solution is that we have the Milne model discussed in Section 22.5.1. The flat spacetime has been foliated using curved spatial hypersurfaces and the scale factor increases linearly with time.

The cases with $\Lambda > 0$ similarly all have the same four-dimensional spacetime background, but one which has been separated into 'time' and 'space' in different ways. This is because, in the FLRW construction, we associate 'time' at the outset with the proper time of fundamental observers and the worldlines of these observers can be assigned in more than one way. This illustrates the fact that coordinates in GR need not be mere labels but are often carefully chosen so that they have useful physical content. In this example each coordinate line of fixed position is not just any old timelike geodesic: it is the worldline of a test particle which could in principle be present, and which acts as one of the fundamental observers of the comoving coordinate system. The hypersurfaces called 'space' are those which are orthogonal to these worldlines. This spacetime is called *de Sitter space* D^4. It has been much studied because of its innate simplicity and symmetry; it is the maximally symmetric space in four (not just three) dimensions for a Lorentzian manifold with positive spatial curvature.

The case $\Lambda < 0$ gives a four-dimensional geometry called *anti de-Sitter space*; it is also interesting but we shall not consider it further. (For an introduction to de-Sitter and anti-de Sitter space, see (Moschella, 2005; Rindler, 2000)).

The most significant feature of the vacuum-only solutions is the exponential growth in a when $\Lambda > 0$. All of (23.46)–(23.48) give $H \to \alpha$ for $ct \gg \sqrt{3/|\Lambda|}$ and consequently, using (23.34),

$$\Omega(t) \to 1 \tag{23.51}$$

(c.f. Fig. 23.4). This means that the evolution is such that spacetime naturally configures itself so as to have the critical energy density and become spatially flat, or almost so. One may say that as it expands under the influence of a cosmological constant, space is filling itself up with energy to just the right degree to give this outcome, approaching more and more closely to the $\Omega_{\Lambda 0} = 1$, $k = 0$ model. The latter is called the *de Sitter model* or *de Sitter universe*. It plays a significant role in the understanding of cosmic inflation which will be discussed in Chapter 26. The quantity $R_\Lambda = \sqrt{3/|\Lambda|}$ is called the *de Sitter radius*.

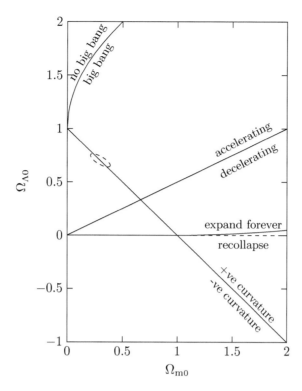

Fig. 23.5 Types of solution for Friedmann-Lemaître (matter and vacuum) models. The dashed ellipse shows the region which is consistent with cosmological observations (c.f. Fig. 25.4).

Matter and vacuum: the Friedmann-Lemaître models. Since the radiation-dominated period is thought to have been of very short duration compared to the total (a mere 47 thousand years out of 13.8 billion), a model setting $\Omega_{r0} = 0$ but incorporating both dust and vacuum can be expected to be quite precise. It should account correctly for most of cosmological history, and give values for quantities such as the age and spatial curvature of the universe to better than 1% precision, if the parameters can be measured accurately enough.

Fig. 23.5 summarizes the behaviour of solutions depending on the values of Ω_{m0} and $\Omega_{\Lambda0}$. For any given Ω_{m0}, $\Omega_{\Lambda0}$ the solution for $a(t)$ can be obtained numerically with little difficulty. The main new feature, compared with the dust-only solutions, is that the expansion can either accelerate or decelerate. The effect of matter is to slow the expansion; the effect of a positive Λ is to accelerate the expansion.

The straight line at $\Omega_{\Lambda0} = 1 - \Omega_{m0}$ on the graph separates the regions of positive and negative curvature index k. The straight line at $\Omega_{\Lambda0} = \Omega_{m0}/2$ is obtained from eqn (25.4) for the *deceleration parameter* described in Chapter 25. The line separating 'expand forever' and 'recollapse' is obtained by examining whether or not the graph of $H(a)$ has a turning point. This line intersects the $k = 0$ line at $\Omega_{\Lambda0} = 0$, which indicates that in the absence of a cosmological constant, the long-term temporal behaviour is in one-to-one correlation with the spatial curvature. When

$\Lambda \neq 0$ this link no longer applies and all combinations are possible. The region marked 'no big bang' indicates solutions which either either decay exponentially to zero as $t \to -\infty$, or which 'bounce', i.e. first contract and then expand monotonically without ever reaching $a = 0$.

The spatially flat case ($\Omega_{\mathrm{r}0} = 0$, $\Omega_{\mathrm{m}0} + \Omega_{\Lambda0} = 1$ leading to $k = 0$) can be solved analytically. Cosmological observations suggest the universe is close to this case, so even if it may not give an exact model of the universe, it gives a very good first approximation. The Friedmann equation (23.37) then reads

$$\frac{\mathrm{d}a}{\mathrm{d}t} = H_0 \left((1 - \Omega_{\Lambda0})a^{-1} + \Omega_{\Lambda0}a^2 \right)^{1/2} \quad \Rightarrow \quad t = \frac{1}{H_0} \int_0^a \frac{x\,\mathrm{d}x}{((1 - \Omega_{\Lambda0})x + \Omega_{\Lambda0}x^4)^{1/2}}. \quad (23.52)$$

The integral can be evaluated by the change of variable $y^2 = x^3|\Omega_{\Lambda0}|/(1 - \Omega_{\Lambda0})$ and one obtains

$$H_0 t = \frac{2}{3\sqrt{|\Omega_{\Lambda0}|}} \begin{cases} \sinh^{-1}\left(\sqrt{a^3|\Omega_{\Lambda0}|/\Omega_{\mathrm{m}0}}\right) & \text{if } \Lambda > 0 \\ \sin^{-1}\left(\sqrt{a^3|\Omega_{\Lambda0}|/\Omega_{\mathrm{m}0}}\right) & \text{if } \Lambda < 0 \end{cases} \quad (23.53)$$

Hence (Heckmann, 1932; Blome, Hoell and Priester, 1997; Rindler, 2006)

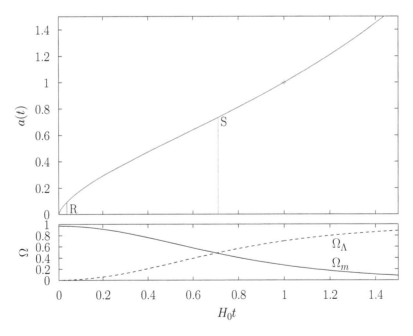

Fig. 23.6 Top: the scale factor as a function of time for a matter+vacuum model, with parameters suited to our universe ($\Omega_{\mathrm{m}0} = 0.3$, $\Omega_{\Lambda0} = 0.7$). The times marked R and S indicate approximately the epoch of reionization (the first stars) and the formation of the solar system. Bottom: $\Omega_{\mathrm{m}}(t)$ and $\Omega_\Lambda(t)$ for this case.

$$a(t) = (\Omega_{m0}/|\Omega_{\Lambda 0}|)^{1/3} \sinh^{2/3}\left(\frac{3}{2}\sqrt{|\Omega_{\Lambda 0}|}H_0 t\right) \qquad \text{if } \Lambda > 0,\ k=0 \qquad (23.54)$$

$$a(t) = (\Omega_{m0}/|\Omega_{\Lambda 0}|)^{1/3} \sin^{2/3}\left(\frac{3}{2}\sqrt{|\Omega_{\Lambda 0}|}H_0 t\right) \qquad \text{if } \Lambda < 0,\ k=0 \qquad (23.55)$$

The dashed ellipse on Fig. 23.5 shows where the parameters of our universe are believed to lie (see Chapter 25). The corresponding evolution is shown in Fig. 23.6. The scale factor first rises sharply and then the expansion slows as all the matter mutually attracts, but eventually the expansion begins to speed up again, owing to the vacuum term, and once this term dominates, it always will (under the assumption of no further physical effects to be accounted for). The expansion asymptotically approaches the exponential expansion indicated in eqn (23.46). By coincidence, it happens that the age of the universe in this model is now quite close to $1/H_0$.

23.3   The physical interpretation of the cosmological constant

Although Friedmann and Lemaître both included Λ in their theoretical studies of the possible dynamics of the cosmos in the 1920s, the cosmological constant was set to one side in most discussions of cosmology in the period 1930–1990. Its status was unclear, and it remains somewhat mysterious to this date. Whereas the value $\Lambda = 0$ might be a statement about the field equation which we must simply accept (until a deeper theory is discovered that can subsume GR), any other value naturally leads to questions about why Λ has that value and not another, and what is the physical nature of whatever it is that gives rise to Λ.

If we interpret the cosmological constant term as a form of stress-energy tensor, then for $\Lambda > 0$ we observe that the vacuum has $\rho_\Lambda > 0$, i.e. a positive energy-density, which is familiar. One then finds that the pressure of the vacuum is negative, which represents tension. Tension is a familiar property of solids and some liquids, but it is odd to think of it somehow existing in the vacuum. There are some hints from quantum field theory that suggest the quantum vacuum might exhibit tension, since what we call 'vacuum' is in fact everywhere filled with a rich collection of quantum fields such as the electromagnetic field, the Dirac field, the quark and gluon fields, and so on, and one can argue that when the room available for all these fields increases, so does their total energy content, even when they are all in their ground state. An increasing energy as a function of volume is physically manifested as a tension. However, all attempts to estimate Λ even roughly from quantum field theory have failed so far. A simple calculation based on merely summing the zero-point energies gets an answer too large by a factor of order 10^{120}—the answer is clearly not that simple!

The next unfamiliar aspect of Λ is that it gives rise to gravitational *repulsion*. Since, on average, the cosmological tension owing to Λ is homogeneous and isotropic, it offers no net non-gravitational force to any 'piece' of either vacuum or matter in the cosmos, so it does not disturb the galaxies away from their geodesic free-fall motion, but it does contribute to the overall *gravitation*, and it does so in a negative sense, i.e. the opposite sense to mass-energy. For example, in the field equation for an isotropic small region, (16.7), the gravitational attraction is proportional to $(\rho + 3p/c^2)$. This is positive whenever $|p| \ll \rho c^2$ and whenever p is positive. However, the energy tensor associated with Λ is proportional to the metric, so in a LIF it has the form $p = -\rho c^2$, and in this case $\rho + 3p/c^2 = -2\rho$. Hence we have negative attraction, which is repulsion. For any region containing some matter, this repulsion is negligible on distance scales of the size of the solar system or a galaxy, but on the cosmic scale it is not. It leads to the exponentially increasing scale factor of eqns (23.46) and (23.54).

We have modelled the cosmological constant by noting that it gives a contribution to T_{ab} described by the equation of state $p = w\rho c^2$ (eqn (23.22)) with $w = -1$. In search of more general possibilities, one may propose a more general equation of state in which w may vary, or may simply have a value other than -1. Observations to date are consistent with w constant and equal to -1, but this remains an active area of investigation.

23.4 Particle horizon and event horizon

The *particle horizon* is a statement about the past light cone of an observer at any particular place in the universe. It expresses useful information about the past evolution of the universe. The cosmic *event horizon* is a statement about the possible future evolution of the universe. It has no relevance in itself to any observation or measurement we can make now, but its discussion can be a convenient way to express some aspect of the overall dynamics. When cosmologists speak simply of 'the horizon' they usually mean the particle horizon.[5]

In an expanding space, the past light cone of any observer has the generic features illustrated by Fig. 23.7. Place the origin of the comoving coordinates at the observer in question. The light cone consists of rays that set off from a comoving coordinate χ at some time t_1, travelling in the negative χ direction, and journeyed towards the origin of the spatial coordinates, so as to arrive 'now' at time t. In order to understand this, it is best not to set $t_1 = 0$ in the first instance. The comoving coordinate distance travelled by such a light ray is given by:

$$\chi = c \int_{t_1}^{t} \frac{dt}{R_0 a} \tag{23.56}$$

and therefore this is also the formula for the coordinate from which the light set out. As t_1 goes to earlier times, this χ grows. In other words as we look longer ago, we are also looking further way. If we now consider the case $t_1 = 0$ then there are two possibilities: either the integral diverges, or it does not. If the integral diverges, it means that we (and therefore also any other observer) can now see light arriving from any comoving coordinate location. If the integral is finite, on the other hand, then there is a particle horizon given by:

Particle horizon

$$\chi_h(t) = \frac{c}{R_0} \int_0^t \frac{dt}{a} = \frac{c}{R_0} \int_0^a \frac{da}{a\dot{a}} \tag{23.57}$$

[5]But beware, the word is often employed to refer to the Hubble distance c/H, which is a different thing.

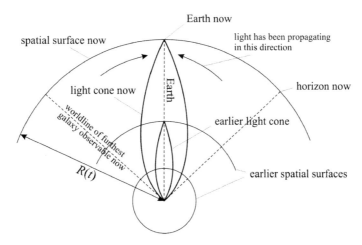

Fig. 23.7 The past light cone of Earth at two different times. The figure shows an 'expanding balloon' diagram, oriented such that Earth's worldline is vertical. The distance outwards from the origin on the diagram is proportional to the scale factor. The two lines bounding a given light cone indicate the worldlines of light arriving at our location from opposite directions. The EDS model has been used to plot illustrative photon worldlines (null geodesics). All the light entering our instruments in the present comes from the surface, not the interior, of the light cone of our present position. In the past the light cone was smaller. Therefore we can now see events which are outside the earlier light cone. In other words the horizon increases with time. (Compare this diagram with Fig. 23.8).

The particle horizon is, then, the largest comoving coordinate distance from which one part of the universe can have influenced another in the time available. The integral with respect to a will be finite if $a\dot{a} \sim a^n$ with $n < 1$; this gives $\dot{a} \sim a^{n-1}$ and $\ddot{a} \sim (n-1)a^{n-2}$ so the condition is $\ddot{a} < 0$. Therefore a sufficient (but not a necessary) condition for the horizon to be finite is that the expansion has been continually decelerating since the beginning.

A warning is needed here. Unlike many other calculations in cosmology, such as the age of the universe or the value of the Hubble parameter, the horizon calculation is sensitive to the behaviour of a at times very close to $t = 0$. Since this involves the pre-GR or Planck era, the true horizon is completely unknown. It has become the standard practice in cosmology to use the phrase 'the horizon' as a shorthand for either 'the horizon for that part of the evolution which took place after the Planck era' or 'the horizon if we trust the Friedmann equation all the way to $a = 0$'. That is, either one cuts off the lower limit of the integral in (23.57) at whatever t corresponds to the time when the thermal energy was of the order of the Planck energy, or else one extends the time-dependence of $a(t)$ given by the Friedmann equation right back to $a = 0$ even though there is no good reason to trust it even approximately at such scales. In the following we will adopt this practice, and then reconsider this issue in Section 26.1.1.

The particle horizon is often quoted as a proper distance by quoting the value of

$$d_\mathrm{h}(t) = a(t)R_0\chi_\mathrm{h}(t) = ac\int_0^t \frac{\mathrm{d}t}{a}. \tag{23.58}$$

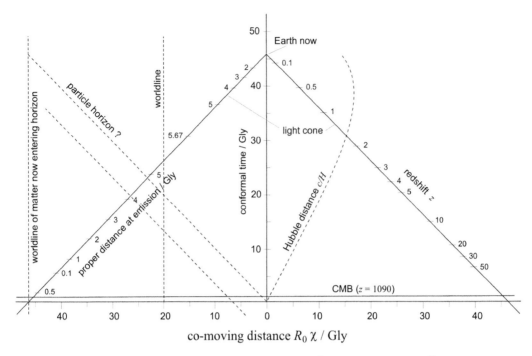

Fig. 23.8 Spacetime diagram of the universe plotted in $\chi, \bar{t}$ coordinates, where $\bar{t}$ is conformal time. Lines of constant χ, which were radial in Fig. 23.7, are vertical on this diagram. The use of $\bar{t}$ has the effect of stretching the time axis in such a way that the past light cone is shown by straight lines. The diagram shows the behaviour for the case $\Omega_{\Lambda 0} = 0.7$, $\Omega_{m0} = 0.3$, $H_0 = 70\,\mathrm{km\,s^{-1}Mpc^{-1}}$. If the universe evolved differently at early times then $\bar{t}$ need not begin at the value zero; consequently the true particle horizon may not be as shown (see Chapter 26). [The idea for this diagram is from Robert Lambourne (Lambourne, 2010)]

At $t =$ the present, this is the distance to the horizon 'now', but of course the galaxies at this distance from us at the present cosmic moment have moved to locations far outside our past light cone so we have no direct knowledge of what they may be doing now. The current value of d_h thus has little observational significance, except as an aid to our understanding of a spacelike surface that we find it convenient to think about. However, the value of d_h at earlier times gives useful information about what the universe could have been like at earlier times. In particular, at any given time t no physical process (with the possible exception of unknown types of process in the pre-GR era) could have made causal connections across distances larger than $d_h(t)$.

If the horizon is finite then observers at different locations in the universe are not all in the same future light cone of any given cosmic event after the pre-GR era. It is important to appreciate that this is possible. When I first encountered this idea it was very surprising to me, and I struggled to understand it. Naively one might imagine that as the universe expanded from some tiny volume, light travelled outwards faster than anything else, so we and our distant alien friends (the observers at fixed comoving coordinates) are all lagging behind the same light front and therefore we must all be enclosed within a single vast light sphere now. This is wrong.

To build a better intuition about light travel in an expanding universe, the balloon picture of Fig. 22.8 can help us once again. All we need to do is refuse to allow the balloon to start as an infinitesimal point. If we start from a balloon of small but non-zero size then it is clear that the light setting off from any one place may or may not be able to travel all the way to the other side of the balloon in any given time—it will depend on how rapidly and when the balloon expands. The same applies to spacetime. The initial spacetime of non-zero size is the outcome of the pre-GR or Planck era.

Another diagram that may be helpful here is shown in Fig. 23.8. This shows the same information as Fig. 23.7 but plotted in terms of coordinates $\chi, \bar{t}$. The coordinate χ (multiplied by R_0) is taken as the horizontal axis, so the expansion of the universe is hidden (all the comoving worldlines are vertical). It can be inferred from the other information also shown, such as the Hubble distance. For the vertical axis the time coordinate is taken as

$$\bar{t} = \int_0^t \frac{\mathrm{d}t}{a(t)}. \tag{23.59}$$

This is called *conformal time*. This has the effect of making the past light cone, and also the growth of the horizon, appear as straight lines on the diagram. We have $\mathrm{d}\bar{t} = \mathrm{d}t/a$ so in terms of conformal time the metric takes the form

$$\mathrm{d}s^2 = a^2(\bar{t}) \left(-c^2 \mathrm{d}\bar{t}^2 + R_0^2 \mathrm{d}\chi^2 + R_0^2 S^2(\chi) \, \mathrm{d}\Omega^2 \right). \tag{23.60}$$

In this equation the factor $a^2(\bar{t})$ produces a conformal transformation, hence the name for $\bar{t}$. Conformal time is convenient for tracking the motion of light in the cosmos, and also for the treatment of some early universe processes (Chapter 26).

The region within the particle horizon is what is commonly meant by the phrase 'the observable universe' but strictly speaking we cannot observe the 'observable universe' in its present state any more than we can see the Sun in its present state. Who knows what the Sun has been doing in the last eight minutes? And who knows what a galaxy seen by us at redshift 3 is now doing? Maybe it was gobbled up by an alien civilization 2 billion years ago. We observe strictly and only a past light cone. Compared to any timescale of cosmic interest, the time during which humanity has been able to collect data is miniscule, so all our reasoning is based on a slim 'snapshot' showing close regions as they are now and distant regions as they were long ago. We can, in principle, see back to any given moment in cosmic time—if the line of sight is not obscured—and presumably what we see is a fair indication of the whole. But—to repeat—it is not the whole. For all we know, outside our past light cone the universe might be filled with exotic things such as wormholes and dragons, or no universe at all! However, the sliver of spacetime that we can see is so rich that it furnishes enough evidence to allow confident extrapolation to other times and places. It is an entirely plausible conjecture that the rest of the universe at any given time is in broad terms similar to the part we can see.

Any galaxy within the horizon can in principle be seen now (it is seen in its condition at some earlier time) and any galaxy outside the horizon cannot (yet) be seen. In comoving coordinate units the horizon propagates outwards at the speed $c/R_0 a$. In units of proper distance and time it propagates outwards at the speed of light, overtaking more and more galaxies as it goes.

The galaxies do not 'pop into view' suddenly, however, because they first appear with huge redshift and in any case at the furthest reach of observation the material at that location would not be in the form of a galaxy but would be a loose gathering of dust or, earlier still, a blob of plasma in a sea of plasma. As cosmic time goes on the image of any given item visible to us gradually becomes less red-shifted and we see later versions of it. In practice the furthest we can 'see' (i.e. detect with our instruments) with currently available technology is to the surface of last scattering from which the CMB comes. Further back than that the universe was opaque to electromagnetic radiation. The situation is roughly comparable to looking at the Sun, where we see its surface but not (directly) the interior. However, by studying waves on the surface of the Sun we can learn about the interior, and this is comparable to learning about the early universe from the CMB. In future it might be possible to detect neutrinos or gravitational waves (or dark matter particles?) from earlier epochs. Then we could peer even to the early moments of the Big Bang, whose (hugely red-shifted) neutrinos and gravitons may even now be arriving at Earth after setting out long ago from some other coordinate location than ours.

The integral in (23.57) depends on $a(t)$ and therefore on the form the expansion takes. In the Einstein-de Sitter (i.e. matter-dominated, spatially flat) model one finds

$$\chi = \frac{2c}{R_0 H_0} \left(\sqrt{a(t)} - \sqrt{a(t_1)} \right) \tag{23.61}$$

$$= \frac{2c}{R_0 H_0} \left(1 - \frac{1}{\sqrt{1+z}} \right) \tag{23.62}$$

where the second version is for $a(t) = 1$, i.e. the present, and z is the redshift of the emitter. This formula answers the general question: *from what comoving coordinate was the light of given redshift emitted?* The horizon is given by the limit $z \to \infty$ and then the formula reads

$$\chi_h = \frac{2c}{R_0 H_0} = \frac{3ct_0}{R_0} \tag{23.63}$$

where we used (23.43). As a prediction for our universe, this calculation is illustrative, not precise. A more precise answer would be obtained by integrating (23.37). An estimate based on more careful modelling yields $R_0 \chi_h \simeq 14.4$ Gpc, or 47 billion light years.

It is convenient to convert the integral over time to an integral over redshift:

$$d_h = ac \int_0^t \frac{dt}{a} = ac \int_\infty^z \frac{1}{a} \frac{dt}{dz} \frac{da}{dz} dz = ac \int_\infty^z \frac{-a}{\dot{a}} dz = ac \int_z^\infty \frac{dz}{H(z)} \tag{23.64}$$

One may then employ (23.36) to provide the formula for $H(z)$ in terms of cosmological parameters:

$$d_h = \frac{ac}{H_0} \int_z^\infty \frac{dz}{[\Omega_{m0}(1+z)^3 + \Omega_{r0}(1+z)^4 + \Omega_{\Lambda 0} + (1 - \Omega_0)(1+z)^2]^{1/2}}. \tag{23.65}$$

This formula is not perfectly precise because it assumes the parameters to remain constant, whereas in fact the 'matter' becomes 'radiation' (i.e. relativistic) when it is hot enough, which modifies the formula at large z, and this can be significant when calculating the horizon even if it is not significant to other issues. Also, in Chapter 26 we will discuss the possibility that $\Omega_{\Lambda 0}$ may have gone through large changes over time.

Another horizon called the *cosmic event horizon* concerns the future evolution of the universe. It is defined as the largest comoving distance from which light emitted now can ever reach a given comoving observer in the future. This horizon moves inwards at the speed of light relative to any local matter in its vicinity, but owing to the expansion of the universe, it may either grow or shrink relative to us. The existence of such a horizon leads to the notion that far away galaxies are 'moving away from us at faster than the speed of light', and so they may be, as measured by ruler distance measured across the universe, per unit cosmic time. It is interesting to note that relativity places no fundamental limit on the timescale of the expansion of space itself: the proper comoving distance between distant coordinate locations can change at any rate. But light moving in either the inwards or the outwards directions sweeps over those distant galaxies at exactly c as determined by local instruments. One may express this as 'nothing can beat light in a race', which is perhaps a less ambiguous assertion than 'nothing can go faster than light'. One may also simply decide not to use the word 'speed' for a change in proper distance in an expanding space, but there is no consensus among physicists on such usage.

Using a calculation based on measured cosmological parameters, the distance to our cosmic event horizon is reckoned to be about 5 Gpc (16 billion light years). This is well within our particle horizon (though the value depends on future cosmological modelling so is debatable). This means that if there is a galaxy 6 Gpc away whose infancy we can see, and which is currently undergoing some interesting process such as consumption by an alien civilization, then our future descendants will never know it, unless they move to another place in the universe.

Needless to say, these cosmic horizons have nothing to do with the event horizons associated with black holes.

23.5 Last scattering

In this chapter we have mostly considered the overall growth of the cosmos, indicated by parameters such as scale factor, redshift, and Hubble parameter. We now have the tools to perform an important calculation, which is to find the temperature at last scattering. This is the process, also called photon decoupling, which was mentioned in Section 22.1. It is the completion of the transition from electron-proton plasma to neutral hydrogen gas which concludes the early part of the universe's history. The temperature at which this occurred tells us the temperature of the background radiation (now the CMB) when it was first emitted. By comparing this with the measured temperature in the present we have one of the major ways in which the whole scheme is calibrated.

We model the scattering of light off electrons at the time of last scattering using ordinary physics. This is the Thomson scattering whose cross-section can be calculated from classical electromagnetism; it is

$$\sigma = \frac{8\pi}{3} \left(\frac{e^2}{4\pi\epsilon_0 m_e c^2} \right)^2 \simeq 6.65 \times 10^{-29} \, \text{m}^2 \tag{23.66}$$

where we have given the value for electrons. The scattering rate at any given place is therefore

$$\Gamma = n_e \sigma c \tag{23.67}$$

where n_e is the number density of electrons in the plasma. Light also scatters off protons but since their mass is much greater their Thomson scattering cross-section is negligible at the precision we are attempting. We shall define 'last scattering' by the condition $\Gamma = H$. Obviously the scattering switches off gradually not abruptly, but it turns out that it is fairly abrupt and this is a good definition. More generally one may note that any process whose rate is below the Hubble parameter is slow enough to be significantly affected by cosmological expansion.

Our goal is to find the temperature. We can write $n_e = x n_B = x \eta n_\gamma$ where x is the ionization fraction defined in (22.4) and η is the baryon-photon ratio defined in (23.26). We obtain n_γ from (22.6), so the condition $\Gamma = H$ can be written

$$x \eta \, 0.243 \left(\frac{k_B T}{\hbar c} \right)^3 \sigma c = H_0 \sqrt{\Omega_{m0} a^{-3}} \tag{23.68}$$

where on the right-hand side we used (23.35) for a matter-dominated situation, which is approximately the case at last scattering. The ionization fraction is given by the Saha equation (22.3), (22.5), and the cosmic redshift formula gives $a = T_{CMB}/T$ where $T_{CMB} = 2.725 \, \text{K}$ is the CMB temperature now. Therefore we have all the information we need. The Saha equation and (23.68) are now a pair of simultaneous equations for x and T. They have to be solved by trial and error or a numerical method; the answer thus obtained is

$$x = 0.007, \quad T = 3060 \, \text{K} \tag{23.69}$$

where we used $\eta = 6 \times 10^{-10}$ and $\Omega_{m0} = 0.3$. Notice that the ionization fraction is quite low at last scattering. This means there are not many electrons around in comparison to hydrogen atoms. A more thorough analysis would model the hydrogen atom more fully and also allow for the influence of Ω_{r0} on H; the result is then 2970 K. This is how the redshift value reported in Table 22.1 is obtained.

Exercises

23.1 Starting from eqns (23.3), (23.4) for the Ricci tensor, obtain the Friedmann equations.

23.2 (i) Confirm that the FLRW metric has determinant $g = -c^2 a^6 r^4 \sin^2\theta (1 - K_0 r^2)^{-1}$. Hence obtain (23.14) from (23.13).

(ii) Obtain (23.14) from the Friedmann equations.

23.3 Obtain (23.18) from (23.15) and (23.17).

23.4 In the three-fluid model, is the Hubble parameter a monotonic function of time?

23.5 Show that in a spatially flat case ($k = 0$), if there is only a single component in the cosmological fluid, with $w \neq -1$, then the solution to the Friedmann equation is $a \propto t^n$ where $n = 2/(3(w+1))$.

23.6 Solve (23.37) in the case $\Omega_{m0} = \Omega_{\Lambda 0} = 0$ and confirm that (23.41) is a solution in the case $\Omega_{r0} = \Omega_{\Lambda 0} = 0$.

23.7 Equations (17.32) and (23.41) both describe a cycloid. Is this a coincidence?

23.8 Use (23.53) to estimate the age of the universe, taking parameter values from Figs 22.2 and 23.5.

23.9 (i) Using the data from Fig. 23.2, estimate the redshift and hence the temperature at matter-radiation equality. How dense was the universe then?
(ii) The temperature and density at the centre of the Sun are 1.5×10^7 K and $150 \, \mathrm{g/cm^3}$. What is the redshift for the events when the universe had this temperature? What was the density for radiation, baryonic matter, and dark matter respectively? (you may assume that the ratio of dark to baryonic matter density is approximately 5.3).
(iii) At what redshift was the density of matter in the universe similar to that of water now?

23.10 Consider a universe which somehow begins at some universal moment $t = 0$, and thereafter is stationary. Suppose that at time t observer O sees A and B, both at distance L but in opposite directions. How large must L be in order for A and B to be unaware of each other's existence at the time when they are seen by O? [*Ans. ct/3*]

23.11 In a dynamic universe, two galaxies A and B are at comoving coordinates χ_A, χ_B with $\chi_B > \chi_A$. Light arrives in the present at Earth (comoving coordinate zero) from both of them. Is it possible that the proper distance from Earth to B when its light set out was smaller than the proper distance from Earth to A when its light set out? [Hint: draw a diagram]

23.12 Suppose light leaves fundamental observer E at time t_e and arrives at observer O at time t_o. Define 'past light cone distance' D_{plc} as the proper distance between O and E at t_e.
(i) Show that in an Einstein–de Sitter universe, $D_{\mathrm{plc}} = 3c(t_o^{1/3} t_e^{2/3} - t_e)$.
(ii) Show that this distance has a maximum value at redshift $z = 1.25$.

23.13 Consider a static manifold in $1+2$ dimensions whose spatial part is a 2-sphere. Small circles, all having the same proper diameter, are scattered in space. Show that, when observed from any one point, the visually observed angular diameter of the circles first decreases and then increases with distance. This can also happen in a non-static flat space. Using (22.41), find the minimum observed angular diameter of an object of proper diameter L in a matter-dominated spatially flat cosmology.

23.14 Show that the particle horizon in the Einstein–de Sitter model grows at three times the speed of light. Show that in a radiation-dominated universe it grows at twice the speed of light and is equal to the Hubble distance c/H.

23.15 Show that the Milne universe has no particle horizon, and state what feature of Fig. 22.10 indicates this result.

23.16 Define the dimensionless conformal time

$$\tilde{t} = \frac{c}{R_0} \int \frac{\mathrm{d}t}{a(t)}$$

Show that for a single-fluid model with equation of state $p = w\rho c^2$ the Friedmann equations can be written as

$$h^2 + k = \frac{8\pi G R_0^2}{3c^2} \rho a^2, \qquad (23.70)$$

$$2h' + (1+3w)(h^2 + k) = 0 \qquad (23.71)$$

where $h = a'/a$ and $a' = \mathrm{d}a/\mathrm{d}\tilde{t}$. Hence show that

$$\frac{d^2 y}{d\bar{t}^2} = -\tfrac{1}{4}(1+3w)^2 ky \qquad (23.72)$$

where $y = a^{(1+3w)/2}$. Find $a(\tilde{t})$ and $t(\tilde{t})$ for the case $\rho = \rho_0/a^4$ (the radiation-dominated universe).

23.17 Show that for a flat universe containing matter and radiation the Friedmann equation can be written

$$\left(\frac{da}{d\bar{t}}\right)^2 \propto (a + a_{\text{eq}}) \qquad (23.73)$$

where $\bar{t}$ is the conformal time $\bar{t} = \int_0^t dt/a$ and a_{eq} is the value of the scale factor at matter-radiation equality. Show that the solution is $a(\bar{t}) = A\bar{t}^2 + B\bar{t}$ and find the constants A, B in terms of $H_0, \Omega_{m0}, a_{\text{eq}}$.

23.18 Olbers' paradox may be summarized as the question 'why is the sky dark at night?' Kepler and others argued that if the universe is large enough and on average uniformly filled with stars, then in any line of sight from Earth there must (at some distance) be the surface of a star. More precisely, the number of stars in a given solid angle must increase in proportion to the square of their distance from Earth. This is just sufficient to counterbalance the inverse square law for the intensity of radiation at given distance from an isotropic source, so the net effect should be that the sky appears uniformly bright everywhere (with possibly some small variations reflecting the different types of star). If one asserts that the universe is static for long enough, then thermodynamic reasoning suggests that even these variations should have been smoothed away. Olbers reasserted this argument more precisely in 1826.

(i) Supposing the universe is static and spatially flat, is Kepler's and Olbers' reasoning correct? Is the reasoning still correct if each star has a finite lifetime?

(ii) Does the paradox hold in a spatially curved but still static universe?

(iii) Does it follow that the darkness of the night sky represents evidence for a non-static universe?

24

The growth of structure

In our study of cosmology so far we have studied the FLRW metric and the Friedmann equations and their solution. This gives the big picture for the curvature and average density of the universe over space and time. Within this background lies all the structure which makes the universe what it is—a fascinating universe, full of interest and complexity. A major component of cosmology is the study of how this structure grows from its primordial beginnings. In this chapter we will introduce the main concepts and the simpler methods applicable to such study.

Gravity characteristically tends to amplify, over time, any small inhomogeneity in the density; pressure tends to smooth things out. In a cloud with small inhomogeneities, we will show that this leads to a characteristic length called the *Jeans length*, given by

$$\lambda_{\mathrm{J}} = c_s \sqrt{\pi/G\rho} \tag{24.1}$$

where ρ is the density of a self-gravitating cloud of matter and c_s is the speed of sound in the cloud, which is related to the pressure and density:

$$c_s = \sqrt{(\partial p/\partial \rho)}. \tag{24.2}$$

Density inhomogeneities with wavelength less than λ_{J} oscillate as sound waves; density inhomogeneities with wavelength greater than λ_{J} grow. In short, large regions collapse, small regions do not. As the universe grows, ρ falls so λ_{J} gets larger, but meanwhile the inhomogeneities grow. Eventually in some regions they are of order ρ: a proto-galaxy has formed. Thereafter the density in such a proto-galactic cloud increases as the cloud collapses, so λ_{J} gets smaller. Then smaller and smaller regions become sufficiently dense to collapse in their turn. Eventually small and dense enough regions coalesce into stars and planets as each local gathering of matter is pulled together by its own gravity (Fig. 24.1).

We have noted in Chapter 22 that the CMB radiation is very smooth, with fractional temperature fluctuations of order 10^{-5}. On the other hand the universe today is very lumpy, with matter concentrated into galaxies; the average matter density within a galaxy is of order 10^6 times the overall average. We would like to understand both how the situation at decoupling came about, and also how matter clumped together since then. It will emerge that one of the important 'dates' is matter-radiation equality, which occurred at $z = 3391$ when the universe was 47,000 years old. Before this the universe was radiation-dominated and fluctuations grew

Relativity Made Relatively Easy: General Relativity and Cosmology. Volume 2. Andrew M. Steane, Oxford University Press. © Andrew M. Steane 2021. DOI: 10.1093/oso/9780192895646.003.0024

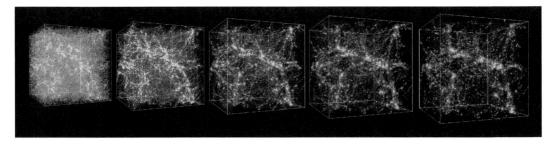

Fig. 24.1 Data from a numerical simulation of structure formation by gravitational collapse. The frames show the evolution of structures in a 43 Mpc box from redshift 30 to the present, in a cold dark matter with dark energy (ΛCDM) model. [Andrey Kravtsov, Kavli Institute for Cosmological Physics, University of Chicago]

Table 24.1 The growth of perturbations to the dark matter density distribution of small and large wavelength λ in the two main regimes of behaviour.

	$\lambda < \lambda_J$	$\lambda > \lambda_J$	λ_J/d_H
radiation-dominated	osc.	a^2	$\sqrt{8\pi/3} \simeq 2.96$
matter-dominated	osc.	a	$\sqrt{8/3}\,\pi c_s/c \ll 1$

one way; afterwards the universe was matter-dominated and fluctuations grew differently. Indeed the main result of this chapter is easy to state and is shown in Table 24.1. In this table a is the cosmological scale factor and $d_H = c/H$ is the Hubble distance. The density contrast (eqn (24.3)) for dark matter grows in proportion to a^2 or a, or not at all, as shown. Baryonic matter behaves similarly, except for the period between matter–radiation equality and recombination (to be discussed). The main information is that the long-wavelength fluctuations grow in proportion to a^2 in a radiation-dominated universe, and in proportion to a in a matter-dominated universe—but the definition of what is a 'long' and a 'short' wavelength also depends on the regime. Under matter-domination most wavelengths of cosmic significance are longer than the Jeans length so most perturbations grow.

Table 24.1 is the destination rather than the start of the argument. To start the argument we introduce some terminology and a remark about initial conditions.

If $\rho(t, x, y, z)$ is a density of some part of the cosmic fluid then we define the *density contrast* δ by

$$\delta \equiv \frac{\rho(t, x, y, z) - \bar{\rho}(t)}{\bar{\rho}(t)} \tag{24.3}$$

where $\bar{\rho}$ refers to the spatial average at the given time. There are perturbations in four main parameters: the density of matter (mostly dark matter), the density of radiation, the velocity, and the spacetime metric. These various types of perturbation may in principle be correlated with one another (which one would expect if one is the cause of the others). Two extreme cases of such correlation are given names. Matter and radiation perturbations correlated such that

$$\delta_{\mathrm{m}} = (3/4)\delta_{\mathrm{r}} \qquad\qquad (24.4)$$

are called *adiabatic*; perturbations anti-correlated such that $\delta\rho_{\mathrm{m}} = -\delta\rho_{\mathrm{r}}$ are called *isocurvature* (because their net effect on spacetime curvature cancels out.) More fully, the adiabatic case is where perturbations at a given time and place $(\tau, \mathbf{r})$ are such that all the quantities (densities, pressure etc.) are the same as those of the average background at some slightly different time $\tau + \delta\tau(\mathbf{r})$. The whole perturbation can then be viewed as different parts of the universe being slightly 'ahead' or 'behind' others, and entropy is conserved. Observations suggest that the primordial perturbations were adiabatic.

Causality.   It is not necessary for two regions to be in causal contact in order for density fluctuations of large length scale to grow. For example, if we consider two entirely unconnected and unrelated clouds of matter at opposite sides of the observable universe, one of which is more dense than the other, then the denser one will collapse more rapidly than the less dense one, and thus a perturbation with characteristic length equal to the diameter of the observable universe will grow. The particle horizon is not relevant to such processes.

However, the particle horizon is relevant to some of the ways in which instabilities can arise. For example, if some quadrupolar perturbation were suddenly to occur at the centre of a cloud, then it will take time for gravity elsewhere in the cloud to react—the cloud will not respond until the local metric at each place has been adjusted via the propagation of gravity waves and pressure waves.

In the present chapter we will not, and we will not need to, consider processes of the kind just described. It will be sufficient to take as initial conditions the assumption that the universe was, in its early state, almost but not quite homogeneous, where by 'early' we mean at any time after the first few minutes and within the first few hours or years or centuries after that. The timescales we are interested in here are thousands to billions of years. This nearly homogeneous universe had small perturbations in its metric and in the densities and velocities of matter and radiation. By calculating the growth of these perturbations and comparing with observations one can infer their initial spectrum. Concerning how those initial conditions arose, it is not known whether some other less homogeneous state was driven towards homogeneity in the Planck era, nor whether it was the other way round, and things started out even more homogeneous. But we note that random perturbations in the metric of spacetime itself would be sufficient to act as causes of all the other fluctuations. We shall return to this issue in Chapter 26.

24.1   The structure equations

We treat the growth of structure by modelling the universe as a fluid which is mostly smooth but has small perturbations in density or velocity or both. One sees immediately that this will only be adequate to treat the growth of these perturbations while they are still small. After that other types of calculation must take over, and ultimately cosmologists also make much use of numerical studies.

There are three basic concepts which lead to equations describing the growth of structure:

1. conservation of the amount of matter, leading to a continuity equation
2. conservation of momentum, leading to an equation for fluid flow
3. gravitation

We will account for the third item almost entirely using the weak-field limit for gravity, which amounts to Newtonian gravity with an adjustment for the impact of pressure (eqn (5.50)). This amounts to doing perturbation theory on a background metric which is taken to be Minkowskian. Such an approach is valid for a universe which is spatially flat to good approximation, which is the case for our universe, except that one ought to check that the dynamics are also correctly handled. We will not provide a full justification for this, but we can immediately justify it for small scales, and motivate it for all scales: the equations we will describe give the same prediction for the average dynamics, called the *Hubble flow*, as do the Friedmann equations and the FLRW metric. Therefore they correctly describe what General Relativity says happens to the background, and they give a correct perturbation theory on that background for small regions. For large regions one requires a fuller analysis; it is found that this justifies our approach.[1]

We treat a cloud of matter or radiation with density $\rho(\mathbf{r}, t)$, velocity $\mathbf{v}(\mathbf{r}, t)$, and gravitational potential $\Phi(\mathbf{r}, t)$. These functions satisfy:

$$\left.\begin{array}{rl} \dfrac{\mathrm{d}\rho}{\mathrm{d}\tau} + \rho(1+w)\boldsymbol{\nabla}\cdot\mathbf{v} = 0 & \textcircled{1} \\[2mm] \rho(1+w)\left(\dfrac{\mathrm{d}\mathbf{v}}{\mathrm{d}\tau} + \boldsymbol{\nabla}\Phi\right) = -\boldsymbol{\nabla}p & \textcircled{2} \\[2mm] \nabla^2\Phi = 4\pi G(\rho_{\mathrm{tot}} + 3p_{\mathrm{tot}}/c^2) & \textcircled{3} \end{array}\right\} \qquad (24.5)$$

These are the continuity equation (14.22), Euler equation (14.24), and the Poisson equation for gravity (5.50), all written in the case of low velocity $v \ll c$ and the linearized gravity limit. We are using a notation in which the bold font indicates 3-vectors, and $\boldsymbol{\nabla}$ is the three-dimensional gradient operator. The total derivative in the first two equations refers to the derivative along the flow,

$$\frac{\mathrm{d}}{\mathrm{d}\tau} = \frac{\partial}{\partial t} + \mathbf{v}\cdot\boldsymbol{\nabla} \qquad (24.6)$$

We have adopted the equation of state $p = w\rho c^2$ and note that in the continuity equation, only the density of the fluid component under discussion enters, whereas all components contribute to gravitation in the Poisson equation. In the three-fluid model one has

$$\nabla^2\Phi = 4\pi G(\rho_{\mathrm{m}} + 2\rho_{\mathrm{r}}) - \Lambda c^2 \qquad (24.7)$$

and for flat space the Friedmann equation (23.7) reads

[1]The full analysis is rather subtle; the equations we adopt turn out to be valid in a certain choice of coordinates, and this is the choice we will implicitly be making.

$$3H^2 = 8\pi G \left(\bar{\rho}_{\mathrm{m}} + \bar{\rho}_{\mathrm{r}} \right) + \Lambda c^2. \tag{24.8}$$

In this chapter we are concerned mostly with the evolution of just one type of material: the matter. But it will be helpful to note what happens to radiation too.

Towards solving the equations, we first express the density, velocity and potential as the sum of a zeroth-order term, plus the rest. We adopt the notation

$$\rho = (1 + \delta)\bar{\rho} \tag{24.9}$$
$$\mathbf{v} = \mathbf{v}_0 + \mathbf{u} \tag{24.10}$$
$$\Phi = \Phi_0 + \phi \tag{24.11}$$

in which $\bar{\rho}, \mathbf{v}_0$ and Φ_0 describe the Hubble flow, and δ, $\mathbf{u}$ and ϕ are perturbations to the Hubble flow. $\delta = (\bar{\rho} - \rho)/\rho$ is the density contrast for the fluid component under discussion, introduced in (24.3). $\mathbf{u}$ is the peculiar velocity at any given place. All these parameters depend on position and time, with the exception of $\bar{\rho}$ which depends on time only.

First we note that $\{\bar{\rho}, \mathbf{v}_0, \Phi_0\}$ is itself a solution of the three equations (24.5). Then, upon substituting (24.9)–(24.11) into (24.5) the zero-order terms cancel, and we can extract equations for the perturbations. It is not necessary to assume that the perturbations are small in order to do this, and we shall not make that assumption in the first instance. After using that $\boldsymbol{\nabla}\bar{\rho} = 0$ and allowing the zero order terms to cancel each other out, the first equation of (24.5) yields (exercise)

$$\textcircled{1} \qquad \left(\frac{\partial}{\partial t} + \mathbf{v}_0 \cdot \boldsymbol{\nabla} \right) \delta + (1 + \delta)(1 + w) \boldsymbol{\nabla} \cdot \mathbf{u} + \mathbf{u} \cdot \boldsymbol{\nabla}\delta = 0. \tag{24.12}$$

Next, in the Euler equation we can write

$$\boldsymbol{\nabla} p = c_s^2 \boldsymbol{\nabla} \rho = c_s^2 \bar{\rho} \boldsymbol{\nabla} \delta \tag{24.13}$$

where c_s is the speed of sound (c.f. (24.2)). After cancelling the zero order terms we have

$$\left(\frac{\partial}{\partial t} + \mathbf{v}_0 \cdot \boldsymbol{\nabla} \right) \mathbf{u} + \mathbf{u} \cdot \boldsymbol{\nabla}(\mathbf{v}_0 + \mathbf{u}) + \boldsymbol{\nabla}\phi = \frac{-c_s^2 \boldsymbol{\nabla}\delta}{(1 + \delta)(1 + w)}. \tag{24.14}$$

The Hubble flow gives

$$\mathbf{v}_0 = \frac{\dot{a}}{a}\mathbf{r} = H\mathbf{r} \tag{24.15}$$

where H is the Hubble parameter; c.f. (23.9). It follows that $(\mathbf{u} \cdot \boldsymbol{\nabla})\mathbf{v}_0 = H(\mathbf{u} \cdot \boldsymbol{\nabla})\mathbf{r} = H\mathbf{u}$. Substituting this into (24.14) we obtain

$$\textcircled{2} \qquad \left(\frac{\partial}{\partial t} + \mathbf{v}_0 \cdot \boldsymbol{\nabla} \right) \mathbf{u} + H\mathbf{u} + (\mathbf{u} \cdot \boldsymbol{\nabla})\mathbf{u} = -\mathbf{f} - \boldsymbol{\nabla}\phi. \tag{24.16}$$

where $\mathbf{f} = c_s^2 \boldsymbol{\nabla}\delta / ((1 + \delta)(1 + w))$ (the force term which came from the gradient of the pressure).

Finally, Poisson's equation (24.7) yields

$$\nabla^2 \phi = 4\pi G(\bar{\rho}_{\mathrm{m}} \delta_{\mathrm{m}} + 2\bar{\rho}_{\mathrm{r}} \delta_{\mathrm{r}}). \tag{24.17}$$

Note that the pressure associated with matter makes a negligible contribution to gravitation here, but its gradient makes a non-negligible contribution to the force on any region of the fluid in the Euler equation.

Equations (24.12), (24.16), (24.17) are all exact within the approximation of adopting (24.5) and setting $w = 0$ for matter in the Poisson equation.

We now wish to adopt a useful change of coordinates. Let

$$\mathbf{x} \equiv \mathbf{r}/a. \tag{24.18}$$

The coordinates $\mathbf{x}$ are called comoving coordinates, because they track the motion of a fundamental observer, or, equally, a fluid element in the Hubble flow. One then finds

$$\left(\frac{\partial}{\partial t}\right)_{\mathbf{x}} = \left(\left(\frac{\partial}{\partial t}\right)_{\mathbf{r}} + \mathbf{v}_0 \cdot \nabla_{\mathbf{r}}\right), \qquad \nabla_{\mathbf{x}} = a\nabla_{\mathbf{r}}$$

where $\nabla_{\mathbf{r}}$ is the operator we have up till now written simply ∇. Using these results, it is straightforward to write (24.12), (24.16), (24.17) in terms of the coordinates $(\mathbf{x}, t)$. We shall do this and immediately drop the subscript $\mathbf{x}$, thus obtaining

---

Structure equations in comoving coordinates

$$\left.\begin{array}{ll} \dfrac{\partial \delta}{\partial t} + \dfrac{1}{a}(1+\delta)(1+w)\nabla \cdot \mathbf{u} + \dfrac{1}{a}\mathbf{u} \cdot \nabla\delta = 0 & \text{\textcircled{1}} \\[2ex] \dfrac{\partial \mathbf{u}}{\partial t} + \dfrac{\dot{a}}{a}\mathbf{u} + \dfrac{1}{a}(\mathbf{u} \cdot \nabla)\mathbf{u} = -\dfrac{\mathbf{f}}{a} - \dfrac{1}{a}\nabla\phi & \text{\textcircled{2}} \\[2ex] \nabla^2\phi - 4\pi Ga^2(\bar{\rho}_{\mathrm{m}}\delta_{\mathrm{m}} + 2\bar{\rho}_{\mathrm{r}}\delta_{\mathrm{r}}) = 0 & \text{\textcircled{3}} \end{array}\right\} \tag{24.19}$$

Keep in mind here that the new operators $\partial/\partial t$ and ∇ are not the same as the old operators $\partial/\partial t$ and ∇.

24.2 Linearized treatment

The first two structure equations are non-linear; their solution in general requires numerical integration. However, it is both useful and insightful to consider the case where the perturbations are small enough that the non-linear terms can be neglected in comparison to the linear terms. This procedure is an example of *first order perturbation theory*; it is also commonly referred to as obtaining *linearized* equations. One finds:

Linearized structure equations in comoving coordinates

$$\left.\begin{aligned}\frac{\partial\delta}{\partial t}+\frac{1+w}{a}\boldsymbol{\nabla}\cdot\mathbf{u}&=0 \qquad\qquad\qquad\quad\text{①}\\ \frac{\partial\mathbf{u}}{\partial t}+\frac{\dot{a}}{a}\mathbf{u}&=-\frac{c_s^2\boldsymbol{\nabla}\delta}{a(1+w)}-\frac{1}{a}\boldsymbol{\nabla}\phi \quad\text{②}\end{aligned}\right\} \qquad (24.20)$$

(and the third equation in (24.19) is already linear). By taking the divergence of the second equation and then using the first equation to eliminate $\boldsymbol{\nabla}\cdot\mathbf{u}$ in favour of $\dot{\delta}$, one obtains (exercise)

Governing equation for density perturbations in linear regime

$$\frac{\partial^2\delta}{\partial t^2}+2H\frac{\partial\delta}{\partial t}-\frac{c_s^2\nabla^2\delta}{a^2}-(1+w)4\pi G\left(\bar{\rho}_\mathrm{m}\delta_\mathrm{m}+2\bar{\rho}_\mathrm{r}\delta_\mathrm{r}\right)=0. \qquad (24.21)$$

---

where δ is the density contrast in whichever component is under consideration.

Our aim is now to explore the solutions to (24.21) in various interesting cases. The main situations of interest are the static background (valid for periods of time small compared to $1/H$) and those where the background evolution (the Hubble flow) is dominated by one of radiation, matter, vacuum.

24.2.1 Static background: the Jeans instability

The static case arises when $H\dot{\delta}$ can be neglected in comparison to the other terms in (24.21). For simplicity we shall assume also that we are dealing with a matter-only fluid. Then one has

$$\frac{\partial^2\delta}{\partial t^2}-\frac{c_s^2\nabla^2\delta}{a^2}-4\pi G\bar{\rho}\,\delta=0. \qquad (24.22)$$

In this situation a can be regarded as a constant, and therefore the equation has plane wave solutions:

$$\delta=Ae^{i(\mathbf{k}\cdot\mathbf{x}-\omega t)}. \qquad (24.23)$$

Upon substituting this into (24.22) one obtains the dispersion relation

$$\omega^2=\frac{c_s^2k^2}{a^2}-4\pi G\rho. \qquad (24.24)$$

The right-hand side is positive for large k, and negative for small k. This means that perturbations of large k (short wavelength) oscillate and those of small k (long wavelength) grow or decay exponentially. The Jeans length (24.1) is the critical wavelength (in $\mathbf{r}$ coordinates) that determines which behaviour obtains. The underlying physics here is that in the absence of gravity one has a wave equation, and indeed it is the equation for sound waves in an ordinary gas. But at large distance scales gravity overwhelms everything else. The exponential growth of fluctuations of large wavelength is called the *Jeans instability*. It occurs for $(k/a) < k_J$, where

$$k_J \equiv \frac{(4\pi G\rho)^{1/2}}{c_s}. \tag{24.25}$$

The time $\tau = (4\pi G\rho)^{-1/2}$ can be interpreted as the timescale of the gravitational collapse of a spherical ball (of any size) if the collapse were unopposed. When this time is less than l/c_s for some region of size l then the transmission of pressure cannot keep up with the collapse; it is all over before the pressure can build up.

24.2.2 Non-static single-component universe

Next let us consider a single-component fluid in a dynamic universe. We consider solutions to the governing equation (24.21) of the form

$$\delta(\mathbf{x}, t) = e^{i\mathbf{k}\cdot\mathbf{x}}\tilde{\delta}(\mathbf{k}, t) \tag{24.26}$$

where $\tilde{\delta}(\mathbf{k}, t)$ is some function of time to be discovered. Since a general function of $\mathbf{x}$ can be expressed in terms of such solutions by Fourier analysis, it suffices to find how each $\tilde{\delta}(\mathbf{k}, t)$ function evolves. The function of time for any given $\mathbf{k}$ is called the *growth factor* for that $\mathbf{k}$. To reduce clutter we shall drop the tilde; as long as one remembers this has been done then no confusion should arise. By substituting (24.26) into (24.21) one finds that the growth factor is a solution of

$$\ddot{\delta}(\mathbf{k}) + 2H\dot{\delta}(\mathbf{k}) + c_s^2\left(\frac{k^2}{a^2} - (1+w)(1+3w)k_J^2\right)\delta(\mathbf{k}) = 0 \tag{24.27}$$

where we suppressed the explicit mention of t in $\delta(\mathbf{k}, t)$ since this is implicit in the fact that we now have a differential equation in time alone. Keep in mind that k_J here is a function of time, but k is not. This equation does not have exponential solutions, but the solutions do separate into oscillatory ones when k is large (small distance scales), and growing or decaying ones when k is small (large distance scales). The separation between the two types of behaviour is indicated by the Jeans length, up to a factor of order 1.

Recall that when a single component dominates, the solution of the Friedmann equation is $a \propto t^n$ where $n = 2/3(w+1)$ (exercise 23.5 of Chapter 22), and therefore

$$H = \frac{\dot{a}}{a} = \frac{2}{3(1+w)}\frac{1}{t}. \tag{24.28}$$

We can also write the Jeans wave vector in terms of the Hubble parameter via the Friedmann equation, which, in a flat universe, leads to

$$k_J(t) = \sqrt{\frac{3}{2}}\frac{H(t)}{c_s(t)}. \tag{24.29}$$

Hence in the limit $k/a \ll k_J$ equation (24.27) becomes

$$3(1+w)\ddot{\delta} + 4\frac{\dot{\delta}}{t} - \frac{2(1+3w)}{t^2} = 0. \tag{24.30}$$

This is a second order differential equation with power-law solutions. By trying the form $\delta(\mathbf{k}, t) = At^n$. One finds the general solution

$$\delta(\mathbf{k}) = At^1 + Bt^{-1} \qquad \text{for } w = 1/3$$
$$\delta(\mathbf{k}) = At^{2/3} + Bt^{-1} \qquad \text{for } w = 0 \tag{24.31}$$

where A and B are constants of integration. In each case, the solution which grows with t is called a *growing mode* and the solution which decays with t is called a *decaying mode*. In practice the initial conditions will lead to a solution consisting of a sum of both kinds of mode for any given $\mathbf{k}$, but since the decaying mode decays it soon becomes negligible compared to the growing mode, so we focus our attention on the latter. The main message of (24.31) is that, for the perturbations whose wavelength is above the Jeans length,

> radiation perturbations in a radiation-dominated era grow as $\delta \sim t \quad \sim a^2$
> matter perturbations in a matter-dominated era grow as $\quad \delta \sim t^{2/3} \sim a$
> and in both cases metric perturbations ϕ are constant (in comoving coordinates)

The final statement is obtained by substituting the behaviours into the comoving Poisson equation (24.20) and using that $\bar{\rho}_m \propto a^{-3}$, $\bar{\rho}_r \propto a^{-4}$.

The other crucial fact is that the Jeans length is itself very different in these regimes. Examine (24.29), and note that in a relativistic gas $c_s = c/\sqrt{3}$ (one can obtain this by noting that if $p = wc^2\rho$ then $\mathrm{d}p/\mathrm{d}\rho = wc^2$ for an adiabatic sound wave). Define the *Hubble distance*

$$d_H = c/H, \tag{24.32}$$

then we have that in a radiation dominated era, the Jeans length and the Hubble distance are similar, but in a matter dominated era, the Jeans length is tiny compared to the Hubble distance. Therefore only modes of very long wavelength, above the Hubble distance, can grow in the radiation dominated era, whereas modes of essentially all sizes can grow in the matter dominated era, as long as the pressure is small. Low pressure is true for cold dark matter at all times under consideration and for baryonic matter after decoupling—more on this later.

Let us now highlight the second result above in order to comment on it:

Growth of perturbations in matter-dominated expanding universe

$$\delta \propto a, \tag{24.33}$$
$$\phi = \text{constant in comoving coordinates} \tag{24.34}$$

These simple formulae give us a valuable insight into the growth of structure in our universe, because the universe has been matter dominated for much of its history. It is a very gentle growth in comparison with the exponential growth that occurs in the case of a static background.

The cosmic microwave background radiation exhibits fractional perturbations of only a few parts in 10^5. It has a redshift of order 1000, and therefore the scale factor has grown by approximately 1000 since the time of last scattering when the CMB was produced. The anisotropies of the CMB are a measure of the density perturbations of ordinary, baryonic matter which interacts with photons. If these density perturbations had grown in proportion to a since the time of last scattering then they would now be of order a few percent and therefore no dense structures (stars, planets) would have formed. Thus the smoothness of the CMB is an indication that something else has contributed to the growth of structure. This is strong evidence for dark matter. The observations are consistent with a model in which dark matter density perturbations were substantially larger than those in baryonic matter at the time of last scattering, and they contributed the primary seeds for the growth of structure of all types.

As the density perturbations grow, eventually they lead to significant amounts of kinetic energy in the collapsing material. Baryonic matter can get rid of this energy through emission of electromagnetic radiation after atoms collide and excite one another, a process called *virialization* because the resulting orbits obey the virial theorem. Dark matter does not have this avenue and therefore clumps less strongly in the non-linear regime. This is why stars and planets are mostly made of baryonic matter and not dark matter. The latter now resides mainly in large diffuse filaments between galaxies and roughly spherical 'halos' permeating and surrounding galaxies.

24.2.3   Matter perturbations in the radiation-dominated era

We have now obtained most of the results that were previewed in Table 24.1, but we have not yet shown what happens to matter before matter domination. To treat this we return to the governing equation (24.21) and obtain

$$\ddot{\delta}_{\rm m}(\mathbf{k}) + 2H\dot{\delta}_{\rm m}(\mathbf{k}) + c_s^2 \frac{k^2}{a^2}\delta_{\rm m} - 4\pi G\left(\bar{\rho}_{\rm m}\delta_{\rm m}(\mathbf{k}) + 2\bar{\rho}_{\rm r}\delta_{\rm r}(\mathbf{k})\right) = 0. \qquad (24.35)$$

First consider distance scales above the Jeans length for matter but below the Jeans length for radiation, which is of the order of the Hubble distance. On these scales the radiation perturbations oscillate and do not grow, so $\delta_{\rm r} \simeq 0$. Also the definition of radiation domination is that $\bar{\rho}_{\rm r} \gg \bar{\rho}_{\rm m}$ so we must find that the $\bar{\rho}_{\rm m}\delta_{\rm m}$ term is small compared to the $H\dot{\delta}_{\rm m}$ term. Hence the behaviour of the matter perturbations is roughly modelled by

$$\ddot{\delta}_{\rm m}(\mathbf{k}) + \frac{1}{t}\dot{\delta}_{\rm m}(\mathbf{k}) \simeq 0. \qquad (24.36)$$

This has solutions $\delta_{\rm m} = $ constant and $\delta_{\rm m} \sim \log t \sim \log a$. This logarithmic growth is much slower than the linear growth in the matter-dominated regime and is roughly equivalent to no growth at all. This is the result reported in Table 24.1. (A more thorough analysis yields $\delta_{\rm m} \propto 1 + (3/2)\rho_{\rm m}/\rho_{\rm r}$, hence the matter perturbation grows by at most 5/2 during the entire radiation-dominated era (exercise 24.4).)

For the modes of large wavelength, above the Hubble distance, (24.35) gives

$$\left.\begin{array}{l} \ddot{\delta}_{\rm r} + \dot{\delta}_{\rm r}/t - \delta_{\rm r}/t^2 = 0, \\ \ddot{\delta}_{\rm m} + \dot{\delta}_{\rm m}/t - (3/4)\delta_{\rm r}/t^2 = 0. \end{array}\right\} \tag{24.37}$$

The solution to the first equation is $\delta_{\rm r} = At + B/t$ as we found before. The second can be solved by combining the two equations. After integrating one finds

$$t\frac{\rm d}{{\rm d}t}\left(\delta_{\rm m} - \frac{3}{4}\delta_{\rm r}\right) = \text{constant} \tag{24.38}$$

This has a logarithmic solution, which tells us that even if $\delta_{\rm m}$ starts out not equal to $\delta_{\rm r}$, it will not depart much from $\delta_{\rm r}$. Hence one concludes that for these modes $\delta_{\rm m} \simeq \delta_{\rm r}$, so the growing mode for long-wavelength matter perturbations grows as $\delta_{\rm m} \sim t \sim a^2$. This is the result reported in the first row of Table 24.1.

24.2.4 Matter and vacuum

The EDS (i.e. matter-dominated) model considered in Section 24.2.2 is a good approximation for the growth of structure during the period when Ω_m was close to 1. This includes for example the so-called 'dark ages' between last scattering and reionization when the first stars formed. The subsequent evolution is well approximated by a model allowing for matter and vacuum. By using (23.30) to write the density ρ in terms of the Hubble parameter and $\Omega_{\rm m}$, one finds

$$\ddot{\delta} + 2H\dot{\delta} - (3/2)H_0^2\Omega_{\rm m0}a^{-3}\,\delta = 0. \tag{24.39}$$

The growing solution satisfies

$$\frac{{\rm d}\ln\delta}{{\rm d}\ln a} \simeq \Omega_m^{0.6}. \tag{24.40}$$

Hence the growth is proportional to a while Ω_m is close to 1, and proportional to a slightly smaller power of a when Ω_m falls somewhat below 1. By inspecting Fig. (23.6) one can observe that Ω_m has been in the range 0.5 to 1 for most of the period during which structure has grown in the matter-dominated era.

Matter in a universe dominated by a cosmological constant. If the growth of the scale factor is dominated by a cosmological constant then $H = \text{constant}$ and

$$\ddot{\delta} + 2H\dot{\delta} \simeq 0. \tag{24.41}$$

The solutions are now constant or decaying exponentially. In other words the accelerated expansion driven by a cosmological constant will completely suppress the growth of structure by gravitational collapse.

24.3   The overall picture

We are now ready to give the broad outline of how structure grew in the universe through the radiation-dominated era for some 47,000 years, and then subsequently through some further billions of years. The structures are conveniently discussed in the comoving coordinates that we have adopted since the end of Section 24.1. In these coordinates the wavenumber k of any given Fourier component of the perturbations stays constant as time goes on and the universe grows, which means the corresponding physical wave is stretching in proportion to a. Meanwhile the Jeans length is growing in proportion to $1/\sqrt{\rho}$, and so is the Hubble distance.

The phrase 'the Jeans length for radiation fluctuations in a radiation-dominated universe' is rather a mouthful. It is useful to note that, during radiation domination,

$$\lambda_{\mathrm{J}} \simeq 2.96 \, d_H \qquad (24.42)$$

so we can use the Hubble distance as a stand-in for the Jeans length in this period. Any perturbation which starts out with a long wavelength and therefore a small k will begin by growing rapidly, as a^2, and then when the Hubble distance (and thus also λ_{J}) grows to match the perturbation's wavelength, the growth of its amplitude will stall during the radiation-dominated era. Subsequently, after matter-radiation equality λ_{J} falls and so the perturbation amplitude will grow again, now in proportion to a (and eventually somewhat less rapidly as the cosmological constant comes into play). This behaviour is illustrated in Fig. 24.2.

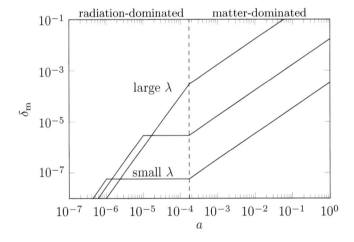

Fig. 24.2 The growth of matter density perturbations of a variety of wavelengths, illustrating the effect of 'stalling' during the radiation-dominated era. The perturbations of small wavelength fall below the Jeans length ('enter the horizon') earlier during radiation domination, and so grow less. The lines show an indicative case, in which perturbations of various wavelengths start out with sizes roughly in proportion to $\sqrt{k}$. The figure shows that the effect of the 'processing' offered by the subsequent dynamics is to reverse this ordering, except at the longest wavelengths.

The terminology widely used in this area is to say that as the Hubble distance grows to meet the distance scale of any given fluctuation, the fluctuation 'enters the horizon'. This terminology is unfortunate, because it confuses the notion of horizon, which is a property of an integral over time, with the notion of a rate of change such as H. Also it obscures the fact that the underlying physics is not to do with cosmic expansion *per se*, but with the Jeans instability.

The overall effect of the 'processing' offered by the dynamics is conveniently expressed by defining a **transfer function** $T(k)$:

$$\delta(\mathbf{k}, t_0) = T(k)\delta(\mathbf{k}, t_i) \tag{24.43}$$

where t_0 is the present and t_i is some initial time. Let us define

$$k_{\text{eq}} \equiv \frac{2\pi}{c}(aH)_{\text{eq}}. \tag{24.44}$$

Modes with larger distance scale than this, i.e. $k < k_{\text{eq}}$, will grow throughout the radiation-dominated era, so

$$T(k) = \left(\frac{a_{\text{eq}}}{a_i}\right)^2 \frac{a_0}{a_{\text{eq}}} \qquad \text{for } k \ll k_{\text{eq}}. \tag{24.45}$$

The amplitude of a mode with shorter wavelength first grows until the mode 'enters the horizon' (i.e. the Jeans length overtakes the wavelength), and then does not grow until a reaches a_{eq}, so we have

$$T(k) = \left(\frac{a_{\text{enter}}}{a_i}\right)^2 \frac{a_0}{a_{\text{eq}}}$$

$$= \left(\frac{a_{\text{enter}}}{a_{\text{eq}}}\right)^2 \times \left(\frac{a_{\text{eq}}}{a_i}\right)^2 \frac{a_0}{a_{\text{eq}}} \qquad \text{for } k \gg k_{\text{eq}}. \tag{24.46}$$

The second way of expressing the result is useful because it enables an easy comparison with formula (24.45). The physical wavenumber k/a matches $k_{\text{J}} \sim H$ when $H \sim k/a$. Using $H \propto 1/a^2$ in the radiation dominated era this implies $a_{\text{enter}}/a_{\text{eq}} = k_{\text{eq}}/k$, so the net result is

$$T(k) = [k\text{-independent growth factor}] \times \begin{cases} 1 & k \ll k_{\text{eq}} \\ k^{-2} & k \gg k_{\text{eq}} \end{cases} \tag{24.47}$$

To get a rough prediction at all values of k it suffices to draw a smooth curve linking these two extremes. A more thorough analysis would allow for the modest growth during radiation domination, and would model the fluid more completely, allowing for neutrinos for example. But it is remarkable how well the above discussion captures the main features of the structure of the universe—see Fig. 24.3. The only thing missing is a statement of the dependence of δ on k in the initial conditions. To be precise, one wants a statement about a suitable statistical property such as the variance of the distribution of δ values at any particular k, which we consider in the next section.

Notice that throughout the discussion we have adopted a model in which dark matter has negligible kinetic energy and negligible pressure. Such a model is termed *cold* dark matter. Hot

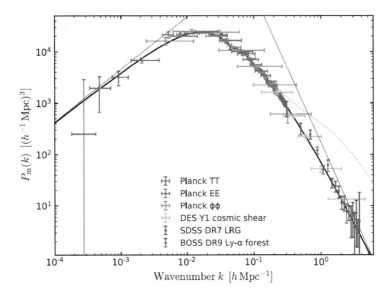

Fig. 24.3 A combination of many experimental observations of structure in the universe, expressed as a power spectrum for matter in the present. The curve is based on an initial power-law distribution followed by detailed numerical modelling of the dynamics. The two straight lines have been drawn with slopes differing by 4, to show how the simple model captures the main global feature (they are not quite the true asymptotes). [Courtesy ESA and the Planck Collaboration, (Planck Collaboration: Y. Akrami *et al.*, 2018)]

dark matter would lead to a different transfer function. The combination of observation and early universe modelling together imply that cold dark matter is largely the correct model. In the early universe there was also a hot neutrino background which somewhat modifies the picture. There are also other details which we shall not discuss.

24.3.1 The power spectrum

We shall now consider the statistical properties of the early universe fluctuations, which provide the initial conditions for the processes described in earlier sections.

A distribution such as $\delta(\mathbf{x}, t)$ fills a wide region of space and has, by definition, zero average at any time, since it is a distribution of departures from the average. The next interesting statistical property is the variance $\langle \delta^2 \rangle$, and its generalization which is the *two-point correlation function*

$$\xi(\mathbf{x}_1, \mathbf{x}_2, t) \equiv \langle \delta(\mathbf{x}_1, t)\delta(\mathbf{x}_2, t) \rangle \tag{24.48}$$

where the bracket indicates an average over all realizations of the fluctuations which are potentially possible for given statistical properties. To clarify this notion, note that for a distribution which shows no preference for one place or direction over another on average, one must find that this is a function of only the difference $|\mathbf{x}_1 - \mathbf{x}_2|$, and then it can be calculated by considering each value of $r = |\mathbf{x}_1 - \mathbf{x}_2|$ and taking a spatial average. The result can be written $\xi(|\mathbf{x}_1 - \mathbf{x}_2|, t)$. It is an indication of how likely it is, on average, to find two dense things (such as galaxies) separated by a given distance.

Now introduce the Fourier transform of the density fluctuations,[2]

$$\delta(\mathbf{k}, t) \equiv \int \mathrm{d}^3 x \, e^{\mathbf{k} \cdot \mathbf{x}} \delta(\mathbf{x}, t). \tag{24.49}$$

Then one finds

$$\langle \delta(\mathbf{k}_1, t) \delta(\mathbf{k}_2, t)^* \rangle = (2\pi)^3 \delta_{\mathrm{Dirac}}(\mathbf{k}_1 - \mathbf{k}_2) \int \mathrm{d}^3 r \, e^{i\mathbf{k} \cdot \mathbf{r}} \xi(r, t). \tag{24.50}$$

This is telling us that fluctuations at different $\mathbf{k}$ are uncorrelated, and the variance at each k is equal to $(2\pi)^3 P(k, t)$ after integrating over the three-dimensional Dirac delta function, where

$$P(k, t) \equiv \int \mathrm{d}^3 r \, e^{i\mathbf{k} \cdot \mathbf{r}} \xi(r, t). \tag{24.51}$$

Since it gives the variance at each k, this quantity is called the **power spectrum** of the fluctuations.

A good guess for the power spectrum of primordial fluctuations is the power-law

$$P(k) = A k^n \tag{24.52}$$

with constants A and n (the **spectral index**) to be found by observation. This is an informed guess, because some modest familiarity with the universe reveals that no one frequency scale stands out, and a power law is the simplest function having no scale-related parameter. $P \sim k^n$ gives $\xi(r) \sim r^{-(n+3)}$ so we expect $n > -3$.

An especially interesting possibility is $n = 1$, and in fact it is found that structure in the universe is consistent with a spectral index close to 1 for the primordial fluctuations. This type of distribution is called a *Harrison–Zel'dovich spectrum*, and it is interesting because there is a significant sense in which it is scale-independent.

To see this, first introduce the dimensionless quantity

$$\Delta(k) \equiv k^3 P(k)/(2\pi^2). \tag{24.53}$$

The three factors of k are needed to get something dimensionless from the function $P(k)$; the factor $2\pi^2$ is conventional. Next observe that Poisson's equation gives $\nabla^2 \phi \propto \delta$ and therefore $k^2 \phi(k) \propto \delta(k)$ where ϕ is the metric perturbation. Therefore the power spectrum of metric perturbations (or gravitational potential perturbations, which is the same thing) satisfies

$$P_\phi(k) \propto k^{-4} P(k) \quad \text{and} \quad \Delta_\phi(k) \propto P(k)/k. \tag{24.54}$$

It follows that if $P(k) \propto k$ then $P_\phi(k) \propto k^{-3}$ and $\Delta_\phi = $ constant, i.e. independent of k. That is, a dimensionless measure of the metric perturbations is scale-invariant (and see exercise 24.5).

---

[2]We here continue to use the same unadorned letter δ for a function and its Fourier transform; it is clear from the context which one is intended.

In the next chapter we will discuss observations which give an observed value $n = 0.97$. It is one of the successes of inflationary theory (the subject of Chapter 26) that it can make a prediction for n and the prediction is consistent with the observed value. On this model, the perturbations originate as natural properties of quantum fields in an expanding spacetime.

24.4   Baryon acoustic oscillations

Baryon acoustic oscillations are oscillations that took place in the plasma which filled the universe during the first 380,000 years, before decoupling. So far we have ignored these oscillations because we have been discussing the dark matter. But in this period the universe was filled with a plasma of photons, ionized ordinary matter and dark matter, and the ionized ordinary (baryonic) matter behaves differently. Owing to its strong coupling to the photons via Thomson scattering, it experiences the same high pressure as the photons, and therefore has a much higher Jeans length than the dark matter. It oscillates on all scales up to approximately the Hubble distance. These oscillations are called *Baryon acoustic oscillations*. They are sound waves with time and distance scales of order 10^5 years and 10^5 lightyears. They are truly the *basso profundo* notes of the universe! The oscillation ceased at recombination because then the matter and photons became largely decoupled.

This process is responsible for a slight departure from randomness in the density and other perturbations at recombination. If we assume that at very early times small perturbations in the curvature and dark matter distribution were distributed on all distance scales, with no particular preference for one distance scale over another, then one might expect the subsequent distribution to have this same feature. However, one would be wrong, because the *temporal* behaviour is not random. The temporal oscillations begin at specific times and impose structure on the position distribution of baryonic matter.

We analyse the density perturbations of baryonic matter in terms of their Fourier components. Each Fourier component starts out with some random amplitude A and with a phase φ of the spatial oscillation which is also random, but the temporal part is not random. The components at long wavelengths do not oscillate, whereas those whose wavelength is below the Hubble distance do. It follows that as the Hubble distance grows, the Fourier components at longer and longer wavelengths start to oscillate at particular, non-random, times. Hence at each wavelength one has a contribution to the distribution having the form

$$A \cos(\mathbf{k} \cdot \mathbf{x} + \varphi) \cos(\omega(t - t_k))$$

where t_k is, approximately, the time when the Hubble distance reaches π/k. This is not a precise statement but is intended to give a qualitative insight into the phenomenon. The frequency and wavelength are related by the speed of sound in the plasma, $\omega = c_s k$.

The acoustic oscillation continues until decoupling, and then it stops, because once the photons and baryons are no longer coupled the pressure drops abruptly. At this stage, the various Fourier components will be at various stages in their temporal oscillation. Some will happen to be at a stationary point, whether a compression or rarefaction, and consequently the distribution of baryons will have a corresponding over-density or under-density on that distance scale.

A full calculation of the BAO process involves extensive numerical modelling and is outside the scope of our discussion. However, the distance scale of the most prominent feature can be calculated straightforwardly.

Consider a region where there happens to be a slight over-density in both the dark matter and the ordinary matter at some very early time. There will result a spherical pressure wave in the ordinary (baryonic) matter propagating outwards from that region at the speed of sound c_s. This continues as long as the photon pressure is maintained, which is until decoupling. After this the slight over-density associated with the spherical wavefront is left 'frozen' in comoving coordinates. There also remains the overdensity in the dark matter at the centre. Therefore the net result is that the matter distribution at recombination has correlated over-density regions separated by a comoving coordinate separation χ_s given by

$$\chi_s = \int_0^{t_{\mathrm{ls}}} \frac{c_s}{R(t)}\mathrm{d}t = \frac{1}{R_0}\int_\infty^{z_{\mathrm{ls}}} \frac{c_s}{a}\frac{\mathrm{d}t}{\mathrm{d}a}\frac{\mathrm{d}a}{\mathrm{d}z}\mathrm{d}z = \frac{1}{R_0}\int_{z_{\mathrm{ls}}}^\infty \frac{c_s}{H(z)}\mathrm{d}z \qquad (24.55)$$

where $t_{\mathrm{ls}} \simeq 380,000$ years is the time of last scattering, c_s is the speed of sound in the plasma (approximately $c/\sqrt{3}$) and we used the FLRW metric in the form (22.16), which gives $c_s\mathrm{d}t = R(t)\mathrm{d}\chi$ for a worldline whose proper speed is c_s.

The quantity

Sound horizon

$$s \equiv R_0\chi_s = \int_{z_{\mathrm{ls}}}^\infty \frac{c_s}{H(z)}\mathrm{d}z \qquad (24.56)$$

---

is called the *sound horizon*; it is the total distance travelled by sound waves when the universe could support them, expressed as a comoving coordinate distance χ_s 'mapped' via R_0 onto the spatial surface at the present cosmic time. It is useful because we can calculate it without knowing R_0, and since $z_{\mathrm{ls}} \simeq 1100$ we do not need to know $\Omega_{\Lambda 0}$ either: the other terms dominate in eqn (23.36). However, one does need both Ω_{m} and Ω_{r}; the latter is not negligible but dominates at early times.

We will show in the next chapter how the outcome of the BAO process is used to deduce cosmological parameters from the distribution of galaxies and from the CMB.

24.5 Galaxy formation

With the advent of fast large-scale computing, it has become possible to run large numerical simulations of the process of structure formation, and this has been used to build up a detailed picture of the sequence of events.

Density perturbations of both baryonic and dark matter grew slowly before decoupling, as we have seen. Dark matter perturbations grew more, partly because there was more dark matter, and partly because they grew more rapidly after matter-radiation equality ($z = 3390$), whereas baryonic perturbations were prevented from growing much by their interactions with photons up until decoupling ($z = 1090$).

After decoupling, both baryonic and dark matter fell towards the bottom of the potential wells provided by the dark matter. Baryonic matter was able to concentrate at the bottom of these wells, because it could get rid of excess kinetic energy via collisions and emission of photons. Dark matter does not do this and consequently remains in larger more diffuse structures.

For any given region with an over-density, gravitational collapse happens most quickly along the smallest diameter, leading to a roughly planar or 'wall' shape. After this the next smallest diameter collapses, resulting in a linear filament, and then the filament contracts to a sphere. Consequently the universe is reckoned now to be filled with walls, filaments and spheres of dark matter, with the visible galaxies lying primarily at the intersections of these filaments.

Numerical simulations suggest that most of the galaxies we now see were formed by many hundreds of mergers. Such mergers are still going on. A fragment of several million solar masses has been discovered currently undergoing integration into our Milky Way galaxy on the other side from us. Our neighbours the Magellanic Clouds may someday do the same. The after-effects of some ancient mergers are expected to be detectable through their lingering effect on the velocity distribution. There has been time for several generations of stars to be born, grow old and die, liberating much of their contents via solar winds and supernovae. Later stars are formed partly from the material ejected from earlier ones, and consequently the chemical composition and structure of stars changes as one moves from one generation to the next. It is found that apparently all galaxies have huge black holes at their centre. It is not yet known whether these formed early on by collapse of huge clouds of gas, or whether they themselves are the result of numerous mergers.

Overall the cosmos is a thoroughly dynamic place. From its early rapid beginnings, to its present stately pace, it is full of 'life', of change and development.

Exercises

24.1 Consider the Newtonian dynamics of a spherical cloud of mass M and initial radius R collapsing under its own gravity. Show that the radius obeys $\ddot{r} = -GM/r^2$ and that the solution can be expressed parametrically as

$$r(\theta) = R\cos^2\theta, \quad t = A\left(\theta + \tfrac{1}{2}\sin 2\theta\right) \tag{24.57}$$

where A is a constant. Show that the cloud collapses completely in a time proportional to $\rho_0^{-1/2}$ where ρ_0 is the initial density. Why does not the collapse time depend on R?

Estimate (very roughly) the collapse time for a galaxy like the Milky Way.

24.2 Show that if the pressure term is negligible then equations 2 and 3 of (24.5) have the solution $\mathbf{v} = H(t)\mathbf{r}$ such that the acceleration equation (23.8) holds.

24.3 Discover whether or not isotropic homogeneous electromagnetic radiation in otherwise empty space undergoes gravitational collapse.

24.4 Consider the behaviour of radiation and cold dark matter in flat space (with negligible cosmological constant). We seek a more thorough treatment of (24.35) than the one given in the main text.
(i) For distance scales large compared to the Jeans length for matter, but small compared to the Jeans length for radiation, justify the assertion $\bar{\rho}_r \delta_r \ll \bar{\rho}_m \delta_m$, and hence obtain

$$\ddot{\delta}_m(\mathbf{k}) + 2H\dot{\delta}_m(\mathbf{k}) - \frac{3}{2}H^2 \frac{\bar{\rho}_m \delta_m}{\bar{\rho}_m + \bar{\rho}_r} = 0. \tag{24.58}$$

(ii) Let $y \equiv \bar{\rho}_m/\bar{\rho}_r$ (thus the situation is radiation-dominated when $y \ll 1$, and matter-radiation equality occurs when $y = 1$.) Show that $y = y_0 a$ and $\dot{y} = yH$.

(iii) With a view to using y as the new time parameter, show that $\dot{\delta} = \delta' yH$, and $\ddot{\delta} = \delta''(yH)^2 + \delta' \ddot{a}y_0$ where the prime indicates derivatives with respect to y. Hence obtain

$$\delta_m'' + \frac{2+3y}{2y(1+y)}\delta_m' - \frac{3}{2y(1+y)}\delta_m = 0.$$

Try the ansatz $\delta_m = A + By$ for constants A and B, and thus find one solution to this equation.[3] Hence find the total growth of matter perturbations during the radiation-dominated era.

24.5 Show that

$$\langle \phi(\mathbf{x})\phi^*(\mathbf{x})\rangle = \frac{1}{(2\pi)^3}\int P_\phi(k)\,\mathrm{d}^3k$$
$$= \int_0^\infty \Delta_\phi(k)\,\mathrm{d}(\ln k). \tag{24.59}$$

Hence $\Delta_\phi(k)$ can be interpreted as the contribution to the metric variance at each place from Fourier components of given wavenumber, per unit logarithmic range of k. The scale invariant case therefore has the property that each range of values in which k increases by the same factor contributes equally to $\langle \phi(\mathbf{x})\phi^*(\mathbf{x})\rangle$.

[3]The second solution is $\delta_m = -3x + (1+3y/2)\log((x+1)/(x-1))$ where $x = \sqrt{1+y}$.

25

Observational cosmology

It is a major aim of observational cosmology to derive accurate values of H_0, $\Omega_{i,0}$ and other parameters from observations of the universe. In order to build a picture and check for internal consistency, many different experimental methods are brought together. We have already mentioned the use of redshift and luminosity measurements for standard candles such as cepheid variables and type 1a supernovae. The main further tools are the precise observation of the cosmic microwave background (CMB) radiation and the study of the distribution of galaxies ('galaxy clustering'). In each case one combines exquisitely precise and voluminous measurements with lengthy theoretical and numerical modelling in order to learn what the observations are telling us.

25.1 Models, statistical and systematic error

It is a common feature of all empirical science that one must grapple with both statistical and systematic uncertainty in the interpretation of data. This is especially true when one is faced with large unknowns, such as the nature of dark matter, and the details of very early-universe processes. Fig. 25.1 presents an illustrative set of data which we will use to make a simple point about the difference between statistical and systematic uncertainty.

The figure shows a set of eleven data points (for some arbitrary physical phenomenon), and three attempts to extract information from the data by fitting a curve using the least squares method. One may say that we are thus using the data in order to learn the values of parameters in a model. Three models are tried:

$$\text{Model 1:} \qquad y = A + Bx$$
$$\text{Model 2:} \qquad y = Bx + Cx^2$$
$$\text{Model 3:} \qquad y = Bx + Dx^3$$

The degree to which the data is consistent with a model can be estimated by examining the *residuals*, that is, the set of differences $y_{\text{data}} - y_{\text{fit}}$. These are plotted in the three lower graphs of Fig. 25.1. In the case of model 1 (a straight line fit) it is found that the departures of the data from the fit are statistically significant. We say that the model 1 *does not provide a good fit*.

Relativity Made Relatively Easy: General Relativity and Cosmology. Volume 2. Andrew M. Steane,
Oxford University Press. © Andrew M. Steane 2021. DOI: 10.1093/oso/9780192895646.003.0025

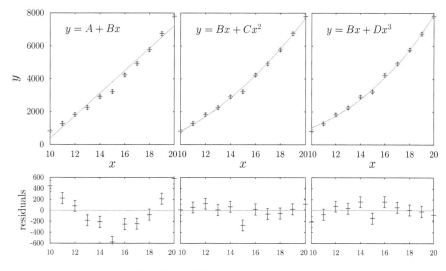

Fig. 25.1 Fitting models to data.

One way to see this is to note that the residuals are not distributed randomly about zero. For the other two models, on the other hand, there is a 'good fit'. The residuals are consistent with random noise at the level of the experimental uncertainty.

In order to assess the precision with which the parameters can be determined from the data, one tries a range of parameter values. As the parameters range further from their best-fit values, the residuals grow, and one determines what range of values may be said to be reasonably likely given the data. For example, in model 2, on the assumption that the fitted curve has the correct functional form (a linear plus a quadratic term), one finds $C = 30 \pm 3$ in this example.

A naive interpretation of this result would be to say that the data has determined the value of C to ten percent precision. But such an interpretation is only valid if one already has strong confidence that model 2 is itself correct. For suppose we consider model 3. In this model we get just as good a fit to the data, and it asserts that the value of the quadratic term in the formula is not 30 nor 27 nor 33 but zero! In order to explore further, one might try the model $y = Bx + Cx^2 + Dx^3$. In this case, if there is no other information available, then we obtain $C = 15 \pm 35$. The reason for the wide range is that a low C can be compensated by a high D, and *vice versa*. On this basis one would say that that the data is consistent with $C = 0$, and if C is non-zero then it is probably within the range -20 to $+50$.

The main lesson of this illustrative example is that when reporting, and assessing reports of, experimental precision, attention must be paid to what type of task is being reported. There is a large difference between using data to extract parameter values for a model which is known to be correct, and using data to distinguish between two conjectured physical processes whose effects are similar (C and D in our example).

Cosmological measurements typically must combine observation with some model of the relevant features of the cosmos, so the precision of any measured value is not limited purely by uncertainty in the data, but also by uncertainty in the degree of correctness of the model. The currently best supported model of the cosmos at large—one which is consistent with a large number of independent pieces of evidence—is called the ΛCDM model, which stands for 'Lambda, cold dark matter'. In this model the dark energy term in the Friedmann equations is non-zero and has the form of a cosmological constant Λ, and the matter content of the universe is largely in the form of dark matter—material which gravitates but has either no or only very small interactions of other kinds—and which is furthermore 'cold' meaning of low kinetic energy so that it does not present much pressure. In the following we will not assume this model at the outset, but we will present some of the pieces of evidence which lead to it. These form part of what is now a large body of scientific work, to which we cannot do justice. Our aim is purely that the reader should come to understand some of the main ideas.

25.2 The age of the universe

If we take the value of the Hubble constant determined from distance and red-shift measurements of stars, galaxies and supernovae, then, as we commented after eqn (23.43), one finds that a matter-only model of the evolution of the universe is unable to give a value of the age of the universe which is consistent with the age of stars and other evidence. The same is true of models which combine matter and radiation but exclude Λ. On this evidence, then, there is a strong suggestion that Λ ought not to be excluded. It is unlikely that measurements of the Hubble constant have been thrown off by some feature of the universe that has not been accounted for, to a degree sufficient to make the matter- and radiation-only models viable.

If we adopt the matter plus vacuum model, we can explore which values of Ω_{m0} and $\Omega_{\Lambda0}$ are consistent with the age of the universe, and which are not, and thus gain an initial indication of the parameter values on the assumption that the model is correct. It is straightforward to integrate (23.37) numerically in order to find the value of t when $a = 1$ (and one may note that the addition of a small non-zero value of Ω_{r0} would not change the answers significantly). Fig. 25.2 shows the result. The figure shows a contour plot of $H_0 t$ as a function of Ω_{m0} and $\Omega_{\Lambda0}$. Values of H_0 determined from redshift studies fall in the range 68 to 72 km/s per Mpc. Stellar models give reasonable confidence that there are stars of age 13 ± 1.5 Gyr, therefore we have that $t > 13$ Gyr, and on a conservative estimate we thus find $H_0 t > 0.9$. Thus the region shown shaded on the figure is ruled out. If we now bring in the fact that even a very rough measurement of spatial curvature suggests Ω_0 is in the range 0.6 to 1.4, then we can further restrict the allowed region to between the two dashed lines. This gives a strong suggestion, then, that $\Omega_{\Lambda0} > 0$. Even without this argument from spatial curvature, it would be hard to maintain $\Omega_{\Lambda0} = 0$ because then the allowed values of Ω_{m0} are small, which would imply that the matter is so thinly spread that there would not have been enough time for it to come together by gravitational collapse so as to form the galaxies that are observed.

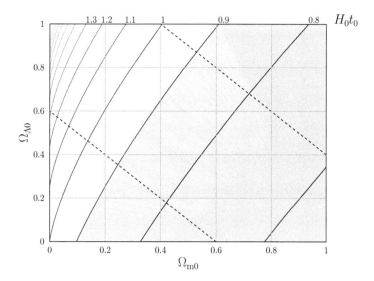

Fig. 25.2 Contours of $H_0 t_0$ obtained by integrating (23.37), as a function of Ω_{m0} and $\Omega_{\Lambda 0}$, taking $\Omega_{r0} = 0$. (The integral yields t as a function of a; t_0 is the value of t when $a = 1$). The shaded region is ruled out as giving too young a universe, and the region outside the dashed lines is ruled out by studies of spatial curvature. Galaxy formation also requires that Ω_{m0} be not too small.

25.3   Hubble parameter and deceleration parameter

The time-dependence of the scale factor $a(t)$ of our universe can in principle be deduced from measurements that determine the redshift and the distance of a large collection of galaxies, via a measure such as luminosity distance. In order to organize the data, a useful first step is to write down a Taylor expansion of $a(t)$ about any convenient moment. The most convenient moment is the present, t_0. So let us define $\delta t = (t_0 - t)$, the 'look-back time'. The sign is chosen so that events which we can observe (i.e. events that occurred at times $t < t_0$) shall have a positive δt. Then the Taylor expansion reads

$$
\begin{aligned}
a(t) &= a(t_0 - \delta t) \\
&= a(t_0) - \delta t \, \dot{a} + \tfrac{1}{2}\delta t^2 \ddot{a} + \cdots \\
&= a(t_0) \left(1 - H_0 \delta t - \tfrac{1}{2} q_0 H_0^2 \delta t^2 + \cdots \right)
\end{aligned}
\tag{25.1}
$$

where we have used the Hubble parameter $H(t) \equiv \dot{a}/a$ defined in (22.1), and introduced the *deceleration parameter*

$$
q(t) \equiv -\frac{\ddot{a}(t)a(t)}{\dot{a}^2(t)}
\tag{25.2}
$$

with $H_0 = H(t_0)$ and $q_0 \equiv q(t_0)$. The deceleration parameter is defined this way because then

$$
H(t) = H_0 + H_0^2(1 + q_0)(t_0 - t) + O((t_0 - t)^2)
\tag{25.3}
$$

and see also eqn (25.10) below. The deceleration parameter quantifies the rate of change of the Hubble parameter. In other words it quantifies the degree to which the expansion is slowing, at any given time. Using the Friedmann equation (23.8) one finds (exercise)

$$q = \tfrac{1}{2}(\Omega_m + 2\Omega_r - 2\Omega_\Lambda). \tag{25.4}$$

Notice that in a homogeneous ΛCDM model q can only be negative if Ω_Λ is non-zero.

If we apply the redshift formula (22.22) to the redshift observed now, at t_0, for a photon which set out at time t then we find

$$z = \frac{a(t_0)}{a(t)} - 1 = \left(1 - H_0\delta t - \tfrac{1}{2}q_0 H_0^2 \delta t^2 + \cdots\right)^{-1} - 1$$

$$\simeq H_0\delta t + \tfrac{1}{2}q_0 H_0^2 \delta t^2 \tag{25.5}$$

(valid for $H_0\delta t \ll 1$). Solving for δt we have

$$\delta t = t_0 - t \simeq \frac{z - (1 + q_0/2)z^2}{H_0}. \tag{25.6}$$

Thus, if we know the values of H_0 and q_0 then we can discover the amount of cosmic time before present at which light of any given redshift was emitted, to reasonable precision for small z.

There are a number of ways to write an exact formula. One useful way is to differentiate the redshift formula, obtaining:

$$\frac{dz}{dt} = -\frac{a_0}{a^2}\dot{a} = -(1+z)H \tag{25.7}$$

from which,

$$t_0 - t = \int_t^{t_0} dt = \int_0^z \frac{dz}{(1+z)H(z)}. \tag{25.8}$$

To evaluate the integral we would need to know H as a function of z, which requires a knowledge of $a(t)$.

The comoving coordinate distance travelled by the light is given by

$$\chi = \int_t^{t_0} \frac{c\,dt}{R(t)} = \int_t^{t_0} \frac{c\,dt}{R_0 a(t)} = \frac{c}{R_0}\int_0^z \frac{dz}{H(z)}. \tag{25.9}$$

In the quadratic approximation introduced above one finds

$$H \simeq H_0\left(1 + (1+q_0)z + \cdots\right), \tag{25.10}$$

$$\chi \simeq \frac{c}{H_0 R_0(1+q_0)}\ln\left(1 + (1+q_0)z\right). \tag{25.11}$$

Luminosity distance measurements can be used to determine H_0 and q_0. To lowest order in z, and using $S(\chi) \simeq \chi$, eqns (25.11) and (22.38) give

> **Magnitude.** In astronomy, *magnitude* is a measure of brightness. The *absolute magnitude* M of a source of luminosity L is defined by
>
> $$M = -2.5 \log_{10}(L/L_0)$$
>
> where $L_0 = 3.0125 \times 10^{28}\,\mathrm{W}$ is a standard value. Thus a lower value of M corresponds to a brighter star (the term arose from the expressions 'of the first magnitude,' 'of the second magnitude,' etc.). The absolute magnitude of the Sun is approximately 4.74. *Apparent magnitude m* is defined by
>
> $$m = -2.5 \log_{10}(F/F_0)$$
>
> where F is the flux (power per unit area) received in a detector on Earth, and F_0 is a standard flux given by $F_0 = L_0/(4\pi d_0^2)$ where $d_0 = 10\,\mathrm{parsec}$. Hence the absolute magnitude is equal to what the apparent magnitude would be if the source were viewed from a distance of ten parsecs. By the definition of luminosity distance d_L, one has $L = 4\pi d_L^2 F$, and therefore
>
> $$m - M = 5 \log_{10}(d_L/d_0)$$
>
> which can also be written $m - M = 5 \log_{10}(d_L/\mathrm{Mpc}) + 25$. The quantity $m - M$ is called the *distance modulus*. The unaided human eye can detect apparent magnitudes below (i.e. brighter than) approximately 6.

$$F \simeq \frac{L}{4\pi}\left(\frac{H_0(1+q_0)}{c(1+z)\ln(1+(1+q_0)z)}\right)^2 \simeq \frac{LH_0^2}{4\pi c^2 z^2}(1 + (q_0-1)z + O(z^2)). \qquad (25.12)$$

where the first verion is accurate for z up to of order 1, and the second version for $z \ll 1$. By using this formula when L values are known, through the use of standard candles, the value of H_0 can be directly obtained from a set of flux and redshift measurements, and in principle the value of q_0 also obtained, though with less precision. A positive q_0 would make the stars look brighter. In practice the available precision is such that observations up to $z \simeq 0.2$ do not constrain q_0 well enough to determine its sign; therefore one requires larger z values for which the second version of (25.12) is too rough. More generally one uses numerical integration to obtain the predicted $F(z)$ for some $H(z)$ given by a cosmological model, and then one uses the observations to constrain the parameters of the model.

25.3.1   Measurements

A Hubble diagram out to modest values of z (e.g. 0.2) allows H_0 to be measured to an accuracy of a few percent. One would like to obtain the deceleration parameter also, but this only becomes a practical possibility when the diagram can be extended to values of redshift of order 1. This was finally accomplished in the late 1990s through the accumulation of Type 1a supernova events at high (i.e. of order 1) redshift. Modern large telescopes allow observations to probe deeply into the sky, covering a large number of galaxies in a single run. This compensates for the rarity of supernova events (their rate is a few per century in a Milky-Way-sized galaxy).

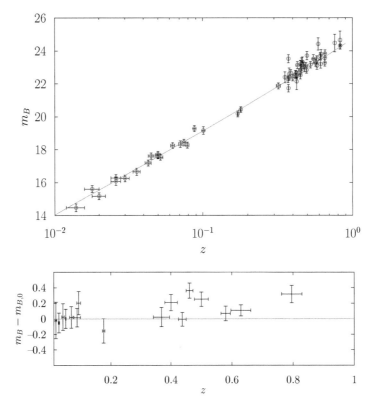

Fig. 25.3 Observations of type 1a supernova from Perlmutter *et al.*, 1999. The data shows the corrected apparent magnitude m of 60 supernova events at a range of distances (the correction being based on the light curve). The line shows eqn (25.14) with $q_0 = 0$ and a value of H_0 chosen to match the low z data. The high z data fall above this line; this is shown more clearly in the residuals plot (lower graph). The data at high z fall above the line to a statistically significant degree, therefore either the expansion of the universe is accelerating, or some unknown process has influenced the luminosity at high z. (The residuals (data minus curve) have been binned so as to reduce the statistical error on the weighted mean for each group. The horizontal error bars indicate the range of z values included in each group.)

Fig. 25.3 shows the observations published in 1999 by one of two leading research teams employing the method of accumulating observations of distant type 1a supernovae.(Perlmutter *et al.*, 1999; Riess *et al.*, 1998; Perlmutter and Schmidt, 2003) The data is reported in terms of the apparent magnitude m (see box) as a function of redshift z. Using (22.39) and (25.12) (which was obtained using a linear approximation to $H(z)$) one finds

$$m = M + 5 \log_{10} \frac{d_L}{d_0} \tag{25.13}$$

$$\simeq M + 5 \log_{10} \frac{c}{H_0 d_0} + 5 \log_{10}(1 + z) + 5 \log_{10} \left(\frac{\ln(1 + (1 + q_0)z)}{1 + q_0} \right) \tag{25.14}$$

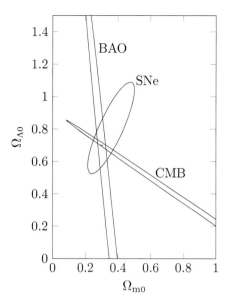

Fig. 25.4 One sigma (68% confidence) limits on the parameters Ω_{m0}, $\Omega_{\Lambda0}$ given by galaxy clustering (BAO), supernova (SNe) and CMB observations, assuming the homogeneous ΛCDM model.

where M is the absolute magnitude of a supernova event at the peak of the light curve. The peak absolute magnitude of a standard Type 1a supernova is $M = -19.3$. A typical supernova brightens during a few days and then glows for a month. The duration of each light curve is used to infer its relation to the standard case and an adjustment is applied to m to account for this.(Suzuki *et al.*, 2012)

The form of (25.14) is such that H_0 provides a vertical offset to the graph of m versus z, whereas q_0 affects the shape of the functional dependence on z. The resulting curve is shown on Fig. 25.3 for the values $H_0 = 2.1 \times 10^{-18}\,\text{s}^{-1}$ and $q_0 = 0$. The high-z data falls somewhat *above* this line, as can be seen more easily in the residuals plot. In other words the supernovae look a little dim compared to what would be the case if H were constant. This implies that q_0 is *negative*, which suggests that the expansion rate is *accelerating*.

If we take $\Omega_{r0} \ll 1$ then (25.4) gives

$$q \simeq \Omega_m/2 - \Omega_\Lambda. \tag{25.15}$$

This and other supernova studies in the range $z = 0.1$ to 1 determined the value $q_0 = -0.45\pm0.1$. Owing to the differing influence of Ω_m and Ω_Λ on $a(t)$, the data is able to determine the difference $\Omega_{\Lambda0} - \Omega_{m0}$ more accurately than the sum $\Omega_{\Lambda0} + \Omega_{m0}$, with the result that the conclusion, expressed as a limit on Ω_{m0} and $\Omega_{\Lambda0}$, at 68% statistical confidence, is as shown in Fig. 25.4. The most significant source of systematic imprecision is obscuration by dust. Ordinary astrophysical dust scatters blue light preferentially, leading to the reddening of the received spectrum; this can be quantified from the observations and hence compensated for (it is found to be almost negligible). There remains the possibility of scattering by some unknown form of 'grey' dust, but one can place upper bounds on such a process since it would also affect other types of

Finding Ω_{r0}. The value of Ω_{r0} can be found in the present from the standard thermodynamic expression for the energy density of thermal radiation, since Ω_{r0} is dominated by the CMB. The energy density u_γ in cavity (black body) radiation is given by $u_\gamma = 4\sigma T^4/c$ where $\sigma = \pi^2 k_{\rm B}^4/(60\hbar^3 c^2)$ is the Stefan–Boltzmann constant, whose value is $\sigma \simeq 5.6704 \times 10^{-8}\,\mathrm{Wm^{-2}K^{-4}}$. At $T = 2.72548\,\mathrm{K}$ this gives $u_\gamma = 0.2606\,\mathrm{MeV\,m^{-3}}$ and therefore $\Omega_{r0} = 5.47 \times 10^{-5}$ if we take $H_0 = 2.18 \times 10^{-18}\,\mathrm{s^{-1}}$. However, in early times one must allow also for neutrinos. It is calculated that the contribution made by neutrinos to the density of the early universe is related to that of photons by

$$u_\nu = 3.046\frac{7}{8}\left(\frac{4}{11}\right)^{4/3} u_\gamma \tag{25.16}$$

After adding this to u_γ one finds $\Omega_{r0} = 9.2 \times 10^{-5}$. To check this, one treats the value of redshift at matter–radiation equality as a fitted parameter in the study of the CMB. The best fit value is $z_{\rm eq} = 3391$. Using $\Omega_{m0}(z_{\rm eq} + 1)^3 = \Omega_{r0}(z_{\rm eq} + 1)^4$ one then finds $\Omega_{r0} = 9.29 \times 10^{-5}$ for $\Omega_{m0} = 0.315$.

astronomical observation. Overall, then, there is good reason to believe that the data indicate the dynamics of the universe, rather than some other effect.

The most significant feature of the supernova study is that it strongly suggests that the expansion of the observable universe is accelerating, irrespective of cosmological models. This could be partially explained without appealing to Λ if the Milky Way happened to be situated in an under-density region in the cosmic fluid, but it is hard to account for the whole of the result this way. In a ΛCDM model, the negative q_0 implies that Ω_Λ is non-zero and positive.

25.4 Baryon acoustic oscillations

The baryon acoustic oscillation (BAO) process was described in Section 24.4. It results in a slight preference in the baryonic matter distribution, at last scattering, for over-densities separated by χ_s given by (24.55), which can be conveniently expressed in terms of the sound horizon $s = R_0\chi_s$, eqn (24.56).

In order to calculate s one requires a cosmological model, but, as we noted in Section 24.4, it is not necessary to know the value of $\Omega_{\Lambda0}$ because at large z the other terms dominate in (23.36). One does need both Ω_m and Ω_r; the latter is not negligible but dominates at early times. The box summarizes two methods to obtain Ω_{r0}. The value of Ω_{m0} is found either from other information, or by using the BAO method to discover it by seeking consistency between observations at several redshifts. One thus obtains $s \simeq 150$ Mpc.

The density perturbations formed the seeds around which gravitational collapse subsequently happened, eventually forming galaxies. The spherical sound-wave process in the original plasma occurred simultaneously for a large number of randomly distributed centres, so one does not

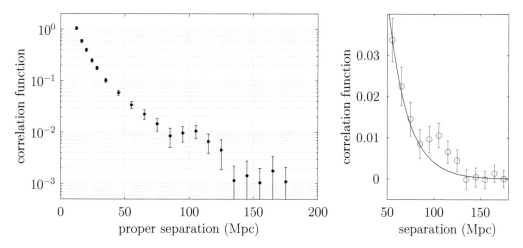

Fig. 25.5 Correlation function of galaxy separations at $z = 0.35$, observed in the Sloan Digital Sky Survey. The right-hand diagram is a close-up of the data near the BAO 'bump' at 110 Mpc. (Eisenstein *et al.*, 2005)

expect to see spheres in the final density distribution, but one does expect to find that pairs of galaxies are more likely to be separated now by 150 Mpc than by 140 Mpc or 160 Mpc, and similar statements apply at earlier epochs, after allowing for the scale factor. In a triumph of observational cosmology, this is what is found in large-scale sky surveys (Fig. 25.5). This observation can be used to determine the spatial curvature of the universe, as follows.

At any given cosmic epoch after last scattering, the correlation length $R(t)\chi_s = as = s/(1+z)$ acts as a standard ruler whose length is known to us because we can calculate s. One first picks some value of z, then one determines from a sky survey the distribution of angular separations of galaxy pairs at that z. Owing to the BAO process there is a bump in the distribution at some $\Delta\theta$ which is thus measured (Fig. 25.5). By substituting this into (22.40) we obtain the angular diameter distance to the galaxies in question:

$$d_A = \frac{as}{\Delta\theta} = \frac{s}{(1+z)\Delta\theta}. \tag{25.17}$$

Next, we employ (22.41):

$$d_A = \frac{R_0 S(\chi)}{1+z} \tag{25.18}$$

where $S(\chi)$ is one of $\sin\chi$, χ, $\sinh\chi$ (see (22.16)), and note that in this formula we are using χ to indicate radial distance from Earth to some other galaxy, not the length of the standard ruler described by χ_s. For small χ we have

$$S(\chi) = \chi - k\chi^2/6 + \dots \tag{25.19}$$

so

$$d_A = \frac{1}{1+z} x \left(1 - \frac{1}{6} K_0 x^2\right) \tag{25.20}$$

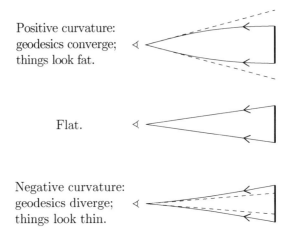

Positive curvature: geodesics converge; things look fat.

Flat.

Negative curvature: geodesics diverge; things look thin.

Fig. 25.6 The influence of spatial curvature on the visual appearance of distant objects. These diagrams give a correct general impression but, N.B., they do not show the change of scale factor with time, which also has to be included when interpreting d_A.

where we introduced $x \equiv R_0\chi$ (called the *comoving radial distance*) and K_0 is the Gaussian curvature of space in the present ($|K_0| = 1/R_0^2$). (Takada and Doré, 2015) Since we already know the value of d_A from the angle measurement indicated in (25.17), we can now determine K_0 if we can measure x. This is determined by the time it takes light to travel from a galaxy at redshift z to Earth. The null geodesic satisfies

$$c\,\mathrm{d}t = R(t)\,\mathrm{d}\chi = R_0 a\,\mathrm{d}\chi = a\,\mathrm{d}x, \tag{25.21}$$

therefore

$$x = \int_t^{t_o} \frac{c}{a}\,\mathrm{d}t = \int_0^z \frac{c}{H(z)}\,\mathrm{d}z \tag{25.22}$$

where the working is similar to that in (24.55), but notice the different limits: in order to evaluate (25.22) for low to moderate values of z, we only need to know how H has been behaving recently (for an estimate one can simply use the value $H \simeq H_0$ for $z \lesssim 1$). Now we can solve (25.17) and (25.22) in order to find a formula for K_0 in terms of the known or measured quantities $s, x, \Delta\theta$:

$$K_0 = \frac{6}{x^2}\left(1 - \frac{s}{x\Delta\theta}\right). \tag{25.23}$$

The basic idea is that if the standard ruler 'looks' a bit fat ($\Delta\theta > s/x$) then space is positively curved; if it looks thin ($\Delta\theta < s/x$) then space is negatively curved—see Fig. 25.6. In the event, the observed angles are neither fat nor thin: the curvature is found to be small and possibly zero:

$$K_0 = (0 \pm 0.1)(H_0/c)^2 \tag{25.24}$$

where we have invoked the deceleration parameter measurement from the previous section in order to constrain $\Omega_{\Lambda 0}$. This is because in a more thorough analysis the BAO method turns out to be rather insensitive to $\Omega_{\Lambda 0}$ but achieves a good precision for Ω_{m0}: see Fig. 25.4.

25.5   The cosmic microwave background radiation

Study of the CMB radiation is the third main method to obtain information about cosmological parameters, and in many respects it is the most precise. For an introduction to the physics of the CMB, see Bucher 2015 (Bucher, 2015).

25.5.1   The observed structure of the CMB

The CMB radiation received at Earth was emitted from points on a huge sphere centred at the comoving coordinate of Earth, and now arriving after travelling for billions of years. This does not imply that Earth is at a special location, since all other locations also have such a sphere around them.

The radiation is highly uniform in direction, and highly regular in frequency composition, matching the spectrum of thermal radiation of temperature 2.72548 ± 0.00057 kelvin to better than one part in 10^4 after an overall dipole term is allowed for. (Fixsen, 2009) Imposed on this uniformity there are small variations in temperature as a function of direction. These variations are random on the smallest angular scales, and structured on larger angular scales. This structure is conveniently analysed by expressing the temperature anisotropies as a sum of spherical harmonics:

$$\Delta T(\theta, \phi) = \sum_{l=0}^{\infty} \sum_{m=-l}^{l} a_{lm} Y_m^l(\theta, \phi) \tag{25.25}$$

This is like a form of Fourier analysis, but in direction rather than distance or time. The constant $l = 0$ term can be ignored. The $l = 1$ (dipole) term is owing to the fact that Earth (and the Milky Way Galaxy) has a non-zero velocity relative to the local cosmic comoving frame. It is the higher order terms that concern us.

Each value of l captures information about a given angular scale. The power in the fluctuations is determined by defining the coefficients

$$C_l = \frac{1}{2l+1} \sum_{m=-l}^{l} |a_{lm}|^2. \tag{25.26}$$

Each C_l is then a measure of the temperature fluctuations on the angular scale approximately $180°/l$.

In the past 30 years there has been a beautifully creative and technically exquisite sequence of balloon and spacecraft instruments used to measure the CMB with increasing sensitivity and angular precision. The major missions were named COBE (1989–1996), Boomerang (1997–2003), WMAP (2001–2010) and Planck (2009–2013). The results of the latter are shown in Figs 25.7, 25.8. The locations and shapes of the peaks in the C_l spectrum hold a wealth of information of cosmological interest. Together with the BAO and supernova measurements they permit the main parameters of the ΛCDM model to be obtained with a precision of order

Fig. 25.7 A map of the microwave temperature across the sky, after subtracting a dipole component and adjusting for dust and sources within the Milky Way. (Courtesy ESA and the Planck Collaboration: Ade, P. A. R. et al., 2014)

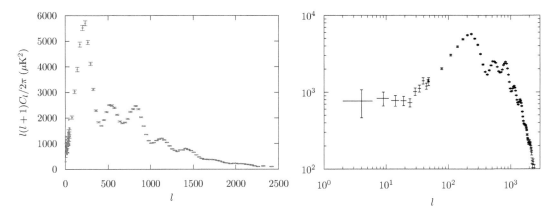

Fig. 25.8 The angular spectrum of CMB anisotropies. The same data is shown first on linear and then on logarithmic scales. Data from Planck mission. (Planck Collaboration: Ade, P. A. R. *et al.*, 2014)

a few percent on the assumption that the model is correct: see Table 25.1. This does not in itself constitute a guarantee that the model is correct, but it strongly constrains what the possibilities are; we shall return to this issue at the end of the chapter.

The main features of the spectrum of CMB fluctuations are:

1. A series of well-defined peaks
2. A specific baseline function on which these peaks are superimposed

Table 25.1 Cosmological parameters inferred from a combination of CMB, galaxy clustering and supernova studies, as reported in (Di Valentino, Melchiorri and Silk, 2015; Planck Collaboration: N. Aghanim *et al.*, 2018). The 12 parameter model allows all the listed parameters to vary in the fits to the data; the 6 parameter model adopts assumed values for six of the parameters (such as $w = -1$). The dimensionless Hubble parameter h is defined as $h = H_0/(100\,\mathrm{km\,s^{-1}/Mpc})$; hence for example the data imply $\Omega_b = 0.0224/0.674^2 = 0.049(5)$. The tensor/scalar ratio r refers to the size of anisotropies in tensor (as opposed to scalar and vector) properties of the CMB. The neutrino-related quantities calibrate models of a relativistic plasma. The last two rows give the density parameters inferred from the first three rows. The models adopted in order to obtain the bounds shown in this table assumed zero spatial curvature so $1 - \Omega_{m0}$ can be taken to be the value of $\Omega_{\Lambda 0}$.

		12 parameter model	6 parameter model
baryon density parameter	$\Omega_{\mathrm{b}} h^2$	0.0224(6)	0.0223(3)
cold dark matter density parameter	$\Omega_{\mathrm{c}} h^2$	0.119(7)	0.119(2)
Hubble parameter (kms^{-1}/Mpc)	H_0	67.4(4.3)	67.5(0.9)
reionization optical depth	τ	0.058(42)	0.066(12)
scalar spectral index	n_{s}	0.968(25)	0.966(8)
amplitude	$\log(10^{10} A_s)$	0.76(9)	0.832(25)
dark energy state parameter	w	$-1.06(14)$	
running of spectral index	$\mathrm{d}n_{\mathrm{s}}/\mathrm{d}\ln(k)$	0 ± 0.02	
tensor/scalar ratio	r	< 0.183	
effective neutrino number	N_{eff}	3.1(6)	
Neutrino mass sum	$\sum m_\nu$	$< 0.854\,eV$	
lensing amplitude	A_{lens}	1.2(2)	
$\Omega_b + \Omega_c$	Ω_{m0}	0.31(3)	0.310(7)
$1 - \Omega_{\mathrm{m0}}$	$\Omega_{\Lambda 0}$	0.69(3)	0.690(7)

In order to interpret the data, one begins with the statistical measures of perturbations in a three-dimensional fluid described in Section 24.3.1. One then has to find the mathematical relation between fluctuations in three dimensions and fluctuations on a spherical surface (one whose thickness is small but non-zero). Finally one needs to determine what the CMB spectrum would be for any given form of the power spectrum $P(k)$ of fluctuations in density, velocity and metric. A lengthy calculation is needed, first to convert the three-dimensional picture to a statement about correlations as a function of angular separation on the surface of last scattering, and then to account for phenomena such as gravitational lensing of the light on its long journey from the surface of last scattering to Earth. These details are outside the scope of our discussion, except that we will describe one part of the calculation, the Sachs–Wolfe effect.

25.5.2 Sachs–Wolfe effect

The Sachs–Wolfe effect is a dynamic process which shifts the observed frequency of radiation emitted from any given surface on which there are non-negligible metric fluctuations. It tells us that the temperature fluctuations observed by ourselves, long after and far away from the surface of last scattering, are not simply a map of the temperature on that surface. Rather, the combined effect of redshift on leaving a gravitational well, and also a slight change in the timing of last scattering, make the ΔT observed by us partly a reflection of the original ΔT, and partly an indication of the size of this further gravitational effect.

If the situation were static, then one would find that a metric perturbation $\Delta\Phi$ would lead to a gravitational potential well of this size, and consequently light emerging from the well has its frequency shifted by

$$\frac{\Delta\nu}{\nu} = \Delta\Phi \tag{25.27}$$

(where we took $c = 1$ and used (13.48) for the case $\Delta\Phi \ll 1$). A positive $\Delta\Phi$ here indicates a higher potential and therefore a blue shift compared to light coming from somewhere at average potential. Since all frequencies get the same fractional change, the observed spectrum remains thermal, but with a temperature change $\Delta T_g = T\Delta\Phi$.

The Sachs–Wolfe effect now modifies this by allowing for the non-static nature of the problem. The metric perturbation $\Delta\Phi$ at any location has persisted with fixed amplitude (in comoving coordinates) during both radiation-domination and matter-domination, for large wavelengths $k < k_{\text{eq}}$, (Section 24.2.2). Consequently local processes such as scattering and the formation of neutral hydrogen have been taking place more rapidly in a region with positive $\Delta\Phi$, with the result that last scattering occurs 'early', a little before the average, by the amount $\Delta t \simeq -\int \Delta\Phi(t)\mathrm{d}t \simeq -t\Delta\Phi$. Consequently the scale parameter at last scattering, in the local region, is smaller than the average by $\Delta a = (\mathrm{d}a/\mathrm{d}t)\Delta t = -(\mathrm{d}a/\mathrm{d}t)t\Delta\Phi$. The subsequent expansion to the value of a in the present at our detectors (i.e. $a = 1$) will therefore introduce a larger redshift. Using the cosmological redshift formula $\nu/\nu_{\text{LS}} = a_{\text{LS}}/a$ where ν is the frequency observed now, we have

$$\frac{\Delta\nu}{\nu_{\text{LS}}} = \frac{\Delta a_{\text{LS}}}{a} \implies \frac{\Delta\nu}{\nu} = \frac{\Delta a_{\text{LS}}}{a_{\text{LS}}} = -\frac{\mathrm{d}a}{\mathrm{d}t}\frac{t}{a}\Delta\Phi \tag{25.28}$$

where the quantities on the right are all to be evaluated at last scattering. Using $a \propto t^{2/3}$ for the matter-dominated universe, which is a good approximation at last scattering, one finds $\Delta\nu = -(2/3)\nu\Delta\Phi$ and thus a temperature reduction, $\Delta T_a = -(2/3)T\Delta\Phi$.

We now combine the two contributions. The result is that if the original temperature had been completely smooth, but there were metric perturbations of size $\Delta\Phi$, then we would observe temperature fluctuations of size

$$\frac{\Delta T_{\text{SW}}}{T} = \frac{\Delta T_g + \Delta T_a}{T} = \frac{1}{3}\Delta\Phi. \tag{25.29}$$

In fact the temperature on the last scattering surface was not completely smooth. For thermal radiation the energy density $\rho_{\text{r}} \propto T^4$ so, in the notation of Chapter 24, $\delta_{\text{r}} = \delta\rho_{\text{r}}/\rho_{\text{r}} = 4\delta T/T$. For adiabatic perturbations (eqn (24.4)) this implies the fractional temperature perturbations at last scattering were given by

$$\frac{\Delta T}{T} = \frac{1}{3}\delta_{\text{m}} \tag{25.30}$$

where $\delta_{\rm m}$ is the density contrast. The relative sizes of the two contributions (25.29) and (25.30) depend on the distance scale of the perturbation, because one depends on the metric perturbations and the other on the density perturbations, and these are related by Poisson's equation, which gives $k^2 \Delta \Phi = 4\pi G a^2 \rho \delta_{\rm m}$ as we noted in the previous chapter. Therefore $\Delta \Phi$ dominates for small k, and $\delta_{\rm m}$ dominates for large k. Hence the anisotropies in the CMB at low l (in practice, $l < 50$ or so) indicate metric perturbations, and the anisotropies at high l indicate density (and velocity) perturbations. It is found that, after this 'processing' via the Sachs–Wolfe effect, one expects $C_l \propto (l(l+1))^{-1}$ for a scale-invariant power spectrum (the Harrison and Zel'dovich of Section 24.3.1), and for this reason the flat region at low l in the graph of $l(l+1)C_l$ is called the 'Sachs–Wolfe plateau'.[1] A small departure from scale-invariance is called 'tilt'.

Another process is called the *late time integrated Sachs–Wolfe (ISW) effect*. This refers to the effect of light falling into wide potential wells (or hills) in the moderately recent past (on cosmic scales), which then become shallower owing to the cosmic expansion. It is found that an accelerating expansion is required for there to be a non-zero net effect. Therefore evidence of late-time ISW in the data is evidence for an accelerating expansion.

25.5.3 Extracting cosmological information

It is found that the CMB C_l values are consistent with a power spectrum $P_\delta(k)$ having the following features:

1. acoustic peaks consistent with Fourier components that have evolved from an initial extension beyond the Hubble distance (to be explained shortly).
2. A nearly scale-invariant underlying spectrum: $n_{\rm s} = 0.968 \pm 0.025$.
3. Gaussian statistics; that is, fluctuation amplitudes which are normally distributed.
4. Fluctuations mainly such that all forms of matter share the same fractional over- or under-density at each point, and furthermore this is correlated with the radiation fluctuations as in (24.4). (In the context of cosmology, such fluctuations are called *adiabatic*.)

The *acoustic peaks* (item 1 in the above list) are the result of the BAO process described in Section 24.4. The first peak, called the *first acoustic peak*, is at $l \simeq 220$, corresponding to an angular scale of $\sim 1°$. This peak, and others, is the signature in the CMB of the acoustic oscillation process. There are several peaks because the BAO process involves oscillation, not just spherical waves. The various peaks correspond to standing waves which completed successive half-integer numbers of oscillations. At high l the observed spectrum is damped exponentially owing to diffusion of the photons on their journey to us. At low $l \lesssim 10$ the statistical uncertainty in the data grows because we have only one sky to observe and there are fewer regions being averaged over at the widest angular scales; this is called *cosmic variance*.

---

[1] This name is slightly misleading because the Sachs–Wolfe effect is not the cause of the metric perturbations; it is the translator of those perturbations into observed frequency perturbations.

It is significant that the peak widths are less than their separations, so that the peaks are clearly distinguished from one another. If one models the growth of density perturbations in the early universe then there are, broadly speaking, two classes of model. In one class the perturbations are already present at a very early stage, on all distance scales including those that are larger than the Hubble distance. This was assumed in Chapter 24. In the other class the perturbations are produced at each distance scale as the Hubble distance reaches that scale. The latter class of model leads to broader peaks and is disfavoured by the data.

The peak positions are used to infer the curvature of space by the following argument.

By calculating the temperature and density conditions for the recombination process and comparing with the CMB temperature now observed, it is deduced that the CMB was produced at redshift $z = 1090$. The proper size (ruler length) of the sound horizon at last scattering was therefore $s/1090 \simeq 0.14\,\mathrm{Mpc}$, where s is given by (24.56). The location of the first acoustic peak in the CMB (i.e. 1 degree) corresponds to a ruler length on the scattering surface $1° \times (\pi/180°)x_{\mathrm{ls}}/1090 \simeq 0.22\,\mathrm{Mpc}$ where

$$x_{\mathrm{ls}} = \int_{t_{\mathrm{ls}}}^{t_0} \frac{c}{a}\mathrm{d}t \simeq 14\,\mathrm{Gpc}. \tag{25.31}$$

The two values (0.14 Mpc and 0.22 Mpc) are not the same because one is the radius of a typical sphere, the other is an average over diameters of cross-sections through spheres. A complete analysis confirms that they are mutually consistent. It is a beautiful confirmation of the ideas that the study of the galaxy distribution thus correlates with tiny variations in the CMB produced billions of years before. The CMB peak can now be used as a standard ruler in order to measure the average curvature of space, just as we discussed in the previous section. Owing to the precision of the data and to the much larger value of z, a greater overall precision is available.

The method hinges on our ability to model the early universe precisely. It might appear that this is very uncertain, but the main feature of interest—the distance scale—can be obtained by quite simple arguments based on the speed of sound in a plasma whose conditions are well within the range of energy and density that we are able to calculate with confidence. The result of the calculation can be summarized as

$$K \simeq \frac{H^2}{c^2}\left(\left(\frac{l_{\mathrm{peak}}}{220}\right)^2 - 1\right) \tag{25.32}$$

which, using (23.29) can also be written

$$\Omega \simeq (220/l_{\mathrm{peak}})^2. \tag{25.33}$$

The observed value is $l_{\mathrm{peak}} = 220$, therefore it can be inferred that the curvature is close to zero. The experimental bounds shown on Fig. 25.4 give $\Omega = 1.006 \pm 0.020$.

A lot of further information is available either directly from the CMB or by combining all the types of observation. After reionization there is a small but non-negligible scattering of the

CMB radiation on its long journey; this is parameterized in terms of the optical depth τ; its value can be inferred from the data and this in turn indicates when reionization happened. The acoustic peaks in the CMB power spectrum also indicate the relative amounts of baryonic (Ω_b) and dark matter (Ω_c). The ratio Ω_b/Ω_c can be inferred from the relative heights of odd and even acoustic peaks (odd peaks come from compression, even ones rarefaction). If there is a high baryon density then the compressions are stronger, emphasizing the odd peaks in comparison to the even peaks. If one relaxes the assumption that dark energy has the form of a cosmological constant, then one can also use the data to constrain the value of the state parameter w in the equation of state for dark energy (assuming it to be of the form (23.17))—see Table 25.1.

The CMB anisotropies are approximately 10% polarized and further information is available from the distribution of polarization. This partial polarization comes about when light scatters off small quadrupolar inhomogeneities and hence it probes tensorial (as opposed to scalar) fluctuations; these indicate the strength of the gravity wave background in the early universe (c.f. appendix D).

The total matter density parameter is $\Omega_m = \Omega_b + \Omega_c$; it can be found by combining all the information from galaxy clustering, supernovae and the CMB (Fig. 25.4). By combining this value with the observed small value for curvature, one then infers, in the ΛCDM model, that the difference $\Omega - \Omega_m$ must be owing to the Ω_Λ term. Hence the dark energy is not observed directly but its presence is inferred and the result is consistent with the supernova observations.

In view of the fact that we have little understanding of what exactly dark matter is, and little knowledge of how dark energy relates to particle physics calculations, it is proper to view the ΛCDM model as indeed a model rather than a clearly delineated physical theory. However, it has held up well under scrutiny and does a good job of capturing large amounts of information about the cosmos in terms of a small number of parameters.

Extensive numerical modelling of the growth of structure in the universe is consistent with the type and amount of dark matter implied by the observations of the CMB and galaxy clustering. Further information is available from measurements of rotation curves and gravitational lensing. Thus the evidential basis for dark matter is very strong. We lack primarily knowledge of what non-gravitational effects, if any, it has, and of the details of its spatial distribution on scales small compared to the size of a galaxy.

Dark energy is harder to study. (Durrer, 2011) The error bounds reported in Table 25.1 do not make allowance for the slight inhomogeneity in the universe, so it is possible that the value of Λ lies well outside those bounds. This possibility is being actively studied, along with other ideas. Nonetheless, studies to date suggest that the value $\Lambda = 0$ is ruled out. Elucidation of the nature of dark energy is one of the main ongoing research aims of contemporary cosmology.

Exercises

25.1 If $H_0 = 70 \, \text{km s}^{-1}/\text{Mpc}$ and the parameters of a ΛCDM model were $\Omega_{m0} = 0.8$, $\Omega_{r0} = 0$, $\Omega_{\Lambda 0} = 0.8$ then how old would the universe be? If the age is 13.8 billion years and $\Omega_{m0} = 0.3$, $\Omega_{r0} = 0$ then what is the value of $\Omega_{\Lambda 0}$? [You may use Fig. 25.2 in the first instance, but for a full answer you could write a computer program to carry out the numerical integration.]

25.2 Find $a(t)$ in a universe with constant deceleration parameter q. [*Ans.* $a \propto t^n$ with $n = 1/(1+q)$]

25.3 Obtain (25.12).

25.4 Obtain the expression for apparent magnitude m in terms of M and L.

25.5 Use the data in Fig. 25.5 to estimate K_0. (Adopt the values $H_0 = 70 \, \text{km s}^{-1}/\text{Mpc}$, $s = 150 \, \text{Mpc}$).

25.6 Show that if there is isotropic thermal radiation of temperature T_0 in an inertial frame S_0 in flat spacetime, then the radiation will also be thermal in any given direction in a frame S moving with respect to S_0, but now with an anisotropic temperature

$$T = \frac{T_0}{\gamma(1 - (v/c)\cos\theta)} \qquad (25.34)$$

where $\mathbf{v}$ is the relative velocity of the frames and θ is the angle between $\mathbf{v}$ and the direction of observation in S. The CMB radiation observed from Earth has $T_{\max} - T_{\min} \simeq$

$6.5 \times 10^{-3} \, \text{K}$ and $(T_{\max} T_{\min})^{1/2} \simeq 2.73 \, \text{K}$. How fast is Earth moving with respect to the frame in which the radiation is isotropic, and how does that frame relate to (i) the source of the radiation, (ii) the average Hubble flow, and (iii) the notion of a preferred reference frame?

25.7 Last scattering took place over a range of times (a modest but non-zero range), therefore different CMB photons originate from a plasma at different temperatures. How then can the radiation be so accurately thermal now, with a single temperature?

25.8 Show that the cross-over between which of Sachs–Wolfe effect and density perturbation gives the dominant contribution to an observed temperature anisotropy occurs for structure wavelengths of the order of the Hubble distance at last scattering. What is the corresponding value of l?

25.9 In what respect are the CMB anisotropies random? In what respect are they not? Explain qualitatively how the non-random aspect of the anisotropies came about.

25.10 Revisit Olbers' paradox (exercise 23.18 of Chapter 23). We *do* receive radiation almost uniformly from the entire sky! But it is invoked as evidence of a non-static not a static cosmology. Why?

26

The very early universe

The last few decades have seen a large effort to explore what processes may have gone on in the very early universe, at temperatures and characteristic energies at or above the grand unification (GUT) scale (10^{29} K, 10^{16} GeV). The aim is to reduce the degree to which cosmology has to depend on simply quoting facts about the early universe. One would rather describe physical processes which link those facts together.

Two observed facts are that the universe is highly homogeneous and either spatially flat or nearly so. The puzzles set by these observations are called 'the horizon problem' and 'the flatness problem'. However, it is not clear that these observations are especially puzzling in comparison to other issues; we will comment on them below. The more important point is that the conditions in the very early universe, at times before $\sim 10^{-35}$ s and temperatures above $\sim 10^{29}$ K have to be *very* special in order to lead to the universe as we observe it to be, and we do not at present have a good understanding of this. The important parameter here is entropy.

Let us consider just the observable universe and make a very broad-brush estimate of what the possibilities are in terms of entropy. We know the observable universe is in a rather rich and interesting state at the moment, but how special is this state? What is the size of the total set of available states? To estimate this, note that if the mass (of order 10^{53} kg) of the observable universe had all been located in a single black hole, then, by using (21.25), the entropy would be of order 10^{100} J/K. This number gives an impression of the size of the state space that is in principle available to the matter of the observable universe. The state space must include

$$\exp(S/k_{\mathrm{B}}) \sim 10^{10^{122}} \tag{26.1}$$

states. This is an unimaginably large number, and out of this vast number only a miniscule fraction of states are anything like the universe we observe.

If we try to construct a parameter space of cosmological possibilities, including such things as density parameters and size of fluctuations, then a conservative estimate of the fraction of possible configurations at the GUT scale that are sufficiently smooth to lead to the observed situation at recombination is (Carroll, 2014b)

$$f(\text{sufficiently smooth at GUT scale}) \simeq 10^{-70\,000\,000}. \tag{26.2}$$

Relativity Made Relatively Easy: General Relativity and Cosmology. Volume 2. Andrew M. Steane,
Oxford University Press. © Andrew M. Steane 2021. DOI: 10.1093/oso/9780192895646.003.0026

This estimate is not intended to be precise; it serves merely to illustrate the fact that any such estimate yields a tiny number. It means that the universe is and has always been in a very special condition, and in our current state of knowledge we can only note this as a physical fact that we are unable to explain in terms of other physical facts. We shall refer to this as the *entropy problem*.

The concept of *inflation* was introduced with a view to elucidating some features of the universe, especially the fact that the universe is sufficiently flat to be long-lived, that it is close to homogeneous over observable distance scales, and that the perturbations to that homogeneity are of a certain kind. Another feature is also noteworthy, going by the name of *magnetic monopole problem*. When one studies quantum field theory at the GUT energy scale it is reasonable to expect a symmetry-breaking process that results in the observed strong and electroweak forces, and such a process is predicted to produce a large density of magnetic monopole particles throughout the universe, but none have ever been observed. This discrepancy can be interpreted either as a failure in our understanding of physics in this energy regime,[1] or as evidence that some further process intervened so as to disperse or destroy these monopoles. In some inflationary scenarios, the universe expanded by a huge factor during or after the production of monopoles, with the result that their density became vastly diluted.

In the following we will outline the broad ideas of inflationary cosmology. This has become a mainstream part of cosmology; textbooks are written about it and the word 'inflation' is often invoked as the precursor to a discussion of other matters, as if it is an agreed thing. However, this is far less settled knowledge than that discussed in the rest of this book. Therefore we shall present the ideas with the type of friendly scepticism which is part of the ordinary scientific mindset. Inflation is an interlude between two great unknowns—the unknown physics of the Planck era, and the unknown physics of the GUT era—and therefore a receptive but questioning attitude is appropriate. The mathematical ideas and methods can become established, and worthy of textbook treatment, whether or not they capture correctly the nature of the universe, but for physics we want a good case that they do. The major success of these ideas is that they lead to a model which correctly predicts the spectrum and statistics of the CMB anisotropies in several different respects. This gives hope that the ideas are on the right track. However, at the time of writing, no version of inflation has produced a model in which the conditions of our universe are predicted or likely. These models do not reduce the degree to which the initial conditions have to be special. In fact inflation not only does not solve the entropy problem; it makes it worse. The initial conditions necessary for getting inflation to start and finish in the right way are extremely fine-tuned—more so, it seems, than those of a Big Bang model which merely quotes initial conditions without inflation. As Carroll puts it, 'inflation "explains" the fine-tuned nature of the early universe by positing an initial condition that is even more fine-tuned.' (Carroll, 2014*b*) This is not a settled matter, however. Baumann, for example, asserts, 'Via inflation, the universe we know and love grew out of generic initial conditions.'(Baumann, 2012) These statements cannot both be right. I think the second one is adopting a restricted sense of the word 'generic'.

---

[1] Accelerator-based experiments have so far explored collision energies up to 10^4 GeV; inflationary cosmology is concerned with processes involving collision energies $\geq 10^{15}$ GeV.

The majority view among experts is that whereas the analysis of an early inflationary period remains somewhat tentative, it is not in serious doubt that such a period occurred. Inflation uses ideas that arise naturally in field theory, and it suggests a mechanism whereby fluctuations can arise that have all the properties inferred from the CMB anisotropies (nearly scale-invariant; Gaussian; adiabatic; and such as to give well-resolved acoustic peaks). Therefore something like this may have happened before the strong and electroweak forces separated. Even if inflation did not happen after the Planck era, its ideas may be capturing some aspects of the Planck era, especially the role of quantum physics in a dynamic spacetime in seeding inhomogeneities.

26.1 The horizon and flatness 'problems'

Since the horizon and flatness problems are often also mentioned in the context of inflation, we shall first discuss those.

26.1.1 The horizon problem

The particle horizon was defined in Section 23.4; it is the distance over which light-speed-limited causal connection has been possible in a given universe with a given history. A quick way to estimate the value of this horizon at the time of last scattering is to use the propagation of sound in the early universe. In our discussion of baryon acoustic oscillations, we gave reason to believe that the correlations observed in the CMB should appear on distance scales equal to the sound horizon. On the assumption that there was not time for light to travel very far before these sound waves started to propagate (an assumption we shall question in a moment), one will find that the horizon at last scattering is larger than the sound horizon at last scattering by approximately a factor $\sqrt{3}$. Hence we find that at last scattering the horizon corresponded to angular separations of order 1 or 2 degrees. But the CMB radiation has a thermal spectrum at very precisely the same temperature across the entire sky. How did regions separated by more than the horizon come to have the same temperature? Furthermore, there are patterns in the polarization of the CMB which are correlated with the temperature patterns over patches of the sky as large as $5°$. How could this occur unless there was causal contact?

This question is called the *horizon problem*. This has often been presented as a great mystery but the answer is straightforward. It is simply that the horizon has not been calculated correctly. One should not assume that a simple cosmology, such as ΛCDM with constant parameters, applies all the way to $a = 0$, since at sufficiently high densities and temperatures classical physics can no longer be trusted. Once one enters the Planck era all classical bets are off. So the 'horizon problem' is like the 'flying bees problem' (the supposed aerodynamic mystery of how can bees manage to fly). In both cases the problem is not so much a problem as a convenient way to draw attention to something which one ought not to overlook: in the bee case that bees flap their wings, in the cosmology case that quantum physics is not the same as classical physics. This suffices to change completely the prediction for causal connection in the very early universe. In order to explore the possibilities, one can propose different functional forms for $a(t)$ as $a \to 0$ in the horizon integral (23.57); two such possibilities are shown as the dashed lines in Fig. 22.4.

This is not intended to be precise physics; it is serving as a way to discover the degree of sensitivity of the horizon to the unknown physics of the pre-GR period. It is found that any power-law dependence $a \propto t^n$ with $n \geq 1$ in the limit $t \to 0$ will lead to $\chi_{\rm h} \to \infty$. This shows that the horizon is very sensitive to the dynamics and any value can be obtained (exercise 26.1). Therefore it would be unsurprising if the physics at work in the quantum gravity era allowed causal connection between comoving coordinate locations of arbitrary separation. Hence there is no horizon problem as such; there is only the already admitted problem that we do not have a good understanding of the pre-GR era.

A behaviour such as $a \propto t^n$ with $n > 1$ would itself be termed 'inflation', so one may say that inflation is here being advocated. The pertinent point is that the 'traditional' Big Bang model should not be understood as a model which ignores the complexity of the Planck era and just blindly applies classical GR all the way to $a = 0$. Rather, all agree that the Planck era is special; what is unknown is what happened in the Planck era, and whether or not inflation along the lines of existing models occurred, either then or later.

With this in mind, the real problem is not whether or not the universe was causally connected, but why it was so homogeneous. If it became firmly established that in fact the entire observable universe was causally connected at an early stage, as seems plausible, then this would not in itself guarantee that the result of that causal connection would be a nearly uniform temperature and density. Gravitation characteristically leads to instability and non-uniformity. The observed uniformity would remain equally as puzzling as it already is.

In conclusion, the horizon problem can be expressed as follows:

1. A naive argument based on classical GR with radiation-dominated early conditions leads to a small horizon, but there is no good reason to apply such an argument to the actual universe because classical physics is inadequate in the early conditions.
2. The scientific puzzle here is the fact that the initial conditions of the universe were special, including a remarkable degree of homogeneity. A physical process which allows the horizon to be large does not necessarily explain this homogeneity.

It is tempting, when pondering the CMB radiation with its near-uniform temperature, to think that the universe somehow relaxed to thermal equilibrium in its very early evolution. But there was not time for that to happen, at the larger scales, by the usual process of energy exchange by radiation and collisions. Therefore the thermal nature of the CMB should not be interpreted purely as a result of thermal relaxation (which in any case does not lead to homogeneity when gravity is involved). The hint offered by GR and quantum field theory is that the conditions came about by another type of process altogether.

26.1.2   The flatness problem

The supposed 'flatness problem' is observed in the fact that any spatially flat solution to the Friedmann equation is dynamically unstable. If one studies the curvature density Ω_k in a ΛCDM cosmology, one finds that it satisfies

$$\Omega_{\rm k} = a^{-2} \frac{H_0^2}{H^2} \Omega_{\rm k0} \tag{26.3}$$

and in the radiation-dominated era $H \sim a^{-2}$ so one has $\Omega_{\rm k} \sim a^2 \Omega_{\rm k0}$. Therefore, in order that $\Omega_{\rm k}$ be small now, it must have been tiny at early times (small a). Now recall that $\Omega_{\rm k} = 1 - \Omega$. What caused Ω to be so close to 1? It seems odd, goes the argument, that the universe should have been so close to its critical density.

In fact this argument is spurious because it does not offer a reason to say why one value of $1 - \Omega$ is more or less to be expected than another. However, it does highlight the interest in discovering, if we can, how this parameter value is connected to other properties, and what sequence of events led to its getting the value it has.

The nearly-flat nature of the large-scale spatial geometry of the cosmos is worthy of remark, but one should not assume that far-from-flat geometries would be typical if one could some-how draw a universe at random from some sort of collection. To see this, one requires a way to assess what probability distribution ought to be associated with the parameter values in question (in the absence of a theory which can specify them). In formal terms, one requires a *state space*, a *course-graining*, and a *measure*. The state space is a space with one dimension per independent parameter of the system under investigation; the course-graining is a way to say which states in a continuous space are like one another; the measure is a function like a probability density which says how much weight should be assigned to each course-grained cell in the space. The construction of these quantities is not a well-developed science in the case of cosmology, but one may make reasonable arguments based on simple models and thus find out whether the case of small curvature gets a large or a small weight in the space of all possible cosmologies. Such a calculation is described by Carroll, who finds that far from being exceptional, low curvature is typical: rather than sufficiently flat universes being rare, they are generic (Carroll, 2014b). On this argument, the surprise would be if the universe were not flat or almost flat.

Such a calculation need not be precise in detail; it is sufficient to our present purpose merely to observe that, for all we know, very low curvature might as well be expected as not. The flatness problem is, then, not that we know flatness to be remarkable, but that we do not know whether it is remarkable or not.

We have not presented the details of the rather technical argument required to show this, but a useful further insight is provided by examining the trajectory of the state of the universe in the parameter space furnished by $\Omega_{\rm m}$ and Ω_Λ in a ΛCDM cosmology: see Fig. 26.1. If we set $\Omega_{\rm r} = 0$ then we can express (23.29) and (23.30) in the form (exercise 26.2)

$$\left.\begin{aligned} \Omega_\Lambda' &= (\Omega_{\rm m} - 2\Omega_\Lambda + 2)\Omega_\Lambda \\ \Omega_{\rm m}' &= (\Omega_{\rm m} - 2\Omega_\Lambda - 1)\Omega_{\rm m} \end{aligned}\right\} \tag{26.4}$$

where the prime indicates the derivative with respect to η, where $\eta \equiv \ln a = -\ln(1+z)$. One may then show that the quantity

$$\frac{\Omega_{\rm m}^2 \Omega_\Lambda}{(1 - \Omega_{\rm m} - \Omega_\Lambda)^3} \tag{26.5}$$

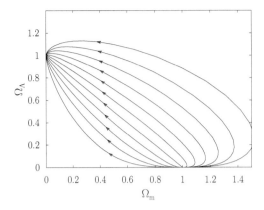

Fig. 26.1 A selection of paths through the Ω_m, Ω_Λ plane, as described by equations (26.4).

is a constant of the motion. The lines on Fig. 26.1 show example solutions to these equations, marked by an arrow in the direction of increasing η (so increasing a, or evolution forward in time). This study shows that when a is small the trajectory does initially evolve away from the line $k = 0$ (which is the line $\Omega_\Lambda = 1 - \Omega_\mathrm{m}$), but subsequently it is drawn back to that line through the Ω_Λ term. Also, no matter which trajectory one picks, it will tend to the point $(\Omega_\mathrm{m}, \Omega_\Lambda) = (1, 0)$ as $a \to 0$, so there is no cause for surprise that this is a property of our universe.

Put simply, the flatness problem was the problem that the universe evolves quickly away from $\Omega = 1$ at early times, but when $\Omega_\Lambda > 0$ it generically does not get far from $\Omega = 1$ in its later evolution, so it is not so surprising to find it near that value after all.

This is not to say that there is no mystery here or nothing to explain. The puzzle could better be described as a *longevity problem*. The density of the universe is high enough to provide interesting phenomena such as stars and planets and life, but not so high as to cause it all to collapse again under its own gravity before life could develop. In this sense the universe is finely balanced. (Adler and Overduin, 2005; Lake, 2005; Helbig, 2012)

26.2 Inflation

The notion of cosmological inflation has two main components: first, the general idea that a cosmological constant term in the Friedmann equation will result in a period of accelerating expansion if it dominates, and this will result in a larger horizon. The second component is the general idea that quantum field theory, on a background spacetime described by GR, offers a rich variety of possibilities for what the early universe may have been like, and amongst these possibilities it is easy to construct models in which quantum fields give rise to metric perturbations of the kind that seed structure formation in the universe.

Let us define the term 'inflation' by:

Definition of inflation

$$\ddot{a} > 0. \tag{26.6}$$

That is, by inflation we mean generically any situation in which the cosmological expansion is accelerating. An equally good (and equivalent) definition would be to say that inflation is the case where the comoving Hubble distance $c/(aH)$ gets smaller as time goes on. In a single perfect fluid model either definition implies the condition

$$p < -\rho c^2/3 \qquad \text{or} \qquad w < -1/3. \tag{26.7}$$

Having noted the role of Higgs fields in the Standard Model of particle physics, let us now posit a scalar field ϕ described by a Lagrangian density

$$\mathcal{L} = -\tfrac{1}{2}g^{\mu\nu}(\partial_\mu\phi)(\partial_\nu\phi) - V(\phi) \tag{26.8}$$

where $V(\phi)$ is some sort of potential energy function which we shall leave unspecified for the moment. (For example, a simple case would be $V = \tfrac{1}{2}m^2\phi^2$ which would lead to the Klein–Gordon equation for ϕ; c.f. Chapter 27). This proposed new field is called the *inflaton* field. If it exists, then it exists even now, undetectable to our instruments because it is now in a very low-lying state and its excitations correspond to particles of high mass (possibly, but we do not know). We shall explore its dynamics using classical field theory in the first instance.

First we find the equation of motion. It is a type of Klein–Gordon equation, but one that incorporates the effect of an expanding space. We do not know the spatial curvature, but if we take it to be zero then this will likely be a good approximation in the later part of the dynamics; in any case it is the simplest case so in the interests of getting some general understanding we assume it. Then the equation of motion in comoving coordinates is (exercise 26.3)

$$\ddot{\phi} + 3H\dot{\phi} - \frac{c^2}{a^2}\nabla^2\phi + \frac{\mathrm{d}V}{\mathrm{d}\phi} = 0. \tag{26.9}$$

The solution of this classical equation, and its quantized cousin, is the main theme of the rest of this chapter.

Next we find the energy tensor; it is given by (28.9):

$$T_{ab} = (\partial_a\phi)(\partial_b\phi) - g_{ab}[\tfrac{1}{2}(\partial_\lambda\phi)(\partial^\lambda\phi) + V(\phi)]. \tag{26.10}$$

Let us compare this to the energy tensor for a perfect fluid. The comoving frame is the one in which any given fluid element is at rest; in this frame one finds[2] (setting $c = 1$)

---

[2] The derivation here requires some care; (Weinberg, 2008) I am grateful to Steven Balbus for pointing this out. One way to justify the result is to show that it is consistent with the equation of motion.

$$\rho_\phi = \tfrac{1}{2}\dot{\phi}^2 - \tfrac{1}{2}\nabla^2\phi + V(\phi),$$
$$p_\phi = \tfrac{1}{2}\dot{\phi}^2 - \tfrac{1}{2}\nabla^2\phi - V(\phi). \tag{26.11}$$

Therefore if ϕ is providing the dominant contribution to the total T_{ab} of some part of (or all of) the universe, then the condition (26.7) for inflation of that part is

---

Inflation condition when ϕ dominates and is homogeneous

$$\dot{\phi}^2 < V(\phi) \tag{26.12}$$

In the limit $\dot{\phi}^2, \nabla^2\phi \ll V(\phi)$ the relation between the pressure and density becomes $p_\phi \simeq -\rho_\phi$ so the inflaton fluid mimics dark energy.

In the first instance we will treat the homogeneous case, where ϕ evolves with time but has no spatial structure so $\vec{\nabla}\phi = 0$. In this case (26.9) becomes[3]

$$\ddot{\phi} + 3H\dot{\phi} + \frac{dV}{d\phi} = 0. \tag{26.13}$$

The form of this equation is reminiscent of a simple scenario in classical mechanics: a particle at position ϕ moving in a potential V, with H providing a damping term sometimes called 'Hubble friction'. H is itself time-dependent, so in order to specify the behaviour of ϕ we require a further equation: the first Friedmann equation (23.7), which gives

$$H^2 = \frac{8\pi G}{3}\left(\tfrac{1}{2}\dot{\phi}^2 + V(\phi)\right) - \frac{k}{R_0^2 a^2} \simeq \frac{8\pi G}{3}\left(\tfrac{1}{2}\dot{\phi}^2 + V(\phi)\right). \tag{26.14}$$

The second version is valid when the curvature term can be neglected.

26.2.1   The slow-roll approximation

Equations (26.13) and (26.14) are a pair of coupled differential equations which determine the behaviour of ϕ and H, and hence a, when ρ_ϕ, p_ϕ provide the dominant contributions to the energy tensor.

In order that the inflation should last long enough to have a large effect, the condition (26.12) has to hold for long enough as the field evolves. This can be achieved if initially the condition is strongly satisfied, i.e.

$$\dot{\phi}^2 \ll V(\phi) \qquad\qquad \text{slow roll} \tag{26.15}$$

Let us now examine the evolution in this *slow roll approximation*. The name refers to the picture in which ϕ is like the position of a particle 'rolling' along the potential curve provided by V.

---

[3] (26.13) can also be derived by substituting (26.11) into the homogeneous continuity equation (23.14).

The condition simplifies (26.14) and also implies that $2\dot\phi\ddot\phi \ll \dot V$ and hence $\ddot\phi \ll \mathrm{d}V/\mathrm{d}\phi$, which simplifies (26.13). Hence we have

$$3\dot\phi H \simeq -\frac{\mathrm{d}V}{\mathrm{d}\phi}, \tag{26.16}$$

$$H^2 \simeq \tfrac{1}{3}V \tag{26.17}$$

where we have adopted units where $8\pi G = 1$. By dividing the first equation by the second and squaring, we obtain

$$\left(\frac{1}{V}\frac{\mathrm{d}V}{\mathrm{d}\phi}\right)^2 \simeq \frac{\dot\phi^2}{H^2} \ll 1 \tag{26.18}$$

(since $H^2 = V/3$ and $\dot\phi^2 \ll V$ under the slow roll assumption). By differentiating this result one finds that it requires $V'' \ll V$. Hence we have identified three requirements for slow roll:

Slow roll conditions

$$\dot\phi^2 \ll V; \quad V' \ll V; \quad V'' \ll V. \tag{26.19}$$

These conditions imply that the potential must be very flat at the initial value of ϕ; this is an example of an assumption of fine tuning.

If the evolution is such that V stays roughly constant during some period, then for that period we have from (26.17) that the Hubble parameter stays constant and therefore the scale factor grows exponentially, as in (23.46)—the de Sitter universe. This exponential growth is a characteristic feature of the theory of inflation, and it leads to a universe in which the total density parameter is close to 1, and therefore the spatial curvature is small; c.f. Fig. 23.4. The simplified argument presented in this chapter assumed spatial flatness at the outset, but if one drops that assumption then inflationary evolution still generically tends towards flatness because the dark-energy-like term in the energy tensor will always dominate after sufficient expansion. Hence the observed near-flatness of the universe now is taken as a modest item of evidence for inflationary cosmology. The curvature predicted at the end of inflation will be extremely small either if inflation persists for long enough, or if the curvature was already quite small at the start of inflation.

A significant feature of exponential growth is that it is *slower* than power-law growth at small t, when the power is less than 1 (Fig. 26.2). Therefore in the early part of an exponential expansion the universe first 'lingers' at small a which gives time for the particle horizon to grow. The universe thus becomes causally connected over a larger range.

So far we have shown that the application of ordinary classical field theory to a simple scalar field can lead to a joint evolution of the field and spacetime in which the latter tends to the condition $\Omega = 1$ and has a large particle horizon. The reason for this is the presence of the term proportional to the metric in the energy tensor (26.10) which leads to the field behaving like a cosmological constant: its total energy (the integral of ρ over volume) grows as space does. The next important ingredient is that this growth does not necessarily persist. The equation of motion is such that the field 'rolls' *down* the potential and therefore, for many forms of V, at

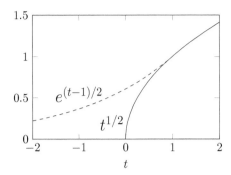

Fig. 26.2 Exponential growth is slower than the alternative.

some point the inflation condition (26.12) will no longer be satisfied and inflation ceases. The friction term guarantees that ϕ will not simply escape such a well, instead it undergoes damped oscillations. There are also models in which inflation may cease in one part of the universe while it does not in another, and models in which the process cannot fully cease but forever spawns newly inflating patches. This idea (called eternal inflation) then runs into the difficulty of it being unclear whether it is truly predictive.[4]

The next part of inflationary cosmology is called *reheating*. It is assumed that the 'inflaton' field ϕ eventually transfers all or part of its energy to other fields—the quark, lepton and other fields of the Standard Model. It is natural to assume that all these fields are coupled and therefore one field can transfer energy to another through decay and pair creation. The timing is such that this decay is rapid, with the characteristic time of the GUT regime, but the inflation itself is more rapid. Timescales of order 10^{-34} s or less are commonly invoked. There is a difficulty, however, which is the requirement for a 'graceful exit'. In order that this scenario can correctly describe the observed universe we require that the inflaton field should pass not only energy but also momentum to the more familiar forms of matter and radiation, and it must do so with just the right pattern to produce the homogeneous expansion (the *Hubble flow*) we observe.

26.2.2 The power spectrum

We now address the question of the power spectrum, and other statistical properties, of the primordial fluctuations. The calculation invokes quantum field theory, and we shall not present it in full, but we shall offer a few highlights, following the treatments of Baumann and of Tong.(Baumann, 2012; Tong, 2019) The overall idea is illustrated in Fig. 26.3, in which we have adopted comoving coordinates and sketched a picture of some patch of the universe. In an inflationary period H is constant, or approximately constant, and therefore the comoving Hubble distance $d_{H,c} = c/(aH)$ is getting smaller. Quantum fields such as the inflaton field have

---

[4]This is a non-trivial metaphysical puzzle. If a theoretical model suggests that most of physical reality is unobservable to us, and the part we do observe is a highly untypical part of the whole, then how is this different from a model which is simply wrong?

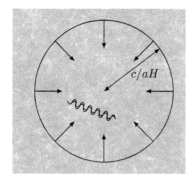
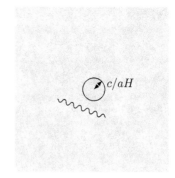

Fig. 26.3 The evolution of the inflaton field and the Hubble distance during inflation, in comoving coordinates. The comoving Hubble distance reduces as a increases. The quantum uncertainty of a field amplitude component of given k first decreases and then stays constant; each comoving wavevector k is constant.

the characteristic feature that their amplitude and phase cannot simultaneously have completely precise values (the Heisenberg Uncertainty Principle). In particular, the field amplitude for any given Fourier component has non-zero variance $\langle \hat{\phi}_{\mathbf{k}} \hat{\phi}_{\mathbf{k}}^{\dagger} \rangle$ where $\hat{\phi}_{\mathbf{k}}$ describes departures from the mean (for each component the mean amplitude is zero). This is the property commonly called 'quantum fluctuation', but one should recall the remarks about this terminology in Section 21.3. The field does not fluctuate in the classical sense of the word; rather, its quantum state is not an eigenstate of field amplitude. More precise terms are *variance* and *correlation*, and indeed it is the correlation function

$$\langle \hat{\phi}_{\mathbf{k_1}} \hat{\phi}_{\mathbf{k_2}}^{\dagger} \rangle \tag{26.20}$$

which is the quantity of interest.

Subsequently, as time goes on the comoving Hubble distance rapidly gets smaller, while the k values are constant, so the correlations at each distance scale rapidly pass 'outside the horizon', i.e. k falls below (aH/c). The correlations with that k are then 'frozen', i.e. unchanging with time, until the expansion begins to decelerate after the end of inflation. We showed in Chapter 24 that in the subsequent radiation-dominated epoch, fractional density perturbations with long distance scales grow by gravitational collapse, while the average density falls, such that metric perturbations are constant. But keep in mind that the behaviour of matter of all kinds is not well known at these extreme energies, that of dark matter least of all. The main idea is that the quantum correlations in the hypothesized inflaton field cause correlations in the metric, and these metric correlations become classical metric perturbations, which serve as the initial conditions for the growth of structure described in Chapter 24.

The movement from quantum correlation to classical perturbation is itself a non-trivial issue, related to the quantum measurement problem. Quantum physics leads to *superpositions* of states of well-defined field amplitude, rather than a classical mixture of states, but the final outcome is a classical mixture of density perturbations. It is like a Schrödinger cat thought-experiment writ large across the sky. Formally, one notes that the quantum behaviour becomes indistinguishable from classical statistical behaviour when field operators commute with conjugate momentum operators; this happens on scales above the Hubble distance (or de Sitter radius—they are the same in de Sitter spacetime) owing to a form of decoherence. There is a strong hint here that the dynamics of the physical world are not completely unitary or deterministic; this is an open question.

Another way of describing the same physics is to invoke the idea of a thermal state. A generic property of quantum field theory in an expanding spacetime is that when the expansion gives rise to a horizon, such as the de Sitter horizon, then the vacuum state has thermal characteristics. It has a temperature, the Gibbons–Hawking or de Sitter temperature, analogous to the Hawking temperature of a black hole. This temperature T is proportional to the Hubble parameter H, and it gives rise to (or is a shorthand for) the presence of the correlations under discussion.

The calculation of the correlation function involves first classical GR perturbation theory, much as in appendix D, and a judicious choice of gauge-invariant property of the metric and inflaton field ϕ together. We shall skip this part and pretend that it is sufficient to examine the inflaton field itself. Here ϕ can be regarded as a stand-in for a gauge-invariant property involving a sum of ϕ and a metric perturbation; it still obeys the equation of motion (26.9). For this calculation we retain the $\nabla^2\phi$ term but for simplicity we will drop the $dV/d\phi$ term; this amounts to a very slow roll which leads to a de Sitter spacetime. For each Fourier component $\phi_\mathbf{k}$ we then have

$$\ddot{\phi}_\mathbf{k} + 3H\dot{\phi}_\mathbf{k} - (k/a)^2\phi = 0 \tag{26.21}$$

where $\mathbf{k}$ is the comoving wavevector (so $\mathbf{k}/a$ is the physical wavevector; c.f. Chapter 24). We would like eventually to quantize this expression, but as it stands the friction term is making this hard to do. We deal with this problem by two changes of variable, first to conformal time and then by rescaling the field amplitude with a factor a. That is to say, first introduce[5]

$$\tau \equiv \int \frac{dt}{a}, \qquad \frac{d\tau}{dt} = \frac{1}{a}. \tag{26.22}$$

Note that for de Sitter spacetime H is constant and then

$$\tau = -1/(aH). \tag{26.23}$$

The conformal time τ begins at $\tau \to -\infty$ when $a \to 0$, and extends to $\tau \to 0_-$ when $a \to \infty$; see Fig. 26.4. In fact we only expect the de Sitter dynamics to continue until inflation ceases at some finite a, and a more precise treatment need not assume an infinite conformal past. The important point is to notice that τ is negative and there is 'plenty of time' (conformal time that is) for the processes under discussion. With this change of variable (26.21) becomes

$$\phi_\mathbf{k}'' - \frac{2}{\tau}\phi_\mathbf{k}' + k^2\phi_\mathbf{k} = 0. \tag{26.24}$$

---

[5]We employ the letter τ for convenience here, but it is not proper time. It is simply a time-related variable that increments more and more rapidly as a gets small.

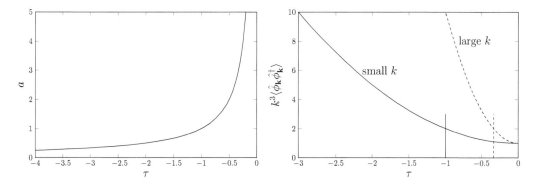

Fig. 26.4 Left: the scale factor as a function of conformal time in a de Sitter universe (arbitrary units). Right: the variance given by (26.32) multiplied by k^3, as a function of conformal time, for two illustrative values of k, in arbitrary units. The vertical lines indicate when $|k\tau| = 1$.

We still have a damped oscillator; we now introduce a rescaling that magically turns it into an undamped oscillator with a time-dependent frequency. Define

$$v_\mathbf{k} \equiv a\phi_\mathbf{k} = -\phi_\mathbf{k}/(H\tau). \tag{26.25}$$

Then after some simple algebra one finds

Classical equation for amplitude of v modes

$$v_k'' + \left(k^2 - \frac{2}{\tau^2} \right) v_k = 0 \tag{26.26}$$

This is the equation of motion for a harmonic oscillator with a time-dependent frequency:

$$\omega_k^2 = k^2 - \frac{2}{\tau^2} \tag{26.27}$$

Starting from $\tau \to -\infty$, ω_k begins almost constant and we have an ordinary oscillator. Then as time goes on $|\tau|$ gets smaller and the frequency falls—imagine a ball oscillating in a bowl that is being opened out. Eventually ω_k goes to zero and then becomes imaginary as the bowl is inverted and then the solutions grow or decay without oscillating. Note that (26.23) tells us that τ is equal to the comoving Hubble distance (often called comoving horizon) and the cross-over between the two types of behaviour takes place when $\tau \simeq 1/k$, i.e. when the mode 'exits the horizon'.

We can solve (26.26) without much difficulty; the general solution is

$$v_k = Ae^{-ik\tau} \left(1 - \frac{i}{k\tau} \right) + Be^{ik\tau} \left(1 + \frac{i}{k\tau} \right) \tag{26.28}$$

where A and B are constants of integration. Observe that for τ large and negative we have ordinary oscillation, but as $|k\tau|$ approaches zero the amplitude blows up. Using (26.25) we find that for the inflaton mode the behaviour is the other way round:

$$\phi_k = -A\frac{H}{k}e^{-ik\tau}(k\tau - i) - B\frac{H}{k}e^{ik\tau}(k\tau + i). \tag{26.29}$$

These modes have a huge amplitude at $\tau \to -\infty$ but finish with ordinary fixed-frequency, fixed-amplitude oscillation as $\tau \to 0$. The last cycle of oscillation begins at $k\tau = -2\pi$; after the mode exits the horizon at $k\tau \simeq -1$ there is almost no further evolution.

There is a difficulty with these huge values in the limit $\tau \to -\infty$; they suggest our model must break down there. The hope is that the treatment of v_k is trustworthy, and we only need to make the connection to ϕ_k at late times when its behaviour is credible.

Our discussion since (26.21) has been entirely classical. Now we would like to quantize the treatment, and the advantage of (26.26) is that it offers a natural way to do it. We promote v_k to a quantum operator

$$v_{\mathbf{k}} \to v_k(\tau)\hat{a}_{\mathbf{k}} + v^*_{-k}(\tau)\hat{a}^\dagger_{-\mathbf{k}} \tag{26.30}$$

where the mode functions v_k are normalized such that the $a_{\mathbf{k}}$ and $a^\dagger_{\mathbf{k}}$ satisfy canonical commutation relations. This is an important part of the analysis but we shall merely quote the answer. It depends on the behaviour of the metric and on the initial conditions; in the case of de Sitter space one finds

$$v_k = \sqrt{\frac{\hbar}{2k}}e^{-ik\tau}\left(1 - \frac{i}{k\tau}\right). \tag{26.31}$$

Finally one computes the correlation function of the field $\hat{\phi}_{\mathbf{k}} = \hat{v}_{\mathbf{k}}/a$:

$$\langle\hat{\phi}_{\mathbf{k_1}}(\tau)\hat{\phi}^\dagger_{\mathbf{k_2}}(\tau)\rangle = (2\pi)^3\delta(\mathbf{k}_1 - \mathbf{k}_2)\frac{|v_k(\tau)|^2}{a^2}$$

$$= (2\pi)^2\delta(\mathbf{k}_1 - \mathbf{k}_2)\frac{\hbar H^2}{2k^3}\left(1 + k^2\tau^2\right) \tag{26.32}$$

Keep in mind that each $\hat{\phi}_{\mathbf{k}}$ here is not the whole inflaton field, it is one Fourier component of the field, or, more strictly, of a gauge-invariant variable related to both the field and the metric.

(26.32) is a central result in inflationary cosmology. At early times, when τ is large and negative, the variance is large (see Fig. 26.4); for late times $\tau \to 0_-$ it becomes constant in time, where here 'late' means after the process has continued enough to make $|k\tau|$ small; the model assumes this occurs before the inflaton falls into a potential well and passes its energy to other fields whose equation of state is the more 'ordinary' $w \simeq 1/3$, leading to the Big Bang.

For $|k\tau| \ll 1$ and constant H (26.32) yields $P_\phi(k) \propto 1/k^3$ which corresponds to a scale-invariant power spectrum (c.f. (24.53)). By allowing that H is not completely constant and by setting τ according to when each mode exits the horizon one finds that the spectral index is predicted to be slightly smaller than one. To lowest order it is

$$n_s = 1 - 2\epsilon \tag{26.33}$$

where

$$\epsilon \equiv -\frac{\dot{H}}{H^2} \; = \; \frac{M_{\mathrm{Pl}}}{2}\left(\frac{V'}{V}\right)^2. \tag{26.34}$$

This is called the *slow roll parameter*; $M_{\mathrm{Pl}} = (\hbar c/8\pi G)^{1/2}$ is the Planck mass.

We have sketched here the highlights of a calculation involving scalar perturbations to the metric (in the terminology of appendix D). One also finds that tensor perturbations (i.e. gravitational waves) are expected, and a prediction of their relative strength in comparison to the scalar modes can be made. It is eagerly hoped to test this prediction when studies of tensor modes of the CMB become more precise.

26.2.3 Lessons of inflationary cosmology

Inflationary cosmology is a theoretical research effort which has made contact with experimental observations, but there remain many unresolved questions. In the course of the past 20 years, many models have been proposed, with names such as old, (R^2), new, chaotic, extended, power-law, hybrid, natural, extranatural, eternal, D-term, F-term, brane, oscillating, trace-anomaly driven, ... These differ in the functional form proposed for V, which may itself depend on parameters such as temperature in addition to ϕ, and the models may differ in other respects, such as invoking a tensor field or more than one scalar field. It was first suggested that the inflation took place towards the lower end of the GUT energy range, but versions taking place closer to the Planck energy have subsequently been proposed.

The illustrative calculation we have presented assumed zero average spatial curvature from the start; clearly if one hopes to argue that flatness is derived not assumed then one must embark on a more general calculation. It is not yet known how to do this for quantum superpositions of highly curved spaces; it would involve a much more confident knowledge of quantum gravity than we have yet attained. The conditions at the start of inflation are either in the Planck era or are the outcome of it. Therefore they are largely unknown. For all we know the particle horizon could already have been infinite before any of the physical models we can apply with confidence became valid!

The chief evidence for the correctness of inflationary cosmology is its success in predicting the size and spectral form of the CMB anisotropies, assuming that the reheating process is sufficiently uniform. The simpler models of inflation predict all the properties listed in Section 25.5.3 (nearly scale-invariant; Gaussian; adiabatic; and such as to give well-resolved acoustic peaks). The observation of these properties is therefore notable indirect evidence for inflation. It does not amount to a confirmation because those same data will also support any other model that can predict them. In particular, the presence of resolved acoustic peaks does not amount to evidence for inflation *per se*, but merely for the presence of metric perturbations on distance scales above the Hubble distance at very early times. Alternative non-inflationary models include some form of 'bounce' cosmology, where before the Big Bang the universe was in a collapsing state, or else simply the assertion that Planck-era physics resulted in a spacetime and matter structure with metric perturbations of the form required to give rise to the observed behaviour. Since a finite-temperature vacuum is a natural aspect of quantum field theory in an expanding space, one would expect it to feature in any model of the very early universe. This would account for the acoustic peaks and other features of the CMB, but not the absence of magnetic monopoles; this absence is then interpreted as empirical counter-evidence to the grand unified theories that predict a large density of monopoles.

Research in this area of cosmology naturally links with efforts to develop a quantum theory of gravity. There are many unknowns, but given the simplicity of the basic concept of inflation, it is reasonable to conjecture that effects like those we have described may have taken place in an epoch when quantum physics and gravity were intimately coupled.

Exercises

26.1 Suppose that in very early times the scale factor grew linearly, as $a = A(t + t_\mathrm{p})$ where A and t_p are constants, and then subsequently it grew as $a = (2H_0 t)^{1/2}$ (the radiation dominated form). Show that the conditions for both a and $\dot{a}$ to be continuous are that $A = (H_0/2t_0)^{1/2}$ and that the switch between the two functional forms occurs at $t = t_\mathrm{p}$. Show also that the particle horizon is infinite in this case.

26.2 Taking $\Omega_r = 0$, obtain (26.4) from (23.29) and (23.30) [Hint: H^2 can be written in terms of Ω_Λ].

26.3 *The equation of motion for ϕ.*
(i) A covariant formulation of field theory is introduced in Chapter 28; all we require here is eqn (28.27). The first term is a covariant divergence of a vector field; use (11.38) to evaluate it for the metric $\mathrm{d}s^2 = -c^2\mathrm{d}t^2 + a^2(t)(\mathrm{d}x^2 + \mathrm{d}y^2 + \mathrm{d}z^2)$ and hence obtain (26.9).
(ii) *The reader unfamiliar with Lagrangian methods in field theory will need to become acquainted with Chapter 28 before attempting this method.* In a flat space we can use Special Relativity, and then the action integral for (26.8) is

$$S = \int \mathrm{d}t\mathrm{d}x^3 \left[\tfrac{1}{2}\dot{\phi}^2 - \tfrac{1}{2}c^2\vec{\nabla}\phi \cdot \vec{\nabla}\phi - V(\phi)\right].$$

Show that if we change coordinates to comoving coordinates, then the integrand becomes

$$a^3(t)\left[\frac{1}{2}\dot{\phi}^2 - \frac{c^2}{2a^2(t)}\vec{\nabla}\phi \cdot \vec{\nabla}\phi - V(\phi)\right].$$

Show that a variation of ϕ gives

$$\delta S = \int \mathrm{d}t\mathrm{d}x^3 a^3(t)\left[\dot{\phi}\delta\dot{\phi} - \frac{c^2\vec{\nabla}\phi \cdot \vec{\nabla}\delta\phi}{a^2(t)} - \frac{\partial V}{\partial \phi}\delta\phi\right]$$

and that this is equal to

$$\int \mathrm{d}t\mathrm{d}x^3 \left[-\frac{\mathrm{d}}{\mathrm{d}t}(a^3\dot{\phi}) + c^2 a\nabla^2\phi - a^3\frac{\partial V}{\partial \phi}\right]\delta\phi \tag{26.35}$$

[Hint: consider $\partial_\mu((\delta\phi)\partial^\mu\phi)$ and integration by parts.] Hence obtain (26.9).

26.4 Given that some low-level noise of all forms would be an expected outcome of early-universe processes, one should expect that stochastic gravity waves in particular were produced. What features of such a gravity wave background can be used to rule out one account or another of their origins?

Part V

27

First steps in classical field theory

We conclude the book with two chapters on classical field theory. (Landau and Lifshitz, 1971) A full discussion of this subject would require a text book in its own right. Our aim here is to introduce some field equations that are close cousins of the wave equation, especially the Dirac equation, and to introduce Lagrangian methods for fields, spinors, and the connection between symmetry and conservation. The discussion is restricted to flat spacetime (Special Relativity) in the whole of the present chapter and most of the next, but our final destination is a treatment of GR using Lagrangian methods. We will adopt the bold font for 3-vectors and index notation for 4-vectors.

Classical field theory deals with the general idea of a quantity that is a function of time and space, which can be used to describe wave-like physical phenomena such as sound and light, or other continuous phenomena such as fluid flow. The word 'classical' is here used in the sense 'not quantum mechanical'. We shall define a field to be *classical* if it satisfies the following criteria:

1. The state of the field at a given time is represented by furnishing, for each point in space, a finite set of numbers (e.g. a single real number or a tensor or a spinor, depending on the type of field).
2. The field can in principle be observed without disturbing it.

By contrast, a quantum field would be described by furnishing at each point in space a set of operators not numbers, and it could not in general be observed without disturbing it. It is important to maintain a tight grip on terminology here, because in many textbooks the equations described in this chapter are first introduced in the context of quantum mechanics. However, I believe it is better to become acquainted with these fields in their classical guise first, and then quantize them afterwards.

All fields treated in this book are classical fields. The electromagnetic field has been introduced as a tensor field. We have sometimes invoked scalar fields for illustrative purposes. In this chapter the idea of a classical spinor field will be introduced. We shall allow that high relative speeds may be involved; to make sure our results satisfy the postulates of Special Relativity we shall only write Lorentz covariant field equations. The language of tensors and spinors makes

Relativity Made Relatively Easy: General Relativity and Cosmology. Volume 2. Andrew M. Steane, Oxford University Press. © Andrew M. Steane 2021. DOI: 10.1093/oso/9780192895646.003.0027

this easy to accomplish. We shall proceed by assuming flat spacetime in this chapter, and then consider GR in the next.

It is assumed that the reader has encountered the Pauli matrices:

$$\sigma_x = \begin{pmatrix} 0 & 1 \\ 1 & 0 \end{pmatrix}, \quad \sigma_y = \begin{pmatrix} 0 & -i \\ i & 0 \end{pmatrix}, \quad \sigma_z = \begin{pmatrix} 1 & 0 \\ 0 & -1 \end{pmatrix}. \tag{27.1}$$

These matrices are Hermitian, traceless, and square to $\mathbf{1}$. They anticommute in pairs (that is, $\sigma_x \sigma_y = -\sigma_y \sigma_x$, etc.), and their commutation relation is

$$[\sigma_i, \ \sigma_j] = 2i\epsilon_{ijk}\sigma_k. \tag{27.2}$$

27.1 Wave equation and Klein–Gordon equation

Notation. In this section the symbol ϕ is used to indicate a classical Lorentz-scalar field. That is, $\phi(t, x, y, z)$ is a function of time and space, whose value is unchanged under a change of reference frame. For example, in the case of a fluid, ϕ could be the proper density; in the case of an electromagnetic field, ϕ could be one of the invariants $E^2 - c^2 B^2$ or $\mathbf{E} \cdot \mathbf{B}$. We shall not take an interest here in any specific physical field, however. Our interest is rather in establishing what are the simplest covariant differential equations that can be written down for scalar fields.

27.1.1 The wave equation

The most familiar non-trivial Lorentz covariant field equation is the scalar wave equation

$$\Box^2 \phi = 0 \qquad \text{or} \qquad -\frac{1}{c^2}\frac{\partial^2 \phi}{\partial t^2} + \nabla^2 \phi = 0. \tag{27.3}$$

where ∇^2 is the Laplacian in three dimensions and we are assuming flat spacetime treated in an inertial frame. The equation has a complete set of plane wave solutions of the form

$$\phi(t, x, y, z) = \phi_0 e^{i(\mathbf{k} \cdot \mathbf{r} - \omega t)}. \tag{27.4}$$

Substituting this solution into the wave equation yields the *dispersion relation*

$$\omega^2 - k^2 c^2 = 0. \tag{27.5}$$

This is reminiscent of the energy-momentum relationship for massless particles, $E^2 - p^2 c^2 = 0$.

27.1.2 Klein–Gordon equation

We can modify the wave equation while preserving Lorentz covariance by introducing a Lorentz scalar s. Possibilities include, for example

$$\Box^2 \phi = s, \tag{27.6}$$

$$\Box^2 \phi = s\phi. \tag{27.7}$$

The first of these is the wave equation with a source term; we studied it in Volume 1 and in Chapters 5 and 7 of the present volume. If s is non-zero it does not in general have plane wave solutions, but the second equation does. If s is independent of time and space then (27.4) is a solution of (27.7) with the dispersion relation

$$\omega^2 - k^2c^2 = sc^2. \tag{27.8}$$

This is reminiscent of the energy-momentum relationship for massive particles, $E^2 - p^2c^2 = m^2c^4$ if we set $s \propto m^2$. Therefore let us name the constant μ^2c^2 to remind us of mass, and we have

---

Klein–Gordon equation

$$(\Box^2 - \mu^2c^2)\phi = 0. \tag{27.9}$$

The dispersion relation now reads

$$\omega^2 - k^2c^2 = \mu^2c^4. \tag{27.10}$$

The Klein–Gordon equation was first applied to physics in the context of quantum theory, and this is the way it is often introduced in text books. However, here we are dealing with a classical field. Do not forget that in this chapter $\phi(t, x, y, z)$ is a scalar quantity that in principle could be measured without disturbing it; it is not a wavefunction. (In quantum theory the constant μ^2c^2 may be written $m^2c^2/\hbar^2$).

Yukawa potential. Let us examine the time-independent solution to the Klein–Gordon equation (the d.c. limit of the plane waves). If ϕ is independent of time then we have

$$\nabla^2\phi = \mu^2c^2\phi. \tag{27.11}$$

Now let us obtain a spherically symmetric solution. When there is no angular dependence, we have

$$\nabla^2\phi = \frac{1}{r}\frac{d^2}{dr^2}(r\phi)$$

and therefore

$$\frac{d^2}{dr^2}(r\phi) = \mu^2c^2(r\phi). \tag{27.12}$$

Thinking of $(r\phi)$ as the function, this is readily solved: one finds the general solution

$$r\phi = Ae^{-\mu cr} + Be^{\mu cr} \tag{27.13}$$

where A, B are constants of integration. We take it that ϕ cannot become infinite at large r, so $B = 0$ and the solution is

$$\phi = A\frac{e^{-\mu cr}}{r}. \tag{27.14}$$

This solution is called the *Yukawa potential*. It is reminiscent of the Coulomb potential, except for the exponential term which makes ϕ fall to zero much more quickly as a function of r

than does the Coulomb potential. Indeed, the exponential term has a *natural length scale* given by $(\mu c)^{-1}$. We say that such a form has a 'finite range'; this does not mean it falls strictly to zero beyond a given distance, but its exponential fall is so rapid that to all intents and purposes it is zero once the distance is large compared to $(\mu c)^{-1}$. This is in contrast to the Coulomb potential which has no such restricted range.

The solution may also be thought of as a 'spherical wave', but one which decays instead of propagates—this is called an *evanescent* wave.

In the mid 1930s Hideki Yukawa was trying to understand the nature of the strong force that holds together nucleons in a nucleus. The wave-particle duality for light was established, though the full quantum field theory for electromagnetism was not. Yukawa supposed that the strong force was mediated by a particle, somewhat analogous to the photon. The essence of his insight was that if one assumed that the mediating particle had a finite rest mass, and proceeded by analogy with the case of photons, then one needed a mass term in the wave equation and then the associated force would have a finite range, as exhibited in equation (27.14). He thus predicted the existence of a previously unknown type of particle, and from the known data on the range of nuclear forces, he was able to predict (approximately) the mass of his 'heavy photon-like particle'. It was subsequently discovered in cosmic ray data and cyclotron experiments in the predicted mass range; it is now called the *meson*. It is now known that there are several types of meson, and it turns out that it is a composite particle (composed of one quark and one antiquark). Yukawa's theory is not correct in detail (it has to replaced by quantum chromodynamics), but it gives a sound qualitative picture and helped guide the way to the correct theory.

Klein–Gordon current. Now return to the general Klein–Gordon equation (27.9) and introduce a real quantity $\mathbf{j}$ defined by

$$\mathbf{j} \equiv i\left(\phi\boldsymbol{\nabla}\phi^* - \phi^*\boldsymbol{\nabla}\phi\right). \tag{27.15}$$

We take an interest in whether this could be a current of a conserved quantity, i.e. a quantity satisfying the continuity equation $\boldsymbol{\nabla} \cdot \mathbf{j} = -\partial\rho/\partial t$ where ρ is a density. You are invited to confirm, by using the Klein–Gordon equation, that

$$\boldsymbol{\nabla} \cdot \mathbf{j} = \frac{i}{c^2}\left(\phi\frac{\partial^2\phi^*}{\partial t^2} - \phi^*\frac{\partial^2\phi}{\partial t^2}\right)$$

and therefore

$$\boldsymbol{\nabla} \cdot \mathbf{j} = -\frac{\partial\rho}{\partial t} \qquad \text{with} \qquad \rho \equiv \frac{i}{c^2}\left(\phi^*\frac{\partial\phi}{\partial t} - \phi\frac{\partial\phi^*}{\partial t}\right). \tag{27.16}$$

Therefore we have a conservation law. The density ρ can be either positive or negative. It could represent, for example, a density of electric charge (with an appropriate constant premultiplying factor to get the physical dimensions right). Eqns (27.15) and (27.16) combine to give the 4-current

$$j^a = (\rho c, \mathbf{j}) = ig^{a\mu}\left(\phi\,\partial_\mu\phi^* - \phi^*\partial_\mu\phi\right). \tag{27.17}$$

27.2 The Dirac equation

The wave equation and the Klein–Gordon equation are both second-order in time. This means that, to obtain a unique solution, it is not sufficient to specify the field ϕ at some initial time; one must also specify its first derivative with respect to time. This implies that ϕ alone cannot serve as a complete specification of the physical situation or 'state'. In many applications, such as to particle physics, this is a drawback. We would prefer an equation that only involves the first derivative with respect to time. We shall approach this in stages.

Massless Dirac equation in two dimensions. The two-dimensional case (i.e. one spatial dimension plus time) is straightforward when the mass is zero. Then eqn (27.9) can be written

$$\left(\frac{\partial}{\partial t} + c\frac{\partial}{\partial x}\right)\left(\frac{\partial}{\partial t} - c\frac{\partial}{\partial x}\right)\phi(x,t) = 0. \tag{27.18}$$

A solution is ensured if we keep either one of the factors. Thus we can obtain a first-order equation

$$\left(\frac{\partial}{\partial t} + c\frac{\partial}{\partial x}\right)\phi(x,t) = 0.$$

This corresponds to the relation $E - cp = 0$ which makes sense for massless particles.

A plane wave solution has the form $\phi = \exp(i(kx - \omega t))$ with $k = \omega c$. Notice that in this case the form $\exp(i(-kx - \omega t))$, which is a solution of the second-order equation, does not satisfy our first-order equation. The first-order equation only has plane wave solutions whose wavefronts propagate towards positive x. A wave described by such a solution in one spatial dimension is called a 'right-mover'. This is a Lorentz-invariant property because we are dealing with waves having phase velocity c. ω and k can be either both positive or both negative.

By retaining the other factor in (27.18) we obtain a first-order equation for 'left-movers'.

Massive Dirac equation in two dimensions. The assumption $\mu = 0$ led to the simplicity of the preceding discussion. When $\mu \neq 0$ we can make some progress by recasting the Klein–Gordon equation into a pair of first-order equations:

$$(\hat{\omega} + \hat{k}_x c)\phi_1 = \mu c^2 \phi_2, \tag{27.19}$$

$$(\hat{\omega} - \hat{k}_x c)\phi_2 = \mu c^2 \phi_1, \tag{27.20}$$

where to reduce clutter we have used

$$\hat{\omega} \equiv i\frac{\partial}{\partial t}, \qquad \hat{\mathbf{k}} \equiv -i\boldsymbol{\nabla}.$$

The factor i is introduced in these definitions so that when acting on a plane wave solution, we will find

$$\hat{\omega}\phi = \omega\phi, \qquad \hat{\mathbf{k}}\phi = \mathbf{k}\phi. \tag{27.21}$$

I know it looks like we are doing quantum mechanics, but I promise you we are not!

You should verify that both ϕ_1 and ϕ_2 satisfy the original equation. Indeed, a common method of solution of such a pair of first-order equations is to obtain the second-order equation and then try to solve it. We have proceeded in the opposite direction.

In the limit $\mu \to 0$, ϕ_1 describes right-movers and ϕ_2 describes left-movers. However, when $\mu \neq 0$ there is no clear distinction between them because a wavefront moving to the right in one reference frame will move to the left in another when the speed is less than c. Therefore both equations are required, and the number of degrees of freedom is roughly doubled. Notice that ϕ_2 is acting like a source in the equation for ϕ_1, and ϕ_1 is acting like a source in the equation for ϕ_2. The two parts are coupled through the 'mass' μ.

We can write the pair of equations for ϕ_1 and ϕ_2 as a single matrix equation for a two-component object called a *spinor*:

$$\left(\hat{\omega} + \sigma_z \hat{k}_x c\right) \psi = \mu c^2 \sigma_x \psi \tag{27.22}$$

where $\psi = \begin{pmatrix} \phi_1 \\ \phi_2 \end{pmatrix}$. This is the two-dimensional Dirac equation.

The spinor is a function of position and time. We started with an equation (Klein–Gordon) for a scalar field, now we have an equation for a spinor field. That is, at each event in spacetime the field is described by a two-component object called a spinor. The Klein–Gordon equation has not completely gone away, because each component of the spinor field satisfies it.

We shall say a little more about spinors after we have introduced some further wave-like equations.

Massless Dirac equation in four dimensions.   In four dimensions, the idea is to attempt once again a factorization of the Klein–Gordon equation $(\hat{\omega}^2 - \hat{k}^2 c^2)\phi = \mu^2 c^4 \phi$. However, we now have a scalar operator $\hat{\omega}$ and a vector operator $\hat{\mathbf{k}}$ so we cannot write $(\hat{\omega} + \hat{\mathbf{k}})(\hat{\omega} - \hat{\mathbf{k}})$ (well, we can write it, but it does not make any mathematical sense). The key to performing the factorization is to introduce a set of quantities σ_i (these will be the Pauli matrices, but pretend for a moment you do not know that) and try

$$\left(\hat{\omega} + \boldsymbol{\sigma} \cdot \hat{\mathbf{k}}c\right)\left(\hat{\omega} - \boldsymbol{\sigma} \cdot \hat{\mathbf{k}}c\right)\phi = 0 \tag{27.23}$$

Since the operators $\hat{\omega}$ and $\hat{\mathbf{k}}$ commute, this evaluates to

$$\left(\hat{\omega}^2 - (\boldsymbol{\sigma} \cdot \hat{\mathbf{k}})^2\right)\phi = 0$$

so we have the Klein–Gordon equation as long as $(\boldsymbol{\sigma} \cdot \hat{\mathbf{k}})^2 = \hat{k}^2$. Now, assuming the three entities σ_i commute with $\hat{k}_j$ we have

$$(\boldsymbol{\sigma} \cdot \hat{\mathbf{k}})^2 = (\sigma_x \hat{k}_x + \sigma_y \hat{k}_y + \sigma_z \hat{k}_z)(\sigma_x \hat{k}_x + \sigma_y \hat{k}_y + \sigma_z \hat{k}_z)$$

$$= \sigma_x^2 \hat{k}_x^2 + \sigma_y^2 \hat{k}_y^2 + \sigma_z^2 \hat{k}_z^2$$

$$+ (\sigma_x \sigma_y + \sigma_y \sigma_x)\hat{k}_x \hat{k}_y + (\sigma_y \sigma_z + \sigma_z \sigma_y)\hat{k}_y \hat{k}_z + (\sigma_z \sigma_x + \sigma_x \sigma_z)\hat{k}_z \hat{k}_x$$

therefore the result is $\hat{k}^2$ as long as the σ_i all square to 1 and anticommute among themselves. The Pauli matrices have this property. Therefore we can represent $\boldsymbol{\sigma}$ in (27.23) by a set of three 2×2 Hermitian matrices, and the wave ϕ becomes a two-component spinor. Each of its components satisfies the Klein–Gordon equation.

Now we throw away one of the factors, to obtain either of the first-order equations

$$\left(\hat{\omega} - \boldsymbol{\sigma} \cdot \hat{\mathbf{k}}c \right) \phi_1 = 0, \tag{27.24}$$

$$\left(\hat{\omega} + \boldsymbol{\sigma} \cdot \hat{\mathbf{k}}c \right) \phi_2 = 0. \tag{27.25}$$

These are the massless Dirac equations in four dimensions, also called the *Weyl equations*. The solutions ϕ_1 and ϕ_2 are Weyl (i.e. two-component) spinor fields. There are two separate equations, and solutions of one are not solutions of the other.

For the plane wave solutions we can replace $\hat{\omega}$ and $\hat{k}$ by $\omega, \mathbf{k}$. In this case the Weyl equations express the dispersion relation for the waves associated with Weyl spinor fields in empty space.

To relate the two types of Weyl spinor, consider the complex conjugate of the first equation, (27.24):

$$\left(\hat{\omega}/c - \sigma_x \hat{k}_x + \sigma_y \hat{k}_y - \sigma_z \hat{k}_z \right) \phi_1^* = 0.$$

Note the sign change of the σ_y term because this matrix is imaginary. Using the anticommutation relations we can write this

$$\sigma_y \left(\hat{\omega}/c + \boldsymbol{\sigma} \cdot \hat{\mathbf{k}} \right) \sigma_y \phi_1^* = 0.$$

The left-hand factor of σ_y can be dropped since this matrix is invertible. We have thus established that if ϕ_1 satisfies (27.24), then $\sigma_y \phi_1^*$ satisfies (27.25).

27.2.1 Massive Dirac equation in four dimensions

In turning to the massive case, we must retain the μ^2 term in the Klein–Gordon equation. We proceed as we did in the two-dimensional case, by writing a pair of coupled equations (c.f. (27.20)), taking advantage of the spin matrix 'trick' for factoring $\hat{k}^2$:

$$\left(\hat{\omega} - \boldsymbol{\sigma} \cdot \hat{\mathbf{k}}c \right) \phi_R = \mu c^2 \chi_L, \qquad \left(\hat{\omega} + \boldsymbol{\sigma} \cdot \hat{\mathbf{k}}c \right) \chi_L = \mu c^2 \phi_R.$$

Here, ϕ_R and χ_L are both two-component spinors. The equation guarantees that each of the four components satisfies the Klein–Gordon equation. Gathering everything together into a matrix, we have:

Dirac equation

$$\begin{pmatrix} -\mu c^2 I & \hat{\omega} I + \boldsymbol{\sigma} \cdot \hat{\mathbf{k}}c \\ \hat{\omega} I - \boldsymbol{\sigma} \cdot \hat{\mathbf{k}}c & -\mu c^2 I \end{pmatrix} \begin{pmatrix} \phi_R \\ \chi_L \end{pmatrix} = 0. \tag{27.26}$$

---

where we included the 2×2 identity matrix I for clarity. This is the Dirac equation in empty space (i.e. without interactions) in four dimensions. $\psi = (\phi_R, \chi_L)$ is a spinor field of a new type called **Dirac spinor**. It has four components (since it is constructed from a pair of two-component entities).

We shall consider the introduction of further terms describing interactions in Section 28.3.

Now introduce a set of four 4×4 matrices, called **Dirac matrices**:

$$\gamma^0 = \begin{pmatrix} 0 & I \\ I & 0 \end{pmatrix}, \quad \gamma^i = \begin{pmatrix} 0 & -\sigma^i \\ \sigma^i & 0 \end{pmatrix}. \tag{27.27}$$

Here we are writing these matrices in a basis that is matched to the form (ϕ_R, χ_L) that we have adopted for the Dirac spinor. This is called the **chiral** basis.

By using the γ^μ matrices the Dirac equation can be written:

Dirac equation, manifestly covariant form

$$(i\gamma^\nu \partial_\nu - \mu c)\, \psi = 0. \tag{27.28}$$

The Einstein summation *is* implied here; the combination $\gamma^\nu \partial_\nu$ is a sum of four terms that results in a 4×4 matrix of differential operators. This form of the equation is suggestive, because it appears to be a manifestly covariant form. To prove that this is a true appearance, it is sufficient to show that $\gamma^\nu \partial_\nu$ is a 'Lorentz scalar' matrix, i.e. a 4×4 matrix which is invariant under Lorentz transformations. This means we need to show that the matrices γ^a transform in the same way as components of a contravariant 4-vector, as the notation implies. The logic is as follows. Suppose we define the Dirac matrices in the first instance by the anti-commutation relations

$$\left\{ \gamma^a, \gamma^b \right\} = -2\eta^{ab} I. \tag{27.29}$$

Then to prove that the Dirac equation is consistent with the postulates of relativity, eqn (27.28) shows that we need to prove that the Dirac matrices transform as the contravariant components of a 4-vector. We accomplish this by asserting *by definition* that the Dirac matrices transform the required way. All that is needed is to show that this assertion does not contradict the anti-commutation relations. It is found that it does not.

In writing down the Dirac equation, (27.26), we used the notation ϕ_R, χ_L for the Dirac spinor. This notation is an indicator of the way the spinor transforms under Lorentz transformations, which we shall now describe.

27.3   Lorentz transformation of spinors

So far we have introduced spinors as collections of complex numbers which can be used to write down solutions of certain wave-like equations. The intention has been to write Lorentz covariant equations, but in order to establish that this is achieved, we need to explore how the spinors themselves change when one passes from one reference frame to another.

A two-component rank 1 spinor ϕ (called Weyl spinor in the context of particle physics) may be defined as a pair of complex numbers which transforms as

$$\phi' = \Lambda\phi \tag{27.30}$$

under a change of reference frame, where Λ is a 2×2 complex matrix with determinant 1, and we have assumed the components of the spinor are arranged as a column vector, such as

$$\boldsymbol{s} = se^{-i\alpha/2} \begin{pmatrix} \cos(\theta/2)e^{-i\phi/2} \\ \sin(\theta/2)e^{i\phi/2} \end{pmatrix}. \tag{27.31}$$

Here we introduced a parameterization of the pair of complex numbers in terms of four real numbers $\{s, \theta, \phi, \alpha\}$, for reasons that will become apparent (and see Fig. 27.1).

We can also introduce another type of spinor, which relates to the first type like the relationship between contravariant and covariant. The two types of spinor transform differently under a given change of frame:

1. Type I, called 'right-handed spinor' or 'positive chirality spinor'

$$\phi_R \rightarrow \Lambda\phi_R = \exp\left(i\frac{\boldsymbol{\sigma}\cdot\boldsymbol{\theta}}{2} - \frac{\boldsymbol{\sigma}\cdot\boldsymbol{\rho}}{2}\right)\phi_R \tag{27.32}$$

2. Type II, called 'left-handed spinor' or 'negative chirality spinor'

$$\phi_L \rightarrow (\Lambda^\dagger)^{-1}\phi_L = \exp\left(i\frac{\boldsymbol{\sigma}\cdot\boldsymbol{\theta}}{2} + \frac{\boldsymbol{\sigma}\cdot\boldsymbol{\rho}}{2}\right)\phi_L \tag{27.33}$$

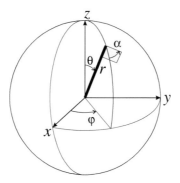

Fig. 27.1 A spinor. The spinor has a direction in space ('flagpole'), an orientation about this axis ('flag'), and an overall sign (not shown). A suitable set of parameters to describe the spinor state, up to a sign, is $(r, \theta, \phi, \alpha)$, as shown. The first three fix the length and direction of the flagpole by using standard spherical coordinates, the last gives the orientation of the flag.

> **Spinor summary**. A rank 1 spinor can be represented by a two-component complex vector, or by a null 4-vector, angle and sign. The spatial part can be pictured as a flagpole with a rigid flag attached.
> The 4-vector is obtained from the 2-component complex vector by
>
> $$U^a = \langle u| \sigma^a |u\rangle \qquad \text{if } \boldsymbol{u} \text{ is a contraspinor ('right-handed')}$$
> $$U_a = \langle \tilde{u}| \sigma^a |\tilde{u}\rangle \qquad \text{if } \tilde{\boldsymbol{u}} \text{ is a cospinor ('left-handed').}$$
>
> Any 2×2 matrix Λ with unit determinant Lorentz-transforms a spinor. Such matrices can be written
> $$\Lambda = \exp\left(i\boldsymbol{\sigma} \cdot \boldsymbol{\theta}/2 - \boldsymbol{\sigma} \cdot \boldsymbol{\rho}/2\right)$$
> where ρ is rapidity. If Λ is unitary the transformation is a rotation in space; if Λ is Hermitian it is a boost.
> If $\boldsymbol{s}' = \Lambda(v)\boldsymbol{s}$ is the Lorentz transform of a right-handed spinor, then under the same change of reference frame a left-handed spinor transforms as $\tilde{\boldsymbol{s}}' = (\Lambda^\dagger)^{-1}\tilde{\boldsymbol{s}} = \Lambda(-v)\tilde{\boldsymbol{s}}$.

The use of the term 'chirality' here is a reference to the notion that the Lorentz transformation is itself a type of rotation in spacetime, but the terminology is unfortunate because it leads to confusion between rotation in spacetime and rotation in space. The 'left'- or 'right'-handedness here is not about the spinor itself in any given frame; it is a reference to how its components transform between inertial frames in relative motion.

In (27.32), (27.33) we have expressed a general 2×2 complex matrix with determinant 1 in terms of exponentials of Pauli matrices and two parameters $\boldsymbol{\theta}$, $\boldsymbol{\rho}$. Once we have given the relationship between spinors and 4-vectors, it will become possible to show that the transformation matrix $\exp(i\boldsymbol{\sigma} \cdot \boldsymbol{\theta}/2)$ represents a rotation and the transformation matrix $\exp(\boldsymbol{\sigma} \cdot \boldsymbol{\rho}/2)$ represents a boost.

Let us introduce
$$\epsilon = \begin{pmatrix} 0 & 1 \\ -1 & 0 \end{pmatrix}. \tag{27.34}$$

For an arbitrary Lorentz transformation
$$\Lambda = \begin{pmatrix} a & b \\ c & d \end{pmatrix}, \qquad ad - bc = 1$$

we have
$$\Lambda^T \epsilon \Lambda = \begin{pmatrix} a & c \\ b & d \end{pmatrix} \begin{pmatrix} c & d \\ -a & -b \end{pmatrix} = \epsilon \tag{27.35}$$

It follows that for a pair of spinors ϕ, ψ the scalar quantity
$$\phi^T \epsilon \psi = \phi_1 \psi_2 - \phi_2 \psi_1$$

is Lorentz-invariant.

The matrix ϵ satisfying (27.35) is called the **spinor Minkowski metric**.

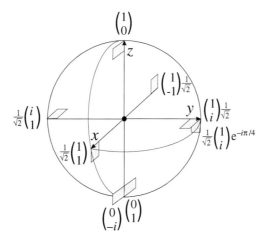

Fig. 27.2 Some example spinors. In two cases a pair of spinors pointing in the same direction but with flags in different directions are shown, to illustrate the role of the flag angle α. Any given direction and flag angle can also be represented by a spinor of opposite sign to the one shown here.

27.3.1   Obtaining 4-vectors from spinors

So far we introduced spinors by specifying that a pair of complex numbers suffices to express a spinor, and we made an assertion about the way these component numbers change under Lorentz transformations. In order to understand the implications we will now propose and prove a connection between spinors and 4-vectors. Every rank 1 right-handed spinor ϕ_R can be associated with a null 4-vector whose contravariant components are given by

---

obtaining a (null) 4-vector from a spinor

$$u^a = \phi_R^\dagger \sigma^a \phi_R \qquad (27.36)$$

---

where σ^i are the Pauli matrices, and we define $\sigma^0 \equiv I$ (the identity matrix). For example, if we substitute (27.31) into this expression then we obtain $u^a = \{r,\ r\sin\theta\cos\phi,\ r\sin\theta\sin\phi,\ r\cos\theta\}$ where $r = |s|^2$.

A similar expression involving a left-handed spinor gives covariant components:

$$u_a = \phi_L^\dagger \sigma^a \phi_L \qquad (27.37)$$

The placement of the index up on the right and down on the left is correct in this expression. The spinor itself has done the job of lowering the index.

The above equations show us that spinors are in some respects like 4-vectors. The connection can be set out more fully by thinking of a spinor as a geometric object consisting of a 'flagpole' with a rigid 'flag' attached (Fig. 27.1). Examples of eqn (27.36) are shown in Fig. 27.2. Under the influence of spatial rotations, the flagpole transforms precisely in the same way as a vector, and the flag changes direction (or not) just as would be the case for a rigid object. So much for the spatial properties. In spacetime, the spinor should be seen as lying in the appropriate direction in the null cone at the event at which it is present.

Proof of (27.36), (27.37). We now wish to show that if $\phi_R \to \Lambda \phi_R$ then $u^a \to \Lambda^a{}_\mu u^\mu$, where the first Λ is a spinor transformation matrix, and $\Lambda^a{}_b$ is a 4-vector Lorentz transformation matrix. We shall make use of

$$e^{\alpha \sigma_j} \equiv \sum_{n=0}^{\infty} \frac{1}{n!} (\alpha \sigma_j)^n = I \cosh(\alpha) + \sigma_j \sinh(\alpha) \tag{27.38}$$

which applies for any scalar real or imaginary α, and can be obtained readily by using that $\sigma_j^2 = I$. We shall treat a Lorentz boost; the argument for a rotation is essentially the same. Under a boost in the x direction, (27.32) asserts that the spinor transforms as

$$\phi_R \to e^{-\rho \sigma_x/2} \phi_R. \tag{27.39}$$

Substituting this into (27.36), we have

$$u^a \to \phi_R^\dagger e^{-\rho \sigma_x/2} \sigma^a e^{-\rho \sigma_x/2} \phi_R. \tag{27.40}$$

Let us treat the components one by one:

$$e^{-\rho \sigma_x/2} \sigma^0 e^{-\rho \sigma_x/2} = e^{-\rho \sigma_x} \quad = I \cosh(\rho) - \sigma_x \sinh(\rho) \tag{27.41}$$

$$e^{-\rho \sigma_x/2} \sigma^1 e^{-\rho \sigma_x/2} = e^{-\rho \sigma_x} \sigma_x = \sigma_x \cosh(\rho) - I \sinh(\rho) \tag{27.42}$$

$$e^{-\rho \sigma_x/2} \sigma^2 e^{-\rho \sigma_x/2} = (Ic - \sigma_x s) \sigma_y (Ic - \sigma_x s)$$
$$= \sigma_y c^2 - \sigma_x \sigma_y sc - \sigma_y \sigma_x cs + \sigma_x \sigma_y \sigma_x s^2$$
$$= \sigma_y (c^2 - s^2) = \sigma_y \tag{27.43}$$

$$e^{-\rho \sigma_x/2} \sigma^3 e^{-\rho \sigma_x/2} = \sigma_z \tag{27.44}$$

where for brevity we used $c = \cosh(\rho/2)$ and $s = \sinh(\rho/2)$, we used $\sigma_x \sigma_y = -\sigma_y \sigma_x$, and the reasoning for the last case (σ_z) is similar to that for σ_y. When one substitutes (27.41)–(27.44) into (27.40) one finds that the components of u^a are transforming exactly as they would for a Lorentz transformation expressed in terms of rapidity ρ. By combining this result with similar reasoning for rotations, one finds that the effect of any member of the Lorentz group is correctly reproduced, and therefore u^a is indeed a 4-vector. (That it is null is easily seen by writing out the components). Equation (27.37) also follows immediately, using the relationship between (27.32) and (27.33). QED.

A good way to understand the relationship between spinors and vectors is to see it in terms of a mapping or *homomorphism* between the Lie groups SU(2) and SO(3). SU(2) is the group of two by two unitary matrices with determinant 1. SO(3) is the special orthogonal group of degree 3, isomorphic to the rotation group. The former is the group of three by three orthogonal real matrices with determinant 1. The latter is the group of rotations about the origin in Euclidean space in three dimensions. The homomorphism is a $2:1$ mapping in which members U and $-U$ of SU(2) both map to member R of SO(3), where $U = \exp(i\boldsymbol{\sigma} \cdot \boldsymbol{\theta})$, $R = \exp(i\mathbf{J} \cdot \boldsymbol{\theta})$ and $\mathbf{J}$ are the generators of rotations in three dimensions:

$$J_x = \begin{pmatrix} 0 & 0 & 0 \\ 0 & 0 & -i \\ 0 & i & 0 \end{pmatrix}, \quad J_y = \begin{pmatrix} 0 & 0 & i \\ 0 & 0 & 0 \\ -i & 0 & 0 \end{pmatrix}, \quad J_z = \begin{pmatrix} 0 & -i & 0 \\ i & 0 & 0 \\ 0 & 0 & 0 \end{pmatrix}. \tag{27.45}$$

It is this homomorphism to which we are alluding when we use the flagpole picture of a spinor.

The Dirac equation involves a pair of spinors, one of each type. This leads to a pair of four-vectors:

$$A^a = \phi_R^\dagger \sigma^a \phi_R, \qquad B_a = \chi_L^\dagger \sigma^a \chi_L.$$

By forming the sum and difference of these 4-vectors, one can obtain the timelike vector $j^a = A^a - B^a$ and the spacelike vector $w^a = A^a + B^a$. It is easy to show that since $\mathbf{A}$ and $\mathbf{B}$ are both null, $\mathbf{j}$ and $\mathbf{w}$ must be orthogonal. In particle physics this is the way the Dirac spinor field encodes a pair of conserved quantities called 4-current and 4-spin density. These quantities can be written using the Dirac matrices (27.27) as

$$j^a = \psi^\dagger \gamma^0 \gamma^a \psi, \tag{27.46}$$

$$w^a = \Psi^\dagger \gamma^0 \gamma^\mu \gamma^5 \Psi, \qquad \gamma^5 \equiv \begin{pmatrix} I & 0 \\ 0 & -I \end{pmatrix} \tag{27.47}$$

where a 5th Dirac matrix has been introduced. In order to prove these results, evaluate the temporal and spatial parts separately and pay attention to the sign change involved when an index is raised or lowered.

We can easily show that this *Dirac current* satisfies the continuity equation:

$$\partial_\alpha j^\alpha = (\partial_\alpha \bar{\psi})\gamma^\alpha \psi + \bar{\psi}\gamma^\alpha \partial_\alpha \psi = i\mu c \bar{\psi}\psi - i\mu c \bar{\psi}\psi = 0 \tag{27.48}$$

where we adopted the notation $\bar{\psi} \equiv \psi^\dagger \gamma^0$ to reduce clutter, and to evaluate the first term we used the 'Dirac adjoint' of the Dirac equation, which reads

$$-i\gamma^\alpha \partial_\alpha \bar{\psi} = \mu c \bar{\psi}. \tag{27.49}$$

We have in this section barely scratched the surface of the mathematics of spinors. The rank 1 objects we have discussed are the building blocks of a whole structure of spinors of any rank, very much like the journey from vectors to tensors. There are several types of rank 2 spinor, one of which can be mapped to a 4-vector of any type. One can translate tensor equations such as Maxwell's equations into spinor language. The spinor may be said to be the most general type of object that can be Lorentz-transformed.

Dirac spinor summary. A Dirac spinor $\psi = (\phi_R, \phi_L)$ is composed of a pair of spinors, one of each handedness.

From the two associated null 4-vectors one can extract two orthogonal non-null 4-vectors

$$j^a = \psi^\dagger \gamma^0 \gamma^a \Psi,$$
$$w^a = \psi^\dagger \gamma^0 \gamma^a \gamma^5 \Psi,$$

where γ^a, γ^5 are the Dirac matrices. With appropriate normalization factors these can represent the 4-velocity and 4-spin of a particle.

Starting from a frame in which $j^i = 0$ (i.e. the rest frame), the result of a Lorentz boost to a general frame can be written

$$\begin{pmatrix} -m & E + \boldsymbol{\sigma} \cdot \mathbf{p} \\ E - \boldsymbol{\sigma} \cdot \mathbf{p} & -m \end{pmatrix} \begin{pmatrix} \phi_R(\mathbf{p}) \\ \chi_L(\mathbf{p}) \end{pmatrix} = 0.$$

This is the Dirac equation. Under parity inversion the parts of a Dirac spinor swap over; the Dirac equation is therefore parity-invariant.

28

Lagrangian mechanics for fields

The classical field equations considered in the previous chapter are listed in Table 28.1, along with two more: the Maxwell equations in empty space, and the Proca equation, which is a generalization of the Maxwell equations to non-zero mass. We have taken the approach of introducing the reader to these equations by steps from the wave equation, so that there is some familiarity to build on. Now we shall describe a more general approach, based on Lagrangian mechanics and least action. The discussion is restricted to flat spacetime up until the final section.

One way to approach Lagrangian mechanics for waves is to consider a simple example such as a string, modelled as a set of masses connected by springs, and develop the particle mechanics of the masses. Suppose the string lies along the x axis. The Lagrangian would involve a sum over all the terms corresponding to different parts of the string, tending to an integral when the continuous limit is taken. The action would then involve an integral over x and t. Generalizing this to 3 spatial dimensions, one sees that the action must be an integral over space and time.

In field theory we still integrate over four dimensions, but we adopt a slightly different approach. We make the field the basic quantity and consider a Lagrangian per unit volume (Lagrangian density) which we take to be a function of the field ϕ and its derivatives $\partial_a \phi$. Then the action is *defined* by

Table 28.1 A collection of classical field equations.

name	equation	type	spin	mass
Wave	$\Box^2 \phi = 0$	scalar	0	0
Klein–Gordon	$\Box^2 \phi = m^2 \phi$	scalar	0	m
Weyl	$\sigma^\mu \partial_\mu \phi = 0$	Weyl spinor	$\frac{1}{2}$	0
Dirac	$i\gamma^\mu \partial_\mu \psi = m\psi$	Dirac spinor	$\frac{1}{2}$	m
Maxwell	$\partial_\alpha \mathbb{F}^{ab} = 0$	tensor	1	0
	$\partial^a \mathbb{F}^{bc} + \partial^b \mathbb{F}^{ca} + \partial^c \mathbb{F}^{ab} = 0$			
Proca	$\partial_\mu \mathbb{F}^{\mu b} = m^2 A^b$	tensor	1	m
	$\mathbb{F}^{ab} = \partial^a A^b - \partial^b A^a$			
	$\partial_\mu A^\mu = 0$			

Relativity Made Relatively Easy: General Relativity and Cosmology. Volume 2. Andrew M. Steane, Oxford University Press. © Andrew M. Steane 2021. DOI: 10.1093/oso/9780192895646.003.0028

$$S = \int_{\mathcal{R}} \mathcal{L}(\phi, \partial_a \phi)\, \mathrm{d}^4 x \qquad (28.1)$$

where $\mathcal{R}$ is some four-dimensional region. The calculus of variations is now employed to find the condition of stationary action ('Least action'), with the boundary of the region providing the fixed end-points. We introduce the notation $\delta\phi$ and $\delta(\partial_a\phi)$ to refer to a small variation of these quantities throughout $\mathcal{R}$ but going to zero change on the boundary. The resulting change in the action is

$$\delta S = \int_{\mathcal{R}} \mathrm{d}^4 x \left[\frac{\partial \mathcal{L}}{\partial \phi} \delta\phi + \frac{\partial \mathcal{L}}{\partial u_\mu} \delta u_\mu \right] \qquad (28.2)$$

where $u_a \equiv \partial_a \phi$ has been introduced for convenience in writing the expression. But notice that $\delta u_\mu = \delta(\partial_\mu \phi) = \partial_\mu \delta\phi$. Therefore the second term in the integral can be integrated by parts. A convenient way to see the method is to note that

$$\partial_\mu \left(\frac{\partial \mathcal{L}}{\partial u_\mu} \delta\phi \right) = \partial_\mu \left(\frac{\partial \mathcal{L}}{\partial u_\mu} \right) \delta\phi + \frac{\partial \mathcal{L}}{\partial u_\mu} \partial_\mu \delta\phi. \qquad (28.3)$$

The left-hand side here is a divergence. Upon integrating over a volume it can be converted into a surface integral by Stokes' theorem, yielding

$$\int_{\partial \mathcal{R}} \left(\frac{\partial \mathcal{L}}{\partial u_\mu} \delta\phi \right) \mathrm{d}\sigma_\mu = \int_{\mathcal{R}} \mathrm{d}^4 x\, \partial_\mu \frac{\partial \mathcal{L}}{\partial u_\mu} \delta\phi + \int_{\mathcal{R}} \mathrm{d}^4 x \frac{\partial \mathcal{L}}{\partial u_\mu} \partial_\mu \delta\phi. \qquad (28.4)$$

Now use that $\delta\phi$ vanishes on the boundary, so the left-hand side is zero. By using this result in (28.2) we obtain

$$\delta S = \int_{\mathcal{R}} \mathrm{d}^4 x \left[\frac{\partial \mathcal{L}}{\partial \phi} - \partial_\mu \frac{\partial \mathcal{L}}{\partial u_\mu} \right] \delta\phi. \qquad (28.5)$$

If this is to be zero for all $\delta\phi$ then we must have:

Euler–Lagrange equation for a local scalar field

$$\frac{\partial}{\partial x^\mu} \left(\frac{\partial \mathcal{L}}{\partial u_\mu} \right) = \frac{\partial \mathcal{L}}{\partial \phi} \qquad \text{where } u_a = \partial_a \phi \qquad (28.6)$$

Comparing this with the equations for particle mechanics, we see that ϕ acts like a 'position' coordinate, and $\partial_a \phi$ acts like a 'velocity'. We have given it the name u_a to make it easier to read some of the results. The above equation is announced with two qualifying adjectives: 'scalar' and 'local'. The result for tensor fields of higher rank is straightforward: just replace ϕ by the appropriate tensor quantity; each component then has its own EL equation. By 'local' we mean that the Lagrangian only depends on at most first-order derivatives of the field. If second-order derivatives also appear then one obtains

$$\frac{\partial}{\partial x^\mu} \left(\frac{\partial \mathcal{L}}{\partial u_\mu} \right) - \frac{\partial^2}{\partial x^\mu \partial x^\nu} \left(\frac{\partial \mathcal{L}}{\partial(\partial_\mu u_\nu)} \right) = \frac{\partial \mathcal{L}}{\partial \phi}. \qquad (28.7)$$

Table 28.2 Lagrangian densities for the fields listed in Table 28.1.

name	Lagrangian in empty space	
wave	$-\frac{1}{2}u_\alpha u^\alpha$	where $u_\mu \equiv \partial_\mu \phi$
real Klein–Gordon	$-\frac{1}{2}u_\alpha u^\alpha - \frac{1}{2}m^2\phi^2$	
complex Klein–Gordon	$-u_\alpha u^{*\alpha} - m^2\phi^*\phi$	
Dirac	$\bar\psi\left(i\gamma^\alpha\partial_\alpha - m\right)\psi$	where $\bar\psi \equiv \psi^\dagger\gamma^0$
Maxwell	$-\frac{1}{4}\mathbb{F}_{\alpha\beta}\mathbb{F}^{\alpha\beta}$	
Proca	$-\frac{1}{4}\mathbb{F}_{\alpha\beta}\mathbb{F}^{\alpha\beta} - \frac{1}{2}m^2 A_\alpha A^\alpha$	

The equations in Table 28.1 can be obtained from the Lagrangians[1] shown in Table 28.2. A change of metric signature will introduce sign changes in some of these equations.[2] The Klein–Gordon Lagrangian is given first for the case of a real valued field, then for the case of a complex valued field where the complex conjugate appears and the factor of 1/2 is not needed if ϕ and ϕ^* are treated as independent variables (see below). The Dirac Lagrangian is given in a form which assumes that ψ and $\bar\psi$ are treated as independent variables. The more symmetric form $\frac{i}{2}\left(\bar\psi\gamma^\alpha\partial_\alpha\psi - (\partial_\alpha\bar\psi)\gamma^\alpha\psi\right) - m\bar\psi\psi$ may alternatively be used. In the case of Maxwell and Proca equations, the Lagrangian has been written in a conveniently succinct form. In order to use it, it may help to notice that $\mathbb{F}_{\alpha\beta}\mathbb{F}^{\alpha\beta} = 2(\partial_\alpha A_\beta)(\partial^\alpha A^\beta) - 2(\partial_\alpha A^\alpha)^2$. You are meant to be impressed by the simplicity of these Lagrangians. Although they could not be called simple, they could not very well be any simpler.

For example, consider the wave equation. We have

$$\mathcal{L} = \frac{1}{-2}u_\lambda u^\lambda \qquad \Rightarrow \frac{\partial\mathcal{L}}{\partial u_a} = -u^a \quad \text{and} \quad \frac{\partial\mathcal{L}}{\partial\phi} = 0$$

(the first calculation is like $(1/2)\nabla r^2 = r\nabla r = \mathbf{r}$). Substituting these into (28.6) yields

$$\partial_\alpha u^\alpha = \partial_\alpha\partial^\alpha\phi = 0$$

which is the wave equation. The Klein–Gordon equation for a real field comes out similarly.

Since a general variation of a complex number consists of independent variation of its real and imaginary parts, a complex field has to be treated in the Euler–Lagrange equations as a sum of two real fields. So, in the Klein–Gordon Lagrangian we can write $\phi = \phi_1 + i\phi_2$ where ϕ_1, ϕ_2 are real, and let $v_a = \partial_a\phi_1$, $w_a = \partial_a\phi_2$, and then the Lagrangian is

$$\mathcal{L} = -u_\alpha u^{*\alpha} - m^2\phi^*\phi = -\left(v_\alpha v^\alpha + w_\alpha w^\alpha + m^2\phi_1^2 + m^2\phi_2^2\right).$$

In the partial derivatives appearing in (28.6) the real and imaginary contributions to the field should be regarded as independent, hence $\partial\mathcal{L}/\partial v_a = -2v^a$, and $\partial\mathcal{L}/\partial\phi_1 = -2\phi_1$ which when substituted into (28.6) yields the Klein–Gordon equation for ϕ_1. A similar calculation yields the same equation for ϕ_2; summing these gives the Klein–Gordon equation for the complex field ϕ. There is a more straightforward approach, however. The Euler–Lagrange equations essentially concern a stationary point in the action when path variations are considered. Instead of regarding the real and imaginary parts of ϕ as independent in these variations, we may equally regard ϕ and ϕ^* as independent. Then the argument reads

$$\frac{\partial \mathcal{L}}{\partial u_a} = -u^{*\,a}, \quad \frac{\partial \mathcal{L}}{\partial \phi} = -m^2\phi^*, \qquad \Rightarrow \partial_\alpha \partial^\alpha \phi^* = m^2 \phi^*$$

which upon complex conjugation gives the Klein–Gordon equation. In this version the canonical momentum conjugate to ϕ is $u^{*\,0} = \dot{\phi}^*$, and this is the version that has been listed in Table 28.2.

For the Dirac equation, we have

$$\mathcal{L} = \bar{\psi}(i\gamma^\alpha \partial_\alpha - m)\psi = i\bar{\psi}\gamma^\alpha u_\alpha - m\bar{\psi}\psi.$$

where the bar notation is

$$\bar{\psi} \equiv \psi^\dagger \gamma^0. \tag{28.8}$$

Treating now $\bar{\psi}$ and ψ as independent variables, we find

$$\frac{\partial \mathcal{L}}{\partial u_\alpha} = i\bar{\psi}\gamma^\alpha, \quad \frac{\partial \mathcal{L}}{\partial \psi} = -m\bar{\psi}, \qquad \Rightarrow -i\partial_\alpha \bar{\psi}\gamma^\alpha = m\bar{\psi},$$

which is the Dirac adjoint of Dirac's equation (c.f. (27.49)). The canonical momentum conjugate to ψ is $i\psi^\dagger \gamma^0 \gamma^0 = i\psi^\dagger$.

28.1   Field energy

For any field, the energy tensor can also be written in terms of the Lagrangian:

Energy-momentum tensor

$$T_{ab} = -g_{a\mu}\frac{\partial \mathcal{L}}{\partial u_\mu}u_b + \mathcal{L}g_{ab} \tag{28.9}$$

Proof. Consider the 00 component (taking $g_{ab} = \eta_{ab}$):

$$T_{00} = \frac{\partial \mathcal{L}}{\partial \dot{\phi}}\dot{\phi} - \mathcal{L}$$

we recognize this as the Hamiltonian density (compare with the expression one finds for particle mechanics, for example), so T^{00} gives the energy density in the field. It follows that T^{0b} is the energy-momentum 4-vector per unit volume, since it is a 4-vector whose zeroth element is energy density, and therefore T^{ab} is the stress-energy tensor. This is not a 'proof' in the sense of a derivation from axioms, rather it is an interpretive posture. We notice that $\mathbf{T}$ has the form to be expected of an energy tensor as long as the Lagrangian describes energy and momentum in the field in the way we expect. One then finds using Noether's theorem (next section) that this tensor is conserved for Lagrangians having translational invariance. Thus it has the right properties. However, we will also mention in eqn (28.34) another way to define an energy-momentum tensor.

28.2   Conserved quantities and Noether's theorem

A basic and very useful idea in particle mechanics is that if the Lagrangian is independent of a coordinate q_i, then the corresponding canonical momentum $\tilde{p}_i$ is a constant of the motion (i.e. its time derivative is zero). This follows immediately from the Euler–Lagrange equations, and we have made much use of it in earlier chapters. The word 'symmetry' is employed in this context, since the absence of a coordinate such as x from the Lagrangian implies that the Lagrangian is invariant when the system is translated along x; $\mathcal{L}$ is said to exhibit a 'symmetry' under spatial translations. We thus have a connection between symmetry with respect to translation in space, and conservation of momentum. More generally, invariance of $\mathcal{L}$ under a change in some combination of the coordinates implies conservation of some corresponding canonical momentum.

Noether generalized this observation in two directions, first to a generalized form for particle mechanics, and then to an important corresponding result in field theory.

The particle mechanics result is generalized simply by developing a convenient notation to describe a set of symmetry transformations. We suppose there are several transformations, say N, and use an index $r = 1, 2, 3, \ldots N$ to keep track of them. The r'th transformation may involve a change in t and a change in some combination of the q_i:

$$
\begin{aligned}
t &\to t' = t + \epsilon \delta_r \\
q_i &\to q_i' = q_i + \epsilon \Delta_{ri}
\end{aligned}
\tag{28.10}
$$

Here ϵ is a parameter describing the size of the change, Δ_{ri} is a matrix and δ_r is a set of numbers. An arbitrary combination of such transformations is

$$
\begin{aligned}
t &\to t' = t + \sum_r \epsilon_r \delta_r \\
q_i &\to q_i' = q_i + \sum_r \epsilon_r \Delta_{ri}
\end{aligned}
\tag{28.11}
$$

where ϵ_r indicates how much of the r'th transformation appears in the combination.

Noether showed that if the Lagrangian is unchanged under such a set of transformations, then the N quantities

$$
\mathcal{H} \delta_r - \sum_i \Delta_{ri} \tilde{p}_i
\tag{28.12}
$$

are constants of the motion. For example, if $N = 1$, $\delta = 1$ and $\Delta = 0$ we have conservation of energy. If $N = 3$, $\delta = 0$ and $\Delta = I$ (the identity matrix) we have conservation of linear momentum. If $N = 4$, $\delta = (1, 0, 0, 0)$, $\Delta = \text{diag}(0, 1, 1, 1)$ we have conservation of both.

In classical field theory, the corresponding results are as follows. Suppose the action is invariant under the set of transformations

$$x^a \rightarrow x^a + \epsilon^\mu \Delta^a_\mu$$

$$\phi \rightarrow \phi + \epsilon^\mu \Phi_\mu$$

(28.13)

Then one may show that the

Noether current

$$J^a_b \equiv -\frac{\partial \mathcal{L}}{\partial u_a} \Phi_b + T^a_\mu \Delta^\mu_b$$

(28.14)

---

is divergenceless, i.e.

Noether's theorem

$$\partial_\mu J^\mu_b = 0.$$

(28.15)

In the simplest case there is a single transformation in the set, so the index b is not needed: then **J** is a 4-vector current. An example is the conservation of charge, as we shall demonstrate shortly. The next case is a set of four transformations so that **J** is a second-rank tensor. For example, if $\Phi = 0$ and $\Delta^a_b = \delta^a_b$, then one sees that J is equal to the energy tensor, and we have the conservation of energy-momentum.

It is useful to have a general formula for the conserved quantity itself, rather than its current density. To this end we define a generalized 'charge' Q^a by

$$Q_a = \int -J^\mu_a d\sigma_\mu$$

(28.16)

where the integral is over a spacelike hypersurface. For example if this hypersurface is $t=\text{const}$ then the integral is the more familiar volume integral $Q_a = \int J^0_a d^3x$. Noether's theorem then implies that

$$\frac{dQ_a}{dt} = 0.$$

(28.17)

For example, for a Lagrangian that does not change under translations in time, the generalized 'charge' is the energy. To satisfy Lorentz covariance, such a Lagrangian will also be invariant under translations along x, for which the conserved 'charge' is x-momentum, etc.

28.2.1   Conservation of electric charge

The conservation of electric charge may be obtained by considering Noether's theorem for the right type of field. A single scalar field does not support the notion of a conserved invariant scalar charge. Once one has achieved symmetry under space and time (thus conservation of momentum and energy) there are no degrees of freedom remaining to exhibit further symmetries, except possibly combinations of energy and momenta.

We therefore consider next a pair of real scalar fields ϕ_1 and ϕ_2. It is convenient to construct from them a single complex field

$$\phi \equiv (\phi_1 + i\phi_2)/\sqrt{2}.$$

We make the Lagrangian have a further symmetry by ensuring that it is real for all combinations of ϕ_1, ϕ_2. For example, for the Klein–Gordon Lagrangian (the simplest non-trivial scalar field), we adopt

$$\mathcal{L} = (\partial_\alpha \phi)(\partial^\alpha \phi^*) + m^2 \phi^* \phi.$$

Such a Lagrangian is invariant under the transformation $\phi \to e^{i\epsilon}\phi$ where ϵ is a real constant. The infinitesimal version of this transformation is

$$\phi \to \phi + i\epsilon\phi, \qquad \phi^* \to \phi^* - i\epsilon\phi^*$$

This falls into the Noether scheme (28.13) as

$$\Delta = 0, \qquad \Phi = i\phi, \qquad \Phi^* = -i\phi^*.$$

The form we gave for Noether's theorem assumed the fields were real, so to apply it here we must treat the field as composed of two real components, both of which contribute to the total current. This is the reason why we exhibit both Φ and Φ^* explicitly. One then finds a sum of two terms in the divergenceless 'Noether current', and it is precisely the current density we previously noted in connection with the Klein–Gordon equation, eqn (27.17).

This mathematical result does not in itself prove that the Noether charge Q_N associated with invariance under a phase change $e^{i\epsilon}$ must be identified with the electric charge q. However, Q_N is a conserved scalar quantity that is not energy or any sort of momentum component, and it can have either sign. It can serve as the mathematical object which models the behaviour of any physical quantity having such properties. Electric charge is one such property. Others are particle number such as baryon or lepton number.

The Dirac Lagrangian is invariant under $\psi \to e^{i\epsilon}\psi$; the associated Noether current is the Dirac current (27.46).

Symmetry and conservation.   The connection between symmetry and conservation is one of the most beautiful and profound results in physics. The whole of Special Relativity could be considered to be a study in how to require symmetry with respect to inertial frame of reference. In the present chapter, this placed a tight restriction on the forms of Lagrangian we allowed ourselves to write down: only scalar, spinor or tensor fields appeared, and the equations were Lorentz covariant. We then make a further very reasonable requirement for an isolated system, that of symmetry with respect to *translation* in space and time. Noether's theorem then comes along and says that we have energy and momentum conservation. Angular momentum conservation follows similarly from symmetry with respect to rotation. Finally we incorporated a new type of symmetry, with respect to a form of rotation of an internal degree of freedom, and we obtained conservation of a further scalar quantity: charge.

Now we can see the avenue for further discovery: invent further internal degrees of freedom (with Lie group theory serving to provide promising candidates), require symmetry in the Lagrangian, and deduce the implications. This has been the main story of fundamental particle physics for the last sixty years, leading to the great insights of electroweak unification and quantum chromodynamics, and hence to many of the components of the Standard Model of particle physics.

28.3   Interactions

So far we discussed Lagrangians of fields in free space. In this section we present some examples to show how interactions between fields are incorporated.

A charge distribution interacting with an electromagnetic field is an example of one field interacting with another. The charge distribution is described by a scalar field; the electromagnetic field is a tensor field. In order to allow our scalar field to describe a conserved charge we let it be a complex field, and incorporate symmetry under the change $\phi \to e^{i\epsilon}\phi$. Then there is a beautiful argument, based on this symmetry and gauge invariance, that directs one to the right theory. This 'gauge symmetry' argument is outside the scope of our discussion however; we shall merely present the conclusion. The Lagrangian describing both fields and their interaction is

$$\mathcal{L} = (\partial_\mu \phi + iqA_\mu \phi)(\partial^\mu \phi^* - iqA^\mu \phi^*) + m^2 \phi^* \phi - \frac{1}{4}\mathbb{F}^{\mu\nu}\mathbb{F}_{\mu\nu}.$$

This result can be understood in a remarkably straightforward way. We have simply the sum of two Lagrangians, one for the scalar (Klein–Gordon) field and one for the electromagnetic field, except that the velocity term of the former has been modified by the vector potential, through the recipe

$$\partial_\mu \to \partial_\mu + iqA_\mu. \tag{28.18}$$

This is a form of covariant derivative—see (10.10) and surrounding text. When the interaction between two fields can be accounted for by adjusting the differential operator in this way, we say that we have *minimal coupling*.

Next we consider the case of a real scalar field interacting with a spinor field. This is used in the Standard Model, for example, to describe the interaction of fermions with the Higgs field. The Lagrangian is a sum of two free-field Lagrangians, plus a scalar interaction term:

$$\mathcal{L} = \tfrac{1}{2}\partial_\alpha \phi \partial^\alpha \phi + \tfrac{1}{2}\mu^2 \phi^2 + \bar{\psi}\left(i\gamma^\alpha \partial_\alpha - m\right)\psi - \lambda \phi \bar{\psi}\psi.$$

Here μ is the mass associated with the scalar field, m that of the spinor field, and λ is a constant characterizing the strength of the coupling.

A Dirac field interacting with an electromagnetic field is described by the **QED Lagrangian**

$$\mathcal{L} = \bar{\psi}\left(i\gamma^\alpha D_\alpha - m\right)\psi - \frac{1}{4}\mathbb{F}^{\alpha\beta}\mathbb{F}_{\alpha\beta}, \tag{28.19}$$

where $D_\mu = \partial_\mu + ie\mathsf{A}_\mu$ and for electrons e is positive (i.e. the charge on an electron is $-e$). The Euler–Lagrange equations yield the equations of motion

$$\left(i\gamma^\alpha D_\alpha - m\right)\psi = 0, \tag{28.20}$$

$$\partial_\alpha \mathbb{F}^{\alpha\mu} = -e\bar{\psi}\gamma^\mu\psi. \tag{28.21}$$

Thus in this case the Dirac current, multiplied by the charge, acts as the source in Maxwell's equations.

An understanding of the above examples, and their quantization, is the first step in getting to grips with the Standard Model of particle physics.

28.4   The Einstein–Hilbert action

In this final section we will make a few remarks on the application of Lagrangian methods in GR, and how Einstein's field equation itself follows from a remarkably simple Lagrangian.

To apply Lagrangian methods in GR, one must revisit all the derivations, paying attention to the need to write generally covariant expressions, for example by replacing ∂_a by ∇_a, though, as usual, this replacement is not sufficient on its own to guarantee correct logic.

We need the action S to be a scalar invariant, so we introduce the *field Lagrangian L* defined by

$$L \equiv \mathcal{L}/\sqrt{-g} \tag{28.22}$$

where $\mathcal{L}$ is the Lagrangian density, and then (28.1) can be written

$$S = \int_\mathcal{R} L\sqrt{-g}\,\mathrm{d}^4x. \tag{28.23}$$

This is convenient because we know $\sqrt{-g}\,\mathrm{d}^4x$ is invariant, so S will be a scalar invariant quantity if L is. Thus L is a scalar invariant but $\mathcal{L}$ is not; in fact $\mathcal{L}$ is here an example of a *scalar density* (c.f. Section 12.6).

One now has the option of regarding L either as a function of a field and its partial derivatives, or as a function of a field and its covariant derivatives. In the former case the EL equation takes the form (28.6) or (28.7) as before. In the latter case one finds

---

Euler–Lagrange equation for $L(\phi, \nabla_a\phi, g_{ab}, \partial_c g_{ab})$

$$\nabla_\mu\left(\frac{\partial L}{\partial u_\mu}\right) = \frac{\partial L}{\partial \phi} \qquad \text{where } u_a = \nabla_a\phi \tag{28.24}$$

A simple example of a Lagrangian for a scalar field is now

$$L = -\tfrac{1}{2} g^{\mu\nu} (\nabla_\mu \phi)(\nabla_\nu \phi) - V(\phi) \tag{28.25}$$

and in this case (28.24) gives

$$-\nabla_\mu (g^{\mu\nu} \nabla_\nu \phi) = -\frac{\mathrm{d}V}{\mathrm{d}\phi} \tag{28.26}$$

which is

$$\nabla^\mu \nabla_\nu \phi - \frac{\mathrm{d}V}{\mathrm{d}\phi} = 0 \tag{28.27}$$

since g_{ab} is covariantly constant. For $V = (1/2)m^2\phi^2$ we thus obtain a generally covariant Klein–Gordon equation. It describes neutral spinless particles which interact purely through gravitational attraction.

Can we now propose a Lagrangian for the gravitational field itself? If we seek an invariant scalar related to spacetime itself, then the simplest non-trivial example is the Ricci scalar R. Thus we are led naturally to the suggestion $L = R$ which gives the *Einstein–Hilbert action*

$$S_\mathrm{H}[g_{ab}] = \int R\sqrt{-g}\, \mathrm{d}^4 x \tag{28.28}$$

(where some fixed region of integration is taken as understood). By now writing $R = g^{\mu\nu} R_{\mu\nu}$ one can determine the effect on the action, to first order, of a small change in the metric tensor:

$$\delta S_\mathrm{H} = \int (\delta g^{\mu\nu}) R_{\mu\nu} \sqrt{-g} + g^{\mu\nu} (\delta R_{\mu\nu}) \sqrt{-g} + R(\delta\sqrt{-g})\, \mathrm{d}^4 x.$$

Consider the second term. Using (15.2) (and some thought) one finds

$$g^{\mu\nu} \delta R_{\mu\nu} = g^{\mu\nu} \left(\nabla_\lambda (\delta \Gamma^\lambda_{\mu\nu}) - \nabla_\nu (\delta \Gamma^\lambda_{\lambda\mu}) \right) = \nabla_\lambda g^{\mu\nu} \delta\Gamma^\lambda_{\mu\nu} - \nabla_\nu g^{\mu\nu} \delta\Gamma^\lambda_{\lambda\mu}.$$

Hence

$$\int g^{\mu\nu} (\delta R_{\mu\nu}) \sqrt{-g}\, \mathrm{d}^4 x = \int \nabla_\lambda \left(g^{\mu\nu} \delta\Gamma^\lambda_{\mu\nu} - g^{\mu\lambda} \delta\Gamma^\nu_{\nu\mu} \right) \sqrt{-g}\, \mathrm{d}^4 x = 0$$

where we invoke Gauss' divergence theorem and let the variation be zero on the boundary. Next, by using (11.33) one finds

$$\delta\sqrt{-g} = -\tfrac{1}{2}(-g)^{-1/2}\delta g = -\tfrac{1}{2}\sqrt{-g}\, g^{\mu\nu} \delta g_{\mu\nu}. \tag{28.29}$$

Hence (28.4) reads

$$\delta S_\mathrm{H} = \int_\mathcal{R} \left(R_{\mu\nu} - \tfrac{1}{2} g_{\mu\nu} R \right) \delta g^{\mu\nu} \sqrt{-g}\, \mathrm{d}^4 x \tag{28.30}$$

and since we require that this should vanish for any $\delta g^{\mu\nu}$ we obtain

$$R_{\mu\nu} - \tfrac{1}{2} g_{\mu\nu} R = 0 \tag{28.31}$$

which is Einstein's field equation in vacuum.

In order to get the full field equation, we now try an action of the form

$$S = \frac{1}{2\kappa} S_{\mathrm{H}} + S_{\mathrm{M}} \tag{28.32}$$

and using this in the variational method, one finds the condition

$$\frac{1}{2\kappa} \left(R_{\mu\nu} - \tfrac{1}{2} g_{\mu\nu} R \right) + \frac{1}{\sqrt{-g}} \frac{\partial S_{\mathrm{M}}}{\partial g^{\mu\nu}} = 0 \tag{28.33}$$

This is Einstein's equation if the energy tensor for matter is

$$T_{ab} = -\frac{1}{\sqrt{-g}} \frac{\partial S_{\mathrm{M}}}{\partial g^{ab}}. \tag{28.34}$$

This is a different form to the one we quoted previously in (28.9). To distinguish them, (28.9) goes by the name *canonical* energy-momentum tensor and then (28.34) is the *dynamical* energy-momentum tensor. It can be shown that they are equivalent for the simplest fields (e.g. fields that do not couple to the metric derivatives.) Just as with the remarks we made concerning the canonical energy-momentum tensor, the chief property we require of T_{ab} is that it be a (symmetric, second-rank) tensor and that it be conserved. It is certainly the former, and it can be shown that it satisfies $\nabla_\mu T^{\mu b} = 0$. In some respects it is easier to work with than (28.9).

29

Conclusion

In the letter to the London Times newspaper on November 28, 1919, Albert Einstein described the theory of relativity and thanked his English colleagues for their understanding and testing of his work. He commented:[1]

> The great attraction of the theory is its logical consistency. If any deduction from it should prove untenable, it must be given up. A modification of it seems impossible without destruction of the whole.

Here Einstein highlights the formal mathematical beauty of General Relativity. Complicated though the equations presented in Chapter 2 were when we first met them, by the time we have absorbed Riemannian geometry and the Hilbert action, they seem all but forced upon us, and in some sense minimal—no more complicated than they need to be. It is possible to formulate a more general class of theories which respect general covariance; in this sense GR can be modified. But in this class GR is arguably the simplest and thus most natural contender; in any case the empirical evidence suggests that GR is the one that describes our universe.

This is not to suggest that this description is perfect, with no imprecision and nothing more to be said. On the contrary, General Relativity as it stands today is almost certainly inapplicable to extreme conditions such as those involving structure at the Planck scale. But as we learn more about such conditions, GR will remain as a guide or a limiting case. As always in science, a more general framework must be consistent with existing knowledge in its domain of applicability.

The next stage in the development of our understanding of space and time will have to include quantum theory from the start. Efforts to do this, in string theory and loop quantum gravity, for example, have already made impressive mathematical progress. However, the mathematical toolbox of quantum field theory in multiple dimensions is very rich, and it has proven hard to 'tame' this richness, by the use of symmetry arguments, for example, to a sufficient degree that the theory gives a clear guide to what the universe is like and what it is not like. It may be that we will need to learn to apply new and unfamiliar types of reasoning in order to constrain the possibilities.

---

[1] From *The Times* (28 Nov 1919), as reprinted in *The Living Age* (3 Jan 1920), **304**, 41–43; https://todayinsci. com/E/Einstein_Albert/EinsteinAlbert-MyTheory.htm

Relativity Made Relatively Easy: General Relativity and Cosmology. Volume 2. Andrew M. Steane, Oxford University Press. © Andrew M. Steane 2021. DOI: 10.1093/oso/9780192895646.003.0029

One such type of reasoning is called 'anthropic reasoning'; this term refers to the general idea that reasoning beings such as ourselves can only reasonably expect to find themselves located in a physical universe that could possibly give rise to reasoning beings. Philosophers and scientists who have pondered this have tended to agree that it captures something worth asserting, but it is controversial precisely what it asserts or how it contributes to our knowledge. Are anthropic arguments explanatory? Or do they merely alert us to interesting features of the universe that we should then try to explain another way?

My own view is that anthropic reasoning is an early glimpse of a way of thinking that we are not yet grasping clearly. It hints at connections between the nature of the physical world at the level of its basic physical constituents—particles, fields, time, space—and the nature of the physical world at the level of its most complex manifestations—reasoning beings. I am sceptical of anthropic reasoning when it appears to be urging us not to ask questions but just to accept the brute fact of the majestic, astounding universe. But I am open to the possibility of new ways of getting insight into the nature of the physical world and our role in it.

It is sometimes asserted that human beings are insignificant because we can only influence one small planet in the cosmos. I do not subscribe to that way of speaking. One can set alongside our limited influence the fact that each of us is unique, and even a single ant or a bee could out-think a galaxy-full of stars alone. We cannot change the whole universe, nor even one star among its billions upon billions, but we can change what the universe will have produced. We can change the part of it where we live. We can create either a beautiful dance or a wasteland. It has been our privilege, in this book, to explore one of the most remarkable intellectual achievements of the human race. It is an achievement of which we are the inheritors and custodians. If all my book succeeds in doing is opening this subject to a few more people who can appreciate and delight in it, and teach it in their turn, then I will be content.

Appendix A

Kepler orbits for binary system

This appendix summarizes the behaviour of a pair of point masses orbiting in their mutual gravity in a Newtonian treatment, and the calculation of the quadrupole moment tensor.

We have a two-body problem with the equations of motion

$$m_1\ddot{\mathbf{r}}_1 = -\frac{Gm_1m_2}{r^2}\hat{\mathbf{r}} \qquad m_2\ddot{\mathbf{r}}_2 = \frac{Gm_1m_2}{r^2}\hat{\mathbf{r}} \tag{A.1}$$

where $\mathbf{r}_1$, $\mathbf{r}_2$ are the positions of the masses m_1, m_2 and $\mathbf{r} \equiv \mathbf{r}_1 - \mathbf{r}_2$.

In the reference frame in which the centre of mass of the system is not moving, one finds $m_1\dot{\mathbf{r}}_1 + m_2\dot{\mathbf{r}}_2 = 0$ and therefore

$$\dot{\mathbf{r}}_1 = (m_2/M)\dot{\mathbf{r}}, \qquad \dot{\mathbf{r}}_2 = -(m_1/M)\dot{\mathbf{r}}, \tag{A.2}$$

where $M = m_1 + m_2$. Hence the total kinetic and potential energies are

$$\text{k.e.} = \tfrac{1}{2}m_1\dot{\mathbf{r}}_1^2 + \tfrac{1}{2}m_2\dot{\mathbf{r}}_2^2 = \frac{1}{2}\frac{m_1m_2}{M}\dot{\mathbf{r}}^2, \qquad \text{p.e.} = -\frac{Gm_1m_2}{r}. \tag{A.3}$$

The total energy is a constant of the motion, equal to

$$E = \text{p.e.} + \text{k.e.} = \frac{-Gm_1m_2}{2a} \tag{A.4}$$

where a is the semi-major axis of the ellipse described by (A.6). Another constant of the motion is the specific angular momentum

$$L \equiv r^2\dot{\phi} = \sqrt{GMa(1-e^2)}. \tag{A.5}$$

The motion itself is one in which each body traces an ellipse with one of the foci at the centre of mass, which is called the *barycentre*. At any moment the line between the bodies passes through the centre of mass and the quantity $m_1\mathbf{r}_1 + m_2\mathbf{r}_2$ is constant. The pair of ellipses is then completely specified if one gives the behaviour of $\mathbf{r}$.

r traces out an ellipse given by

$$r = \frac{a(1-e^2)}{1 + e\cos\phi} \tag{A.6}$$

where e is the eccentricity, which is defined by this equation, and (r,ϕ) are plane polar coordinates in the plane of the orbit. r is smallest when $\phi = 0$; this is the moment of closest approach of the two bodies, called *periastron*; r is largest when $\phi = \pi$; this is the moment of greatest distance, called *apastron*. a is the semi-major axis of the ellipse. (When considering the solar system, the terms *perihelion* and *aphelion* may also be adopted). It can be deduced that

$$a = \frac{r_{\text{per}} + r_{\text{ap}}}{2}, \qquad e = \frac{r_{\text{ap}} - r_{\text{per}}}{r_{\text{ap}} + r_{\text{per}}} = 1 - \frac{2}{(r_{\text{ap}}/r_{\text{per}}) + 1}. \tag{A.7}$$

The period of the motion is

$$P = 2\pi\sqrt{\frac{a^3}{G(m_1 + m_2)}}. \tag{A.8}$$

The relation $P^2 \propto a^3$ is *Kepler's third law*.

Using the orbit equation and the constant angular momentum, one can find the velocity $\dot{\mathbf{r}}$, whose components are

$$v_r = \dot{r} = (GM/L)e\sin\phi, \qquad v_\phi = r\dot{\phi} = (GM/L)(1 + e\cos\phi). \tag{A.9}$$

Using now rectangular coordinates (x, y) to locate each of the two bodies, one finds

$$\mathbf{r}_1 = \frac{m_2}{r}M(\cos\phi, \sin\phi), \qquad \mathbf{r}_2 = \frac{m_1}{r}M(-\cos\phi, -\sin\phi), \tag{A.10}$$

and the quadrupole moment tensor is

$$I_{ij} = \frac{1}{2}\frac{m_1 m_2}{(m_1 + m_2)}r^2 \begin{pmatrix} 1 + \cos 2\phi & \sin 2\phi & 0 \\ \sin 2\phi & 1 - \cos 2\phi & 0 \\ 0 & 0 & 0 \end{pmatrix}. \tag{A.11}$$

This can be written $I = \frac{1}{2}\mu r^2 A$ where A is the matrix, and then

$$\frac{1}{\mu}\frac{\mathrm{d}^3 I}{\mathrm{d}t^3} = (3\dot{r}\ddot{r} + r\dddot{r})A + 3(\dot{r}^2 + r\ddot{r})\dot{A} + 3r\dot{r}\ddot{A} + \tfrac{1}{2}r^2\dddot{A}. \tag{A.12}$$

From (A.6) and (A.5) one finds

$$3\dot{r}\ddot{r} + r\dddot{r} = -eLl^2(1 + e\cos\phi)^2\sin\phi, \qquad \dot{r}^2 + r\ddot{r} = eLl(e + \cos\phi)$$

$$r\dot{r} = eL(1 + e\cos\phi)^{-1}\sin\phi, \qquad r^2 = Ll^{-1}(1 + e\cos\phi)^{-2}$$

where $l = L/(a(1 - e^2))^2$. The final result is

$$
\begin{aligned}
\ddot{I}_{xx} &= \alpha(1 + e\cos\phi)^2[2\sin 2\phi + 3e\sin\phi\cos^2\phi] \\
\ddot{I}_{yy} &= -\alpha(1 + e\cos\phi)^2[2\sin 2\phi + e\sin\phi(3\cos^2\phi + 1)] \\
\ddot{I}_{xy} &= -\alpha(1 + e\cos\phi)^2[2\cos 2\phi + e\cos\phi(3\cos^2\phi - 1)]
\end{aligned}
\tag{A.13}
$$

where

$$
\alpha^2 = \frac{4G^3 m_1^2 m_2^2 M}{a^5(1 - e^2)^5}.
\tag{A.14}
$$

Further information on binary systems in a post-Newtonian approximation can be found in Damour *et al.* 1985 (Damour and Deruelle, 1985).

Appendix B

The 2-sphere and the 3-sphere

B.1   The 2-sphere

The 2-sphere is the two-dimensional manifold of constant positive curvature; it can be realized as the surface of a sphere in three-dimensional Euclidean space.

Here are three versions of the metric of the 2-sphere, corresponding to the three coordinate choices shown in Fig. B.1a:

$$\text{metric 1:} \qquad \mathrm{d}s^2 = R^2\mathrm{d}\theta^2 + R^2\sin^2\theta\,\mathrm{d}\phi^2 \tag{B.1}$$

$$\text{metric 2:} \qquad \mathrm{d}s^2 = \frac{\mathrm{d}r^2}{1 - Kr^2} + r^2\mathrm{d}\phi^2 \tag{B.2}$$

$$\text{metric 3:} \qquad \mathrm{d}s^2 = \frac{1}{(1 + k^2\rho^2)^2}\left(\mathrm{d}\rho^2 + \rho^2\mathrm{d}\phi^2\right) \tag{B.3}$$

where $K = 1/R^2 = 4k^2$. Note that R is here being used for the radius of our sphere; it is not the Ricci scalar!

Consider the length of a curve from $(\theta, \phi) = (0,0)$ to $(\theta_0, 0)$ at fixed ϕ. By symmetry, this curve is geodesic. Using metric 1 and eqn (8.27) the length of the curve is

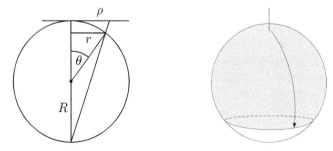

Fig. B.1 Left: a choice of coordinates for the 2-sphere. Right: The shaded region is a circle on the 2-sphere which has large radius and area but small circumference.

$$\int_0^{\theta_0} R\mathrm{d}\theta = R\theta_0.$$ (B.4)

We can observe from this that all the points at $\theta = \theta_0$ are at the same geodesic distance $a \equiv R\theta_0$ from the origin, therefore these points form a circle of radius a. The circumference of this circle is

$$\int_0^{2\pi} R\sin\theta_0\, \mathrm{d}\phi = 2\pi R \sin\theta_0 = 2\pi R \sin(a/R)$$ (B.5)

where to compute the length we used the metric again. For example, a circle of small radius has circumference $2\pi a$, the circle of radius $(\pi/2)R$ has circumference $2\pi R$ and circles of radius larger than this have a smaller circumference, tending to zero at $a = \pi R$.

The area of the circle of radius a is calculated by using (8.33):

$$\int_{\theta=0}^{\theta_0} \int_{\phi=0}^{2\pi} R^2 \sin\theta\, \mathrm{d}\theta\mathrm{d}\phi = 2\pi R^2 (1 - \cos(a/R)).$$ (B.6)

This increases monotonically as a increases from 0 to πR, where it reaches its maximum value $4\pi R^2$; this is the total area of whole manifold (the 2-sphere).

If instead one adopts the coordinates (r, ϕ) and therefore metric 2, one finds that the circle of coordinate radius r_0 has metric (ruler) radius

$$a = \int_0^{r_0} (1 - Kr^2)^{-1/2}\mathrm{d}r = K^{-1/2} \sin^{-1} \sqrt{K}\, r_0$$ (B.7)

The circumference of this circle is (from the metric) $2\pi r_0$, and by using (B.7), this is equal to $2\pi K^{-1/2} \sin(\sqrt{K}\, a)$ which agrees with our previous result (B.5) since $K = 1/R^2$. This illustrates the way in which the metric can be used to obtain relationships that are independent of the choice of coordinates.

B.1.1 Connection and curvature tensor

We can obtain the connection coefficients by using the results presented in Table 17.1 (eqns (17.3)–(17.10)) or by direct calculation from (2.8) which is not too onerous for two dimensions. For metric 1, the only non-zero coefficients are

$$\Gamma^\theta_{\phi\phi} = -\sin\theta\cos\theta, \qquad \Gamma^\phi_{\theta\phi} = \cot\theta.$$ (B.8)

Using eqn (2.13) for the curvature tensor, we find

$$R^\theta_{\phi\theta\phi} = \partial_\theta\Gamma^\theta_{\phi\phi} - \partial_\phi\Gamma^\theta_{\phi\theta} + \Gamma^\theta_{\theta\lambda}\Gamma^\lambda_{\phi\phi} - \Gamma^\theta_{\phi\lambda}\Gamma^\lambda_{\phi\theta}$$ (B.9)

where λ is a dummy index to be summed over, and the others are not. The second and third terms vanish and what remains is

$$R^\theta_{\phi\theta\phi} = \partial_\theta(-\sin\theta\cos\theta) - \Gamma^\theta_{\phi\phi}\Gamma^\phi_{\phi\theta} = \sin^2\theta. \tag{B.10}$$

To obtain the Ricci tensor, we would also like to know $R^\phi_{\theta\phi\theta}$. We can obtain it by first lowering the 1st index, then using a symmetry of the curvature tensor, then raising the index again:

$$R^\phi_{\theta\phi\theta} = \frac{1}{R^2\sin^2\theta}R_{\phi\theta\phi\theta} = \frac{1}{R^2\sin^2\theta}R_{\theta\phi\theta\phi} = \frac{1}{R^2\sin^2\theta}R^2 R^\theta_{\phi\theta\phi} = 1 \tag{B.11}$$

Hence we find

$$R_{\theta\theta} = R^\theta_{\theta\theta\theta} + R^\phi_{\theta\phi\theta} = 1, \qquad\qquad R_{\phi\phi} = R^\theta_{\phi\theta\phi} + R^\phi_{\phi\phi\phi} = \sin^2\theta, \tag{B.12}$$

therefore

$$[R_{ab}] = \begin{pmatrix} 1 & 0 \\ 0 & \sin^2\theta \end{pmatrix} = K g_{ab}. \tag{B.13}$$

More generally, in two dimensions one always has

$$R_{abcd} = K\left(g_{ac}g_{bd} - g_{ad}g_{bc}\right) \tag{B.14}$$

where K is the Gaussian curvature at any given point, and therefore (exercise)

$$R_{ab} = K g_{ab}, \qquad R_{0101} = K g, \qquad K = R^\lambda_\lambda/2 \tag{B.15}$$

where $g = \det(g_{ab})$ as usual.

B.2 The 3-sphere

The 3-sphere is the three-dimensional manifold of constant positive curvature. It can be realized as the surface of a 4-sphere in four-dimensional Euclidean space (such an object is the locus of points satisfying $w^2 + x^2 + y^2 + z^2 = R^2$ where (w, x, y, z) are Cartesian coordinates and R is a constant). Here we will not need to employ that embedding. We can establish coordinates and a metric by suitably adjusting the previous study of the 2-sphere. Let

$$d\Omega^2 \equiv d\theta^2 + \sin^2\theta\, d\phi^2 \tag{B.16}$$

then we can obtain the metric of the 3-sphere by replacing $d\phi^2$ by $d\Omega^2$ in either of (B.2) or (B.3):

$$\text{metric 2:} \qquad ds^2 = \frac{dr^2}{1 - Kr^2} + r^2 d\Omega^2 \tag{B.17}$$

$$\text{metric 3:} \qquad ds^2 = \frac{1}{(1 + K\rho^2/4)^2}\left(d\rho^2 + \rho^2 d\Omega^2\right) \tag{B.18}$$

To see that this is so, it suffices to note that by this choice we have a metric which gives spherical symmetry and which also reproduces the metric of the 2-sphere in the coordinate plane $\theta = \pi/2$. By symmetry it must also do this for all planes through the origin, therefore the Gaussian curvature is indeed the same for all such planes, and we already know that each such plane is itself a surface of constant curvature K.

Consider now the set of points at $r = r_0$ using metric 2. These lie on the surface of a 2-sphere. The radius of this 2-sphere is given by (B.7). Its surface area is easily obtained since the metric of the submanifold with constant r is the one we have called metric 1, i.e. the 2-sphere described by (B.1) with R replaced by r_0, and we already know that the total surface area of such a 2-sphere is

$$S = 4\pi r_0^2 = 4\pi R^2 \sin^2(a/R). \tag{B.19}$$

where we used $R = K^{-1/2}$. For example, for small a we have the familiar $4\pi a^2$ and for $a = (\pi/2)R$ we have $S = 4\pi R^2$. Spheres of radius larger than this have smaller surface area, decreasing until $a = \pi R$ where the surface area goes to zero *but the volume does not!* This is like the circle on the 2-sphere which encloses the whole 2-sphere yet has a vanishing circumference: see Fig. B.1b.

The volume of the sphere within r_0 is calculated by using (8.33) which gives

$$V = \int_0^{r_0} \int_0^{\pi} \int_0^{2\pi} \frac{1}{\sqrt{1 - Kr^2}} r^2 \sin\theta \, dr d\theta d\phi = \frac{2\pi}{K\sqrt{K}} \left(\sin^{-1} \frac{r_0}{R} - \frac{r_0}{R} \sqrt{1 - \frac{r_0^2}{R^2}} \right)$$

$$= \pi R^2 \left(2a - R\sin(2a/R) \right) \tag{B.20}$$

where the last version expresses the result in terms of the ruler radius a (= proper radius) of the sphere in question. Note, the sphere whose surface area and volume we have just calculated is not the whole manifold; it is a sphere contained in the manifold (just as we can find circles on the 2-sphere). In order to find the volume of the whole manifold, we have to interpret carefully because there is a coordinate singularity at $r = R$ in metric 2. To avoid this singularity we can adopt metric 3, and then one finds that the total volume of the manifold (the 3-sphere) is

$$V = 4\pi \int_0^{\infty} \frac{\rho^2}{(1 + K\rho^2/4)^3} d\rho = 2\pi^2 R^3 \tag{B.21}$$

Returning now to metric 2, one sees that at $r_0 = R$, and therefore $a = (\pi/2)R$, eqn (B.20) gives half the volume of the complete manifold. The coordinate system (r, θ, ϕ) is non-singular only up to half the complete 3-sphere, for the same reason that the coordinate system (r, ϕ) only covered one hemisphere of the 2-sphere, see Fig. B.1. (One may admit values of a above $(\pi/2)R$, all the way up to πR, but it is not allowed to consider values of r_0 above R, therefore one must first give a mathematical justification before applying (B.20) to the case $a > (\pi/2)R$.)

B.2.1   Connection and curvature

Using the results presented in Table 17.1 (eqns (17.3)–(17.10)), or by direct calculation from (10.24), we find (B.8) and also

$$\Gamma^r_{rr} = B'/2B \quad = Kr(1 - Kr^2)^{-1} \tag{B.22}$$

$$\Gamma^r_{\theta\theta} = -r/B \quad = -r(1 - Kr^2) \tag{B.23}$$

$$\Gamma^r_{\phi\phi} = -(r\sin^2\theta)/B = -(r\sin^2\theta)(1 - Kr^2) \tag{B.24}$$

$$\Gamma^\theta_{r\theta} = \Gamma^\phi_{r\phi} = 1/r \tag{B.25}$$

One can obtain the Ricci tensor either by using (2.2) or by using (17.11) while carefully allowing for the fact that we have three dimensions not four, so the connection coefficients involving t do not contribute. One thus finds

$$R_{ab} = 2Kg_{ab}. \tag{B.26}$$

This confirms that the 3-sphere has the property $R_{ab} = (N-1)Kg_{ab}$ of the maximally symmetric space mentioned in eqn (15.87).

All of the above gives the merest glimpse of the properties of the 3-sphere, which is itself a rather rich and beautiful mathematical object.

B.3 Hyperbolic space

Let us also briefly mention the spaces of constant negative curvature. Much of the above still applies, but now with a negative value for K, and k^2 is replaced with $-k^2$ in (B.3). The **Poincaré disc** is a model of two-dimensional hyperbolic geometry in which the space is mapped to a disc of coordinate radius $1/k$ in the ρ, ϕ coordinates (Fig. B.2). In this case the range of the ρ coordinate is finite, but the space is infinite because the proper distance from any given point to the edge of the disc is infinite. Geodesics take the form of arcs of circles on the disc which meet its edge at right angles. In the famous image *Circle Limit III* by M.C. Escher, interpreted as a Poncaré disc, all the fish have the same proper size, and all the white lines are geodesics.

One can also construct a finite hyperbolic space by defining a suitable chosen finite region and then identifying opposite sides, as one might a construct a torus.

Fig. B.2 Triangles on the Poincaré disc. The lines are geodesics; the triangles all have the same area.

Appendix C

Differential operators as vectors

In the mathematical analysis of manifolds, it has become standard practice for mathematicians to make an assertion which will sound odd at first to a physicist. It is the assertion that a certain form of differential operator is itself a vector, and a vector is a differential operator. 'But', says the physicist, 'how can an operator be the same as a vector? One is an operator (a mathematical operation waiting to happen) whereas the other is a vector (a quantity having properties of length and direction).' The mathematician will reply, 'yes, but if I show you the definition I have in mind, then you will agree that this operator behaves in many significant respects like a vector, so we may as well say that it is a vector. Equally, the basis vectors are operators, and therefore so are all vectors.' In this appendix we briefly present this concept (it is not adopted in the rest of the book).

The basic idea is to note that for any manifold we can define a scalar function on the manifold and by differentiating it get a one-form, and also we can draw curves and use a derivative along a curve to get a vector. Thus the two types of vectorial object can both be seen as the result of differential operations. The full analysis establishes an *isomorphism* between vectors and differential operators applied to scalar fields on manifolds.

Consider the combination

$$ t^\mu \frac{\partial}{\partial x^\mu} = t^1 \frac{\partial}{\partial x^1} + t^2 \frac{\partial}{\partial x^2} + \cdots + t^n \frac{\partial}{\partial x^n} \tag{C.1} $$

for some set of components t^a. When applied to a scalar field ϕ, this combination gives the rate of change of the field in a direction which is defined by the numbers $(t^1, t^2, \cdots, t^n)$. It is called a *directional derivative*. But we can call it by any name we like. In the context of differential geometry, mathematicians have agreed on the name *tangent vector*. Our aim here is merely to show why this terminology is coherent.

Suppose we have two coordinate systems x^a and x'^a. By the rules of differential calculus, and, N.B., making no appeal to any concept of what a vector may be, we obtain the chain rule:

$$ \partial_a = \frac{\partial x'^\mu}{\partial x^a} \partial'_\mu, \qquad \partial'_a = \frac{\partial x^\mu}{\partial x'^a} \partial_\mu. \tag{C.2} $$

Now notice that these equations are the same as those which describe the transformation of the covariant components of a vector, eqn (9.30), as we have defined vectors up till now. Take the point of view that we have not yet defined the term 'vector' in the context of differentiable manifolds in general; we have simply noticed a common feature of two mathematical results. We are now in a position to take the type of step commonly employed in mathematical analysis: we *define* what we mean by the term 'vector' so that it satisfies whatever are the properties of vector-like quantities that we think most important. Here, we say the notion that a vector has a length and direction is secondary, and the primary property is stated this way:

Definition C.1 *Vector $\equiv$ that which has a list of components, such that the covariant components transform as given by the chain rule for partial differentiation.*

This definition makes no attempt to offer us a nice visual picture (such as a little arrow), but it is mathematically precise and only uses mathematical operations that are well-defined for any differentiable manifold.

Now let us take another look at the idea of a tangent vector. The change in a scalar function ϕ along some curve $x^a(u)$ is given by

$$\frac{\mathrm{d}\phi}{\mathrm{d}u} = \frac{\partial\phi}{\partial x^\mu}\frac{\mathrm{d}x^\mu}{\mathrm{d}u} = \frac{\mathrm{d}x^\mu}{\mathrm{d}u}\frac{\partial\phi}{\partial x^\mu}$$

and since this is valid for any ϕ, we can write

$$\frac{\mathrm{d}}{\mathrm{d}u} = t^\mu\frac{\partial}{\partial x^\mu}$$

where $t^a \equiv \mathrm{d}x^a/\mathrm{d}u$ are a set of components. The operator $\mathrm{d}/\mathrm{d}u$ here is called a *directional derivative* because it is a derivative along a particular direction: along the curve $x^a(u)$. If we now insist that the scalar function is indeed invariant under coordinate transformations, then we must find that the components t^a transform as the contravariant components of a vector.

In the approach using differential operators, one asserts that the operators ∂_a are the coordinate basis vectors $\mathbf{e}_{(a)}$, and the coordinate basis vectors $\mathbf{e}_{(a)}$ are the differential operators ∂_a. This can seem like a very bizarre statement when one is not used to this way of thinking, but if one applies it to the discussion laid out in Chapter 9 one finds that the line of argument survives, and some results fall out more neatly.

If we think of a vector field $\mathbf{X}$ as a directional derivative, then $\mathbf{X}\phi$ signifies the rate of change of scalar field ϕ along $\mathbf{X}$. This rate of change is a single real number at any given point, so it is itself a scalar, but it is a scalar which refers to a rate of change in a direction singled out by a vector.

The *Lie bracket* or commutator of a pair of vector fields is defined under the understanding that we have adopted the directional derivative definition of 'vector'. The Lie bracket of two vector fields $\mathbf{X}, \mathbf{Y}$ is then

$$[\mathbf{X},\ \mathbf{Y}] \equiv \mathbf{X}\mathbf{Y} - \mathbf{Y}\mathbf{X}. \tag{C.3}$$

It might appear from the definition that the Lie bracket must yield a second-order differential operator, but it turns out that this is not so: the second-derivative terms cancel, when the Lie bracket is applied to any scalar field, leaving just a first-order derivative, and furthermore it is one which can be expressed as a directional derivative. Hence we deduce that the Lie bracket of two vector fields is itself a vector field. We will not provide a general proof of this, but we shall prove it in a case which both illustrates the concept and also goes a long way towards constructing a general proof. So, let us consider a vector field $\mathbf{X}$ which is directed along the ith direction, at every point, for some i. That is, $\mathbf{X} = a\partial_i$ where a is the size of the field, and is itself a function of all the coordinates. Suppose also that some other field $\mathbf{Y}$ is along the j'th direction, for some j, thus $\mathbf{Y} = b\partial_j$. Then, for any scalar field ϕ,

$$[\mathbf{X}, \mathbf{Y}]\phi = \mathbf{X}(\mathbf{Y}\phi) - \mathbf{Y}(\mathbf{X}\phi) = a\partial_i(b\partial_j\phi) - b\partial_j(a\partial_i\phi)$$
$$= a(\partial_i b)\partial_j\phi + ab\partial_i\partial_j\phi - b(\partial_j a)\partial_i\phi - ba\partial_j\partial_i\phi$$
$$= \left(a(\partial_i b)\partial_j - b(\partial_j a)\partial_i\right)\phi$$

This is a directional derivative having components in the i and j directions; the i component of size $-b\partial_j a$ and the j component of size $a\partial_i b$. Generalizing to $\mathbf{X} = X^\mu\partial_\mu$ and $\mathbf{Y} = Y^\mu\partial_\mu$, one finds

$$[\mathbf{X}, \mathbf{Y}] = \left((X^\mu\partial_\mu Y^\lambda) - (Y^\mu\partial_\mu X^\lambda)\right)\partial_\lambda \tag{C.4}$$

and in the absence of torsion this is equal to $\nabla_\mathbf{X}\mathbf{Y} - \nabla_\mathbf{Y}\mathbf{X}$. The expression can also be written component-wise as

$$[\mathbf{X}, \mathbf{Y}]^a = X^\mu\partial_\mu Y^a - Y^\mu\partial_\mu X^a. \tag{C.5}$$

We can use the Lie bracket to make a distinction between a coordinate basis and a non-coordinate basis. When writing vectors in terms of components, any set of linearly independent vector fields can be chosen as a basis, and they need not be obtainable from derivatives of a coordinate system. In a coordinate system the coordinate basis vectors are $\{\partial_a\}$ and one has $[\partial_a, \partial_b] = 0$ because partial derivatives commute. One finds that a set of vector fields can form a coordinate system if and only if they commute, i.e. have vanishing Lie brackets.

A reason why one may prefer the differential operator approach is that it deals at all times with quantities that are well-defined 'within' or 'on' the manifold, whereas when we claim that a vector is a directed line segment then we must make appeal to the tangent space where the vector lies. According to the new definition, this traditional picture does not tell us what a vector *is*, but it does allow that the traditional picture is a useful aide to the imagination.

The view adopted for the present book is that, although we acknowledge that the directional derivative offers a precise and useful perspective, so does the view of vectors as directed line segments. There is nothing wrong with adopting the directed line segment picture and using it correctly, and there is much right with the large amounts of intuitive insight it provides.

Appendix D

General equations of the linearized theory

In this appendix we give further information on the linearized theory, also called perturbation theory, extending the discussion of Chapter 5.[1] Note that one can treat perturbation theory in terms of the departures of the exact metric from any chosen background form, not just the Minkowski form, and this is necessary in cosmology, for example. However, here we continue to discuss perturbations on a flat background as in eqn (5.1).

Any symmetric second-rank 4-tensor can be written in terms of a scalar, a 3-vector and a symmetric second-rank 3-tensor. If we introduce a second scalar then we can also impose that the second-rank 3-tensor shall be traceless. To be specific, let us adopt this policy for the perturbation tensor h_{ab}, by introducing Φ, Ψ, w_i, s_{ij} defined by

$$[h_{ab}] = \begin{pmatrix} -2\Phi & w_1 & w_2 & w_3 \\ w_1 & 2(s_{00} - \Psi) & 2s_{12} & 2s_{13} \\ w_2 & 2s_{21} & 2(s_{11} - \Psi) & 2s_{23} \\ w_3 & 2s_{31} & 2s_{32} & 2(s_{33} - \Psi) \end{pmatrix} \tag{D.1}$$

where Ψ is chosen such that

$$\Psi = -\frac{1}{6} h^j_j \tag{D.2}$$

and therefore $s^j_j = 0$. Here s_{ij} is called the *strain*.

The fields Φ, Ψ, w_i, s_{ij} are introduced simply to organize the task of calculating h_{ab}, and to give insight into the results. Note that we have not yet specified a gauge so there remains some redundancy at this stage: more than one set of fields may describe the same region of the spacetime and the phenomena therein.

In terms of Φ, w_i and h_{ij} the Christoffel symbols are

---

[1] The discussion here follows Carroll, who in turn followed E. Bertschinger.

$$\Gamma^0_{a0} = \partial_a \Psi, \qquad\qquad \Gamma^0_{jk} = \tfrac{1}{2}\left(-\partial_j w_k - \partial_k w_j + \partial_0 h_{jk}\right)$$

$$\Gamma^i_{00} = \partial_i \Psi + \partial_0 w_i, \qquad \Gamma^i_{j0} = \tfrac{1}{2}\left(\partial_j w_i - \partial_i w_j + \partial_0 h_{ij}\right)$$

$$\Gamma^i_{jk} = \tfrac{1}{2}\left(\partial_j h_{ki} + \partial_k h_{ji} - \partial_i h_{jk}\right). \tag{D.3}$$

In order to treat the worldline of a massive particle we now introduce symbols $E, \vec{p}, \gamma, \vec{u}$ which are like their versions in Special Relativity:

$$mv^a \equiv m\frac{\mathrm{d}x^a}{\mathrm{d}\tau} \equiv \begin{pmatrix} E \\ \vec{p} \end{pmatrix} \equiv m\begin{pmatrix} \gamma \\ \gamma\vec{u} \end{pmatrix} \tag{D.4}$$

(The ideas will also apply to a null geodesic if one drops m and replaces τ by some other affine parameter). Note in particular that since $\vec{p} = \gamma m\vec{u} = m\mathrm{d}\vec{x}/\mathrm{d}\tau$ and $\gamma = \mathrm{d}t/\mathrm{d}\tau$, we have $\vec{u} = \mathrm{d}\vec{x}/\mathrm{d}t$. One then finds that the geodesic equation gives[2]

$$\frac{\mathrm{d}E}{\mathrm{d}t} = -E\left[\frac{\partial\Phi}{\partial t} + 2\vec{u}\cdot\vec{\nabla}\Phi - \frac{1}{2}u^j u^k\left(\nabla_j w_k + \nabla_k w_j - \frac{\partial h_{jk}}{\partial t}\right)\right], \tag{D.5}$$

$$\frac{\mathrm{d}\vec{p}}{\mathrm{d}t} = E\left[\vec{E}_g + \vec{u}\times\vec{B}_g - 2\frac{\partial h_{ij}}{\partial t}u^j - \frac{1}{2}u^j u^k\left(\nabla_j h_{ki} + \nabla_k h_{ji} - \nabla_i h_{jk}\right)\right] \tag{D.6}$$

where $\vec{E}_g$, $\vec{B}_g$ are given by

$$\vec{E}_g = -\vec{\nabla}\Phi - \frac{\partial\vec{w}}{\partial t},$$
$$\vec{B}_g = \vec{\nabla}\times\vec{w}. \tag{D.7}$$

These are the generalizations of the results of Section 6.1 to the case of arbitrary sources (within the approximation of the linearized treatment on a flat background).

Now let us consider curvature and the field equation.

In terms of Φ, w_i, h_{ij} the Riemann tensor is

$$R_{0j0l} = \partial_j\partial_l\Phi + \partial_0\partial_{(j}w_{l)} - \tfrac{1}{2}\partial_0\partial_0 h_{jl}$$
$$R_{0jkl} = \partial_j\partial_{[k}w_{l]} - \partial_0\partial_{[k}h_{l]j}$$
$$R_{ijkl} = \partial_j\partial_{[k}h_{l]i} - \partial_i\partial_{[k}h_{l]j} \tag{D.8}$$

and the Einstein field equations are

$$\nabla^2\Psi = 4\pi G T_{00} - \tfrac{1}{2}\partial_j\partial_k s^{jk}, \tag{D.9}$$

$$(\delta_{jk}\nabla^2 - \partial_j\partial_k)w^k = -16\pi G T_{0j} + 4\partial_0\partial_j\Psi + 2\partial_0\partial_k s_j^{\ k}, \tag{D.10}$$

$$(\delta_{ij}\nabla^2 - \partial_i\partial_j)(\Phi - \Psi) = 8\pi G T_{ij} - 2\delta_{ij}\partial_0^2\Psi + \partial_0\partial_{(i}w_{j)}$$
$$- \delta_{ij}\left(\partial_0\partial_k w^k + \partial_k\partial_l s^{kl}\right) + \Box^2 s_{ij} - 2\partial_k\partial_{(i}s_{j)}^{\ k}. \tag{D.11}$$

This form brings out the important observation that of the four fields in play (Φ, Ψ, w_i, s_{ij}) only the strain s_{ij} is a truly free degree of freedom. At any moment in time, the others are

---

[2]The combination of vector and index notation in (D.6) is unambiguous; one simply needs to sum over dummy indices and deduce that the index (i) which remains is that of $\vec{p}$, and it does not matter if it is up or down on the rhs here because we are dropping terms which are second or higher order in h_{ab}.

determined by the values of T_{ab}, s_{ij} and its derivatives, and some boundary conditions which can be taken at spatial infinity. For, (D.9) has no time derivatives; therefore Ψ is fixed, at any moment, by T_{00} and spatial derivates of s_{ij}. Similarly, (D.10) involves no time derivative of w^i so it too is determined, at any moment, by the rhs, and the same goes for Φ in eqn (D.11). In summary, once one has a specified energy tensor throughout spacetime, then s_{ij} is not fully determined by the field equation (there may or may not be gravitational waves, for example), but after s_{ij} is specified then so are the other fields.

Upon quantization, only propagating tensor fields give rise to particles. The second-rank tensor field s_{ij} gives rise to a particle of spin 2.

In Chapter 5 we showed that the field equation reduces to a wave equation when one adopts the Lorenz gauge. In this appendix we have not yet restricted to any particular gauge. One finds that under the gauge transformation $h_{ab} \to h_{ab} + \partial_a \xi_b + \partial_b \xi_a$ the fields transform as

$$\Phi \to \Phi + \partial_0 \xi^0, \qquad\qquad w_i \to w_i + \partial_0 \xi^i - \partial_i \xi^0,$$
$$\Psi \to \Psi - \tfrac{1}{3}\partial_i \xi^i, \qquad\qquad s_{ij} \to s_{ij} + \partial_{(i}\xi_{j)} - \tfrac{1}{3}\partial_k \xi^k \delta_{ij}. \qquad (D.12)$$

By choosing ξ^i such that it satisfies

$$\left(\nabla^2 + \tfrac{1}{3}\partial_i \partial_k \xi^k\right) = -2\partial_k s^{ki} \qquad (D.13)$$

one can arrange that s_{ij} is spatially transverse:

$$\partial_i s^{ij} = 0 \qquad (D.14)$$

(this is 'transverse' in the sense that if we examine the Fourier transform then a zero divergence implies the tensor is orthogonal to the wave vector). By further choosing ξ^0 to satisfy

$$\nabla^2 \xi^0 = \partial_i w^i + \partial_0 \partial_i \xi^i \qquad (D.15)$$

one can arrange that $\vec{\nabla} \cdot \vec{w} = 0$ so this field is also transverse. Such a gauge is called the **transverse gauge**. One can see immediately that in this gauge (D.9)–(D.11) simplify to

---

Field equation in transverse gauge

$$\nabla^2 \Psi = 4\pi G T_{00} \qquad\qquad (D.16)$$
$$\nabla^2 w_j = -16\pi G T_{0j} + 4\partial_0 \partial_j \Psi, \qquad\qquad (D.17)$$
$$(\delta_{ij}\nabla^2 - \partial_i \partial_j)(\Phi - \Psi) = 8\pi G T_{ij} - 2\delta_{ij}\partial_0^2 \Psi + \partial_0 \partial_{(i} w_{j)} + \Box^2 s_{ij} \qquad (D.18)$$

Scalar, vector and tensor modes.   Finally, let us note that any vector field can be expressed as a sum of a curl-free and a divergenceless part:

$$\vec{w} = \vec{\nabla}\lambda + \vec{\nabla} \times \vec{\chi} \qquad (D.19)$$

where λ is a scalar field and $\vec{\chi}$ is a vector field (not a unique one, since it could be replaced by $\vec{\chi} + \vec{\nabla}\omega$ for any ω with no effect on $\vec{w}$.) These two parts of $\vec{w}$ are also called the *longitudinal* and *transverse* parts; the longitudinal part being curl-free.

It is less well known, but equally true, that any traceless symmetric tensor field can be decomposed into a transverse part having no divergence, a *solenoidal* part whose divergence is transverse, and a longitudinal part whose divergence is curl-free:

$$s_{ij} = s_{\perp ij} + \partial_{(i}\zeta_{j)} + \left(\partial_i\partial_j - \tfrac{1}{3}\delta_{ij}\nabla^2\right)\theta \qquad (\text{D.20})$$

where $\partial_i s_\perp^{ij} = 0$ and $\partial_i\zeta^i = 0$. Such a decomposition can be adopted in all circumstances, and in particular in any gauge. In the transverse traceless gauge only the $s_\perp^{ij}$ contribution is non-zero.

Through this sequence of steps we have succeeded in expressing the full tensor field h_{ab} in terms of four scalars $\Phi, \Psi, \lambda, \theta$, two transverse vectors $\vec{\chi}, \vec{\zeta}$ (which being transverse have two degrees of freedom each) and one transverse-traceless tensor $s_\perp^{ij}$ (having two degrees of freedom— Chapter 7). When, in cosmology for example, people refer to scalar, vector and tensor *modes* of a 2nd-rank 4-tensor field, it is usually to this set of fields that they refer.

Appendix E

Gravitational energy

We seek to obtain (5.55) from (5.54).

First let us obtain the Christoffel symbols and Ricci tensor to second order in the metric perturbation h_{ab}. We have

$$g_{ab} = \eta_{ab} + h_{ab} \tag{E.1}$$

$$\Gamma^a_{bc} = \tfrac{1}{2} g^{a\lambda}(\partial_b g_{c\lambda} + \partial_c g_{b\lambda} - \partial_\lambda g_{bc}), \tag{E.2}$$

both exact. Hence

$$\Gamma^a_{bc} = \tfrac{1}{2}(\eta^{a\lambda} - h^{a\lambda} + O(h^2))H_{\lambda bc} \;=\; \tfrac{1}{2}H^a_{bc} - \tfrac{1}{2}h^{a\lambda}H_{\lambda bc} + O(h^3) \tag{E.3}$$

where

$$H_{abc} \equiv \partial_b h_{ca} + \partial_c h_{ba} - \partial_a h_{bc} \;=\; 2\partial_{(b}h_{c)a} - \partial_a h_{bc} \tag{E.4}$$

and indices are raised using η^{ab}. That is to say, the quantity H_{abc} is exactly equal to $2\Gamma_{abc}$, whereas $H^a_{bc} = \eta^{a\lambda}H_{\lambda bc}$ which is not equal to $2\Gamma^a_{bc}$.

Next we examine the expression (2.2) for the Ricci tensor, and write down the second-order terms:

$$R^{(2)}_{bc} = \partial_\mu \overset{2}{\Gamma}{}^\mu_{cb} - \partial_c \overset{2}{\Gamma}{}^\mu_{\mu b} + \overset{1}{\Gamma}{}^\mu_{\mu\lambda} \overset{1}{\Gamma}{}^\lambda_{cb} - \overset{1}{\Gamma}{}^\mu_{c\lambda} \overset{1}{\Gamma}{}^\lambda_{\mu b}, \tag{E.5}$$

where the integer $1, 2$ indicates the order of the term in question. One finds

$$\partial_\mu \overset{2}{\Gamma}{}^\mu_{cb} - \partial_c \overset{2}{\Gamma}{}^\mu_{\mu b} = \tfrac{1}{2}[(\partial_c h^{\mu\lambda})(\partial_b h_{\mu\lambda})] - \tfrac{1}{2}(\partial_\mu h^{\mu\lambda})H_{\lambda bc}$$
$$+ \tfrac{1}{2}h^{\mu\lambda}\left(\partial_b\partial_c h_{\mu\lambda} + \partial_\mu\partial_\lambda h_{bc} - \partial_\lambda\partial_c h_{b\mu} - \partial_\mu\partial_b h_{c\lambda}\right) \tag{E.6}$$

$$H^\mu_{\mu\lambda}H^\lambda_{cb} = \left(\partial_\mu h^\mu_\lambda + \partial_\lambda h^\mu_\mu - \partial^\mu h_{\mu\lambda}\right)H^\lambda_{cb} \;=\; (\partial^\lambda h)H_{\lambda bc} \tag{E.7}$$

$$H^\mu_{c\lambda}H^\lambda_{\mu b} = H_{\mu c\lambda}\left(\partial^\mu h^\lambda_b - \partial^\lambda h^\mu_b\right) + [(\partial_c h_{\lambda\mu})(\partial_b h^{\lambda\mu})]$$
$$= 2(\partial_\lambda h^\mu_c)\partial_\mu h^\lambda_b - 2(\partial_\mu h_{c\lambda})\partial^\mu h^\lambda_b + [(\partial_c h_{\lambda\mu})(\partial_b h^{\lambda\mu})] \tag{E.8}$$

and therefore

$$R^{(2)}_{bc} = \tfrac{1}{4}(\partial_b h_{\mu\lambda})(\partial_c h^{\mu\lambda}) + \tfrac{1}{2}h^{\mu\lambda}\left(\partial_b \partial_c h_{\mu\lambda} + \partial_\mu \partial_\lambda h_{bc} - \partial_\lambda \partial_c h_{b\mu} - \partial_\mu \partial_b h_{c\lambda}\right)$$
$$- \tfrac{1}{2}(\partial_\mu h^{\mu\lambda} - \tfrac{1}{2}\partial^\lambda h)H_{\lambda bc} + \tfrac{1}{2}(\partial_\mu h_{c\lambda} - \partial_\lambda h_{\mu c})\partial^\mu h_b^\lambda. \tag{E.9}$$

We now take the average of this expression over a small region. The size L of this region was discussed after eqn (5.54). By taking such an average we gain a great simplification of the equations, at the cost of ignoring nonlinear corrections to gravitational wave amplitudes. But when those amplitudes are already small the corrections to them are negligible.

The simplification arises because whenever a function oscillates about some fixed value, then the average of the gradient of the function must be zero (but the average of the square of the gradient, or of the second derivative, need not be zero), as long as the average is taken over a length- and time-scale larger than that of the oscillations. To be precise, this average gradient is negligible compared to the other quantities. In the present case,

$$\langle \partial_\lambda (h_{ab}\partial_\mu h_{cd}) \rangle = O(L|K|^{1/2}h^2) \simeq 0 \tag{E.10}$$

where L is the size of the region being averaged over, and K is the Gaussian curvature of the background (i.e. zero if the background is flat). We shall make repeated use of this in order to convert products of first derivatives into second derivatives, and *vice versa*:

$$\langle (\partial_\lambda h_{ab})\, \partial_\mu h_{cd} \rangle \simeq -\langle h_{ab}\, \partial_\lambda \partial_\mu h_{cd} \rangle. \tag{E.11}$$

Hereafter we shall freely use this as an equality without explicitly indicating the degree of approximation. We also keep in mind that R_{ab} and h_{ab} are symmetric, and we can freely see-saw dummy indices because raising and lowering is done with η^{ab}, η_{ab} throughout the rest of the analysis.

In order to organize the effort to simplify (E.9), note that indices on h appear in six patterns:

$$(\lambda\mu, \lambda\mu); \ (\lambda\mu, bc); \ (\lambda\mu, b\lambda \text{ or } c\mu); \ (^\mu_\mu, \ldots); \ (b\lambda, c\lambda); \ (b\lambda, c\mu).$$

By using (E.11) one can combine the two terms of the first type (those involving $h^{\lambda\mu}$, $h_{\lambda\mu}$), and one finds the terms of the second and third types cancel. Hence one obtains

$$\left\langle R^{(2)}_{bc} \right\rangle = \frac{1}{4}\left\langle -(\partial_b h_{\mu\lambda})(\partial_c h^{\mu\lambda}) + ④ + ⑤ + ⑥ \right\rangle \tag{E.12}$$

where

$$④ = (\partial^\lambda h)H_{\lambda bc} \tag{E.13}$$
$$⑤ = (\partial_\mu h_{c\lambda})\partial^\mu h_b^\lambda + (b \leftrightarrow c) \tag{E.14}$$
$$⑥ = -\left[(\partial_\lambda h_{\mu c})\partial^\mu h_b^\lambda + (b \leftrightarrow c)\right] \tag{E.15}$$

where we used that the last term in (E.9) is symmetric in b and c (which one should confirm since it is not manifest in (E.9)). We have

$$\langle (\partial^\lambda h)H_{\lambda bc} \rangle = \langle (\partial^\lambda h)(\partial_b h_{c\lambda} + (b \leftrightarrow c)) + h\Box^2 h_{bc} \rangle \tag{E.16}$$

and the average of term ⑤ gives $-\langle h_{c\lambda}\square^2 h_b^\lambda + (b \leftrightarrow c)\rangle$. We use the first order equation (5.28) to express $\square^2 h_{bc}$ in terms of T_{bc} via the field equation:

$$\square^2 h_{bc} = -\frac{16\pi G}{c^4}\left(T_{bc} - \tfrac{1}{2}\eta_{bc}T\right) - \partial_b\partial_c h + \left[\partial_b\partial_\nu h_c^\nu + (b \leftrightarrow c)\right] + O(h^2), \tag{E.17}$$

$$\square^2 h = \frac{8\pi G}{c^4}T + \partial_\lambda\partial_\nu h^{\lambda\nu} + O(h^2). \tag{E.18}$$

Hence, by using this to replace $\square^2 h_{bc}$, $\square^2 h_b^\lambda$, $\square^2 h_c^\lambda$ and combining all the parts of (E.12), one finds

$$\left\langle R_{bc}^{(2)}\right\rangle = \frac{1}{4}\left\langle -(\partial_b h_{\mu\lambda})(\partial_c h^{\mu\lambda}) + 2(\partial_\lambda h^{\lambda\mu})\partial_{(b}h_{c)\mu} - 2(\partial_\lambda h)\partial_{(b}h_{c)}^\lambda + (\partial_b h)\partial_c h \right.$$
$$\left. - \frac{16\pi G}{c^4}\left(hT_{bc} + h_{bc}T - \tfrac{1}{2}\eta_{bc}hT - 2h_{\lambda(b}T_{c)}^\lambda\right)\right\rangle \tag{E.19}$$

in which one must not forget the factor $1/2$ in the definition of the bracket notation introduced in (12.13).

By contracting this expression, and then using (E.17), (E.18) to replace terms in $\square^2 h^{\mu\lambda}$ and $\square^2 h$, one obtains

$$\left\langle R^{(2)}\right\rangle = \frac{4\pi G}{c^4}\left\langle h_{\lambda\mu}T^{\lambda\mu}\right\rangle. \tag{E.20}$$

Next we obtain the Einstein tensor:

$$G_{ab} \equiv R_{ab} - \tfrac{1}{2}Rg_{ab}$$
$$= R_{ab} - \tfrac{1}{2}g_{ab}g^{\mu\nu}R_{\mu\nu}$$
$$= R_{ab} - \tfrac{1}{2}(\eta_{ab} + h_{ab})(\eta^{\mu\nu} - h^{\mu\nu} + O(h^2))R_{\mu\nu}$$
$$= R_{ab} - \tfrac{1}{2}(\eta_{ab} + h_{ab})(\eta^{\mu\nu}R_{\mu\nu} - h^{\mu\nu}R_{\mu\nu} + O(h^3)) \tag{E.21}$$

since R_{ab} is itself of order h. It follows that the first and second order contributions to G_{ab} are

$$G_{ab}^{(1)} = R_{ab}^{(1)} - \tfrac{1}{2}\eta_{ab}R^{(1)}, \tag{E.22}$$
$$G_{ab}^{(2)} = R_{ab}^{(2)} - \tfrac{1}{2}\eta_{ab}R^{(2)} - \tfrac{1}{2}h_{ab}R^{(1)} + \tfrac{1}{2}\eta_{ab}h^{\mu\nu}R_{\mu\nu}^{(1)}. \tag{E.23}$$

where the Ricci scalar is being obtained at each order by using $\eta^{\mu\nu}$ not $g^{\mu\nu}$. The field equation tells us that $R_{ab}^{(1)} = (8\pi G/c^4)(T_{ab} - \tfrac{1}{2}\eta_{ab}T)$ and $R^{(1)} = -(8\pi G/c^4)T$, so we find

$$G^{(2)} = R_{ab}^{(2)} - \tfrac{1}{2}\eta_{ab}R^{(2)} + \frac{4\pi G}{c^4}(\bar{h}_{ab}T + \eta_{ab}h^{\mu\nu}T_{\mu\nu}) \tag{E.24}$$

where we introduced $\bar{h}_{ab} \equiv h_{ab} - \tfrac{1}{2}\eta_{ab}h$.

Finally, the stress-energy pseudo-tensor for the gravitational field in the linearized theory, t_{ab}, is defined by (5.54): $t_{ab} = -(c^4/8\pi G)\langle G_{ab}^{(2)}\rangle$. Using (E.24), (E.19), (E.20), after some effort to replace h_{ab} by $\bar{h}_{ab} + \tfrac{1}{2}\eta_{ab}h$, one obtains (5.55).

Appendix F

Causality and the Cauchy problem in General Relativity

The task of finding a solution to a partial differential equation that satisfies a given boundary condition, such as an initial condition, is called a Cauchy problem. Ordinarily an equation of motion in physics should constrain the motion sufficiently that only one outcome can result from a given initial condition. It is not self-evident that the Einstein equation has this property, therefore we would like to look into it. If there were more than one solution for given initial conditions, it would imply that physical evolution is not determined—there would be more than one possible effect that might result from a single cause. This should not be thought to be a sheer impossibility; there is nothing logically incoherent about indeterminism.[1] It is simply that one would not expect it as part of the nature of gravitation in ordinary circumstances (in the absence of classical chaos, for example).

Studies of the Cauchy problem in GR largely bear this out, in that a spacetime without certain pathological properties will yield a unique future from a given initial condition; we will outline the proof below. A spacetime which does not suffer from the pathologies, and therefore satisfies causality, is said to be *globally hyperbolic*. This odd-sounding name comes from the classification of partial differential equations. In the present context it can be defined:

Definition F.1 *A spacetime is globally hyperbolic if: (1) there are no closed timelike curves and (2) given any two events, the intersection of the future of one and the past of the other is compact.*

We shall define the term *compact* in a moment. The 'future of' and 'past of' mentioned here refer to the surface and the interior of the relevant light cones. The content of this definition is that in a globally hyperbolic spacetime one can find a spacelike surface which intersects every inextendible timelike line, and we will show that the initial-value problem based on initial conditions specified on such a surface has a unique solution. Here, a line is 'inextendible' simply

---

[1] It is not known whether the physical world is deterministic. Neither quantum nor classical models of physics are able to settle it. The former can be interpreted in more than one way (the measurement problem) and the latter predicts exponentially diverging trajectories in phase space in chaotic systems, with the result that indeterminism cannot be ruled out empirically in such cases.

if it does not stop at some finite point—it keeps on going to infinity. The surface is then called a Cauchy surface.

The definition uses the term 'compact' in a technical sense which we need to spell out:

Definition F.2 *A region in a manifold is* compact *if it is* bounded *and* closed. *The region is* bounded *if for any geodesic in the region, the affine parameter is bounded; this is equivalent to the condition that the region does not include any points at infinity. The region is* closed *if it contains its limit points.*

Boundedness is reasonably straightforward; the notion of closure is more subtle. 'Closed' is used here in a sense like that in which a group may be closed, but we are dealing with continuous sets. The term is used to specify how the boundary of the region S relates to S. One may not say, of a compact region, that 'it extends to A without including A' where A is some location, specified, for example, by a set of real-valued coordinates. If A is on the boundary, then S is only compact if $A \in S$. A region that extends infinitesimally close to a singularity cannot be compact, because the singularity cannot itself be said to be part of the manifold.

So much for definitions. The main upshot of the above is to warn us that it will be problematic to evolve the Einstein field equation either forward in time towards the future light cone of a singularity, or backward into the past light cone of a singularity, and closed timelike curves will also present difficulties.

Let us now investigate the problem of finding a solution to the equation $G_{ab} = 8\pi G T_{ab}$ (setting $c = 1$), based on some given initial condition, in the absence of pathologies such as singularities and closed timelike curves. Of course there is no single preferred direction to call 'time', but we can pick a spacelike hypersurface Σ and let our zeroth coordinate, t, have a given value on that surface. We then guess that in order to specify initial conditions it will also be necessary to provide the value of $g_{ab}|_\Sigma$ and the time derivative $\partial_t g_{ab}|_\Sigma$ on the hypersurface, since the field equation is second order in time (spatial derivatives can be extracted from the given g_{ab} should we need them). However, we have $\nabla_\mu G^{\mu b} = 0$, which means that the time derivatives of g_{ab} cannot be chosen arbitrarily when specifying the initial conditions; there are *constraints*. In order to see this, write $\nabla_\mu G^{\mu b} = 0$ in the form

$$\partial_0 G^{0b} = -\partial_i G^{i\nu} - \Gamma^\mu_{\mu\nu} G^{\nu b} - \Gamma^b_{\mu\nu} G^{\mu\nu}. \tag{F.1}$$

Observe that there are no *third*-order time derivatives on the right-hand side; it follows that there are none on the left-hand side. Therefore, although the Einstein tensor involves second-order derivatives of the metric, the specific components G^{0b} do not involve second-order time derivatives (a similar conclusion could be obtained about derivatives of G^{ib} with respect to x^i, but they do not concern us). It follows that of the ten independent components of Einstein's equation, the four given by

$$G^{0b} = 8\pi G T^{0b} \tag{F.2}$$

do not give information that can be used to evolve the initial data $\{g_{ab}, \partial_t g_{ab}\}_\Sigma$. Rather, these components serve as **constraints** on the initial data. For example, in vacuum the metric tensor has to have a form such that $G^{0b} = 0$ at $t = 0$. Once we have furnished such a metric tensor, the field equation will suffice to guarantee that G^{0b} remains zero thereafter (this is easy to show from the Bianchi identity).

We thus discover that we only have six equations describing the dynamical evolution:

$$G^{ij} = 8\pi G T^{ij}. \tag{F.3}$$

This is 6 equations in 10 unknowns $g_{ab}(t)$ (recall that g_{ab} is symmetric so has 10 independent components), so there is a 4-fold ambiguity. Considered as a set of differential equations, one would say that the problem is underdetermined. Thus it may seem as if determinism is not respected in GR. But in fact it is. The 4-fold ambiguity is simply the freedom that must exist, because we are free to choose coordinates throughout spacetime. The situation is similar to that of solving electromagnetic problems via the 4-potential. In order to 'pin down' a specific solution one may adopt a gauge, safe in the knowledge that it will not matter which gauge is picked. The type of coordinate freedom we meet in GR is itself a type of gauge freedom, as we noted in Chapter 5, and can be handled by adopting a gauge such as Lorenz gauge. This is the same gauge we met in the weak field limit in (5.39). Now we write a more general form using the covariant d'Alembertian:

Lorenz gauge condition

$$\nabla^\lambda \nabla_\lambda x^a = 0 \qquad \text{which may also be written} \quad \partial_\lambda \left(\sqrt{|g|}\, g^{\lambda b} \right) = 0 \qquad \text{(F.4)}$$

---

where the second version is obtained in exercise 16.2 of Chapter 16.

This condition represents a differential equation for the four metric components g^{0b}; once we have these we can solve the field equations for the six remaining components g^{ij} and thus we have shown that we can expect a unique solution. We have glossed over the fact that the gauge condition may only be able to be applied over a finite region; one would then have to divide spacetime up into patches described by different coordinate systems.

The conclusion is that as long as there is a Cauchy surface which can 'see' all of spacetime, in the sense specified by the globally hyperbolic condition, then Einstein's field equation does yield a well-defined initial value problem. An initial (or final) state is specified by the spacelike components of the metric and their time derivatives, all furnished on a spacelike hypersurface and such that the constraints (F.2) are satisfied. The spacelike components of Einstein's equation then suffice to evolve the metric forward in time, up to a coordinate ambiguity which can be resolved by a gauge choice. The condition of global hyperbolicity ensures that no event can be influenced by a region of spacetime that was not either specified in the initial conditions or derivable from them.

There remains the fact that some scenarios in GR suggest that spacetimes can come about which are not globally hyperbolic. The situation of a black hole which completely evaporates may be of this kind; this is the *black hole information paradox*.

References

Adler, R. J. and Overduin, J. M. (2005). The nearly flat universe. *Gen. Relativ. Gravit.*, **37**, 1491.

Balbus, Steven (2019). Notes on general relativity and cosmology.

Baumann, Daniel (2012). Tasi lectures on inflation. *arXiv:0907.5424v2*.

Begelman, Mitchell C. (2003). Evidence for black holes. *Science*, **300**(5627), 1898–1903.

Blau, Matthias (2018). Lecture notes on General Relativity.

Blome, H.-J., Hoell, J., and Priester, W. (1997). Kosmologie. In *Lehrbuch der Experimentalphysik, Vol 8* (ed. Bergmann-Schaefer), pp. 311–427. W. de Gruyter, Berlin.

Brynjolfsson, Erling J. and Thorlacius, Larus (2008, sep). Taking the temperature of a black hole. *Journal of High Energy Physics*, **2008**(09), 066–066.

Bucher, Martin (2015). Physics of the cosmic microwave background anisotropy.

Carroll, Sean (2014a). *Spacetime and Geometry*. Pearson, Harlow.

Carroll, Sean M. (1997). Lecture notes on general relativity. *arXiv:gr-qc/9712019v1*.

Carroll, Sean M. (2014b). *In What Sense Is the Early Universe Fine-Tuned?* Harvard University Press (forthcoming).

Crispino, Luís C. B., Higuchi, Atsushi, and Matsas, George E. A. (2008, Jul). The Unruh effect and its applications. *Rev. Mod. Phys.*, **80**, 787–838.

Crowell, Benjamin (2009, 2018). *General Relativity*.

Damour, Thibault and Deruelle, Nathalie (1985). General relativistic celestial mechanics of binary systems. i. the post-newtonian motion. *Annales de l'I.H.P. Physique théorique*, **43**(1), 107–132.

Dey, Ramit, Liberati, Stefano, and Pranzetti, Daniele (2017). The black hole quantum atmosphere. *Physics Letters B*, **774**, 308–316.

Di Valentino, Eleonora, Melchiorri, Alessandro, and Silk, Joseph (2015, Dec). Beyond six parameters: Extending ΛCDM. *Phys. Rev. D*, **92**, 121302.

d'Inverno, Ray (1992). *Introducing Einstein's Relativity*. Oxford University Press, Oxford.

Dirac, P. A. M. (1996). *General Theory of Relativity*. Princeton University Press, Princeton.

Durrer, Ruth (2011). What do we really know about dark energy? *Philosophical Transactions of the Royal Society A*, **369**, 5102–5114.

Eaton, John W., Bateman, David, Hauberg, Søren, and Wehbring, Rik (2018). *GNU Octave version 4.4.0 manual: a high-level interactive language for numerical computations*.

Eisenstein, Daniel J., Zehavi, Idit, Hogg, David W., Scoccimarro, Roman, Blanton, Michael R., Nichol, Robert C., Scranton, Ryan, Seo, Hee-Jong, Tegmark, Max, Zheng, Zheng, Anderson, Scott F., Annis, Jim, Bahcall, Neta, Brinkmann, Jon, Burles, Scott, Castander, Francisco J., Connolly, Andrew, Csabai, Istvan, Doi, Mamoru, Fukugita, Masataka, Frieman, Joshua A., Glazebrook, Karl, Gunn, James E., Hendry, John S., Hennessy, Gregory, Ivezić,

Zeljko, Kent, Stephen, Knapp, Gillian R., Lin, Huan, Loh, Yeong-Shang, Lupton, Robert H., Margon, Bruce, McKay, Timothy A., Meiksin, Avery, Munn, Jeffery A., Pope, Adrian, Richmond, Michael W., Schlegel, David, Schneider, Donald P., Shimasaku, Kazuhiro, Stoughton, Christopher, Strauss, Michael A., SubbaRao, Mark, Szalay, Alexander S., Szapudi, Istvan, Tucker, Douglas L., Yanny, Brian, and York, Donald G. (2005, nov). Detection of the baryon acoustic peak in the large-scale correlation function of SDSS luminous red galaxies. *The Astrophysical Journal*, **633**(2), 560–574.

Fabian, A. C., Vaughan, S., Nandra, K., Iwasawa, K., Ballantyne, D. R., Lee, J. C., Rosa, A. De, Turner, A., and Young, A. J. (2002). A long hard look at MCG–6-30-15 with XMM-Newton. *Mon. Not. R. Astron. Soc.*, **335**, L1–L5.

Fixsen, D. J. (2009, nov). The temperature of the cosmic microwave background. *The Astrophysical Journal*, **707**(2), 916–920.

Fulling, S. A. (2005). Review of some recent work on acceleration radiation. *J. Mod. Opt.*, **52**, 2207–2213.

Fulton, Thomas and Rohrlich, Fritz (1960). Classical radiation from a uniformly accelerated charge. *Annals of Physics*, **9**, 499–517.

Giddings, Steven B. (2016). Hawking radiation, the Stefan–Boltzmann law, and unitarization. *Physics Letters B*, **754**, 39–42.

Gou, Lijun, McClintock, Jeffrey E., Reid, Mark J., Orosz, Jerome A., Steiner, James F., Narayan, Ramesh, Xiang, Jingen, Remillard, Ronald A., Arnaud, Keith A., and Davis, Shane W. (2011, nov). The extreme spin of the black hole in Cygnus X-1. *The Astrophysical Journal*, **742**(2), 85.

Greenwood, Eric and Stojkovic, Dejan (2009, sep). Hawking radiation as seen by an infalling observer. *Journal of High Energy Physics*, **2009**(09), 058–058.

Heckmann, O. (1932, 07). Die Ausdehnung der Welt in ihrer Abhängigkeit von der Zeit. *Veroeffentlichungen der Universitaets-Sternwarte zu Goettingen*, **2**, 180.

Helbig, Phillip (2012, 03). Is there a flatness problem in classical cosmology? *Monthly Notices of the Royal Astronomical Society*, **421**(1), 561–569.

Hobson, M. P., Efstathiou, G. P., and Lasenby, A. N. (2006). *General Relativity: An Introduction for Physicists*. Cambridge University Press, Cambridge.

Lake, Kayll (2005, May). The flatness problem and Λ. *Phys. Rev. Lett.*, **94**, 201102.

Lambourne, Robert (2010). *Relativity, Gravitation and Cosmology*. Cambridge University Press, Cambridge.

Landau, L. D. and Lifshitz, E. M. (1971). *The Classical Theory of Fields*. Pergamon, Oxford. (1st edition (Russian) 1941).

Lightman, Alan P., Press, William H., Price, Richard H., and Teukolsky, Saul A. (1975). *Problem Book in Relativity and Gravitation*. Princeton University Press, Princeton, N.J.

LIGO Scientific Collaboration and Virgo Collaboration (2016, Feb). Observation of gravitational waves from a binary black hole merger. *Phys. Rev. Lett.*, **116**, 061102.

Misne, C. W., Thorne, K. S., and Wheeler, J. A. (1975). *Gravitation*. W. H. Freeman and Co.

Moschella, Ugo (2005). The de Sitter and anti-de Sitter sightseeing tour. *Séminaire Poincaré*, **1**, 1–12.

Nussbaumer, H. (2014). Einstein's conversion from his static to an expanding universe. *European Physics Journal — History*, **39**, 37–62.

Padmanabhan, T. (2010). Thermodynamical aspects of gravity: new insights. *Rep. Prog. Phys.*, **73**.

Peebles, Phillip James Edwin (1993). *Principles of Physical Cosmology*. Princeton University Press.

Perlmutter, S., Aldering, G., Goldhaber, G., Knop, R. A., Nugent, P., Castro, P. G., Deustua, S., Fabbro, S., Goobar, A., Groom, D. E., Hook, I. M., Kim, A. G., Kim, M. Y., Lee, J. C., Nunes, N. J., Pain, R., Pennypacker, C. R., Quimby, R., Lidman, C., Ellis, R. S., Irwin, M., McMahon, R. G., Ruiz-Lapuente, P., Walton, N., Schaefer, B., Boyle, B. J., Filippenko, A. V., Matheson, T., Fruchter, A. S., Panagia, N., Newberg, H. J. M., Couch, W. J., and The Supernova Cosmology Project (1999, jun). Measurements of Ω and Λ from 42 high-redshift supernovae. *The Astrophysical Journal*, **517**(2), 565–586.

Perlmutter, Saul and Schmidt, Brian P. (2003). Measuring cosmology with supernovae. In *Supernovae and Gamma-Ray Bursters. Lecture Notes in Physics* (ed. W. K.W.), Volume 598. Springer, Berlin, Heidelberg.

Planck Collaboration: Ade, P. A. R. *et al.* (2014). Planck 2013 results. i. overview of products and scientific results. *A&A*, **571**, A1.

Planck Collaboration: N. Aghanim *et al.* (2018). Planck 2018 results. VI. cosmological parameters. *arXiv:1807.06209*.

Planck Collaboration: Y. Akrami *et al.* (2018). Planck 2018 results. I. overview and the cosmological legacy of Planck. *arXiv:1807.06205*.

Poisson, Eric (2004). *A Relativist's Toolkit*. Cambridge University Press, cambridge.

Rätzel, Dennis, Wilkens, Martin, and Menzel, Ralf (2016, jan). Gravitational properties of light—the gravitational field of a laser pulse. *New Journal of Physics*, **18**(2), 023009.

Riess, Adam G., Casertano, Stefano, Yuan, Wenlong, Macri, Lucas, Anderson, Jay, MacKenty, John W., Bowers, J. Bradley, Clubb, Kelsey I., Filippenko, Alexei V., Jones, David O., and Tucker, Brad E. (2018*a*, mar). New parallaxes of galactic cepheids from spatially scanning the hubble space telescope: Implications for the hubble constant. *The Astrophysical Journal*, **855**(2), 136.

Riess, Adam G., Casertano, Stefano, Yuan, Wenlong, Macri, Lucas, Bucciarelli, Beatrice, Lattanzi, Mario G., MacKenty, John W., Bowers, J. Bradley, Zheng, WeiKang, Filippenko, Alexei V., Huang, Caroline, and Anderson, Richard I. (2018*b*). Milky way cepheid standards for measuring cosmic distances and application to Gaia DR2: Implications for the Hubble constant. *The Astrophysical Journal*, **861**, 126.

Riess, Adam G., Filippenko, Alexei V., Challis, Peter, Clocchiatti, Alejandro, Diercks, Alan, Garnavich, Peter M., Gilliland, Ron L., Hogan, Craig J., Jha, Saurabh, Kirshner, Robert P., Leibundgut, B., Phillips, M. M., Reiss, David, Schmidt, Brian P., Schommer, Robert A., Smith, R. Chris, Spyromilio, J., Stubbs, Christopher, Suntzeff, Nicholas B., and Tonry, John (1998, sep). Observational evidence from supernovae for an accelerating universe and a cosmological constant. *The Astronomical Journal*, **116**(3), 1009–1038.

Rindler, W. (1997). The case against space dragging. *Physics Letters A*, **233**, 25–29.

Rindler, Wolfgang (2000). Finite foliations of open FRW universes and the point-like big bang. *Physics Letters A*, **276**, 52–58.

Rindler, W. (2006). *Relativity: Special, General, and Cosmological: 2nd ed.* Oxford U.P., Oxford.

Schödel, R., Ott, T., Genzel, R., Hofmann, R., Lehnert, M., Eckart, A., Mouawad, N., Alexander, T., Reid, M. J., Lenzen, R., Hartung, M., Lacombe, F., Rouan, D., Gendron, E., Rousset, G., Lagrange, A.-M., Brandner, W., Ageorges, N., Lidman, C., Moorwood, A. F. M.,

Spyromilio, J., Hubin, N., and Menten, K. M. (2002). A star in a 15.2-year orbit around the supermassive black hole at the centre of the milky way. *Nature*, **419**, 694–696.

Schutz, Bernard (2009). *A First Course in General Relativity, 2nd edition*. Cambridge University Press, Cambridge.

Scully, Marlan O. (1979, Jun). General-relativistic treatment of the gravitational coupling between laser beams. *Phys. Rev. D*, **19**, 3582–3591.

Steane, A. M. (2012). *Relativity Made Relatively Easy*. Oxford U.P., Oxford.

Straumann, Norbert (1989, 05). Evidence for black holes. *Particle Physics and Astrophysics - Current viewpoints*, **-1**, 171–183.

Stuchlík, Z. and Hledík, S. (1999, Jul). Some properties of the schwarzschild–de sitter and schwarzschild–anti-de sitter spacetimes. *Phys. Rev. D*, **60**, 044006.

Suzuki, N., Rubin, D., Lidman, C., Aldering, G., Amanullah, R., Barbary, K., Barrientos, L. F., Botyanszki, J., Brodwin, M., Connolly, N., Dawson, K. S., Dey, A., Doi, M., Donahue, M., Deustua, S., Eisenhardt, P., Ellingson, E., Faccioli, L., Fadeyev, V., Fakhouri, H. K., Fruchter, A. S., Gilbank, D. G., Gladders, M. D., Goldhaber, G., Gonzalez, A. H., Goobar, A., Gude, A., Hattori, T., Hoekstra, H., Hsiao, E., Huang, X., Ihara, Y., Jee, M. J., Johnston, D., Kashikawa, N., Koester, B., Konishi, K., Kowalski, M., Linder, E. V., Lubin, L., Melbourne, J., Meyers, J., Morokuma, T., Munshi, F., Mullis, C., Oda, T., Panagia, N., Perlmutter, S., Postman, M., Pritchard, T., Rhodes, J., Ripoche, P., Rosati, P., Schlegel, D. J., Spadafora, A., Stanford, S. A., Stanishev, V., Stern, D., Strovink, M., Takanashi, N., Tokita, K., Wagner, M., Wang, L., Yasuda, N., and and, H. K. C. Yee (2012, jan). The Hubble Space Telescope cluster supernova survey. V. Improving the dark-energy constraints above $z > 1$ and building an early-type-hosted supernova sample. *The Astrophysical Journal*, **746**(1), 85.

Takada, Masahiro and Doré, Olivier (2015, Dec). Geometrical constraint on curvature with BAO experiments. *Phys. Rev. D*, **92**, 123518.

Tantau, Till (2013). *The TikZ and PGF Packages*.

Tolman, Richard C., Ehrenfest, Paul, and Podolsky, Boris (1931, Mar). On the gravitational field produced by light. *Phys. Rev.*, **37**, 602–615.

Tong, David (2019). Cosmology; university of cambridge part ii mathematical tripos.

Unruh, W. G. (1976, Aug). Notes on black-hole evaporation. *Phys. Rev. D*, **14**, 870–892.

Wald, Robert M. (1984). *General Relativity*. University of Chicago Press, Chicago.

Weinberg, Steven (2008). *Cosmology*. Oxford University Press, Oxford.

Weisberg, J. M., Nice, D. J., and Taylor, J. H. (2010, sep). Timing measurements of the relativistic binary pulsar PSR B1913+16. *The Astrophysical Journal*, **722**(2), 1030–1034.

Index